Recent Titles in This Series

107 **B. M. Makarov, M. G. Goluzina, A. A. Lodkin, and A. N. Podkorytov,** Selected problems in real analysis, 1992
106 **G.-C. Wen,** Conformal mappings and boundary value problems, 1992
105 **D. R. Yafaev,** Mathematical scattering theory: General theory, 1992
104 **R. L. Dobrushin, R. Kotecký, and S. Shlosman,** Wulff construction: A global shape from local interaction, 1992
103 **A. K. Tsikh,** Multidimensional residues and their applications, 1992
102 **A. M. Il'in,** Matching of asymptotic expansions of solutions of boundary value problems, 1992
101 **Zhang Zhi-fen, Ding Tong-ren, Huang Wen-zao, and Dong Zhen-xi,** Qualitative theory of differential equations, 1992
100 **V. L. Popov,** Groups, generators, syzygies, and orbits in invariant theory, 1992
99 **Norio Shimakura,** Partial differential operators of elliptic type, 1992
98 **V. A. Vassiliev,** Complements of discriminants of smooth maps: Topology and applications, 1992
97 **Itiro Tamura,** Topology of foliations: An introduction, 1992
96 **A. I. Markushevich,** Introduction to the classical theory of Abelian functions, 1992
95 **Guangchang Dong,** Nonlinear partial differential equations of second order, 1991
94 **Yu. S. Il'yashenko,** Finiteness theorems for limit cycles, 1991
93 **A. T. Fomenko and A. A. Tuzhilin,** Elements of the geometry and topology of minimal surfaces in three-dimensional space, 1991
92 **E. M. Nikishin and V. N. Sorokin,** Rational approximations and orthogonality, 1991
91 **Mamoru Mimura and Hirosi Toda,** Topology of Lie groups, I and II, 1991
90 **S. L. Sobolev,** Some applications of functional analysis in mathematical physics, third edition, 1991
89 **Valeriĭ V. Kozlov and Dmitriĭ V. Treshchëv,** Billiards: A genetic introduction to the dynamics of systems with impacts, 1991
88 **A. G. Khovanskiĭ,** Fewnomials, 1991
87 **Aleksandr Robertovich Kemer,** Ideals of identities of associative algebras, 1991
86 **V. M. Kadets and M. I. Kadets,** Rearrangements of series in Banach spaces, 1991
85 **Mikio Ise and Masaru Takeuchi,** Lie groups I, II, 1991
84 **Đáo Trọng Thi and A. T. Fomenko,** Minimal surfaces, stratified multivarifolds, and the Plateau problem, 1991
83 **N. I. Portenko,** Generalized diffusion processes, 1990
82 **Yasutaka Sibuya,** Linear differential equations in the complex domain: Problems of analytic continuation, 1990
81 **I. M. Gelfand and S. G. Gindikin, Editors,** Mathematical problems of tomography, 1990
80 **Junjiro Noguchi and Takushiro Ochiai,** Geometric function theory in several complex variables, 1990
79 **N. I. Akhiezer,** Elements of the theory of elliptic functions, 1990
78 **A. V. Skorokhod,** Asymptotic methods of the theory of stochastic differential equations, 1989
77 **V. M. Filippov,** Variational principles for nonpotential operators, 1989
76 **Phillip A. Griffiths,** Introduction to algebraic curves, 1989
75 **B. S. Kashin and A. A. Saakyan,** Orthogonal series, 1989
74 **V. I. Yudovich,** The linearization method in hydrodynamical stability theory, 1989
73 **Yu. G. Reshetnyak,** Space mappings with bounded distortion, 1989

(Continued in the back of this publication)

TRANSLATIONS OF MATHEMATICAL MONOGRAPHS

VOLUME 107

B. M. Makarov, M. G. Goluzina, A. A. Lodkin, and A. N. Podkorytov

Selected Problems in Real Analysis

American Mathematical Society

Translated from the Russian by H. H. McFaden

1991 *Mathematics Subject Classification.* Primary 26-01, 28-01.

Library of Congress Cataloging-in-Publication Data

Izbrannye zadachi po matematicheskomu analizu. English.
Selected problems in real analysis/B. M. Makarov...[et al.].
p. cm.—(Translations of mathematical monographs, ISSN 0065-9282; v. 107)
Includes bibliographical references.
ISBN 0-8218-4559-4 (alk. paper)
1. Functions of real variables. 2. Mathematical analysis. I. Makarov, B. M. II. Title.
III. Series.
QA331.5.I9313 1992 92-15594
515′.8—dc20 CIP

Printed in the United States of America
The paper used in this book is acid-free and falls within the guidelines
established to ensure permanence and durability. ♾
This publication was typeset using $\mathcal{AMS}$-TEX,
the American Mathematical Society's TEX macro system.

10 9 8 7 6 5 4 3 2 02 01 00 99 98

Contents

Foreword

This problem book is intended first and foremost for students wishing to deepen their knowledge of mathematical analysis, and for lecturers conducting seminars in university mathematics departments. It is somewhat different from the usual problem books in the greater difficulty of the problems, which include a number of well-known theorems in analysis. Despite this, no special preparation is required to solve the problems in Chapters I–VII and in §1 of Chapter X, and many of them are accessible even to first-year students in the second semester. All the facts needed to solve these problems are contained in the standard university analysis texts, in particular, in the books of Zorich [7], Kudryavtsev [16], Rudin [23], and Fikhtengol′ts [29]. The problems in Chapters VIII and IX and §2 of Chapter X require a somewhat higher level of preparation of the reader and presuppose that he is familiar with the fundamental concepts of measure theory. The corresponding facts can be found in the last chapter of the cited text of Rudin and, in more complete form, in the books [14] of Kolmogorov and Fomin and [4] of Vulikh.

The contents of the first seven chapters, which include about two thirds of all the problems, do not go outside the framework of the classical topics of analysis (functions, derivatives, integrals, asymptotics). Both here and in the subsequent chapters we make no attempt at maximal generality and, when we have to choose between a more general and less general formulation of a problem, often show preference to the latter. Chapters VIII–X are less traditional for a problem book in analysis. Besides the tastes of the authors, the program of the mathematical analysis course adopted at Leningrad University served as a criterion in choosing the material for these chapters. Problems going outside its framework and relating to the "theory of functions of a real variable" (in spite of the arbitrariness of this division) are not included in the book. For example, we do not use many attractive problems whose solution is based on the Lebesgue theorem on differentiation of an integral with respect to a variable upper limit. Also, almost no reflections of problems connected with complex analysis are found. We refer the readers interested in this circle of questions to the widely known collection of Pólya and Szegő [21] and to the book of Titchmarsh [27].

We tried to combine problems dealing with separate topics or methods into cycles within whose confines it would be possible to exhaust, step by step, various circles of questions with sufficient thoroughness. Partly for this reason we were not able to avoid a certain lack of homogeneity in the degree of difficulty of neighboring problems, which can grow markedly in the course of a single cycle. Therefore, it is not rare that more difficult problems give way to relatively simple problems, and the reader who has not solved some problem must not feel disheartened and can hope in full for success in the solution of subsequent problems.

Brief but often detailed solutions of most of the problems are given in the second part of the book. However, we advise the reader not to be in a hurry to use this part of the book and thus miss the chance of devising a better solution than the one presented there.

The literature in analysis and, in particular, textbooks and problem collections, contains an enormous amount of material, and we think that only a few of the problems presented can pretend to be original. We saw it as our goal first and foremost to try to systematize and introduce into everyday practice problems contained (sometimes in implicit form) in almost inaccessible sources (especially for students) and in the mathematical folklore. The experienced reader will notice that along with the traditional material are borrowings from "Matematicheskoe Prosveshchenie," "American Mathematical Monthly," and the collections [19], [22], [24]–[26], [38], and others. Problems are accompanied by references to the literature only in exceptional cases.

The general editing of this problem book was done by B. M. Makarov.

We express sincere gratitude to our friends and colleagues A. B. Aleksandrov, D. A. Vladimirov, E. D. Gluskin, Yu. G. Dutkevich, V. V. Zhuk, K. P. Kokhas′, M. Yu. Lyubich, G. I. Natanson, A. V. Osipov, A. I. Plotkin, O. I. Reĭnov, B. A. Samokish, S. V. Khrushchev, and D. V. Yakubovich, whose frequent advice and numerous critical remarks were of great help to us. We are also obligated to them for a number of elegant problems.

The following system was used for numbering problems and for references. Within a single chapter the problems are labelled by two numbers, the first denoting the section and the second the problem in that section. In referring to a problem in another chapter we first indicate the chapter number (a Roman numeral). For example, problem VII.2.5 is problem 2.5 in Chapter VII.

We will be grateful to all readers for their opinions and comments.

The authors

Notation

$\mathbb{N}$ is the set of natural numbers;
$\mathbb{Z}$ is the set of integers;
$\mathbb{Q}$ is the set of rational numbers;
$\mathbb{R}$ is the set of real numbers;
$\overline{\mathbb{R}}$ is the extended set of real numbers;
$\mathbb{C}$ is the set of complex numbers;
$\mathbb{R}^n$ is the additive n-dimensional space;
$\varnothing$ is the empty set;
$A \times B$ is the direct (Cartesian) product of sets A and B;
$\operatorname{card}(A)$ is the cardinality of a set A;
$f \circ g$ is the composition of mappings (functions) f and $g: (f \circ g)(x) = f(g(x))$.
$f(A)$ is the image of a set A under a mapping f;
$f^{-1}(A)$ is the complete inverse image of a set A under a mapping f;
$\overline{A}$ is the closure of a subset A of the space $\mathbb{R}^n$;
$B(x, r)$ is the open ball of radius r about a point x;
$B^n(r)$ is the ball $B(0, r)$ in $\mathbb{R}^n$;
B^n is the ball $B^n(1)$;
S^{n-1} is the unit sphere about zero in $\mathbb{R}^n$;
$\langle a, b\rangle$ stands for any one of the four kinds of intervals ($[a, b]$, (a, b), $[a, b)$, or $(a, b]$);
$f\downarrow$, $(f\uparrow)$ denote the character of monotonicity of a function f (nonincreasing, nondecreasing);
$f \downarrow A$, $(f \uparrow A)$ denote the equalities $\lim_{x\to a} f(x) = A$ $(A \in \overline{\mathbb{R}})$ for a nonincreasing (nondecreasing) function f;
$f(x) = O(g(x))$ for $x \in A$ (or on the set A) means that $|f(x)| \le C|g(x)|$ for all $x \in A$, where C is some positive number;
$f(x) = O(g(x))$ as $x \to a$ (or $f(x) \underset{x\to a}{=} O(g(x))$) means that $f(x) = O(g(x))$ on some neighborhood of the point a;
$f(x) \sim g(x)$ as $x \to a$ (or $f(x) \underset{x\to a}{\sim} g(x)$) means that $f(x) = \varphi(x)g(x)$, where $\varphi(x) \to 1$ as $x \to a$;

$f(x) = o(g(x))$ as $x \to a$ (or $f(x) \underset{x\to a}{=} o(g(x))$) means that $f(x) = \varphi(x)g(x)$, where $\varphi(x) \to 0$ as $x \to a$;

$\{a_n\}_{n\geq c}$ is a sequence (a mapping defined on the set of integers $\geq c \in \mathbb{R}$);

$\sum_{n\geq c} a_n$ denotes a series and its sum;

$f_n(x) \rightrightarrows f(x)$ on A denotes uniform convergence of a sequence $\{f_n\}_{n\geq c}$ of functions to a function f on a set A;

In the notation connected with sequences, namely, $\lim_{n\to\infty} a_n$, $a_n \sim b_n$ as $n \to \infty$, etc., we often omit the expression $n \to \infty$ and write $\lim a_n$, $a_n \sim b_n$, etc., let $\{a_n\}$ mean $\{a_n\}_{n\geq 1}$, and let $\sum a_n$ mean $\sum_{n\geq 1} a_n$;

$C(X)$ is the set of functions continuous on a set X;

$C^r(\Delta)$ is the set of functions r times continuously differentiable on an interval $\Delta \subset \mathbb{R}$ $(0 \leq r \leq +\infty)$;

$\mathrm{Lip}_\alpha(\Delta)$ is the set of functions f satisfying the condition $f(x) - f(y) = O(|x-y|^\alpha)$ on the square $\Delta \times \Delta \subset \mathbb{R}^2$ (a Lipschitz condition on the interval $\Delta \subset \mathbb{R}$ with exponent $\alpha > 0$);

Lip_α is the set $\mathrm{Lip}_\alpha(\mathbb{R})$;

$\mathscr{L}^0(A)$ is the set of Lebesgue-measurable functions defined on a (measurable) subset A of the space $\mathbb{R}^n$ and finite almost everywhere;

$\mathscr{L}^r(A)$ is the set of functions f in $\mathscr{L}^0(A)$ such that the function $|f|^r$ is integrable on the set A;

$\mathscr{L}^\infty(A)$ is the set of functions f in $\mathscr{L}^0(A)$ such that the essential supremum of $|f|$ is finite;

$\operatorname{ess\,sup}_A f$ is the essential supremum of a function f in $\mathscr{L}^0(A)$;

λ_n is Lebesgue measure on the space $\mathbb{R}^n$;

λ is the measure λ_1;

α_n is the volume (n-dimensional Lebesgue measure) of the ball B^n;

$[x]$ is the integer part of a real number x;

$x \bmod y$ stands for the number $x - [x/y]y$ $(x, y \in \mathbb{R},\ y > 0)$;

a.e. is an abbreviation for "almost everywhere";

Problems

CHAPTER I

Introduction

§1. Sets

A *binary sequence* is defined to be a sequence "consisting of 0's and 1's," that is, a sequence $\varepsilon = \{\varepsilon_k\}$, where $\varepsilon_k = 0$ or 1 for any $k \in \mathbb{N}$. The set of all possible binary sequences will be denoted by Ξ.

In speaking of sets whose elements, in turn, are sets, we use the expression "*system of sets*."

1.1. Let $\mathscr{P}(\mathbb{N})$ be the system of all possible subsets of $\mathbb{N}$. Prove that:

a) the sets Ξ and $\mathscr{P}(\mathbb{N})$ have the same cardinality;

b) the sets Ξ and $\Xi \times \Xi$ have the same cardinality.

1.2. Prove that the set Ξ has the cardinality of the continuum.

1.3. Prove that the sets $\mathbb{R}^2$ and $\mathbb{R}^3$ have the cardinality of the continuum.

1.4. Prove that the set of all sequences of real numbers has the cardinality of the continuum.

1.5. Prove that the set of continuous functions on an interval $[a, b]$ has the cardinality of the continuum.

1.6. Determine whether there is a system $\mathfrak{A}$ of subsets of $\mathbb{N}$ satisfying the following conditions:

a) $\mathfrak{A}$ has the cardinality of the continuum;

b) $\operatorname{card}(A \cap B) < +\infty$ for any $A, B \in \mathfrak{A}$.

1.7. Determine whether there is a system $\mathfrak{A}$ of subsets of $\mathbb{N}$ satisfying the following conditions:

a) $\mathfrak{A}$ has the cardinality of the continuum;

b) for any number t and any sets $A, B \in \mathfrak{A}$ the inequality $|a - b| < t$ is valid only for finitely many of the points $a \in A$ and $b \in B$.

1.8. Determine whether there is a system $\mathfrak{A}$ of subsets of $\mathbb{N}$ satisfying the following conditions:

a) $\mathfrak{A}$ has the cardinality of the continuum;

b) the system $\mathfrak{A}$ is linearly ordered by inclusion, that is, of any two sets in $\mathfrak{A}$, one is contained in the other.

1.9. Let $\mathscr{F}(\mathbb{N}) = \{A \subset \mathbb{N} \mid \operatorname{card}(A) < +\infty\}$ be the system of all finite subsets of $\mathbb{N}$. Prove that:

a) the system $\mathscr{F}(\mathbb{N})$ is countable;

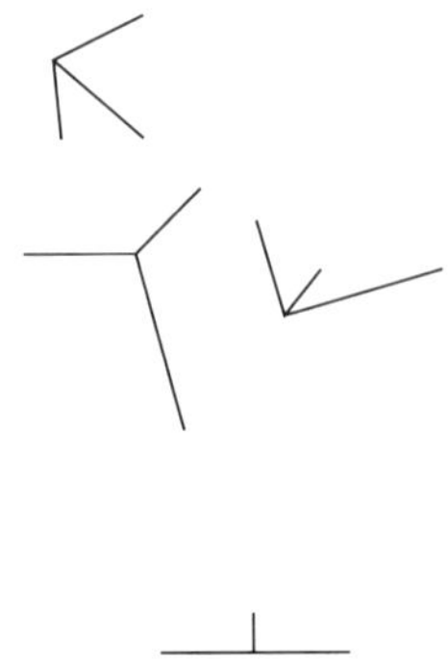

FIGURE 1

b) there exists a function $\varphi : \mathscr{F}(\mathbb{N}) \to \mathbb{N}$ such that $\varphi(A) \le \varphi(B)$ if $A \subset B$ $(A, B \in \mathscr{F}(\mathbb{N}))$.

1.10. a) Define an equivalence relation in the set $\mathbb{R}_+ = (0, +\infty)$ by taking $x \sim y$ if $x/y \in \mathbb{Q}$. Prove that the intersection of each equivalence class with any (nonempty) open interval contained in $\mathbb{R}_+$ is nonempty.

b) Define an equivalence relation on the circle $S^1 = \{z \in \mathbb{C} \mid |z| = 1\}$ by taking $z \sim \zeta$ if $z/\zeta = e^{2\pi i\theta}$, where $\theta \in \mathbb{Q}$. Prove that the set of limit points of any equivalence class coincides with S^1.

1.11. Let $S^1 = \{z \in \mathbb{C} \mid |z| = 1\}$. Sets $A, B \subset S^1$ are said to be *congruent* if there exists a number $\alpha \in \mathbb{R}$ such that $B = \{ze^{i\alpha} \mid z \in A\}$, that is, B "is obtained from A by a rotation through the angle α."

Prove that there exists a sequence $\{E_n\}$ of pairwise disjoint and congruent sets such that $S^1 = \bigcup_{n\ge1} E_n$.

1.12. Prove that on the plane it is possible to arrange a continuum of disjoint figures five, but at most countable many figures eight.

1.13. A bird track is defined to be a set on the plane that is the union of three segments with a common endpoint (the vertex of the track) and lying on distinct rays (Figure 1). Prove that on the plane it is possible to arrange an at most countable set of disjoint bird tracks.

1.14. By a T-shaped figure we understand a union of two mutually perpendicular segments with the midpoint of one an endpoint of the other. Obtain an upper estimate for the number N of disjoint T-shaped figures formed by segments of unit length and contained in a square with sides of length a.

1.15. Prove that in space it is possible to arrange at most countably many disjoint "hoops" (cylindrical rings) of fixed radius (the thickness of the hoops is equal to zero).

1.16. After losing all competitions with Balda, the devils (there are infinitely many of them) decided to develop sports skills and organized sport classes. In each class there were only finitely many devils, but there were so many classes that in any infinite company of demons it was possible to find at least two registered in the same class. Prove that with the exception of finitely many lazy demons each devil is registered in infinitely many classes.*

1.17. Suppose that a sequence $\{A_n\}$ of finite subsets of $\mathbb{N}$ thickly covers $\mathbb{N}$, that is, for any infinite set $B \subset \mathbb{N}$ there is an index m such that $\operatorname{card}(B \cap A_m) \geq 2$. Prove that:

a) the natural numbers belonging only to finitely many of the sets A_n form a finite set;

b) there exists an infinite set $E \subset N$ such that $E \setminus \bigcup_{n \in \Delta_k} A_n \subset \{1, 2, \dots, k-1\}$ for any $k \in E$, where $\Delta_k = \{n \in \mathbb{N} \mid k \in A_n\}$.

1.18. If $\{A_n\}$ and $\{B_n\}$ are two sequences of finite sets, each thickly covering $\mathbb{N}$ (see problem 1.17), then there exist indices p and q such that $\operatorname{card}(A_p \cap B_q) \geq 2$.

1.19. Suppose that $E \subset \mathbb{R}$ and $(x+y)/2 \in E$ for any $x, y \in E$.

a) Is it true that $E \supset [500, 1000]$ if $E \supset [0, 1]$ and $1992 \in E$?

b) Prove that if $\operatorname{Int} E \neq \varnothing$, then E is an interval.

1.20. Find all limit points of the following sets:

a) $\{n^{-1} + m^{-1} \mid n, m \in \mathbb{N}\}$, b) $\{m + n\sqrt{2} \mid m, n \in \mathbb{Z}\}$,

c) $\{\sqrt{m} - \sqrt{n} \mid m, n \in \mathbb{N}\}$, d) $\left\{ \dfrac{\sqrt{m} - \sqrt{n}}{\sqrt{m} + \sqrt{n}} \,\middle|\, m, n \in \mathbb{N} \right\}$.

1.21. Let $E \subset (0, +\infty)$, $E \neq \varnothing$. Prove that if $x/2 \in E$ and $\sqrt{x^2 + y^2} \in E$ for any $x, y \in E$, then $\overline{E} = [0, +\infty)$.

1.22. Let $E \subset \mathbb{R}^n$. Prove that:

a) if the family $\{B_\alpha\}_{\alpha \in A}$ of open balls is such that $E \subset \bigcup_{\alpha \in A} B_\alpha$, then there exists an at most countable set $A_0 \subset A$ such that $E \subset \bigcup_{\alpha \in A_0} B_\alpha$;

b) there exists an at most countable subset of E whose closure contains E.

1.23. a) A set is said to be *discrete* if any point in it is isolated. Prove that every discrete set on the plane is at most countable and its closure cannot have interior points.

b) A point a in a set $E \subset \mathbb{R}$ is said to be *semi-isolated* if there exists a number $\varepsilon > 0$ such that at least one of the intervals $(a - \varepsilon, a)$ and $(a, a + \varepsilon)$ does not contain points of E.

Prove that the set of semi-isolated points of any set $E \subset \mathbb{R}$ is at most countable.

1.24. Let $E \subset \mathbb{N}$, $\operatorname{card} E = +\infty$. Prove the existence of a number $a > 1$ such that infinitely many of the numbers $[a^k]$ $(k \in \mathbb{N})$ are contained in E.

1.25. Let G be an open subset of $\mathbb{R}$ that is unbounded above. Does there exist a positive number x_0 such that the set G contains infinitely many points of the form nx_0 $(n \in \mathbb{N})$?

*This refers to a poem of Pushkin well known in Russia: "The tale of the priest and his worker Balda (Сказка о попе и работнике его Балде)."

1.26. Let $\{G_n\}$ be a sequence of open subset of $\mathbb{R}$ that are unbounded above. Prove that there is a number $x_0 > 0$ such that each of the sets G_n contains infinitely many points of the form mx_0 $(m \in \mathbb{N})$.

The *Cantor set* K is defined to be the intersection of the sets $K_n \subset \mathbb{R}$ $(n \in \mathbb{N})$ defined as follows. The set K_1 is obtained by removing the middle third $(1/3, 2/3)$ of $[0, 1]$. In other words, K_1 is the union of the two closed intervals

$$\Delta_0 = [0, 1/3] \quad \text{and} \quad \Delta_1 = [2/3, 1],$$

which we call the intervals of first rank.

The set K_2 is obtained by removing the middle thirds of the intervals of the first rank, that is, the union of the four closed intervals

$$\Delta_{00} = [0, 1/9], \quad \Delta_{01} = [2/9, 1/3], \quad \Delta_{10} = [2/3, 7/9], \quad \Delta_{11} = [8/9, 1],$$

which we call the intervals of second rank.

The subsequent construction continues by induction: the set K_{n+1} is obtained by removing the middle thirds of the (closed) intervals of nth rank. It is convenient to "enumerate" the intervals of nth rank with the help of the indices $\varepsilon_1, \ldots, \varepsilon_n$, where ε_j takes the values 0 and 1. The indices of the intervals of first and second ranks have already been indicated, and the rest of the indexing is done by induction. Suppose that the intervals of nth rank have already been supplied with indices, and $\Delta_{\varepsilon_1 \cdots \varepsilon_n}$ is one of them. The two intervals of $(n+1)$st rank obtained after removing the middle third of $\Delta_{\varepsilon_1 \cdots \varepsilon_n}$ are "enumerated" as follows: the left-hand one is denoted by $\Delta_{\varepsilon_1 \cdots \varepsilon_n 0}$, and the right-hand one by $\Delta_{\varepsilon_1 \cdots \varepsilon_n 1}$. The set K_{n+1} is the union of all possible closed intervals of $(n+1)$st rank.

1.27. Prove that

a) K has the cardinality of the continuum;

b) K is closed and does not have isolated points;

c) the sum of the lengths of the open intervals making up the set $[0, 1] \setminus K$ is equal to 1.

1.28. Let K be the Cantor set.

a) Prove that a number t belongs to K if and only if it is representable in the form $t = \sum 2\varepsilon_j 3^{-j}$, where ε_j is equal to 0 or 1.

b) Describe the sets

$$K + K = \{s + t \mid t, s \in K\} \quad \text{and} \quad K - K = \{s - t \mid s, t \in K\}.$$

1.29. Suppose that a set $E \subset \mathbb{R}$ has the property that for any points $x, y \in E$, $x < y$, there exists a point $z \in E$ such that $x < z < y$. Does the closure of E necessarily contain an interior point?

1.30. Construct a discrete set on the plane whose closure has the cardinality of the continuum. Does there exist a discrete set on the line with the property? (See problem 1.23 for the definition of a discrete set.)

1.31. Let K be the intersection of the sets $K_n \subset \mathbb{R}$ $(n \in \mathbb{N})$ defined as follows. The set K_1 is obtained by removing from the closed interval $K = [a, b]$ a nonempty open interval $\delta = (p, q)$ whose endpoints are different from a and b. In other words, K_1 is the union of the two closed intervals $\Delta_0 = [a, p]$ and $\Delta_1 = [q, b]$, which we call the intervals of first rank.

The set K_2 is obtained after removing from Δ_0 and Δ_1 open intervals δ_0 and δ_1 whose endpoints are different from those of Δ_0 and Δ_1, respectively. The set $\Delta_\varepsilon \backslash \delta_\varepsilon$ $(\varepsilon = 0, 1)$ consists of two closed intervals, the left-hand one of which we denote by $\Delta_{\varepsilon 0}$, and the right-hand one by $\Delta_{\varepsilon 1}$. Thus, K_2 is the union of four closed intervals Δ_{00}, Δ_{01}, Δ_{10}, Δ_{11}, which we call the intervals of second rank. The subsequent construction is continued by induction. Suppose that we have constructed the set K_n made up of the closed intervals of nth rank. It is convenient to "enumerate" the intervals of nth rank with the help of the indices $\varepsilon_1, \dots, \varepsilon_n$, where ε_j can take the value 0 or 1. The indices of the intervals of first and second ranks have already been indicated, and the subsequent indexing is carried out as follows. In constructing the (closed) intervals of $(n+1)$st rank we remove from each closed interval $\Delta_{\varepsilon_1 \cdots \varepsilon_n}$ of nth rank a nonempty open interval $\delta_{\varepsilon_1 \cdots \varepsilon_n}$ whose endpoints are different from those of $\Delta_{\varepsilon_1 \cdots \varepsilon_n}$. The difference $\Delta_{\varepsilon_1 \cdots \varepsilon_n} \backslash \delta_{\varepsilon_1 \cdots \varepsilon_n}$ consists of two intervals of $(n+1)$st rank, the left-hand one of which we denote by $\Delta_{\varepsilon_1 \cdots \varepsilon_n 0}$, and the right-hand one by $\Delta_{\varepsilon_1 \cdots \varepsilon_n 1}$. The set K_{n+1} is the union of all the intervals of $(n+1)$st rank. If the set $K = \bigcap_{n=1}^{\infty} K_n$ does not have interior points, then it is called a *generalized Cantor set*.

Prove that:

a) K has the cardinality of the continuum;

b) K is closed and does not have isolated points;

c) K is a generalized Cantor set if and only if $l_n \to 0$, where l_n is the maximal length of the intervals of nth rank.

1.32. Let $\{\Delta_{\varepsilon_1 \cdots \varepsilon_n}\}$, where $n \in \mathbb{N}$ and ε_j can take the values 0 or 1, be a family of nonempty bounded intervals satisfying the following conditions:

1) $\Delta_{\varepsilon_1 \cdots \varepsilon_n} \supset \Delta_{\varepsilon_1 \cdots \varepsilon_n 0} \cup \Delta_{\varepsilon_1 \cdots \varepsilon_n 1}$;

2) for each $n \in \mathbb{N}$ the intervals $\Delta_{\varepsilon_1 \cdots \varepsilon_n}$ and $\Delta_{\varepsilon_1' \cdots \varepsilon_n'}$ are disjoint if $(\varepsilon_1, \dots, \varepsilon_n) \neq (\varepsilon_1', \dots, \varepsilon_n')$.

Let $G_n = \bigcup_{\varepsilon_1, \dots, \varepsilon_n \in \{0,1\}} \Delta_{\varepsilon_n \cdots \varepsilon_n}$. Prove that the set $A = \bigcap_{n=1}^{\infty} G_n$ has the cardinality of the continuum.

1.33. Prove that a nonempty open interval cannot be represented as the union of a sequence of disjoint closed sets.

1.34. Prove that the plane cannot be covered by a sequence of closed disks without common interior points.

1.35. Suppose that $-\infty \leq a < b \leq +\infty$ and $(a, b) \subset \bigcup_{n=1}^{\infty} E_n$. Prove that the closure of at least one of the sets E_n has an interior point.

1.36. Prove that the set of irrational numbers is not the union of a sequence of closed sets.

1.37. Let $\nu = \{n_k\}$ be a strictly increasing sequence of positive integers, and let $E_\nu = \{\sum \varepsilon_k 2^{-n_k} | \varepsilon_k = 0 \text{ or } 1\}$.

a) Prove that E_ν is closed and does not have isolated points.

b) Prove that the following assertions are equivalent:

1) E_ν does not contain interior points;

2) $n_{k+1} > 1 + n_k$ infinitely many times;

3) the expansion $t = \sum \varepsilon_k 2^{-n_k}$ (where $\varepsilon_k = 0$ or 1) is unique.

§2. Inequalities

2.1. Prove that for any finite sequence $\{a_k\}_{k=1}^n$ of real numbers there is an index $m \in \{0, \dots, n\}$ such that

$$\left| \sum_{1 \le k \le m} a_k - \sum_{m < k \le n} a_k \right| \le \max_{1 \le k \le n} |a_k|$$

(for $m = 0$ the first sum is taken to be zero, and for $m = n$ the second).

2.2. Let $\{a_k\}_{k=1}^n$ be a finite sequence of positive numbers, and let $M = \max_{1 \le k \le n} a_k$ and $m = \min_{1 \le k \le n} a_k$. Prove that

a)

$$2n\sqrt{\frac{m}{M}} \le \sum_{1 \le k \le n} \frac{a_k}{M} + \sum_{1 \le k \le n} \frac{m}{a_k} \le n\left(1 + \frac{m}{M}\right);$$

b)

$$n^2 \le \sum_{1 \le k \le n} a_k \sum_{1 \le k \le n} \frac{1}{a_k} \le n^2 \frac{(m + M)^2}{4mM}.$$

2.3. Let $\{a_k\}_{k=1}^n$ be a finite sequence of nonnegative numbers with $S = a_1 + a_2 + \cdots + a_n < 1$. Prove the inequalities:

a) $1/(1 - S) \ge \prod_{1 \le k \le n}(1 + a_k) \ge 1 + S$;

b) $1/(1 + S) \ge \prod_{1 \le k \le n}(1 - a_k) \ge 1 - S$.

The left-hand inequalities are strict if $S > 0$. The right-hand inequalities are strict if at least two of the numbers $\{a_k\}_{k=1}^n$ are positive.

2.4. Let $\{a_k\}_{k=1}^n$ be a finite sequence of real numbers with $a_k > -1$ for $k = 1, \dots, n$. Prove that:

a) if $S = a_1 + \cdots + a_n \ge 0$, then

$$\prod_{1 \le k \le n} (1 + a_k) \le 1 + \frac{S}{1!} + \cdots + \frac{S^n}{n!},$$

with equality possible only in the cases when $n = 1$ or $a_1 = \cdots = a_n = 0$;

b) if

$$\sigma = \frac{a_1}{1 + a_1} + \cdots + \frac{a_n}{1 + a_n} \ge 0,$$

then

$$1+\frac{\sigma}{1!}+\cdots+\frac{\sigma^n}{n!} \le \prod_{1\le k\le n}(1+a_k),$$

with equality possible only in the case when $a_1=\cdots=a_n=0$.

2.5. Let $\{p_n\}$ be the sequence of prime numbers, in increasing order ($p_1=2$). Prove the inequalities:

a) $(1-\frac{1}{2})^{-1}(1-\frac{1}{3})^{-1}\cdots(1-\frac{1}{p_m})^{-1} > \sum_{1\le n\le p_m} 1/n$;

b) $1+\sum_{1\le n\le m} 1/p_n > \ln\ln p_m$.

2.6. For finite sequences $\{a_k\}_{k=1}^n$ and $\{b_k\}_{k=1}^n$ of real numbers let the expressions $\{\hat{a}_k\}_{k=1}^n$ and $\{\hat{b}_k\}_{k=1}^n$ (respectively, $\{\check{a}_k\}_{k=1}^n$ and $\{\check{b}_k\}_{k=1}^n$) denote the nondecreasing (respectively, nonincreasing) rearrangements of those sequences. Prove the inequalities:

$$\sum_{1\le k\le n}\check{a}_k\hat{b}_k \le \sum_{1\le k\le n} a_kb_k \le \sum_{1\le k\le n}\hat{a}_k\hat{b}_k.$$

2.7. Let $\{a_k\}_{k=1}^n$ and $\{b_k\}_{k=1}^n$ be finite sequences of real numbers. Prove the inequalities:

$$\sum_{1\le k\le n}\check{a}_k\hat{b}_k \le \frac{1}{n}\sum_{1\le k\le n} a_k \sum_{1\le k\le n} b_k \le \sum_{1\le k\le n}\hat{a}_k\hat{b}_k.$$

In the solution of the following problems the *Abel transformation* can turn out to be useful:

$$\sum_{1\le k\le n} a_kb_k = a_nB_n + \sum_{1\le k\le n}(a_k-a_{k+1})B_k,$$

where $B_k=b_1+\cdots+b_k$ $(k=1,\dots,n)$. It is especially convenient to use this equality in those cases when the sequence $\{a_k\}_{k=1}^n$ is monotone.

2.8. Let $\{a_k\}_{k=1}^n$ and $\{b_k\}_{k=1}^n$ be finite sequences of real numbers, with the sequence $\{a_k\}_{k=1}^n$ nonincreasing and nonnegative. Prove the inequalities:

a) $a_1\min\limits_{1\le k\le n}(b_1+\cdots+b_k) \le \sum_{1\le k\le n} a_kb_k \le a_1\max\limits_{1\le k\le n}(b_1+\cdots+b_k)$;

b) $|\sum_{1\le k\le n} a_kb_k| \le a_1\max_{1\le k\le n}|b_1+\cdots+b_k|$;

c) $m\sum_{1\le k\le n} a_k \le \sum_{1\le k\le n} a_kb_k \le M\sum_{1\le k\le n} a_k$,

where $m=\min_{1\le k\le n}\frac{1}{k}(b_1+\cdots+b_k)$ and $M=\max_{1\le k\le n}\frac{1}{k}(b_1+\cdots+b_k)$. Here M cannot be replaced by a smaller factor, nor m by a larger factor.

2.9. Let $\{a_k\}_{k=1}^n$ and $\{b_k\}_{k=1}^n$ be finite nonnegative sequences, with $\{a_k\}_{k=1}^n$ nonincreasing. Consider an index $m\in\{1,2,\dots,n\}$ such that $m\max_{1\le k\le n} b_k \ge \sum_{1\le k\le n} b_k$. Prove that:

$$\sum_{1\le k\le n} a_kb_k \le \left(\max_{1\le k\le n} b_k\right)\sum_{1\le k\le m} a_k.$$

If $m \max_{1\le k\le n} b_k = \sum_{1\le k\le n} b_k$, then

$$\left(\max_{1\le k\le n} b_k\right) \sum_{n-m<k\le n} a_k \le \sum_{1\le k\le n} a_k b_k .$$

2.10. Let $\{a_k\}_{k=1}^n$ be a finite sequence of real numbers. Prove that:

a)

$$0 \le \frac{1}{n} \sum_{1\le k\le n} a_k^2 - \left(\frac{1}{n}\sum_{1\le k\le n} a_k\right)^2 \le \frac{\Delta^2}{12}(n^2-1),$$

where $\Delta = \max_{1\le k<n} |a_k - a_{k+1}|$;

b) if $\{a_k\}_{k=1}^n$ is monotone, then

$$\frac{\delta^2}{12}(n^2-1) \le \frac{1}{n} \sum_{1\le k\le n} a_k^2 - \left(\frac{1}{n}\sum_{1\le k\le n} a_k\right)^2 ,$$

where $\delta = \min_{1\le k<n} |a_k - a_{k+1}|$.

In what cases do these inequalities become equalities?

2.11. Let $\{a_k\}_{k=1}^n$ be a nondecreasing convex (or nonincreasing concave) sequence of real numbers. Prove the inequalities

$$\frac{1}{3n^2} \sum_{1\le k<n} (n-k)^3 (a_{k+1}-a_k)^2 \le \frac{1}{n} \sum_{1\le k\le n} a_k^2 - \left(\frac{1}{n}\sum_{1\le k\le n} a_k\right)^2$$
$$\le \frac{1}{3n^2} \sum_{1\le k<n} (k+1)^3 (a_{k+1}-a_k)^2 .$$

If $\{a_k\}_{k=1}^n$ is nonincreasing and convex (or nondecreasing and concave), then the inequalities should be reversed.

2.12. Suppose that $\{a_k\}_{k=1}^n$ and $\{b_k\}_{k=1}^n$ are finite sequences of nonnegative numbers. Prove that:

a) if $\min_{b\le k\le n}(b_k - a_k) = \Delta \ge 0$, then $\sqrt[n]{b_1\cdots b_n} \ge \Delta + \sqrt[n]{a_1\cdots a_n}$;

b) $\sqrt[n]{(a_1+b_1)\cdots(a_n+b_n)} \ge \sqrt[n]{a_1\cdots a_n} + \sqrt[n]{b_1\cdots b_n}$.

2.13. Prove that the functions $x \mapsto (1+1/x)^x$ and $x \mapsto (1+1/x)^{x+1}$ are strictly monotone on each of the intervals $(-\infty, -1)$ and $(0, +\infty)$.

2.14. Prove that: a) $(n/e)^n < n!$ for $n \in \mathbb{N}$; b) $n! < n(n/e)^n$ for $n \ge 11$. These inequalities imply that $n! = (n/e)^n n^{\alpha_n}$, $0 < \alpha_n < 1$. More precise representations of factorials are considered in problems II.2.10 and II.2.11.

2.15. Prove the following inequalities for any $x \in [0, 1]$:

a) $2 \le (1+x)^p + (1-x)^p \le 2^p \quad (p \ge 1)$;

b) $(1+x)^p + (1-x)^p \le 2^{p-1}(1+x^p) \quad (p \ge 2)$;

c) $(1+x)^p + (1-x)^p \le 2(1+x^{p/(p-1)})^{p-1} \quad (p \ge 2)$;

(Which of the inequalities b) and c) is stronger than the other?)

d) $(1+x)^2 + (p-1)(1-x)^2 \le 4^{1-1/p}(1+x^p)^{2/p} \quad (1 \le p \le 2)$.

2.16. Let $n \in \mathbb{N}$ and $t \in [0, n]$. Prove the inequalities:
a) $0 \le e^{-t} - (1 - t/n)^n \le t^2 e^{-t}/n$ for $n \ge 1$;
b) $t^2 e^{-t}/n^2 \le e^{-t} - (1 - t/n)^n$ for $n \ge 2$;
c) $e^{-t} - (1 - t/n)^n \le te^{-t}/2\sqrt{n}$ for $n \ge 36$.
d) Prove that

$$0 \le \left(1 + \frac{t}{n}\right)^{-n} - e^{-t} \le \frac{t^2}{n} e^{-t}$$

for $n \ge 1$ and $t \in [0, \sqrt{n}]$.

2.17. Suppose that $n \in \mathbb{N}$ and $t \in [0, n]$. Prove the inequalities

$$0 \le e^{-t^2/2n} - e^t \left(1 - \frac{t}{n}\right)^n \le \frac{1}{\sqrt{en}}.$$

The following problem considers relations supplementing the classical inequalities

$$\sin x < x < \tan x, \qquad \ln(1 + x) < x < e^x - 1,$$
$$\arctan x < x < \arcsin x.$$

2.18. Prove the inequalities:
a) $(1 + x) \ln^2(1 + x) < x^2$ for $x > -1$, $x \ne 0$;
b) $x/(1 + 2x/\pi) < \arctan x$ for $x > 0$;
c) $x^2 < \ln(1 + \tan^2 x) < (\sin x) \tan x$ for $x \in (0, \pi/2)$;
d) $3x - x^3 < 2 \sin(\pi x/2)$ for $x \in (0, 1)$;
e) $x^3 < (\sin^2 x) \tan x$ for $x \in (0, \pi/2)$.
f) Is the inequality $x^{3+\varepsilon} < (\sin x)^{2+\varepsilon} \tan x$ true on $(0, \pi/2)$ for some fixed positive number ε?

2.19. Prove the inequalities:
a) $(e^x - 1) \ln(1 + x) > x^2$ for $x > 0$;
b) $(\tan x) \arctan x > x^2$ for $x \in (0, \pi/2)$;
c) $((1 + x)^p - 1)((1 + x)^{1/p} - 1) > x^2$ for $x > 0$ and $p > 0$, $p \ne 1$;
(What happens for $p < 0$?)
d) $((1 - x)^p - 1)(1 - (1 + x)^{1/p}) > x^2$ for $x \in (0, 1)$ and $p < -1$.
(What happens for $p \in (-1, 0)$?)

2.20. Prove the inequalities

$$\frac{2}{\pi} \left(\sin \frac{\pi}{2} x\right) \arcsin x < x^2 < (\sin x) \arcsin x$$

for $x \in (0, 1)$.

Problems 2.21–2.27 concern integral inequalities which are analogues of the summation inequalities already considered. They can be proved by two paths: either by taking a limit in a summation inequality, or by modifying the proof used in the discrete case. We remark also that by passing to the limit, the inequalities in problems 2.23 and 2.24 can be carried over to the case of improper integrals over an interval of the form $[a, +\infty)$.

2.21. Suppose that the function f is integrable on $[a, b]$ and $m = \inf f > 0$, and let $M = \sup f$. Then the following inequalities hold:

a)

$$2\sqrt{\frac{m}{M}}(b-a) \leq \frac{1}{M}\int_a^b f(x)\,dx + m\int_a^b \frac{dx}{f(x)} \leq \left(1+\frac{m}{M}\right)(b-a);$$

b)

$$(b-a)^2 \leq \int_a^b f(x)\,dx \int_a^b \frac{dx}{f(x)} \leq (b-a)^2\frac{(m+M)^2}{4mM}.$$

2.22. Suppose that the functions f and g are monotone on $[a, b]$. If the monotonicity is of a single type, then

$$\int_a^b f(x)g(a+b-x)\,dx \leq \frac{1}{b-a}\int_a^b f(x)\,dx \int_a^b g(x)\,dx \leq \int_a^b f(x)g(x)\,dx.$$

If the monotonicity is of different types, then the inequalities are reversed.

2.23. Suppose that the functions f and g are integrable on $[a, b]$, $f\downarrow$, and $f \geq 0$. For $x \in [a, b]$ let $G(x) = \int_a^x g(t)\,dt$. Then the following inequalities hold:

a) $f(a)\inf_{a\leq x\leq b} G(x) \leq \int_a^x f(x)g(x)\,dx \leq f(a)\sup_{a\leq x\leq b} G(x)$;

b) $|\int_a^b f(x)g(x)\,dx| \leq f(a)\sup_{a\leq x\leq b}|G(x)|$;

c)

$$\left(\inf_{a<x\leq b}\frac{G(x)}{x-a}\right)\int_a^b f(x)\,dx \leq \int_a^b f(x)g(x)\,dx \leq \left(\sup_{a<x\leq b}\frac{G(x)}{x-a}\right)\int_a^b f(x)\,dx.$$

2.24. Suppose that the functions f and g are integrable on $[a, b]$, $f\downarrow$, and $0 \leq g \leq 1$. Let $c = \int_a^b g(x)\,dx$. Then

$$\int_{b-c}^b f(x)\,dx \leq \int_a^b f(x)g(x)\,dx \leq \int_a^{a+c} f(x)\,dx.$$

2.25. Suppose that $f \in C^1([a,b])$, $\delta = \min_{[a,b]} |f'|$, and $\Delta = \max_{[a,b]} |f'|$. Then

$$\frac{1}{12}(b-a)^2\delta^2 \le \frac{1}{b-a}\int_a^b f^2(x)\,dx - \left(\frac{1}{b-a}\int_a^b f(x)\,dx\right)^2 \le \frac{1}{12}(b-a)^2\Delta^2.$$

Equality holds only for linear functions.

2.26. If a continuously differentiable function f is nondecreasing and convex (or nonincreasing and concave) on $[0, 1]$, then

$$\frac{1}{3}\int_0^1 (1-x)^3(f'(x))^2\,dx \le \int_0^1 f^2(x)\,dx - \left(\int_0^1 f(x)\,dx\right)^2 \le \frac{1}{3}\int_0^1 x^3(f'(x))^2\,dx.$$

For nondecreasing concave (or nonincreasing convex) functions the inequalities are reversed. Equality is possible only for a linear function.

2.27. Suppose that the functions f and g are positive and integrable on $[a, b]$. Then the following inequalities hold:

a)

$$\exp\left(\frac{1}{b-a}\int_a^b \ln f(x)\,dx\right) \le \frac{1}{b-a}\int_a^b f(x)\,dx$$

(an analogue of the Cauchy inequality on the arithmetic mean and the geometric mean);

b)

$$\exp\left(\frac{1}{b-a}\int_a^b \ln f(x)\,dx\right) + \exp\left(\frac{1}{b-a}\int_a^b \ln g(x)\,dx\right) \le \exp\left(\frac{1}{b-a}\int_a^b \ln(f(x)+g(x))\,dx\right)$$

(an analogue of Minkowski's inequality; see problem 2.12b)).

2.28. Let K be the set of all nonnegative and nonincreasing functions f on $[a, b]$ such that $af(a) = bf(b)$ and $\int_a^b f(x)\,dx = 1$. Prove that for any functions f and g in K

$$\int_a^b \max(f(x), g(x))\,dx \le \frac{2\sqrt{b}}{\sqrt{a}+\sqrt{b}}.$$

For what functions is equality attained?

§3. Irrationality

3.1. Let $x \in \mathbb{R}$. Prove the equivalence of the following assertions:

a) $x \notin \mathbb{Q}$;

b) there exist infinitely many irreducible fractions p/q $(q \in \mathbb{N},\ p \in \mathbb{Z})$ such that $|x - p/q| < 1/q^2$.

c) there exists a sequence $\{p_k/q_k\}_{k\geq 1}$ of irreducible fractions ($q_k \in \mathbb{N}$, $p_k \in \mathbb{Z}$) such that $q_k \to +\infty$ and $x - p_k/q_k = o(1/q_k)$ as $k \to +\infty$.

In the following problems we use the notation $\delta(x) = x - [x]$ for the fractional part of a real number x.

3.2. Suppose that $x \in \mathbb{R}\backslash\mathbb{Q}$. Prove that the sequence $\{\delta(nx)\}$ is dense in $(0, 1)$.

3.3. a) Prove that the sequences $\{\sin n^2\}$ and $\{\sin 4^n\}$ do not have limits.

b) Prove that if the sequence $\{\sin(\theta 2^n)\}$ has a limit, then it is equal to zero and $\theta = \pi p 2^{-q}$, where $q \in \mathbb{N}$ and $p \in \mathbb{Z}$.

3.4. Suppose that $x \in \mathbb{R}$. Prove the equivalence of the following statements: a) $x \in \mathbb{Q}$; b) the set $\{\delta(n^{1992}x) | n \in \mathbb{N}\}$ is finite.

3.5. Show by examples that the sequence $\{\delta(10^n x)\}$ can be dense in $(0, 1)$, and that it can be nowhere dense.

3.6. Find the set of limits of all convergent subsequences of the sequence

$$\text{a) } \{\sin n^{2/3}\}; \quad \text{b) } \{\sqrt[4]{n} \sin \sqrt{n}\}; \quad \text{c) } \{\sin \pi n^{3/2}\};$$
$$\text{d) } \{\sin \ln n\}; \quad \text{e) } \{\sin(\pi n \ln n)\}.$$

3.7. Find the set of limits of all convergent subsequences of the sequence:

a) $\{\delta(nx)\}$, where $x \in \mathbb{Q}$; b) $\{\delta(nx)\}$, where $x \in \mathbb{R}\backslash\mathbb{Q}$;

c) $\{\delta(n^\alpha)\}$, where $\alpha \in (0, 1)$; d) $\{\delta(n^{5/2})\}$.

3.8. Does there exist a real number x such that $\delta(x^n) \in [1/3, 2/3]$ for all n?

3.9. Suppose that $f \in C([0, 1])$ and $x \in \mathbb{R}\backslash\mathbb{Q}$. Prove that

$$\frac{1}{n} \sum_{1\leq k\leq n} f(\delta(kx)) \to \int_0^1 f(t)\, dt$$

as $n \to +\infty$.

3.10. Suppose that $x = (x_1, \dots, x_m) \in \mathbb{R}^m\backslash\mathbb{Q}^m$. Prove that there exist integers $p_1, \dots, p_m$ and a positive integer q such that

$$|x_i - p_i/q| < q^{-(1+1/m)} \qquad (i = 1, 2, \dots, m),$$

and the denominator q can be taken as large as desired.

3.11. Let $\alpha_n = \min_{m\in\mathbb{Z}} |\sqrt{2} - m/n|$, $n \in \mathbb{N}$. Prove that:

a) $(2n)^{-2} < \alpha_n < (2n)^{-1}$;

b) if a strictly increasing sequence $\{n_j\}$ of indices is such that $\alpha n_j \leq C/n_j^2$ $(j \in \mathbb{N})$, where C is a fixed number, then $\{n_j\}$ increases at least like a geometric progression: $n_{j+1}/n_j \geq 1 + 1/(8C)$ (thus, the lower estimate in the inequality in a) is coarse for most indices).

In problems 3.12–3.18 $\{\varepsilon_k\}$ is an arbitrary sequence of $+1$'s and -1's, and $\{n_k\}$ is a strictly increasing sequence of positive integers.

3.12. Prove that the sum of the series $\sum \varepsilon_k 2^{-n_k}$ is irrational if $\overline{\lim}(n_{k+1} - n_k) = +\infty$.

3.13. Let $\underline{\lim}(n_1 n_2 \cdots n_k / n_{k+1}) = 0$. Prove that:

a) the sum of the series $\sum (-1)^k / n_k$ is irrational;

b) if $n_{k+1}/n_k \geq Q > 1$ for $k \in \mathbb{N}$, then the sum of the series $\sum \varepsilon_k / n_k$ is irrational.

3.14. Suppose that $\overline{\lim}\ \sqrt[2^k]{n_k} = +\infty$. Prove that:

a) the sum of the series $\sum (-1)^k / n_k$ is irrational;

b) if $n_{k+1}/n_k \geq Q > 1$ for $k \in \mathbb{N}$, then the sum of the series $\sum \varepsilon_k / n_k$ is irrational.

3.15. Suppose that α is an algebraic number of degree at most n (that is, α is a root of an algebraic polynomial of degree n with integer coefficients). Prove that if α is irrational, then there exists a number $c_\alpha > 0$ such that $|\alpha - p/q| > c_\alpha q^{-n}$ for all $p \in \mathbb{Z}$ and $q \in \mathbb{N}$.

3.16. Suppose that $\overline{\lim}\ \sqrt[3^k]{n_k} = +\infty$. Prove that:

a) the number $\sum (-1)^k / n^k$ is not a quadratic irrational;

b) if $n_{k+1}/n_k \geq Q > 1$ for $k \in \mathbb{N}$, then the number $\sum \varepsilon_k / n_k$ is not a quadratic irrational.

3.17. Prove that the numbers $\sum \varepsilon_k 2^{-k!}$ (the Liouville numbers) are transcendental.

3.18. Suppose that $\overline{\lim}\ \sqrt[k!]{n_k} > 1$. Prove that:

a) the sum of the series $\sum (-1)^k / n_k$ is a transcendental number;

b) if $n_{k+1}/n_k \geq Q > 1$ for $k \in \mathbb{N}$, then the sum of the series $\sum \varepsilon_k / n_k$ is a transcendental number.

3.19. Using the expansions $e = \sum_{n \geq 0} 1/n!$ and $e^{-1} = \sum_{n \geq 0} (-1)^n / n!$, prove that:

a) the number e is irrational;

b) $Ae^2 + Be + C \neq 0$ if A, B, and C are integers not all equal to zero.

3.20. Irrationality on the number π.

a) Let

$$I_n = \frac{1}{n!} \int_{-\pi/2}^{\pi/2} \left(\frac{\pi^2}{4} - t^2 \right)^n \cos t \, dt \qquad (n = 0, 1, 2, \dots).$$

Prove that $I_n = P_n(\pi^2)$, where P_n is an algebraic polynomial of degree at most n with integer coefficients.

b) Derive from this that π^2 (and hence also π) is an irrational number.

3.21. Irrationality of the values of the exponential at rational points.

a) Let

$$J_n(x) = \frac{1}{n!} \int_{-x}^{x} (x^2 - t^2)^n e^t \, dt \qquad (n = 0, 1, 2, \ldots).$$

Prove that $J_n(x) = A_n(x)e^x + B_n(x)e^{-x}$, where A_n and B_n are algebraic polynomials of degree n with integer coefficients.

b) Derive from this that $e^r \notin \mathbb{Q}$ if $r \in \mathbb{Q}$, $r \neq 0$.

3.22. Irrationality of the values of the tangent at rational points.

a) Let

$$I_n(x) = \frac{1}{n!} \int_{-x}^{x} (x^2 - t^2)^n \cos t \, dt \qquad (n = 0, 1, 2, \ldots).$$

Prove that $I_n(x) = C_n(x) \cos x + S_n(x) \sin x$, where C_n and S_n are algebraic polynomials of degree at most n with integer coefficients.

b) Derive from this that $\tan r \notin \mathbb{Q}$ if $r \in \mathbb{Q}$, $r \neq 0$.

CHAPTER II

Sequences

§1. Computation of limits

1.1. Let $x_n = \frac{1}{\sqrt{n}} |\sum_{1 \le k \le n} (-1)^k \sqrt{k}|$. Find the limit $\lim x_n$.

1.2. Let $x_n = \sqrt{n}(n-1)!!/n!!$. Find $\sup x_n$, $\inf x_n$, $\overline{\lim}\, x_n$, and $\underline{\lim}\, x_n$.

1.3. Find the limits:

a) $\lim \sum_{0 \le k \le 2n} 2^{-k} \cos\sqrt{k/n}$; b) $\lim \sum_{0 \le k \le 2n} 2^{-nk/(n+k)}$.

1.4. Compute the limit of the sequence $\{x_n\}$, where

$$x_n = \frac{\sqrt{n^{3n}}}{n!} \prod_{1 \le k \le n} \sin\left(\frac{k}{n^{3/2}}\right).$$

1.5. Let $a \in \mathbb{R}$ and $x_n = \sum_{1 \le k \le n} ((a+k)/n)^n$. Find the limit $\lim x_n$.

1.6. Let $S_n = \sum_{1 \le k \le n} (k/n)^n$.

a) Find the limit $\lim S_n$.

b) Determine the character of monotonicity of the sequence $\{S_n\}$.

c) Is it true that $S_n \le 2$ for any $n \in \mathbb{N}$?

1.7. Compute the limit $\lim \frac{1}{n} \sqrt[n]{(n+1)(n+2)\cdots(n+n)}$.

1.8. Suppose that the sequence $\{k_n\}$ of positive integers is such that $k_n/n \to p$, $k_n \le n$. Compute the limit $\lim \frac{1}{n} \ln C_n^{k_n}$.

1.9. Find the limits:

a) $\lim \sum_{0 \le k \le 2n} \frac{k}{k^2+n^2}$; b) $\lim \frac{1}{n^2} \sum_{0 \le k \le n} \sqrt{k(n-k)}$;

c) $\lim \sum_{0 \le k < 2n} \frac{k}{k+n^2}$; d) $\lim \sum_{n \le k \le 2n} \sin\frac{\pi}{k}$;

e) $\lim \sum_{1 \le k < n} \frac{1}{\sqrt{k(n-k)}}$.

1.10. Find the limits:

a) $\lim \sin(2\pi e n!)$; b) $\lim n \sin(2\pi e n!)$;

c) $\lim n^2 \sin(2\pi e n!)$; d) $\lim n^p \sin(\pi(\sqrt{2}+1)^n)$.

1.11. Suppose that the nondecreasing positive sequence $\{x_n\}$ has finite limit a. Prove that the series $\sum(a-x_n)$ converges if and only if $x_1x_2\cdots x_n \sim Ca^n$ for some $C>0$.

1.12. Let $a>0$ and let $x_n = \sqrt{a+\sqrt{a+\cdots+\sqrt{a}}}$ (n square roots). Prove that: a) $x_n \to l_a = \frac{1}{2}(1+\sqrt{1+4a})$; b) $l_a - x_n \sim C_a/(2l_a)^n$ for some $C_a>0$. What is the value of C_2?

1.13. Let $x_n = \sqrt[3]{6+\sqrt[3]{6+\cdots+\sqrt[3]{6}}}$ (n cubic roots). Prove that $2-x_n \sim C/(12)^n$ for some $C>0$.

1.14. Let $p>1$ and $x_n = \sqrt[p]{1+\sqrt[p]{1+\cdots+\sqrt[p]{1}}}$ (n roots). Prove that the sequence $\{x_n\}$ converges to the positive root of the equation $x^p-x-1=0$.

1.15. Suppose that $p>1$ and $\{c_n\}$ is a nonnegative sequence. Let $x_n = \sqrt[p]{c_1+\sqrt[p]{c_2+\cdots+\sqrt[p]{c_n}}}$ for $n\in\mathbb{N}$. Prove that the sequence $\{x_n\}$ converges if and only if the sequence $\{\frac{1}{p^n}\ln c_n\}$ is bounded above.

1.16. Prove that:

a) $3=\sqrt{1+2\sqrt{1+3\sqrt{1+4\sqrt{1+\cdots}}}}$;

b) $3=\sqrt{5+\sqrt{6+2\sqrt{7+3\sqrt{8+4\sqrt{9+\cdots}}}}}$;

c) $\cos x = \sqrt{\sin^2\frac{x}{2}+\cos\frac{3x}{2}\sqrt{\sin^2\frac{x}{4}+\cos\frac{3x}{4}\sqrt{\sin^2\frac{x}{8}+\cdots}}}$ for $|x|\le\pi/2$;

d) $9 = \sqrt{3u_2u_4+u_4\sqrt{3u_4u_6+u_6\sqrt{3u_6u_8+\cdots}}}$, where $\{u_n\}$ is the *sequence of Fibonacci numbers*: $u_1=u_2=1$, and $u_{n+1}=u_n+u_{n-1}$ for $n\ge 2$;

e) $2x^2 = \sqrt{2xT_0T_1+T_1\sqrt{2xT_1T_2+T_2\sqrt{2xT_2T_3+\cdots}}}$ for $x\ge1$, where $\{T_n\}$ is the sequence of values of the *Chebyshev polynomials* at the point x, namely, $T_0=1$, $T_1=x$, $T_{n+1}=2xT_n-T_{n-1}$ for $n\ge1$.

1.17. Let $x_n = \sqrt{1+\sqrt{2+\cdots+\sqrt{n}}}$. Prove that $x_n\to a\in\mathbb{R}$, and $\sqrt[n]{a-x_n}\sim\frac{1}{2}\sqrt{e/n}$.

1.18. Prove that $x_n\to0$ if $x_{n+1}-(1/2)x_n\to0$.

1.19. a) Prove that $\overline{\lim}((a_1+a_{n+1})/a_n)^n\ge e$ for any positive sequence $\{a_n\}$.

b) For any number $a\ge e$ determine a positive sequence $\{a_n\}$ such that $((a_1+a_{n+1})/a_n)^n\to a$.

1.20. Let $\{a_n\}$ be a sequence of real numbers such that $a_{n+m}\le a_n+a_m$. Prove that the limit $\lim(a_n/n)\in\overline{\mathbb{R}}$ exists and equals $\inf(a_n/n)$.

1.21. Suppose that the sequence $\{x_n\}$ of real numbers is such that $x_{n+1}-x_n\to0$. Prove that the set of limits of its convergent subsequences is the interval with endpoints $\underline{\lim}\,x_n$ and $\overline{\lim}\,x_n$.

1.22. a) Suppose that the sequence $\{x_n\}$ of real numbers is such that $x_{n+1} + x_n \to 0$. Prove that the set of limits of convergent subsequences of this sequence is either infinite or contains at most two points.

b) Suppose that the sequence $\{x_n + x_{n+1} + \cdots + x_{n+100}\}$ is convergent. Prove that the set of limits of convergent subsequences of the sequence $\{x_n\}$ can be represented as a union $\Delta_1 \cup \Delta_2 \cup \cdots \cup \Delta_{101}$, where the Δ_i are closed intervals in $\overline{\mathbb{R}}$.

1.23. Suppose that the sequence $\{x_n\}$ of real numbers is such that $x_n - x_{n-2} \to 0$. Prove that $(1/n)x_n \to 0$.

1.24. a) Give an example of a bounded divergent sequence $\{x_n\}$ such that $x_{n+1} - x_n \to 0$.

b) Does there exist a bounded sequence $\{x_n\}$ such that $x_{n+1} - x_n \to 0$, but the sequence $\{(x_1 + \cdots + x_n)/n\}$ does not have a limit?

1.25. Let $\{x_n\}$ be a sequence of real numbers such that the limit $\lim_{n\to+\infty} x_{[C^n]}$ exists for any $C > 1$. Prove that $\{x_n\}$ has a limit.

§2. Averaging of sequences

2.1. Let $\{x_n\}$ be a sequence of real numbers. Prove that $\frac{1}{n}(x_1 + \cdots + x_n) \to L$ if $x_n \to L$. The converse is true only if $\sum_{1<k\leq n} k(x_k - x_{k-1}) = o(n)$.

2.2. Suppose that $a_n \geq 0$ and $A_n = \sum_{1\leq k\leq n} a_k \to +\infty$. For an arbitrary sequence $\{x_n\}$ of real numbers let $\tilde{x}_n = \frac{1}{A_n}\sum_{1\leq k\leq n} a_k x_k$. Prove that

$$\underline{\lim}\, x_n \leq \underline{\lim}\, \tilde{x}_n \leq \overline{\lim}\, \tilde{x}_n \leq \overline{\lim}\, x_n .$$

In particular, $\tilde{x}_n \to a$ if $x_n \to a$ (for $a_n = 1$ this implies the first assertion in problem 2.1).

2.3. Suppose that A is an infinite subset of $\mathbb{N}$, $A = \{n_1, n_2, \ldots\}$ and let $p_n = \operatorname{card}\{m \in A | m \leq n\}$. If the limit $\theta(A) = \lim \frac{1}{n}p_n$ exists, then it is called the *density of the set* A (or the *density of the sequence* $\{n_k\}$). Prove that:

a) $\theta(A) = \lim(k/n_k)$;

b) for any number $\alpha \in [0, 1]$ there exists a set $A \subset \mathbb{N}$ such that $\theta(A) = \alpha$.

2.4. Let $\{x_n\}$ be a sequence of real numbers. Prove that:

a) if $x_n \to a \in \mathbb{R}$, then $\frac{1}{n}\sum_{1\leq k\leq n} |x_k - a| \to 0$;

b) if $\frac{1}{n}\sum_{1\leq k\leq n} |x_k - a| \to 0$ for some $a \in \mathbb{R}$, then there exists a sequence $\{n_k\}$ of density 1 (see problem 2.3) such that $x_{n_k} \to a$;

c) if $\{x_n\}$ is bounded, and $x_{n_k} \to a$ for some sequence $\{n_k\}$ of density 1, then $\frac{1}{n}\sum_{1\leq k\leq n} |x_k - a| \to 0$.

2.5. Let $0 < a_n < 1$ for $n \in \mathbb{N}$.

a) Show by examples that the existence of one of the limits

$$\lim \frac{1}{n}(a_1 + \cdots + a_n) \quad \text{or} \quad \lim \frac{1}{n}(a_1^2 + \cdots + a_n^2)$$

does not imply the existence of the other.

b) Suppose that $(a_1 + \cdots + a_n)/n \to a$ and $(a_1^2 + \cdots + a_n^2)/n \to b$. Prove that $a^2 \le b \le a$ and that b can take any values between a^2 and a.

2.6. Suppose that $y_{n-1} < y_n$ and $\lim y_n = +\infty$. If the sequence $\{x_n\}$ of real numbers is such that the limit

$$\lim \frac{x_n - x_{n-1}}{y_n - y_{n-1}} = l \in \overline{\mathbb{R}}$$

exists, then $x_n/y_n \to l$ (a *theorem of Stoltz*, a discrete analogue of L'Hôpital's rule). Formulate and prove a similar assertion for indeterminants of the type $0/0$.

2.7. Prove that:

a) $\displaystyle\sum_{1\le k\le n} \frac{1}{k} \sim \ln n$; b) $\displaystyle\sum_{1\le k\le n} k^\alpha \sim \frac{n^{\alpha+1}}{\alpha+1} \quad (\alpha > -1)$;

c) $\displaystyle\sum_{k\ge n} k^\alpha \sim -\frac{n^{\alpha+1}}{\alpha+1} \quad (\alpha < -1)$; d) $\displaystyle\sum_{1\le k\le n} \frac{\ln k}{k^\alpha} \sim \frac{\ln n}{1-\alpha} n^{1-\alpha} \quad (\alpha < 1)$;

e) $\displaystyle\sum_{k\ge n} \frac{\ln k}{k^\alpha} \sim \frac{\ln n}{\alpha - 1} n^{1-\alpha} \quad (\alpha > 1)$.

2.8. Prove that:

a) $\displaystyle\sum_{1\le k\le n} k^{\alpha k} \sim n^{\alpha n} \quad (\alpha > 0)$; b) $\displaystyle\sum_{k\ge n} k^{\alpha k} \sim n^{\alpha n} \quad (\alpha < 0)$;

c) $\displaystyle\sum_{1\le k\le n} a^k k! \sim a^n n! \quad (a > 0)$; d) $\displaystyle\sum_{1\le k\le n} (k!)^{-\frac{\alpha}{k}} \sim \frac{e^\alpha}{1-\alpha} n^{1-\alpha} \quad (\alpha < 1)$;

e) $\displaystyle\sum_{k\ge n} (k!)^{-\frac{\alpha}{k}} \sim \frac{e^\alpha}{\alpha-1} n^{1-\alpha} \quad (\alpha > 1)$.

The results in problems 2.9 and 2.10 play an important role in mathematical analysis. They will be used in the solution of many of the subsequent problem.

2.9. Prove that $\sum_{1\le k\le n} \frac{1}{k} = \ln n + \gamma + o(1)$ for some number γ. The number $\gamma (\gamma = 0.5772\ldots)$ is called the *Euler constant*.

2.10. Prove that $\ln(n!) = \sum_{1\le k\le n} \ln k = (n + \frac{1}{2}) \ln n - n + C + o(1)$. With the help of the Wallis formula show that $C = \frac{1}{2}\ln(2\pi)$, which implies the *Stirling formula*: $n! \sim \sqrt{2\pi n}(n/e)^n$.

2.11. Prove that for all $n \in \mathbb{N}$

$$\sqrt{2\pi n}\left(\frac{n}{e}\right)^n e^{1/(12n+1)} < n! < \sqrt{2\pi n}\left(\frac{n}{e}\right)^n e^{1/12n}.$$

In particular,

$$n! = \sqrt{2\pi n}\left(\frac{n}{e}\right)^n \left(1 + \frac{1}{12n} + O\left(\frac{1}{n^2}\right)\right).$$

2.12. Prove that as $n \to +\infty$:

a) $\sum_{1 \le k \le n} k^p = \frac{1}{1+p} n^{1+p} + \alpha_p + o(1)$ if $p \in (-1, 0]$;

b) $\sum_{1 \le k \le n} k^p = \frac{1}{1+p} n^{1+p} + \frac{1}{2} n^p + \beta_p + o(1)$ if $p \in (0, 1]$;

c) $\sum_{1 \le k \le n} k^p = \frac{1}{1+p} n^{1+p} + \frac{1}{2} n^p + \frac{p}{12} n^{p-1} + \gamma_p + o(1)$ if $p \in (1, 2]$.

2.13. Prove that as $n \to +\infty$:

a) $\sum_{1 < k \le n} \frac{1}{k} \ln^p k = \frac{\ln^{1+p} n}{1+p} + C + o(1)$ if $p > -1$;

b) $\sum_{1 < k \le n} \frac{1}{k \ln k} = \ln(\ln n) + C' + o(1)$;

c) $\sum_{1 \le k \le n} \frac{1}{\sqrt[k]{k!}} = e \ln n + C'' + o(1)$.

2.14. Prove that:

a) $\sum_{1 \le k \le n} k \ln k = \frac{1}{2}\left(n^2 + n + \frac{1}{6}\right) \ln n - \frac{1}{4} n^2 + C_1 + o(1)$;

b) $$\sum_{1 \le k \le n} \ln k! = \frac{1}{2}\left(n^2 + 2n + \frac{1}{2}\right) \ln n - \frac{3}{4} n^2 + \left(\frac{1}{2} \ln 2\pi - 1\right) n + C_2 + o(1).$$

§3. Recursive sequences

3.1. Let $x_0 = a$, $x_1 = b$, and $x_{n+2} = \frac{1}{3}(x_n + 2x_{n+1})$ $(n \in \mathbb{N})$. Prove that the limit $\lim x_n$ exists, and find it.

3.2. Let $x_0 = a$ and $x_1 = b$. Find the limit $\lim x_n$ in the following cases:

a) $x_n = (1 - \frac{1}{n})x_{n-1} + \frac{1}{n} x_{n-2}$ $(n = 2, 3, \ldots)$;

b) $x_{n+1} = (1 - \frac{1}{2n})x_n + \frac{1}{2n} x_{n-1}$ $(n \in \mathbb{N})$.

3.3. Let $x_1 \in \mathbb{R}$ and $x_{n+1} = (2 + \frac{2}{n})x_n - 1$.

a) For what values of x_1 does the sequence $\{x_n\}$ diverge? What is the asymptotic behavior of $\{x_n\}$?

b) For what values of x_1 does $\{x_n\}$ converge? What is $\lim x_n$?

3.4. Suppose that $b \in \mathbb{R}$, $a_n \to a \in \mathbb{R}$, and $x_{n+1} = a_n + bx_n$. Prove that:

a) $x_n \to a/(1-b)$ if $|b| < 1$;

b) $x_n \to a/(1-b)$ if $|b| > 1$ and $x_1 + \sum a_k/b^k = 0$;

c) $x_n \sim Sb^{n-1}$ if $|b| > 1$ and $S = x_1 + \sum a_k/b^k \ne 0$.

3.5. Suppose that the sequence $\{x_n\}$ and the number $p \in \mathbb{R}$ are such that $x_{n+1} \leq px_n + (1-p)x_{n-1}$ for any $n = 2, 3, \dots$. Prove that if $p > 0$, then the limit $\lim x_n \in \overline{\mathbb{R}}$ exists, but for $p \leq 0$ this assertion is false for some sequences.

3.6. Suppose that the positive sequence $\{x_n\}$ and the number $p \in \mathbb{R}$ are such that $x_{n+1} \leq x_n^p x_{n-1}^{1-p}$ for $n = 2, 3, \dots$. Prove that if $p \in (0, 2)$, then $\{x_n\}$ converges, while for any other values of p this assertion is false for some sequences.

3.7. Suppose that the sequence $\{x_n\}$ of nonnegative numbers is such that $x_n \leq \frac{1}{n^2}(x_{n-1} + x_{n-2})$ for $n = 3, 4, \dots$. Prove that $x_n = O(1/n!)$.

3.8. Suppose that $k \in \mathbb{N}$, $0 < \beta < \alpha/k$, and $\{x_n\}$ is a nonnegative sequence. Prove that:

$$\text{a) if } x_n = O\left(\frac{x_{n-1} + \dots + x_{n-k}}{\ln^\alpha n}\right), \text{ then } x_n = O((\ln n)^{-\beta n});$$

$$\text{b) if } x_n = O\left(\frac{x_{n-1} + \dots + x_{n-k}}{n^\alpha}\right), \text{ then } x_n = O(n^{-\beta n});$$

$$\text{c) if } x_n = O\left(\frac{x_{n-1} + \dots + x_{n-k}}{e^{\alpha n}}\right), \text{ then } x_n = O(e^{-\beta n^2/2});$$

$$\text{d) if } x_n = O\left(\frac{x_{n-1} + \dots + x_{n-k}}{(n!)^\alpha}\right), \text{ then } x_n = O(n^{-\beta n^2/2}).$$

3.9. Suppose that

$$k \in \mathbb{N}, \quad A_n \uparrow +\infty, \quad n\left(1 - \frac{\ln A_{n-1}}{\ln A_n}\right) \to \theta \geq 0,$$

and $\{x_n\}$ is a nonnegative sequence such that $x_n = O(\frac{1}{A_n}(x_{n-1} + \dots + x_{n-k}))$. Prove that $x_n = O(A_n^{-\beta n})$ if $0 < \beta < 1/(k(1+\theta))$.

3.10. Suppose that $k \in \mathbb{N}$, $A_n \uparrow +\infty$, $A_1 > 1$, and $\{x_n\}$ is a nonnegative sequence such that $x_n \leq \frac{1}{A_n}(x_{n-1} + \dots + x_{n-k})$ for $n > k$. Prove that $x_n = O(e^{S_n}(A_1 \cdots A_n)^{-1/k})$, where $S_n = \sum_{1 \leq j \leq n} 1/\sqrt[k]{A_j - 1}$. In particular, $x_n = O((1+\varepsilon)^n (A_1 \cdots A_n)^{-1/k})$ for any $\varepsilon > 0$; but if the series $\sum 1/\sqrt[k]{A_n}$ converges, then $x_n = O((A_1 \cdots A_n)^{-1/k})$.

3.11. Suppose that $k \in \mathbb{N}$, $\alpha > 0$, and $\{x_n\}$ is a nonnegative sequence. Prove that:

a) if $x_n \leq (x_{n-1} + \dots + x_{n-k})/(\ln n)^\alpha$, then

$$x_n = O\left(\left(\frac{1+\varepsilon}{\ln n}\right)^{\alpha n/k}\right)$$

for any $\varepsilon > 0$;

b) if $x_n \leq (x_{n-1} + \cdots + x_{n-k})/n^\alpha$, then

$$x_n = \begin{cases} O\left((n!)^{-\alpha/k} \exp\left(\frac{k}{\alpha-k} n^{(k-\alpha)/k}\right)\right) & \text{for } 0 < \alpha < k, \\ O(1/(n-1)!) & \text{for } \alpha = k, \\ O((n!)^{-\alpha/k}) & \text{for } \alpha > k; \end{cases}$$

c) if $x_n \leq (x_{n-1} + \cdots + x_{n-k})/e^{\alpha n}$, then $x_n = O(e^{-\alpha n(n+1)/2k})$;

d) if $x_n \leq (x_{n-1} + \cdots + x_{n-k})/(n!)^\alpha$, then

$$x_n = O\left(\left(\frac{e^{n/2}}{n!}\right)^{\alpha n/2k}\right).$$

CHAPTER III

Functions

§1. Continuity and discontinuities of functions

1.1. Describe the sets of functions $f\colon \mathbb{R} \to \mathbb{R}$ having one of the following properties ($\varepsilon, \delta, x_1, x_2 \in \mathbb{R}$):

a) $\forall\varepsilon \quad \exists\delta > 0 \quad (|x_1 - x_2| < \delta \Rightarrow |f(x_1) - f(x_2)| < \varepsilon)$;
b) $\forall\varepsilon > 0 \quad \exists\delta \quad (|x_1 - x_2| < \delta \Rightarrow |f(x_1) - f(x_2)| < \varepsilon)$;
c) $\forall\varepsilon > 0 \quad \exists\delta > 0 \quad (x_1 - x_2 < \delta \Rightarrow |f(x_1) - f(x_2)| < \varepsilon)$;
d) $\forall\varepsilon > 0 \quad \forall\delta > 0 \quad (|x_1 - x_2| < \delta \Rightarrow |f(x_1) - f(x_2)| < \varepsilon)$;
e) $\forall\varepsilon > 0 \quad \exists\delta > 0 \quad (|x_1 - x_2| < \delta \Rightarrow |f(x_1) - f(x_2)| > \varepsilon)$;
f) $\forall\varepsilon > 0 \quad \exists\delta > 0 \quad (|x_1 - x_2| < \varepsilon \Rightarrow |f(x_1) - f(x_2)| < \delta)$;
g) $\forall\varepsilon > 0 \quad \exists\delta > 0 \quad (|f(x_1) - f(x_2)| > \varepsilon \Rightarrow |x_1 - x_2| > \delta)$;
h) $\exists\varepsilon > 0 \quad \forall\delta > 0 \quad (|x_1 - x_2| < \delta \Rightarrow |f(x_1) - f(x_2)| < \varepsilon)$;
i) $\forall\varepsilon > 0 \quad \exists\delta > 0 \quad (x_1 - x_2 < \delta \Rightarrow f(x_1) - f(x_2) < \varepsilon)$.

1.2. Let the function f be defined on $\mathbb{R}$. Prove that the following properties of f are equivalent:

a) f is continuous on $\mathbb{R}$;
b) the set $f^{-1}(G) = \{x \in \mathbb{R} | f(x) \in G\}$ is open for any open set $G \subset \mathbb{R}$;
c) the set $f^{-1}(F) = \{x \in \mathbb{R} | f(x) \in F\}$ is closed for any closed set $F \subset \mathbb{R}$;
d) for any $c \in \mathbb{R}$ the sets $f^{-1}((-\infty, c))$ and $f^{-1}((c, +\infty))$ are open;
e) for any $c \in \mathbb{R}$ the sets $f^{-1}((-\infty, c])$ and $f^{-1}([c, +\infty))$ are closed.
f) Is the condition "for any $c \in \mathbb{R}$ the set $f^{-1}(\{c\})$ is closed" sufficient for the continuity of f?

1.3. Prove that any function defined on a set $E \subset \mathbb{R}$ has an at most countable set of points of disconunity of the first kind.

1.4. Let E_L and E_R be countable subsets of $[0,1]$ with $E_L \cap E_R = \varnothing$. Define a function on $[0,1]$ that is continuous only from the left at points of E_L, continuous only from the right at points of E_R, and continuous at the remaining points of $[0,1]$.

1.5. Let $\mathscr{D}_L(f)(\mathscr{D}_R(f))$ be the set of points at which a function f on $\mathbb{R}$ is discontinuous from the left (respectively, from the right). Is it true that if one of these sets is at most countable, then the other is also at most countable?

1.6. Let f_0 be the Dirichlet function:

$$f_0(x) = \begin{cases} 1 & \text{for } x \in \mathbb{Q}, \\ 0 & \text{for } x \in \mathbb{R}\setminus\mathbb{Q}. \end{cases}$$

a) Prove that $f_0(x) = \lim_{m\to\infty} \lim_{n\to\infty} \cos^{2n}(2\pi x m!)$ for any $x \in \mathbb{R}$.

b) Does there exist a sequence $\{f_n\}$ of continuous functions on $\mathbb{R}$ such that $f_0(x) = \lim_{n\to\infty} f_n(x)$ for any $x \in \mathbb{R}$?

1.7. Prove that the set of points of discontinuity of an arbitrary function on an interval $\Delta \subset \mathbb{R}$ is an F_σ-set, that is, the union of a sequence of closed sets. Is this true if the function is defined on an arbitrary metric space?

1.8. Define on $\mathbb{R}$ a function with a given set E of points of discontinuity if:

a) E is a closed set;

b) $E = \bigcup_{n=1}^{\infty} E_n$, where E_n is a closed set for any $n \in \mathbb{N}$.

1.9. Define on the interval $(0, 1)$ a function such that any nonempty open interval $\Delta \subset (0, 1)$ contains a continuum of its points of continuity and a continuum of its points of discontinuity.

1.10. Prove that a function f defined on an interval $\langle a, b\rangle$ is continuous if and only if:

a) f has the Cauchy property (that is, the image of each interval $\langle p, q\rangle \subset \langle a, b\rangle$ is an interval);

b) the set $f^{-1}(\{y\})$ is closed for any $y \in \mathbb{R}$.

1.11. Prove that a function on an interval $[a, b]$ is continuous if and only if its graph is connected and closed.

1.12. Let $E \subset \mathbb{R}$, and let f be a bounded function on E that has a local extremum at each nonisolated point of E. Prove that the set $f(E)$ is then at most countable. Is this true if E is a separable metric space? An arbitrary metric space?

1.13. Describe all the continuous functions on an interval $[a, b]$ having a local extremum at each point of (a, b).

1.14. Suppose that a function f is defined and one-to-one on (a, b) and continuous at a point $c \in (a, b)$.

a) Is f^{-1} necessarily continuous at the point $f(c)$? Does f^{-1} always have one-sided limits at $f(c)$?

b) Is f^{-1} continuous at the point $f(c)$ if f is strictly monotone?

1.15. Let $f \in C(\langle \alpha, \beta\rangle)$. Prove that f is one-to-one if and only if it is strictly monotone.

1.16. Suppose that a positive function f is defined and differentiable on $[a, +\infty)$. Prove that if $\inf_{x\ge a} f'(x)/f(x) > 0$, then:

a) $f(x) = o(f((1+\delta)x))$ as $x \to +\infty$ for any $\delta > 0$;

b) $f^{-1}(z) \sim f^{-1}(\varepsilon z)$ as $z \to +\infty$ for any $\varepsilon > 0$.

1.17. Prove that if $f \in C([0, +\infty))$ and the limit $\lim_{n\to\infty} f(nx)$ exists for any $x \geq 0$, then the limit $\lim_{x\to+\infty} f(x)$ exists. Prove this if the limit $\lim_{n\to\infty} f(nx)$ exists only for points x in some nonempty closed set without isolated points.

1.18. Define on $[0, +\infty)$ a function f satisfying the following conditions:

a) $\lim_{n\to\infty} f(nx) = 0$ for any $x \in [0, +\infty)$;

b) f is bounded on any finite interval and has discontinuities of the first kind only;

c) the set of limit points of f at infinity fills $\mathbb{R}$ (compare with problem 1.17).

1.19. Define on $[0, +\infty)$ a function f satisfying the following conditions:

a) $\lim_{n\to\infty} f(nx) = 0$ for any $x \in [0, +\infty)$;

b) the set of values of f on any nonempty open interval $\Delta \subset [0, +\infty)$ fills $\mathbb{R}$ (compare with problem 1.17).

1.20. Prove that if $f \in C([0, +\infty))$ and $f(x+h)-f(x) \to 0$ as $x \to +\infty$ for any $h \in \mathbb{R}$, then $f(x+h)-f(x) \rightrightarrows 0$ as $x \to +\infty$ on any finite interval, and hence f is uniformly continuous on $[0, +\infty)$.

1.21. Let $\delta > 0$ and $f \in C([0, 1])$. The graph of f is said to have a horizontal chord of length δ if there is a point $x \in [0, 1-\delta]$ such that $f(x) = f(x+\delta)$. Prove that if $f(0) = f(1)$, then for any $n \in \mathbb{N}$ there is a horizontal chord of length $1/n$. Show that for chords of a different length this is not true in general.

1.22. Prove that if $f \in C(\mathbb{R})$ and $f(x+h) - 2f(x) + f(x-h) \to 0$ as $h \to +\infty$ for any $x \in R$, then f is a linear function.

1.23. A function $f\colon \mathbb{R} \to \mathbb{R}$ is said to be Cèsaro-continuous if the condition $\frac{1}{n}(x_1 + \cdots + x_n) \to x_o$ implies $-(f(x_g) + \cdots + f(x_n)) \to f(x_0)$. Describe all functions that are Cèsaro-continuous.

1.24. Let $f(x) = \Pi_{k\geq 0}(1 + x^{4^k})$ for $0 \leq x < 1$. Prove that:

a) there exist $c_1 > 0$ and $c_2 > 0$ such that $c_1 \leq \sqrt{1-x}f(x) \leq c_2$ for all $x \in [0, 1)$;

b) the function $\sqrt{1-x}f(x)$ is not monotone on the interval $[0, 1)$;

c) $\sqrt{1-x}f(x)$ does not have a limit as $x \to 1-0$.

1.25. Does there exist a continuous function on $[0, 1]$ such that all sets of constancy are countable (have the cardinality of $\mathbb{N}$)?

1.26. Suppose that the function f is defined and separately continuous on the square $Q = [0, 1] \times [0, 1]$. Prove that there is a sequence $\{f_n\}$ of functions continuous on Q such that $f_n(x, y) \to f(x, y)$ for any $x, y \in [0, 1]$ (that is, f is a function of the first Baire class on Q).

1.27. Let f be defined and separately continuous on $\mathbb{R}^2$. Prove that if f vanishes on a dense subset of $\mathbb{R}^2$ (that is, a set whose closure coincides with $\mathbb{R}^2$), then $f \equiv 0$.

§2. Semicontinuous functions

Let $\Delta \subset \mathbb{R}$ be an arbitrary interval. A function f defined on Δ and possibly taking the value $-\infty$ $(+\infty)$ is said to be *lower* (*upper*) *semicontinuous* if $f(a) \le \underline{\lim}_{x\to a} f(x)$ $(f(a) \ge \overline{\lim}_{x\to a} f(x))$ for any point $a \in \Delta$.

2.1. Suppose that a function f is defined on $\mathbb{R}$. Prove that the following properties of f are equivalent (compare with problem 1.2):

a) f is lower semicontinuous;
b) the set $f^{-1}((c, +\infty))$ is open for any $c \in \mathbb{R}$;
c) the set $f^{-1}((-\infty, c])$ is closed for any $c \in \mathbb{R}$;
d) if $x_n \to a$, then $f(a) \le \sup_{n\in\mathbb{N}} f(x_n)$;
e) for any point $a \in \mathbb{R}$ and any number $\varepsilon > 0$ there is a neighborhood V of a such that $f(x) \ge f(a) - \varepsilon$ for $x \in V$.

2.2. Prove that:

a) the characteristic function of a set $E \subset \mathbb{R}$ is lower semicontinuous if and only if E is open;
b) the function $f: \mathbb{R} \to \mathbb{R}$ defined by

$$f(x) = \begin{cases} 0 & \text{if } x \text{ is an irrational number,} \\ 1/n & \text{if } x = m/n \text{ where } n \in N,\ m \in Z, \\ & \text{and } m/n \text{ is irreducible,} \end{cases}$$

is upper semicontinuous.

2.3. Prove that if a function is lower semicontinuous on $[a, b]$ and takes only finite values, then it attains the smallest value on $[a, b]$ (and hence is bounded below). Does it have to be bounded above?

2.4. Let g be an arbitrary function defined on an interval $\Delta \subset \mathbb{R}$, and let $f(x) = \underline{\lim}_{y\to x} g(y)$ $(x \in \Delta)$. Prove that f is lower semicontinuous.

2.5. Let $\{f_\alpha\}_{\alpha\in A}$ be a family of lower semicontinuous functions defined on an interval $\Delta \subset \mathbb{R}$, and let $f(x) = \sup_{\alpha\in A} f_\alpha(x) < +\infty$ for any $x \in \Delta$. Prove that f is lower semicontinuous. Is this true for the function $h = \inf_{\alpha\in A} f_\alpha$? Is h lower semicontinuous if the family $\{f_\alpha\}_{\alpha\in A}$ is finite? Countable?

2.6. Prove that a function $f: [a, b] \to \mathbb{R}$ is lower semicontinuous if and only if $f = \sup\{g \mid g \in C([a, b]),\ g \le f\}$.

§3. Continuous and differentiable functions

3.1. Let $\Delta_h^1 f(x) = \Delta_h f(x) = f(x+h) - f(x)$ and $\Delta_h^{m+1} f(x) = \Delta_h(\Delta_h^m f(x))$. Prove that a function $f \in C(\mathbb{R})$ is a polynomial of degree at most m if and only if $\Delta_h^{m+1} f(x) = 0$ for any $x, h \in \mathbb{R}$.

3.2. Let $f \in C^1([0, +\infty))$. Prove that $f'(x) + f(x) \to L \in \mathbb{R}$ as $x \to +\infty$ if and only if $f(x) \to L$ as $x \to +\infty$ and f' is uniformly continuous on $[0, +\infty)$.

3.3. Suppose that f is differentiable on $[a, b]$. Prove that if $f'(a)\cdot f'(\beta) < 0$, then there exists a point $c \in (a, b)$ such that $f'(c) = 0$.

3.4. Suppose that a function f is differentiable on $[a, b]$. Prove that if the set $(f')^{-1}(x) = \{y \in [a, b] | f'(y) = x\}$ is closed for any $x \in \mathbb{R}$, then $f \in C^1([a, b])$.

3.5. Suppose that a function f is defined on $[a, b]$. Prove that $f \in C^1([a, b])$ if and only if the ratio $\frac{1}{h}(f(x+h) - f(x))$ tends to a finite limit uniformly on $[a, b]$ as $h \to 0$.

3.6. Let $f \in C((a, b))$, and suppose that for any $x \in (a, b)$ the limit $\lim_{h\to 0} \frac{1}{2h}(f(x+h) - f(x-h)) = g(x)$ exists and is finite.

a) Prove that if $g \geq 0$ on (a, b), then f is increasing, and if $g \equiv 0$, then f is constant.

b) Prove that if $g \in C((a, b))$, then $f \in C^1((a, b))$.

3.7. Suppose that functions f and g are defined on (a, b) and satisfy the condition that for any $x \in (a, b)$ there exists a positive number $\delta_x > 0$ such that $f(x+h) - f(x-h) = 2hg(x)$ for $0 < h < \delta_x$. Prove that if f is differentiable, then it is a polynomial of degree at most two. Is this true if f is continuous? If both f and g are continuous on (a, b)?

3.8. Suppose that $F \subset \mathbb{R}$ is a closed set without interior points. Prove that there exists a strictly increasing function $f \in C^1(\mathbb{R})$ such that $f'(x) = 0$ if and only if $x \in F$.

3.9. Suppose that $f \in C([0, 1])$. Determine whether the following statements are equivalent:

a) f is constant on $[0, 1]$;

b) $\forall x \in (0, 1)\ \exists \{x_n\}\colon x_n \to x,\ x_n > x$,
$$\frac{f(x_n) - f(x)}{x_n - x} \to 0;$$

c) $\forall x \in (0, 1)\ \exists \{x_n\}\colon x_n \to x,\ x_n \neq x$,
$$\frac{f(x_n) - f(x)}{x_n - x} \to 0.$$

Suppose that f is defined on the interval $\langle a, b\rangle \subset \mathbb{R}$. The *variation* $\mathrm{Var}_a^b(f)$ of a function f on $\langle a, b\rangle$ is defined to be the supremum of all possible sums $\sum_{1\leq k<n} |f(x_{k+1}) - f(x_k)|$, where $x_k < x_{k+1}$, $x_k \in \langle a, b\rangle$, $k = 1, \dots, n$. The function f is called a *function of bounded variation* on $\langle a, b\rangle$ if $\mathrm{Var}_a^b(f) < +\infty$. The symbol $V(\langle a, b\rangle)$ denotes the set of all functions having bounded variation on $\langle a, b\rangle$.

3.10. Let $f \in V([a, b])$, and let $g(x) = \mathrm{Var}_a^x(f)$ $(x \in (a, b])$, $g(a) = 0$. Prove that if f is continuous at a point $c \in [a, b]$, then so is g.

3.11. Let $f \in C^1([a, b])$. Prove that $\mathrm{Var}_a^b(f) = \int_a^b |f'(x)|\, dx$.

3.12. Suppose that $1 < \alpha < 2$, let $f(x) = x^2 \cos(x^{-\alpha})$ for $0 < x \leq 1$, and let $f(0) = 0$. Prove that $f \in V([0, 1])$ and f is differentiable on $[0, 1]$, but the function $g(x) = \mathrm{Var}_a^x(f)$ is not differentiable at zero. Does $g'(0)$ exist for $\alpha = 1$?

3.13. Suppose that $\alpha, \beta > 0$, and let $f(x) = x^\alpha \cos(x^{-\beta})$ for $0 < x \leq 1$ and $f(0) = 0$. Prove that:

a) $f \in V([0, 1])$ if and only if $\alpha > \beta$;
b) $f \in \mathrm{Lip}_1([0, 1])$ if and only if $\alpha \geq \beta + 1$;
c) if $\alpha < \beta + 1$, then $f \in \mathrm{Lip}_\gamma([0, 1])$, where $\gamma = \alpha/(\beta + 1)$.

3.14. Let $0 < \gamma < 1$. Construct a function $f \in \mathrm{Lip}_\gamma([0, 1])$ not having bounded variation on any nondegenerate interval $\Delta \subset [0, 1]$.

3.15. Find functions $f, g \in C([0, 1])$ such that:
a) $f \in V([0, 1])$ and $f \notin \mathrm{Lip}_\alpha([0, 1])$ for any $\alpha > 0$;
b) $g \notin V([0, 1])$ and $g \in \mathrm{Lip}_\alpha([0, 1])$ for any $\alpha < 1$.

3.16. Suppose that a function $f\colon \mathbb{R} \to \mathbb{R}$ is continuous and 2π-periodic, and let

$$\varphi(x) = \int_{-\pi}^{\pi} \frac{f(x-t)}{\sqrt{|t|}}\, dt \qquad (x \in \mathbb{R}).$$

Prove that $\varphi \in \mathrm{Lip}_{1/2}$; more precisely, prove that $|\varphi(u) - \varphi(v)| \leq 8|u - v|^{1/2} \max_{x\in\mathbb{R}} |f(x)|$.

For $x \in [0, 1]$ let

$$f(x) = \sup\left\{ \sum_{k\geq 1} \frac{\varepsilon_k}{2^k} \,\middle|\, \sum_{k\geq 1} \frac{2\varepsilon_k}{3^k} \leq x,\ \varepsilon_k = 0, 1 \right\}.$$

The graph of f is pictured in Figure 2. The function f is called the *Cantor function* (and its graph the "*Cantor staircase*").

3.17. Prove that:
a) if $x = \sum_{k\geq 1} 2\varepsilon_k/3^k$, where $\varepsilon_k = 0$ or 1, then $f(x) = \sum_{k\geq 1} \varepsilon_k/2^k$;

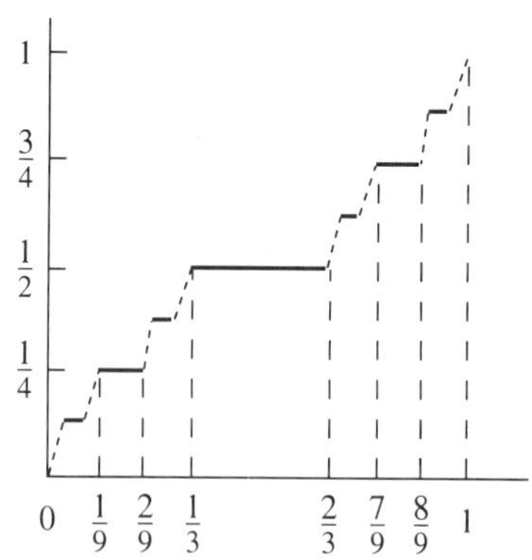

FIGURE 2

b) $f \in C([0, 1])$;

c) on the middle third of each interval $\Delta_{\varepsilon_1 \cdots \varepsilon_n}$ defined in the construction of the Cantor set (see §I.1) the function f is constant and takes the value $\sum_{1 \le k \le n} \frac{\varepsilon_k}{2^k} + \frac{1}{2^{n+1}}$.

3.18. Find the length of the graph of the Cantor function.

3.19. For what $\alpha > 0$ is the Cantor function in the class $\mathrm{Lip}_\alpha([0, 1])$?

3.20. Let K be a generalized Cantor set, and $\Delta_{\varepsilon_1 \cdots \varepsilon_n}$ the closed intervals of nth rank appearing in the definition of K (see the definition of a generalized Cantor set in problem I.1.31). Further, suppose that $a = \inf K$, $b = \sup K$, and $\delta_{\varepsilon_1 \cdots \varepsilon_n}$ are the open intervals removed from the closed intervals $\Delta_{\varepsilon_1 \ldots \varepsilon_n}$ in constructing the closed intervals of $(n+1)$st rank. Define the function φ on $[a, b]$ as follows: $\varphi(a) = 0$, $\varphi(x) = 1/2$ if $x \in [a, b] \setminus (\Delta_0 \cup \Delta_1)$;

$$\varphi(x) = \sum_{1 \le k \le n} \frac{\varepsilon_k}{2^k} + \frac{1}{2^{n+1}} \quad \text{if } x \in \delta_{\varepsilon_1 \cdots \varepsilon_n};$$

$\varphi(x) = \sup\{\varphi(t) | a < t < x,\ t \notin K\}$ if $x \in K$, $a < x$. The function φ will be called the Cantor function corresponding to the set K.

a) Prove that φ is increasing and continuous on $[a, b]$.

b) Prove that

$$\varphi(x_0 + h) - \varphi(x_0 - h) > 0$$

for any point $x_0 \in K$ and any number $h > 0$ (φ is increasing at any point of K).

3.21. Suppose that $F \subset \mathbb{R}$ is a closed set without isolated points. Prove that there exists a function $f \in C(\mathbb{R})$ such that:

a) $f'(x) = 0$ for any $x \in \mathbb{R} \setminus F$;

b) $f(x+h) - f(x-h) > 0$ for any $x \in F$ and any $h > 0$.

A function f on an interval Δ containing a point c is said to belong to the class $\mathrm{Lip}(\alpha, c)$ $(\alpha > 0)$ if there exist positive numbers L and δ such that $|f(x) - f(c)| \le L|x-c|^\alpha$ for any point x in the intersection $\Delta \cap (c-\delta, c+\delta)$.

3.22. Suppose that $\alpha > 0$, $f \in C^1(\Delta)$, and $N_0 = \{x \in \Delta \,|\, f'(x) = 0\}$ is the set of critical points of f. Prove that:

a) if $f' \in \mathrm{Lip}(\alpha, c)$ for any point $c \in \Delta$, then

$$N_0 = \{x \in \Delta \,|\, f \in \mathrm{Lip}(1+\alpha, x)\};$$

b) if f is twice differentiable on Δ, then

$$N_0 = \{x \in \Delta \,|\, f \in \mathrm{Lip}(2, x)\}.$$

§4. Continuous mappings

4.1. Suppose that $P(z) = a_0 + a_1 z + \cdots + a_m z^m$ $(z \in \mathbb{C})$. Prove that the image $P(F)$ of any closed set $F \subset \mathbb{C}$ is again a closed set. Is this true for an arbitrary polynomial in two variables?

4.2. Let $\Gamma_0 = \{z \in \mathbb{C} \mid |z| = 1,\ z \neq -1\}$ and $\varphi(z) = \operatorname{Im} z/(1 + \operatorname{Re} z)$. Prove that φ is a one-to-one mapping of Γ_0 onto $\mathbb{R}$. Find φ^{-1}.

4.3. Does there exist a partition of $[0, 1]$ into two disjoint sets A and B such that each can be mapped onto the other in a one-to-one bicontinuous manner (homeomorphically)?

4.4. Define functions φ and ψ on $[0, 1]$ as follows. Let $t \in [0, 1]$, $t = \sum 3^{-k}\alpha_k$, where $\alpha_k = 0, 1, 2$. Then $\varphi(t) = \sum 3^{-k}\beta_k$, where: $\beta_1 = \alpha_1$; $\beta_2 = \alpha_3$ if α_2 is even, and $\beta_2 = 2 - \alpha_3$ if α_2 is odd; $\beta_n = \alpha_{2n-1}$ if $\alpha_2 + \alpha_4 + \cdots + \alpha_{2n-2}$ is even, and $\beta_n = 2 - \alpha_{2n-1}$ if $\alpha_2 + \alpha_4 + \cdots + \alpha_{2n-2}$ is odd $(n \geq 2)$. Similarly, $\psi(t) = \sum 3^{-k}\gamma_k$, where: $\gamma_1 = \alpha_2$ if α_1 is even, and $\gamma_1 = 2 - \alpha_2$ if α_1 is odd; $\gamma_n = \alpha_{2n}$ if $\alpha_1 + \alpha_3 + \cdots + \alpha_{2n-1}$ is even, and $\gamma_n = 2 - \alpha_{2n}$ if $\alpha_1 + \alpha_3 + \cdots + \alpha_{2n-1}$ is odd $(n \geq 1)$.

a) Prove that φ and ψ are well defined and continuous.

b) Let $f\colon [0, 1] \to \mathbb{R}^2$ be the mapping with coordinate functions φ and ψ ($f(t) = (\varphi(t), \psi(t))$ for $t \in [0, 1]$). Prove that f maps $[0, 1]$ onto the square $[0, 1] \times [0, 1]$. The mapping f is described schematically in Figure 3. Each of the intervals Δ_k, $k = 1, \ldots, 9$, is mapped onto the square with the same index (Figure 3(a)). Figure 3(b) shows the circuit of the unit square when the interval is partitioned into 9^2 parts.

c) Prove that the system

$$\begin{cases} \varphi(t) = \xi, \\ \psi(t) = \eta \end{cases}$$

of equations has at most four solutions for any $\xi, \eta \in [0, 1]$. Find ξ and η such that this system has four solutions.

d) Prove that the sets of constancy of φ and ψ (that is, the sets $\varphi^{-1}(\{c\})$ and $\psi^{-1}(\{c\})$, $c \in \mathbb{R}$) do not have isolated points.

e) Prove that $\varphi, \psi \in \mathrm{Lip}_{1/2}([0, 1])$, but $\varphi, \psi \notin \mathrm{Lip}_\alpha([0, 1])$ if $\alpha > 1/2$.

4.5. A continuous mapping of a closed interval onto a square is called a Peano curve. Let u and v be the coordinate functions of an arbitrary Peano curve defined on $[0, 1]$. Prove that if $u, v \in \mathrm{Lip}_\alpha([0, 1])$, then $\alpha \leq 1/2$. (Thus, the smoothness of the coordinate functions of the Peano

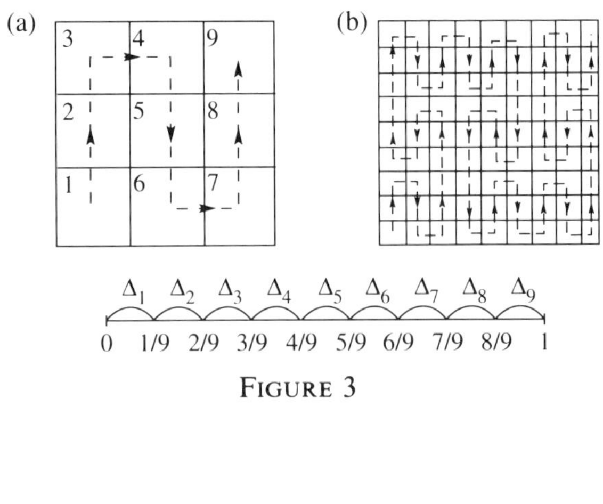

FIGURE 3

curve constructed in the preceding problem is the best possible.)

4.6. Prove that any two generalized Cantor sets are homeomorphic, that is, a one-to-one bicontinuous correspondence can be established between them.

§5. Functional equations

5.1. Describe all the functions $f \in C(X)$ satisfying the equation $f(2x) = f(x)$ for all $x \in X$ in the following cases:

a) $X = \mathbb{R}$; b) $X = (0, +\infty)$.

5.2. Suppose that the function $f: \mathbb{R} \to \mathbb{R}$ satisfies the equation $f(x+y) = f(x) + f(y)$ $(x, y \in \mathbb{R})$. Prove that:

a) $f(rx) = rf(x)$ $(r \in \mathbb{Q}, x \in \mathbb{R})$;

b) if f is bounded above on some nonempty open interval, then it is linear, that is, $f(x) = ax$ $(x \in \mathbb{R})$, where $a = f(1)$;

c) if f is discontinuous at at least one point, then its graph is dense in $\mathbb{R}^2$.

5.3. Describe all the monotone functions $f: (0, +\infty) \to \mathbb{R}$ satisfying the equation $f(xy) = f(x) + f(y)$ $(x, y > 0)$.

5.4. Describe all the functions $f \in C(\mathbb{R})$ satisfying the equation $f(xy) = xf(y) + yf(x)$ for any $x, y \in \mathbb{R}$.

5.5. Find all functions $f: \mathbb{R} \to \mathbb{R}$ satisfying the system of equations $f(x+y) = f(x) + f(y)$, $f(xy) = f(x)f(y)$ for $x, y \in \mathbb{R}$.

5.6. Let $n \in \mathbb{N}$, $n > 1$. Describe all the functions $f: \mathbb{R} \to \mathbb{R}$ such that $f(x + y^n) = f(x) + (f(y))^n$ for any $x, y \in \mathbb{R}$.

5.7. Suppose that for some $\sigma > 0$ the function $f \in C(\mathbb{R})$ satisfies the condition $|f(x) + f(y) - f(x+y)| \le \sigma$ for any $x, y \in \mathbb{R}$. Prove that f is representable as the sum of a linear function and a function not exceeding σ in absolute value.

5.8. Suppose that the function $f \in C(\mathbb{R})$ satisfies the condition $f(x) + f(y) = f(\sqrt{x^2 + y^2})$ for any $x, y \in \mathbb{R}$. Prove that $f(x) = ax^2$ for any $x \in \mathbb{R}$, where $a = f(1)$.

5.9. Suppose that the function $f \in C(\mathbb{R})$ satisfies the conditions $f \not\equiv 0$ and $f(x)f(y) = f(\sqrt{x^2 + y^2})$ for any $x, y \in \mathbb{R}$. Prove that $f(x) = e^{ax^2}$ for any $x \in \mathbb{R}$ (a is a fixed number).

5.10. Let $a > b > 0$. Describe all the functions $f \in C((0, +\infty))$ such that the difference $f(ax) - f(bx)$ does not depend on $x \in (0, +\infty)$.

5.11. a) Describe all the functions $\varphi \in C((0, 1])$ satisfying the condition $\varphi(x^2) = 2\varphi(x)$ for $0 < x \le 1$.

b) Describe all the nondecreasing functions $f \in C((0, 1))$ satisfying the condition $f(x^2) = (f(x))^2$ for $0 < x < 1$.

5.12. Find a continuous strictly increasing function $f: \mathbb{R} \to \mathbb{R}$ such that $f(x^2) = (f(x))^4$ for any $x \in \mathbb{R}$.

5.13. Let $\overline{B} = \{z \in \mathbb{C} \mid |z| \le 1\}$. Find all the continuous one-to-one mappings f of the disk $\overline{B}$ onto itself satisfying the condition $f(z^2) = (f(z))^2$ for all $z \in \overline{B}$.

5.14. Prove that there exists a unique monotone function $f\colon [0, 1] \to \mathbb{R}$ satisfying the conditions $f(x) = 2f(x/3)$ and $f(x) + f(1 - x) = 1$ for any $x \in [0, 1]$.

5.15. Find all the functions $f \in C^2(\mathbb{R}^2)$ satisfying the conditions $f(x, y) = f(y, x)$ and $f(x, y) + f(y, z) + f(z, x) = 3f(\frac{x+y+z}{3}, \frac{x+y+z}{3})$ for any x, y, $z \in \mathbb{R}$.

5.16. For which continuous functions $f\colon \mathbb{R} \to \mathbb{R}$ does there exist a function $g\colon \mathbb{R}^2 \to \mathbb{R}$ such that $f(xy) = g(x, f(y))$ $(x, y \in \mathbb{R})$?

Suppose that G and G' are arbitrary groups (with the multiplicative notation for the group operation). A mapping $\varphi\colon G \to G'$ is called a homomorphism if it preserves the group operation, that is, if $\varphi(xy) = \varphi(x)\varphi(y)$ for any x, $y \in G$. If S^1 is the group of complex numbers with modulus 1, then a homomorphism of G into S^1 is called a *character* of the group G.

5.17. Find the general form of a character for the following groups:

a) $\mathbb{Z}$; b) $\mathbb{Z} \times \mathbb{Z}$; c) $\mathbb{Z}_m$(the group of residues modulo m).

5.18. Find the general form of a continuous character on the following (additive) groups: a) $\mathbb{R}$; b) $\mathbb{R}^n$; c) $\mathbb{R} \times \mathbb{Z}$; d) $\mathbb{C}$. The same for the following (multiplicative) groups: e) S^1; f) $\mathbb{R}_+ = (0, +\infty)$; g) $\dot{\mathbb{C}} = \mathbb{C}\setminus\{0\}$.

5.19. Describe all the continuous homomorphisms φ of a group G into a group G' in the following cases (the notation introduced in the preceding problem is used): a) $G = \mathbb{R}$, $G' = \mathbb{R}$; b) $G = \mathbb{R}$, $G' = \mathbb{R}_+$; c) $G = \mathbb{R}_+$, $G' = \mathbb{R}$; d) $G = \mathbb{R}_+$, $G' = \mathbb{R}_+$; e) $G = \mathbb{R}$, $G' = S^1$; f) $G = \dot{\mathbb{C}}$, $G' = \dot{\mathbb{C}}$.

CHAPTER IV

Series

§1. Convergence

1.1. Does the series[1] $\sum \varepsilon_n/n$ converge, where $\varepsilon_n = 0$, if the decimal expression for n contains the digit 9, and $\varepsilon_n = 1$ otherwise?

1.2. Do the following three series converge:

$$\text{a) } \sum |\cos 2^n|/n; \quad \text{b) } \sum |\sin(n + \ln n)|/n; \quad \text{c) } \sum |\sin n^2|/n?$$

1.3. Prove that the series $\sum \cos(b \ln n)/n^a$ converges only for $a > 1$.

1.4. Determine the values of $p > 0$ for which the following series converge:

$$\text{a) } \sum_{n \geq 2} \frac{(\ln n)^{-2p}}{n^p + p^{-(\ln n)^{1/p}}};$$

$$\text{b) } \sum \frac{(-1)^{[\sqrt{n}]}}{n^p};$$

$$\text{c) } \sum \arccos^p \frac{\nu(n)}{\nu(n+1)},$$

where $\nu(n)$ is the number of digits in the decimal expression for n.

1.5. Determine whether the following series converge:

$$\text{a) } \sum \frac{\tau(n)}{n^2}; \qquad \text{b) } \sum \frac{1}{p_n};$$

$$\text{c) } \sum \frac{\varphi(n)}{n^2}; \qquad \text{d) } \sum \frac{1}{np_n - (n-1)p_{n-1}}.$$

Here $\tau(n)$ is the number of divisors of n, p_n is the nth prime number, and $\varphi(n)$ is the number of positive integers $\leq n$ that are relatively prime to n.

1.6. Prove that for $a > 1$ the series

$$\sum \frac{1}{n^a} \frac{1}{|\sin(\pi\sqrt{2}n)|}$$

converges.

---

[1] Recall that $\sum a_n$ stands for the series $\sum_{n=1}^{\infty} a_n$.

1.7. Determine whether the following double series converge:

a) $\sum_{n,m\geq 1} \frac{1}{n^2+m^2}$; b) $\sum_{n,m\geq 1} \frac{\min(n,m)}{n^3+m^3}$;

c) $\sum_{n,m\geq 1} \frac{\max(n,m)}{n^4+m^4}$; d) $\sum_{n,m\geq 1} \frac{nm}{n^5+m^5}$;

e) $\sum_{n,m\geq 1} \frac{\varepsilon(n,m)}{n^2+m^2}$; f) $\sum_{n,m\geq 1} \frac{\mathrm{GCD}(n,m)}{n^3+m^3}$;

g) $\sum_{n,m\geq 1} \frac{\mathrm{LCM}(n,m)}{n^4+m^4}$; h) $\sum_{n,m\geq 1} \frac{1}{(\mathrm{LCM}(n,m))^p}$, $p>1$.

Here $\mathrm{LCM}(n, m)$ is the least common multiple of the numbers n and m, $\mathrm{GCD}(n, m)$ is their greatest common divisor, and $\varepsilon(n, m) = 1$ when n and m are relatively prime and 0 otherwise.

1.8. Verify that the sum of the series

$$1+\frac{1}{2}+\frac{1}{2}-1+\frac{1}{4}+\frac{1}{4}-\frac{1}{2}+\frac{1}{4}+\frac{1}{4}-\frac{1}{2}$$
$$+\frac{1}{8}+\frac{1}{8}-\frac{1}{4}+\frac{1}{8}+\frac{1}{8}-\frac{1}{4}+\frac{1}{8}+\frac{1}{8}-\frac{1}{4}+\frac{1}{8}+\frac{1}{8}-\frac{1}{4}+\cdots$$

(the group of terms $\frac{1}{2^{n+1}}+\frac{1}{2^{n+1}}-\frac{1}{2^n}$ is repeated 2^n times) is equal to 1. Find a rearrangement after which the sum becomes equal to -1.

§2. Properties of numerical series connected with monotonicity

2.1. Suppose that $a_n > 0$, $a_n \uparrow$. Prove that the series $\sum \arccos^2 a_n/a_{n+1}$ converges if and only if the sequence $\{a_n\}$ is bounded.

2.2. Let $a_n \downarrow 0$. Prove that:

a) the series $\sum a_n$ and $\sum 2^n a_{2^n}$ converge or diverge simultaneously;

b) if $\sum a_n = +\infty$, then $\sum \min(a_n, 1/n) = +\infty$;

c) if $\sum a_n/n = +\infty$, then $\sum \frac{1}{n}\min(a_n, 1/\ln n) = +\infty$.

The Abel transformation for series is used in the solution of many of the subsequent problems: if sequences $\{a_n\}$ and $\{b_n\}$ are such that the sequence $\{a_n B_n\}$ converges (where $B_n = b_1 + \cdots + b_n$), then the series $\sum a_n b_n$ and $\sum (a_n - a_{n+1}) B_n$ converge or diverge simultaneously. In the case when the series $\sum b_n$ converges, another variant of this assertion is useful: if the sequence $\{a_n \beta_n\}$ converges (where $\beta_n = b_n + b_{n+1} + \cdots$), then the series $\sum a_n b_n$ and $\sum (a_n - a_{n+1})\beta_{n+1}$ converge or diverge simultaneously. These assertions are easy to prove by applying the Abel transformation (see Chapter I, §2) to the sums $\sum_{1\leq n\leq N} a_n b_n$. The Abel transformation is especially convenient in the case of a monotone sequence $\{a_n\}$.

2.3. Let $a_n \downarrow 0$. Prove that $\sum a_n < +\infty$ if and only if $a_n = o(1/n)$ and $\sum (a_n - a_{n+1}) n < +\infty$.

2.4. Let $a_n \downarrow 0$. Prove that $\sum a_n/n < +\infty$ if and only if $a_n = o(1/\ln n)$ and $\sum (a_n - a_{n+1}) \ln n < +\infty$.

2.5. Let $p > 1$ and $a_n \downarrow 0$. Prove that the series $\sum a_n^p$ converges if the series $\sum a_n^{p-1}/n^{1/p}$ converges. Is the monotonicity essential?

2.6. Let $a_n \downarrow 0$. Prove that:

a) if the series $\sum a_n x_n$ converges, then $a_n \sum_{1 \le k \le n} x_k = o(1)$;

b) if $a_n > 0$ for all $n \in \mathbb{N}$, then there exists a nonnegative sequence $\{x_n\}$ such that $a_n \sum_{1 \le k \le n} x_k = o(1)$ and $\sum a_n x_n = +\infty$. Is the monotonicity of $\{a_n\}$ essential in a) and b)?

2.7. Suppose that $a_n \downarrow 0$ and the series $\sum x_n$ converges. Prove that the series $\sum a_n x_n$ converges, and $\sum_{k \ge n} a_k x_k = o(a_n)$.

2.8. Suppose that $p > 0$, $q, Q \in \mathbb{N}$, and $Q > 1$. Let $\{a_n\}$ be a nonnegative sequence, and let $A_n = a_1 + \cdots + a_n$. Prove the equivalence of the following assertions:

$$\text{a) } \sum a_n/n^p < +\infty; \quad \text{b) } \sum A_{n^q}/n^{1+pq} < +\infty; \quad \text{c) } \sum A_{Q^n}/Q^{pn} < +\infty.$$

2.9. Prove that the series $\sum x_n/n^{1+\varepsilon}$ converges if $\varepsilon > 0$ and the sequence $\{x_n\}$ is such that $\frac{1}{n}\sum_{1 \le k \le n} x_k \to 0$. Is the convergence always absolute? Does the series $\sum x_n/n$ always converge?

2.10. Suppose that $a_n \downarrow 0$, $b_n \downarrow 0$, and $\sum a_n = \sum b_n = +\infty$. Does the series $\sum \min(a_n, b_n)$ always diverge?

2.11. Suppose that $a_n \ge 0$ and $S_n \to \infty$, where $S_n = a_1 + \cdots + a_n$. Prove that: a) $\sum a_n/S_n = +\infty$; b) $\sum a_n/S_n^{1+\varepsilon} < +\infty$ for any $\varepsilon > 0$.

2.12. Suppose that $a_n \ge 0$ and $\sum a_n < +\infty$, and let $\alpha_n = \sum_{k \ge n} a_k$. Prove that: a) $\sum a_n/\alpha_n = +\infty$; b) $\sum a_n/\alpha_n^{1-\varepsilon} < +\infty$ for $\varepsilon > 0$; c) $\sum a_n/e^{\alpha_n} < 1$. d) Is it always true that $\sum a_n/\sqrt{\alpha_{n+1}} < +\infty$?

2.13. Suppose that $p \ge 1$, $a_n \ge 0$, and $r_n = \sum_{k \ge n} a_k^p < +\infty$. Prove that: a) $\sum a_n \le 4 \sum \sqrt[p]{r_n/n}$; b) if $r_n = O(n^{-\alpha})$ for some $\alpha > 0$, then $\sum_{k \ge n} a_k^q = O(n^{1-(\alpha+1)q/p})$ for any $q > p/(\alpha + 1)$.

2.14. Let $\{a_n\}$ be a positive monotone sequence with $a_n \ne 1$ for all n and with $a_n \to a < +\infty$. Prove that the series

$$\sum \frac{1}{n} \frac{(a_{n-1}/a_n) - 1}{1 - \sqrt[n]{a_n}}$$

diverges if $a = 0$ or $a = 1$, and converges otherwise. What happens if $a = +\infty$?

2.15. Let f be a positive strictly increasing function on $[1, +\infty)$, with $f(x) \to +\infty$ as $x \to +\infty$. Prove that the series $\sum 1/f(n)$ and $\sum \frac{1}{n^2} f^{-1}(n)$ converge only simultaneously (f^{-1} is the function inverse to f).

2.16. Suppose that f decreases to zero on the interval $[0, +\infty)$, the sequence $\{a_n\}$ is nonnegative and bounded, and $\sum a_n = +\infty$. Prove that the series $\sum f(n)$ and $\sum a_n f(a_1 + \cdots + a_n)$ converge or diverge simultaneously.

2.17. Suppose that $a_n \downarrow 0$ and let $\sum a_n = A$. For an arbitrary set $\nu \subset \mathbb{N}$ the sum of the partial series corresponding to ν is defined to be $A(\nu) = \sum_{n\in\nu} a_n$ (by definition, $A(\varnothing) = 0$). Prove that the following assertions are equivalent:

a) the sums of all partial series fill the interval $[0, A]$;

b) $a_n \le a_{n+1} + a_{n+2} + \cdots$ for all indices n.

2.18. Does there exist a series $\sum a_n$ such that $a_n \downarrow 0$ and the sums of all its partial series (see 2.17) fill the Cantor set?

2.19. Suppose that $a_n \downarrow 0$ and $\sum a_n = +\infty$. Prove that there exists a sequence $\{n_k\}$ of indices such that $n_k < n_{k+1}$, $a_{n_k} < 1/k$, and $\sum a_{n_k} = +\infty$. Can the monotonicity of $\{a_n\}$ be dropped?

2.20. a) Prove that if a series $\sum a_n$ converges, then there exists a sequence $\lambda_n \uparrow +\infty$ such that the series $\sum a_n\lambda_n$ converges.

b) Suppose that the series $\sum a_n$ diverges, and $\lambda_n \to +\infty$. Does the series $\sum \lambda_n a_n$ always diverge? What happens if $\lambda_n \uparrow +\infty$?

2.21. Let $\{t_n\}$ be a sequence of positive numbers.

a) Prove that the series

$$\sum \frac{1}{n}\frac{1+t_{n+1}}{t_n}$$

diverges.

b) Is it true that the series $\sum a_n(1+t_{n+1})/t_n$ diverges if $a_n \downarrow 0$ and $\sum a_n = +\infty$?

§3. Various assertions about series

3.1. Does there exist a sequence $\{a_n\}$, $a_n \ne 0$, such that the series $\sum a_n$ and $\sum 1/(n^2 a_n)$ converge? Can $\{a_n\}$ be selected to be positive?

3.2. Suppose that the series $\sum a_n$ converges. Can the series $\sum a_n^3$ diverge? Can $\sum a_n^{2001}$ diverge?

3.3. Suppose that $a_n > 0$ and $\sum a_n < +\infty$. Prove that

$$\text{a) } \sum (a_n)^{1-1/n} < +\infty; \qquad \text{b) } \sum \frac{a_n}{\ln(1+n)} \ln\frac{1}{a_n} < +\infty.$$

3.4. Suppose that $0 < a_n < 1$. Prove that if the series $\sum a_n/(\ln a_n)$ converges, then so does the series $\sum a_n/\ln(1+n)$. Is the converse true if $a_n \downarrow 0$? Can the monotonicity be dropped?

3.5. Suppose that $a_n \ge 0$ and $\sum a_n/n < +\infty$. Prove that there exists a sequence $\{n_k\}$ of numbers with density 1 (see problem II.2.3) such that $a_{n_k} \to 0$.

3.6. Prove that the following assertions are equivalent:

a) $\sum |a_n| < +\infty$;

b) for any sequence $\{n_k\}$ of numbers with density 0 (see problem II.2.3) the series $\sum a_{n_k}$ converges.

3.7. Suppose that the series $\sum a_n$ converges absolutely and $\sum_{n\geq 1} a_{kn} = 0$ for any $k \in \mathbb{N}$. Prove that $a_n = 0$ for all n.

3.8. Prove that if the series $\sum a_n x_n$ converges for any sequence $\{x_n\}$ tending to zero, then $\sum |a_n| < +\infty$.

3.9. Prove that the following two properties of the sequence $\{\lambda_n\}$ are equivalent:

a) $\sum |\lambda_n - \lambda_{n+1}| < +\infty$;

b) if the series $\sum a_n$ converges, then so does the series $\sum \lambda_n a_n$.

3.10. Suppose that the function φ is monotone and bounded on $[0, +\infty)$. Prove that $\sum \varphi(\varepsilon n) a_n \underset{\varepsilon\to+0}{\to} \varphi(+0)\sum a_n$ for any convergent series $\sum a_n$.

3.11. Construct a positive sequence $\{a_n\}$ with the following properties: a) $a_n \to 0$; b) $\sum a_n = +\infty$; c) if $a_{n_k} \downarrow 0$, then $\sum a_{n_k} \leq 2$.

3.12. Construct a positive sequence $\{a_n\}$ with the following properties: a) $a_n \to 0$; b) $\sum a_n = +\infty$; c) if an increasing sequence $\{n_k\}$ of indices is such that $\sup_{k\geq j} a_{n_k}/a_{n_j} < +\infty$, then the series $\sum a_{n_k}$ converges.

3.13. Let $\{a_n\}$ be a sequence of positive numbers, and let $A_n = a_1 + \cdots + a_n$ and $\alpha_n = \frac{1}{a_n} + \frac{1}{a_{n+1}} + \cdots$. Prove that the following six assertions are equivalent:

a) $A_n = O(a_n)$; b) $\alpha_n = O(1/a_n)$;

c) $A_n/A_{n+1} \leq q_1 < 1$ for all $n \in \mathbb{N}$;

d) $\alpha_1 < +\infty$ and $\alpha_{n+1}/\alpha_n \leq q_2 < 1$ for all $n \in \mathbb{N}$;

e) $a_1^p + \cdots + a_n^p = O(a_n^p)$ if $p > 0$;

f) $\dfrac{1}{a_n^p} + \dfrac{1}{a_{n+1}^p} + \cdots = O(\dfrac{1}{a_n^p})$ if $p > 0$.

If a)–f) hold, then so does the assertion g) for any $Q > 1$ there exists an index $L = L_Q$ such that $a_{n+L}/a_n \geq Q$ for $n \in \mathbb{N}$, that is, the sequence $\{a_n\}$ can be decomposed into L subsequences $\{a_n^{(l)}\} = \{a_{nL+l}\}$ $(l = 0, 1, \ldots, L-1)$, each increasing no more slowly than a geometric progression: $a_{n+1}^{(l)}/a_n^{(l)} \geq Q$. Prove that if the sequence $\{a_n\}$ is monotone, then g) $\Rightarrow$ a). Is this assertion true for nonmonotone sequences?

3.14. Let $\{a_n\}$ be a sequence of positive numbers, and let $A_n = a_1 + \cdots + a_n$ and $\alpha_n = \frac{1}{a_n} + \frac{1}{a_{n+1}} + \cdots$. Prove that the following seven assertions are equivalent:

a) $A_{n-1} = o(a_n)$; b) $\alpha_{n+1} = o\left(\dfrac{1}{a_n}\right)$;

c) $A_{n-1} = o(A_n)$; d) $\alpha_1 < +\infty$ and $\alpha_{n+1} = o(\alpha_n)$;

e) $a_1^p + \cdots + a_{n-1}^p = o(a_n^p)$ if $p > 0$;

f) $\dfrac{1}{a_{n+1}^p} + \dfrac{1}{a_{n+2}^p} + \cdots = o\left(\dfrac{1}{a_n^p}\right)$ if $p > 0$;

g) $a_{n-1} = o(a_n)$.

If a)–g) hold, then: h) $\sqrt[n]{a_n} \to +\infty$. Show by examples that h) $\nRightarrow$ a), not even for monotone sequences; but if $\sqrt[n]{a_n} \uparrow +\infty$, then a)–g) hold.

3.15. Suppose that $p > 1$ and $a_n \geq 0$ and let $A_n = (a_1 + \cdots + a_n)/n$ $(n \in \mathbb{N})$. Prove that

a) $\sum A_n^p \leq (p/(p-1)) \sum a_n A_n^{p-1}$;
b) $\sum A_n^p \leq (p/(p-1))^p \sum a_n^p$.

3.16. Suppose that $a_n > 0$ for all $n \in \mathbb{N}$, and the series $\sum a_n$ converges. Prove that the series $\sum \sqrt[n]{a_1 \cdot a_2 \cdots a_n}$ converges, while the series $\sum (a_1 + \cdots + a_n)/n$ diverges.

3.17. Let f be a function defined on $\mathbb{R}$. Prove that the following properties of f are equivalent:

a) $f(x) = O(x)$ as $x \to 0$;
b) for any absolutely convergent series $\sum a_n$ the series $\sum f(a_n)$ converges absolutely;
c) for any absolutely convergent series $\sum a_n$ the series $\sum f(a_n)$ converges.

3.18. Let f be a function defined on $\mathbb{R}$. If the series $\sum f(a_n)$ converges for any convergent series $\sum a_n$, then $f(x) = Cx$ in some neighborhood of zero.

3.19. Suppose that Q is a countable subset of $(0, 1)$ with $\inf Q = 0$ and $\sup Q = 1$. Prove that for any number $x \in (0, 1)$ it is possible to number the set $Q = \{q_1, q_2, \dots\}$ in such a way that $x = \sum q_n/2^n$.

§4. Computation of sums of series

Compute the sums of the series in problems 4.1–4.11 (the sums in 4.7–4.9 can be computed in terms of the Euler constant—see problem II.2.9).

4.1. $\sum 3^n \sin^3(x/3^n)$.
4.2. $\sum (-1)^n (\cos^3(3^n x))/3^n$.
4.3. $\sum_{n\geq 1} 1/\binom{n+r}{1+r}$ $(r \in \mathbb{N})$.
4.4. $\sum (n!)^2 z^n/(2n+1)!$ $(|z| < 4)$.
4.5. $\sum_{n\geq 1} \frac{1}{n} a_n(m)$, where $m \in \mathbb{N}$, $m > 1$, $a_n(m) = 1$ if n is not divisible by m, and $a_n(m) = 1 - m$ if n is divisible by m.

4.6. $\sum n(\cosh 2n - 1)e^{-n^2}$. 4.7. $\displaystyle\sum_{n\geq 2}\left(\frac{1}{n}\sum_{m\geq 2} m^{-n}\right)$.

4.8. $\sum \frac{(-1)^n}{n} \ln n$. 4.9. $\sum \frac{(-1)^n}{n}[\log_2 n]$.

4.10. $\sum \frac{1}{n} \int_{2\pi n}^{\infty} \frac{\sin x}{x}\, dx$. 4.11. $\sum \frac{1}{n} \int_{\pi n}^{\infty} \frac{\sin x}{x}\, dx$.

4.12. Prove that

$$\sum_{n>N} \frac{1}{n^2} = \frac{4}{\pi} \frac{(2N)!!}{(2N-1)!!} \int_0^{\pi/2} x^2 \cos^{2N} x \, dx.$$

In particular, $1 + \frac{1}{2^2} + \frac{1}{3^2} + \cdots + \frac{1}{n^2} + \cdots + = \frac{\pi^2}{6}$.

4.13. Let $s > 1$. Prove that:

$$\text{a) } \sum \frac{1}{n^s} = \lim_{n\to\infty} (1 - 2^{-s})^{-1}(1 - 3^{-s})^{-1} \cdots (1 - p_n^{-s})^{-1};$$

$$\text{b) } \sum \frac{\lambda(n)}{n^s} = \lim_{n\to\infty} (1 + 2^{-s})^{-1}(1 + 3^{-s})^{-1} \cdots (1 + p_n^{-s})^{-1},$$

where $p_1 = 2$, $p_2 = 3$, ... are the prime numbers, numbered in increasing order, $\lambda(n) = 1$ if the number of prime divisors of n (counting multiplicity) is even, and $\lambda(n) = -1$ otherwise ($\lambda(1) = 1$).

4.14. Prove that

$$\prod_{n\geq 1} \left(1 - \frac{1}{p_n^2}\right) = \frac{6}{\pi^2}.$$

4.15. Prove that for $|t| < 1$

$$\text{a) } \sum \frac{t^n}{1-t^n} = \sum \tau(n) t^n; \qquad \text{b) } \sum \frac{nt^n}{1-t^n} = \sum \sigma(n) t^n,$$

where $\tau(n)$ is the number of divisors and $\sigma(n)$ the sum of the divisors of n.

§5. Function series

5.1. For what $p, q > 0$ and $x, y \in \mathbb{R}$ does the series

$$\sum \left| \cos \frac{x}{n^p} + \sin \frac{y}{n^{2p}} \right|^{n^q}$$

converge?

5.2. Suppose that $\lambda_n \to 0$. Does the series $\sum \lambda_n e^{-|x-n|}$ converge uniformly on $\mathbb{R}$? Estimate $\sup_x |\sum \lambda_n e^{-|x-n|}|$ under the assumption that $|\lambda_n| \leq 1$ for any $n \in \mathbb{N}$.

5.3. Does the series

$$\sum \frac{x^n}{(1+x)(1+x^2)\cdots(1+x^n)}$$

converge uniformly on $[0, +\infty)$? Estimate

$$\sup_{x\geq 0}\sum \frac{x^n}{(1+x)(1+x^2)\cdots(1+x^n)}.$$

5.4. Does the series $\sum \min(xn!, 1/xn!)$ converge for $x > 0$? Does it converge uniformly on $(0, +\infty)$? Estimate $\sup_{x>0}\sum \min(xn!, 1/xn!)$.

5.5. Suppose that $a_n \in \mathbb{R}$, $b_n > 0$, $a_n = O(b_n)$, and $\sum b_n = +\infty$, and assume that the series $B(t) = \sum_{n\geq 0} b_n t^n$ converges for $|t| < 1$. Prove that the series $A(t) = \sum_{n\geq 0} a_n t^n$ converges for $|t| < 1$, and

$$\underline{\lim}\,\frac{a_n}{b_n} \leq \underline{\lim}_{t\to 1-0}\frac{A(t)}{B(t)} \leq \overline{\lim}_{t\to 1-0}\frac{A(t)}{B(t)} \leq \overline{\lim}\,\frac{a_n}{b_n}$$

(compare with problem II.2.2). In particular, if $a_n \sim cb_n$, $c \neq 0$, then $A(t) \sim cB(t)$ as $t \to 1-0$; if $a_n = o(b_n)$, then $A(t) = o(B(t))$ as $t \to 1-0$.

5.6. Prove that if the series $\sum_{n\geq 0} a_n$ converges, then $\sum_{n\geq 0} a_n t^n \to \sum_{n\geq 0} a_n$ as $t \to 1-0$ (*Abel's theorem*).

5.7. If the partial sums S_n of the series $\sum_{n\geq 0} a_n$ are such that the limit $\lim((S_1+\cdots+S_n)/n) = l$ exists and is finite, then the series $A(t) = \sum_{n\geq 0} a_n t^n$ converges for $|t| < 1$, and $A(t) \to l$ as $t \to 1-0$. Thus, the *Abel method* for summation of series (computation of the limit $\lim_{t\to 1-0}\sum_{n\geq 0} a_n t^n$) is stronger than the arithmetic mean method (the *Cesàro method*).

5.8. Suppose that $b_n \geq 0$, $\sum b_n = +\infty$, and the series $B(t) = \sum_{n\geq 0} b_n t^n$ converges for $|t| < 1$. If the sequence $\{a_n\}_{n\geq 0}$ is such that the limit $\lim(a_0+\cdots+a_n)/(b_0+\cdots+b_n) = l$ exists and is finite, then the series $A(t) = \sum_{n\geq 0} a_n t^n$ converges for $|t| < 1$, and $A(t)/B(t) \to l$ as $t \to 1-0$.

5.9. Suppose that the sequence $\{a_n\}$ is such that $a_n > -1$ for $n \in \mathbb{N}$, and $\prod(1+a_n) = A \in (0, +\infty)$. Prove that for $|t| < 1$ the product $\prod(1+a_n t^n) = A(t)$ converges, and $A(t) \to A$ as $t \to 1-0$ if $\sum a_n^2 < +\infty$. Can the condition $\sum a_n^2 < +\infty$ be dropped?

5.10. Let $\alpha > 1/2$. Prove that:

a) the series $\sum n^{-\alpha}\sin(\pi x\sqrt{n})$ converges for any $x \in \mathbb{R}$;

b) for $1/2 < \alpha < 1$ the convergence is uniform on $[a, A]$, $0 < a < A$, but not uniform on $[0, A]$;

c) for $\alpha = 1$ the convergence is uniform on $[0, A]$, $A > 0$.

5.11. Let $f(x) = \sum \frac{1}{n}\sin(x/4^n)$. Prove that

$$0 < \overline{\lim}_{x\to+\infty} |f(x)|/\ln\ln x < +\infty, \qquad \underline{\lim}_{x\to+\infty} |f(x)| = 0.$$

5.12. Let $f(x) = \sum a_n|\sin nx|$, where $a_n \geq 0$ and $\sum a_n < +\infty$. Prove that $f \in \mathrm{Lip}_1$ if and only if $\sum na_n < +\infty$.

5.13. Suppose that the sequence $\{P_n\}$ of polynomials satisfies the conditions:

a) the degree of P_n is at most $m \in \mathbb{N}$ for all $n \in \mathbb{N}$;

b) the limit $P(x) = \lim P_n(x)$ exists and is finite for all $x \in [\alpha, \beta]$, $\alpha < \beta$.

Prove that P is a polynomial of degree at most m, and $P_n \rightrightarrows P$ on any finite interval.

5.14. Suppose that $-\infty \le a < b \le +\infty$, $f_n \in C((a, b))$, and $f_n \ge 0$ $(n \in \mathbb{N})$, and let $f = \sum f_n$. Prove that if f is continuous on (a, b), then

$$\int_a^b f(x)\,dx = \sum \int_a^b f_n(x)\,dx.$$

§6. Trigonometric series

6.1. Let $\{A_n\}$ and $\{\varphi_n\}$ be numerical sequences, with $A_n \to +\infty$. Prove that in any nonempty open interval there is a point x such that $\overline{\lim}_{n\to+\infty} \cos(A_n x + \varphi_n) = 1$.

6.2. Prove that $a_n, b_n \to 0$ if $a_n \cos nx + b_n \sin nx \to 0$ as $n \to +\infty$ for any $x \in [\alpha, \beta]$, $\alpha < \beta$.

6.3. Find a set $E \subset [0, 2\pi]$ with the cardinality of the continuum and a sequence $B_n \to +\infty$ such that $\sin B_n x \rightrightarrows 0$ on E.

6.4. Prove that $\sum(|a_n| + |b_n|) < +\infty$ if $\sum |a_n \cos nx + b_n \sin nt| \le \text{const}$ on $[\alpha, \beta]$, $\alpha < \beta$.

6.5. Prove that for $x \in (0, \pi)$

$$S_n(x) = \sin x + \dots + \sin nx = O(\min(n, 1/x));$$
$$C_n(x) = \cos x + \dots + \cos nx = O(\min(n, 1/x)).$$

6.6. Compute the limits

$$\lim_{A\to+\infty} \frac{1}{\ln A} \int_0^{\pi/2} \frac{|\sin Ax|}{\sin x}\,dx \quad \text{and} \quad \lim_{A\to+\infty} \frac{1}{\ln A} \int_0^{\pi/2} \frac{\sin^2 Ax}{\sin x}\,dx.$$

What is the asymptotic behavior of the integrals $L_n = \int_{-\pi}^{\pi} |C_n(x)|\,dx$ and $\widetilde{L}_n = \int_{-\pi}^{\pi} |S_n(x)|\,dx$?

6.7. Prove that the series $\sum \frac{1}{\sqrt{n}} \exp(in\pi(x - \ln[\sqrt{n}]))$ diverges for any $x \in \mathbb{R}$.

6.8. Compute the sums of the following series:

$$\text{a) } \sum \frac{\sin nx}{n}; \qquad \text{b) } \sum \frac{\cos nx}{2n-1};$$
$$\text{c) } \sum (-1)^n \frac{\sin nx}{n}; \qquad \text{d) } \sum \frac{\sin nx}{2n-1}.$$

6.9. Prove that:

$$\text{a) } \frac{\pi}{2}|\sin x| = 1 - 2\sum \frac{\cos 2nx}{4n^2-1}; \qquad \text{b) } \frac{\pi}{2}|\cos x| = 1 - 2\sum (-1)^n \frac{\cos 2nx}{4n^2-1}.$$

In problems 6.10–6.21 the sequence $\{\lambda_n\}$ decreases to zero. The Abel transformation (see §2) can prove to be useful in solving these problems.

6.10. Prove that the series

$$\sum \lambda_n \cos nx \quad \text{and} \quad \sum \lambda_n \sin nx$$

converge uniformly on any closed interval not containing points $2\pi k$ $(k \in \mathbb{Z})$.

6.11. Prove that for $x \in (0, \pi)$

$$\sum_{n\geq N} \lambda_n \sin nx = O\left(\frac{\lambda_N}{x}\right), \qquad \sum_{n\geq N} \lambda_n \cos nx = O\left(\frac{\lambda_N}{x}\right).$$

6.12. Prove that the series

$$\sum_{n\geq 1} \lambda_n \sin nx \sin mx \quad \text{and} \quad \sum_{n\geq 1} \lambda_n \cos nx \sin mx$$

$(m \in \mathbb{Z})$ converge uniformly on $\mathbb{R}$.

6.13. Prove that if $g(x) = \sum \lambda_n \sin nx$ for $x \in \mathbb{R}$, then

$$\lambda_m = \frac{2}{\pi} \int_0^\pi g(x) \sin mx\, dx.$$

6.14. Prove that the series $\sum \lambda_n \sin nx$ converges uniformly on $\mathbb{R}$ if and only if $\lambda_n = o(1/n)$.

6.15. Prove that the partial sums of the series $\sum \lambda_n \sin nx$ are uniformly bounded if and only if $\lambda_n = O(1/n)$.

6.16. Let $\lambda_n = O(1/n)$ and suppose that the function φ is absolutely integrable on $[a, b]$. Prove that the series $\sum \lambda_n \varphi(x) \sin nx$ can be integrated termwise on $[a, b]$. Using the result of problem 6.8a), compute the sum of the series $\sum n^{-2}$.

6.17. Let $f(x) = \sum \lambda_n \cos nx$. Prove that $\int_0^\pi |f(x)|\, dx < +\infty$ if $\sum \frac{1}{n}\lambda_n < +\infty$.

6.18. Let $g(x) = \sum \lambda_n \sin nx$. Prove that $\int_0^\pi |g(x)|\, dx < +\infty$ if and only if $\sum \frac{1}{n}\lambda_n < +\infty$.

6.19. Suppose that $\sum \frac{1}{n}\lambda_n < +\infty$, $f(x) = \sum \lambda_n \cos nx$, $g(x) = \sum \lambda_n \sin nx$, and $\varphi \in C([0, \pi])$. Prove that the series $\sum \lambda_n \varphi(x) \cos nx$ and $\sum \lambda_n \varphi(x) \sin nx$ can be integrated termwise on $[0, \pi]$. In particular,

$$\int_0^\pi f(x) \cos mx\, dx = \frac{\pi}{2}\lambda_m \qquad (m \in \mathbb{N}).$$

6.20. Prove that for $x \in (0, \pi)$:

a) $\sum_{1\leq k\leq n} (\sin kx)/k > 0$;

b) $\sum_{1\leq k\leq n} (\cos kx)/k \geq -1$.

6.21. a) Suppose that $g(x) = \sum \frac{\lambda_n}{n} \sin nx$. Prove that $g \geq 0$ on $[0, \pi]$.

b) Let $f(x) = \lambda_0/2 + \sum \lambda_n \cos nx$. Prove that if the sequence $\{\lambda_n\}_{n\geq 0}$ is convex and $\lambda_n \to 0$, then $f \geq 0$ on $\mathbb{R}$ and $\int_0^\pi f(x)\, dx < +\infty$.

6.22. Let $S_n(x)$ be the nth partial sum of the series $f(x) = \sum(\sin nx)/n$, and let $M_n = \max_x S_n(x)$. Prove that

$$M_n = S_n\left(\frac{\pi}{n+1}\right) \uparrow G = \int_0^\pi \frac{\sin u}{u}\,du > \frac{\pi}{2} = f(+0)$$

(the *Gibbs phenomenon*).

6.23. Let $f(x) = \sum_{|k|\le N} c_k e^{ikx}$ end $\alpha, m > 0$. Prove that

$$|c_k| \le \frac{M}{2}\left(\frac{\pi}{|k|}\right)^\alpha$$

for $k = \pm 1, \dots, \pm N$ if $|f(x) - f(y)| \le M|x-y|^\alpha$ $(x, y \in \mathbb{R})$.

6.24. Let $f(x) = \sum 2^{-n}\cos 4^n x$. Prove that $f \in \mathrm{Lip}_{1/2}$, but $f \notin \mathrm{Lip}_\alpha$ for $\alpha > 1/2$.

6.25. Let $f(x) = \sum A^{-n}\sin 3^n x$, where $A \in (1, 3)$. Prove that $f \in \mathrm{Lip}_\alpha$, where $\alpha = \log_3 A$, but $f \notin \mathrm{Lip}_\beta$ for $\beta > \alpha$.

6.26. Let $f(x) = \sum 3^{-n}\sin 3^n x$. Prove that $f \in \mathrm{Lip}_\alpha$ for any $\alpha \in (0, 1)$. Is it true that $f \in \mathrm{Lip}_1$?

6.27. Let $f(x) = \sum 2^{-n}\sin(2\pi n!\,x)$. Prove that:

a) $f \notin \mathrm{Lip}_\alpha$ for any $\alpha \in (0, 1]$;

b) f is nondifferentiable at every point.

6.28. Prove that the following three properties of a trigonometric series

$$\sum_{n\ge 0}(a_n\cos(2^n x) + b_n\sin(2^n x))$$

are equivalent:

a) $\sum_{n\ge 0}(|a_n| + |b_n|) < +\infty$;

b) the partial sums $S_N(x) = \sum_{0\le n\le N}(a_n\cos(2^n x) + b_n\sin(2^n x))$ are uniformly bounded on some interval $[a, b]$, $a < b$;

c) the partial sums $S_N(x)$ are uniformly bounded on $\mathbb{R}$.

6.29. Consider the two sequences $\{P_k\}_{k\ge 0}$ and $\{Q_k\}_{k\ge 0}$ of algebraic polynomials (*Rudin-Shapiro polynomials*) defined recursively as follows: $P_0 = Q_0 = 1$, $P_{n+1}(z) = P_n(z) + z^{2^n}Q_n(z)$, $Q_{n+1}(z) = P_n(z) - z^{2^n}Q_n(z)$ $(n = 0, 1, 2, \dots)$. Prove that:

a) the polynomials P_n and Q_n have degree $2^n - 1$;

b) all the coefficients of the polynomials P_n and Q_n are equal to ± 1;

c) if $|z| = 1$, then $|P_n(z)|^2 + |Q_n(z)|^2 = 2^{n+1}$.

6.30. Let R_N $(N = 0, 1, 2, \dots)$ be the natural projection of the set of all algebraic polynomials onto the set of polynomials of degree at most N (R_N annihilates the monomials of degree greater than N and does not change the polynomials of degree at most N). Prove the following inequalities for $|z| = 1$ and for any N, $n \in \mathbb{N}$;

$$|R_N(P_n)(z)| \le 10\sqrt{N}, \qquad |R_N(Q_n)(z)| \le 10\sqrt{N}.$$

6.31. Prove that there exists a series $\sum_{k\geq 0}\varepsilon_k e^{ik\theta}$ with $\varepsilon_k = \pm 1$ whose partial sums $S_m(\theta)$ satisfy the inequalities:

$$\text{a)}\ \max_{0\leq\theta\leq 2\pi}|S_m(\theta)| \leq 10\sqrt{m}; \qquad \text{b)}\ \int_0^{2\pi}|S_m(\theta)|\,d\theta \geq \sqrt{m/3}.$$

6.32. Using the sequence $\{\varepsilon_k\}_{k\geq 0}$ in problem 6.31, we see that there exists a continuous function f of the form $f(\theta) = \sum_{k\geq 0} a_k e^{ik\theta}$ such that $\sum |a_k|^p = +\infty$ for any $p < 2$.

6.33. Suppose that the sequence $\{\varepsilon_n\} \subset \mathbb{C}$ satisfies the condition (cf. problem 6.31)

$$\max_{\theta\in\mathbb{R}}\left|\sum_{1\leq k\leq n}\varepsilon_k e^{ik\theta}\right| = O(\sqrt{n}).$$

Prove that for $\alpha > 0$:

a)

$$\max_{\theta\in\mathbb{R}}\left|\sum_{1\leq k\leq n}\frac{\varepsilon_k}{k^{1/2-\alpha}}e^{ik\theta}\right| = O(n^{\alpha});$$

b)

$$\max_{\theta\in\mathbb{R}}\left|\sum_{k\geq n}\frac{\varepsilon_k}{k^{1/2+\alpha}}e^{ik\theta}\right| = O(n^{-\alpha});$$

c) the function

$$f(\theta) = \sum_{k\geq 1}\frac{\varepsilon_k}{k^{1/2+\alpha}}e^{ik\theta}$$

is in the class Lip_α for $\alpha \in (0, 1)$.

6.34. Prove that there exists a function $f \in \mathrm{Lip}_{1/2}$ of the form $f(\theta) = \sum_{k\geq 1} c_k e^{ik\theta}$ such that $\sum_{k\geq 1}|c_k| = +\infty$ (cf. problem IX.4.13).

CHAPTER V

Integrals

§1. Improper integrals of functions of a single variable

In problems 1.1–1.11 investigate the convergence of the corresponding integrals.

1.1. $\displaystyle\int_0^{+\infty} \frac{|\sin x|\,dx}{e^{x^2\sin^2 x}}$. 1.2. $\displaystyle\int_0^{+\infty} \frac{dx}{1+x^p\sin^2 x}$.

1.3. $\displaystyle\int_1^{\infty} \frac{x^q\,dx}{1+x^p|\sin x|^r}$. 1.4. $\displaystyle\int_1^{+\infty} \frac{x^q\,dx}{e^{x^p|\sin x|^r}}$.

1.5. $\displaystyle\int_1^{+\infty} \frac{|\sin x^p|\,dx}{x^q}$. 1.6. $\displaystyle\int_1^{+\infty} |\sin(x^p+\ln^q x)|\frac{dx}{x}$.

1.7. $\displaystyle\int_1^{+\infty} \frac{dx}{x^q|\sin x|^p}$. 1.8. $\displaystyle\int_1^{+\infty} x^p|\sin x|^{x^q}\,dx$.

1.9. $\displaystyle\int_0^{\infty} \cos(x^3-x)\,dx$. 1.10. $\displaystyle\int_0^{+\infty} \sin(x^p+ax+b)\,dx$, $p>1$.

1.11. $\displaystyle\int_0^{+\infty} \sin(x\ln x)\,dx$.

1.12. Determine whether the convergence of the integral $\int_1^{+\infty} f(x)\,dx$ implies that of the following integrals:

$$\text{a)}\ \int_1^{+\infty} f^3(x)\,dx; \qquad \text{b)}\ \int_1^{\infty} |f(x)|\frac{dx}{x^2}?$$

1.13. Let $f(x)=x^{-\varepsilon}\int_0^x t^{-2}\ln|1-t^2|\,dt$ $(x>0)$. For what $\varepsilon>0$ does the integral $\int_0^{+\infty} f(x)\,dx$ converge?

1.14. Compute the *Frullani integral* $\int_0^{+\infty}(f(ax)-f(bx))/x\,dx$, where f is continuous on $[0,+\infty)$ and satisfies one of the following conditions:

a) the integral $\int_1^{+\infty} f(x)/x\,dx$ converges;

b) $f(x+T)=f(x)$ for some $T>0$ and all $x\geq 0$;

c) the limit $l=\lim_{x\to+\infty} f(x)$ exists and is finite.

1.15. Suppose that the integral $\int_0^{+\infty} f(x)\,dx$ converges. Prove that:

a) the integral $\int_0^{+\infty} e^{-\varepsilon x}f(x)\,dx$ converges for any $\varepsilon>0$;

b) $\int_0^{+\infty} e^{-\varepsilon x} f(x)\,dx \to \int_0^{+\infty} f(x)\,dx$ as $\varepsilon \to 0$ (cf. Abel's theorem—problem IV.5.6).

In problems 1.16–1.38 compute the integrals. Some of them are expressed in terms of the Euler constant γ (see problem II.2.9).

1.16. $\displaystyle\int_0^{+\infty} \frac{1}{x^2}\ln|1-x^2|\,dx.$ 1.17. $\displaystyle\int_0^{\pi/2} \ln(\sin x)\,dx.$

1.18. $\displaystyle\int_0^1 \frac{\ln x}{1-x}\,dx.$ 1.19. $\displaystyle\int_0^1 \frac{\ln x}{1+x}\,dx.$

1.20. $\displaystyle\int_0^{+\infty} \frac{x^2\,dx}{\cosh^2 x}.$ 1.21. $\displaystyle\int_0^{+\infty} \frac{\sin x}{x} e^{-ax}\,dx \ (a>0).$

1.22. $\displaystyle\int_0^{+\infty} \frac{\sin x}{x}\,dx.$ 1.23. $\displaystyle\int_0^{+\infty} \frac{x-\sin x}{x^3}\,dx.$

1.24. $\displaystyle\int_0^{+\infty} \frac{e^{-x}-\cos x}{x}\,dx.$

1.25. $\displaystyle\int_{-\infty}^{\infty} \frac{(e^{iax}-1)(e^{ibx}-1)}{x^2}\,dx \ (a, b \in \mathbb{R}).$

1.26. $\displaystyle\int_0^{+\infty} \frac{\ln x}{e^x}\,dx.$ 1.27. $\displaystyle\int_0^{+\infty} \frac{\ln^2 x}{e^x}\,dx.$

1.28. $\displaystyle\int_0^1 \frac{1-\cos x}{x}\,dx - \int_1^{+\infty} \frac{\cos x}{x}\,dx.$

1.29. $\displaystyle\int_0^{+\infty} \frac{\cos(x^p)-\cos(x^q)}{x}\,dx \ (p, q>0).$

1.30. $\displaystyle\int_0^{+\infty} \frac{\exp(-x^p)-\exp(-x^q)}{x}\,dx \ (p, q>0).$

1.31. $\displaystyle\int_0^1 \frac{x^2\,dx}{(1-x)\ln^2(1-x)}.$ 1.32. $\displaystyle\int_0^{+\infty} \frac{\ln x}{1+e^x}\,dx.$

1.33. Using the result of problem I.2.16a), prove that

$$\int_0^{\sqrt{n}} e^{-x^2}\,dx = \sqrt{n}\frac{(2n)!!}{(2n+1)!!} + o(1).$$

With the help of the Stirling formula find the value of the *Euler-Poisson integral* $\int_{-\infty}^{+\infty} e^{-x^2}\,dx$.

1.34. $\int_0^{+\infty} \frac{e^{-x} - e^{-2x}}{x^p} \, dx$, where $p = \frac{k}{2} - 1$, $k = 0, 1, 2, 3, 4, 5$.

1.35. $\int_0^{+\infty} \frac{e^{-x^2}}{(x^2 + \frac{1}{2})^2} \, dx$.

1.36. $\int_0^{+\infty} e^{-(x^2 + a^2/x^2)} \, dx$.

1.37. a) $\int_0^{+\infty} \frac{\sin x}{\sqrt{x}} \, dx$; b) $\int_0^{+\infty} \frac{\cos x}{\sqrt{x}} \, dx$.

1.38. $\int_0^{+\infty} \pi(x)/(x^3 - x) \, dx$, where $\pi(x)$ is the number of prime numbers not exceeding x.

§2. Computation of multiple integrals

In problems 2.1–2.7 it is required to compute the given integrals.

2.1. $\int_0^{+\infty} \int_0^{+\infty} |\ln x - \ln y| e^{-(x+y)} \, dx \, dy$.

2.2.

a) $$\int_{[0,1]^n} \max(x_1, \dots, x_n) \, dx_1 \cdots dx_n;$$

b) $$\int_{[0,1]^n} \min(x_1, \dots, x_n) \, dx_1 \cdots dx_n;$$

c) $$\int_{[1,+\infty)^n} \frac{dx_1 \cdots dx_n}{x_1 \cdots x_n (\max(x_1, \dots, x_n))^\varepsilon} \quad (\varepsilon > 0).$$

2.3. $\int_{\mathbb{R}^n} e^{-\|x\|^2/2} \, dx$.

2.4. a) $\iint_{\mathbb{R}^2} |ax + by| e^{-(x^2+y^2)/2} \, dx \, dy$;

b) $\int_{\mathbb{R}^n} |(x, a)|^p / e^{-\|x\|^2/2} \, dx$ $(a \in \mathbb{R}^n,\ p > -1)$.

2.5. $\iiint_{\|y\| \leq 1} \|x - y\|^{-1} \, dy_1 \, dy_2 \, dy_3$ $(x \in \mathbb{R}^3)$.

2.6. $\iiiint_{x^2+y^2+u^2+v^2 \leq 1} e^{x^2+y^2-u^2-v^2} \, dx \, dy \, du \, dv$.

2.7. $\int_{(Ax,x) \leq 1} e^{(Ax,x)} \, dx$ $(x \in \mathbb{R}^4$, A a positive-definite matrix).

2.8. Let $A = \{x = (x_1, x_2, x_3, x_4) \in \mathbb{R}^4 \mid \sqrt{x_2^2 + x_3^2 + x_4^2} \leq x_1\}$. For what $t \in \mathbb{R}^4$ is the integral $K(t) = \int_A e^{-(x,t)} \, dx$ finite? What is its value?

In problems 2.9–2.13 we understand a random choice of points to mean that the probability of choosing a point belonging to some set is proportional to its measure (length, area, volume).

2.9. Two points are chosen randomly in the interval $[a, b]$. Find the mean value M of the distance between them. What is the probability P that this distance is greater than M?

2.10. Three numbers are chosen randomly on the interval $[0, a]$. What is the probability that they are the lengths of the sides of some triangle?

2.11. Two points u and v are chosen randomly on the interval $[-a, a]$.

a) What is more probable: that the roots of the equation $z^2 + uz + v = 0$ lie on the real axis (probability $P_1(a)$), or that the roots of this equation do not lie on the real axis (probability $P_2(a)$)? What do the probabilities $P_1(a)$ and $P_2(a)$ tend to as $a \to +\infty$?

b) What is the probability that the biquadratic equation $z^4 + uz^2 + v = 0$ has both real and complex roots?

2.12. Two points are chosen randomly in the unit disk. What is the mean value S of the area of the disk constructed with the segment joining them as a diameter?

2.13. Two points are chosen randomly on a circle of radius R. Find:

a) the mean value L of the length of the chord joining them;

b) the mean value α of the angle $(\leq \pi)$ formed by the radii drawn to these points.

2.14. Let $p < 1$ and let

$$\Phi(x, y) = \iint_{u^2+v^2\leq 1} ((x-v)^2 + (y-v)^2)^{-p/2}\, du\, dv,$$

$$\Psi(x, y) = \iint_{u^2+v^2\leq 1} \ln((x-u)^2 + (y-v)^2)\, du\, dv.$$

Prove that $\Phi, \Psi \in C^1(\mathbb{R}^2)$ and find Ψ.

2.15. Let $A_n = \{x = (x_1, \dots, x_n) \in \mathbb{R}^n \mid \sum_{1\leq k\leq n} x_k/k \leq 1,\ x_1, \dots, x_n \geq 0\}$. Prove that for any $t \in \mathbb{R}$

$$\int \cdots \int_{A_n} e^{t(x_1+\cdots+x_n)}\, dx_1 \cdots dx_n = \left(\frac{e^t - 1}{t}\right)^n.$$

2.16. Prove that for $m, n \in \mathbb{N}$ with $m \leq n$ and for $a > 0$

$$\int_{[0,1]^n} \frac{dx}{(a + x_1 + \cdots + x_n)^m} = \frac{(-1)^{m-1}}{(m-1)!(n-m)!}\Delta_n(a^{n-m}\ln a),$$

where $\Delta_1 f(a) = \Delta f(a) = f(a+1) - f(a)$ and $\Delta_n f(a) = \Delta(\Delta_{n-1} f(a))$.

2.17. Let a_k $(k = 0, 1, \dots, n)$ be positive numbers.

a) Prove that for any nonnegative continuous function f on $[0, +\infty)$

$$\int \cdots \int_{[0,1]^n} f(a_1 + (a_2 - a_1)x_1 + (a_3 - a_2)x_1x_2 + \cdots + (a_0 - a_n)x_1 \cdots x_n) \times x_1^{n-1} x_2^{n-2} \cdots x_{n-1}\, dx_1 \cdots dx_n$$
$$= \int \cdots \int_{Q_n} f(a_0 + (a_1 - a_0)x_1 + (a_2 - a_0)x_2 + \cdots + (a_n - a_0)x_n)\, dx_1 \cdots dx_n,$$

where $Q_n = \{(x_1, \dots, x_n) \in \mathbb{R}^n \mid \sum_{1\le k\le n} x_k \le 1,\ x_1 \ge 0, \dots, x_n \ge 0\}$.

b) Compute the integrals

$$\int \cdots \int_{[0,1]^n} \frac{x_1^{n-1} x_2^{n-2} \cdots x_{n-1}\, dx_1 \cdots dx_n}{(a_1 + (a_2 - a_1)x_1 + (a_3 - a_2)x_1x_2 + \cdots + (a_0 - a_n)x_1 \cdots x_n)^{n+1}},$$

$$\int \cdots \int_{Q_n} \frac{dx_1 \cdots dx_n}{(a_0 + (a_1 - a_0)x_1 + (a_2 - a_0)x_2 + \cdots + (a_n - a_0)x_n)^{n+1}}.$$

2.18. Prove that:

a) if f is continuous and nonnegative on $\mathbb{R}$, then

$$\frac{1}{\pi} \iint_{\mathbb{R}^2} f(ax+by) \frac{dx\,dy}{(1+x^2)(1+y^2)} = \int_{-\infty}^{+\infty} f((|a|+|b|)u) \frac{du}{1+u^2} \quad (a, b \in \mathbb{R});$$

b) for $0 < p < 1$ and $a_k \in \mathbb{R}$ $(k = 1, 2, \dots, n)$

$$\frac{1}{\pi^n} \int_{\mathbb{R}^n} |(a, x)|^p \frac{dx}{(1+x_1^2)(1+x_2^2)\cdots(1+x_n^2)} = \left(\cos \frac{\pi}{2} p\right)^{-1} \left(\sum_{1\le k\le n} |a_k|\right)^p.$$

2.19. Let $f \in C^2([0,1] \times [0,1])$. Find the limit

$$\lim_{n\to+\infty} n^2 \left(\int_0^1 \int_0^1 f(x, y)\, dx\, dy - \frac{1}{n^2} \sum_{1\le k\le n} \sum_{1\le j\le n} f\left(\frac{k-1/2}{n}, \frac{j-1/2}{n}\right)\right).$$

CHAPTER VI

Asymptotics

§1. Asymptotics of integrals

1.1. Suppose that $-\infty < a < b \leq +\infty$, $f, g \in C([a, b))$, and $g > 0$. Prove that:

a) if $\int_a^b g(t)\,dt = +\infty$ and $f(t) \sim g(t)$ as $t \to b-0$, then $\int_a^x f(t)\,dt \sim \int_a^x g(t)\,dt$ as $x \to b-0$;

b) if $\int_a^b g(t)\,dt = +\infty$ and $f(t) = o(g(t))$ as $t \to b-0$, then $\int_a^x f(t)\,dt = o(\int_a^x g(t)\,dt)$ as $x \to b-0$;

c) if $\int_a^b g(t)\,dt < +\infty$ and $f(t) \sim g(t)$ as $t \to b-0$, then $\int_x^b f(t)\,dt \sim \int_x^b g(t)\,dt$ as $x \to b-0$;

d) if $\int_a^b g(t)\,dt < +\infty$ and $f(t) = o(g(t))$ as $t \to b-0$, then $\int_x^b f(t)\,dt = o(\int_x^b g(t)\,dt)$ as $x \to b-0$.

1.2. Find the asymptotics as $t \to 1-0$ of the following integrals:

a) $\mathscr{J}(t) = \int_0^t (1-x^a)^{-1}\,dx \quad (a \neq 0)$;

b) $\mathscr{K}(t) = \int_0^1 \sqrt{(1-x^2)(1-t^2x^2)}^{\,-1}\,dx$;

c) $\mathscr{L}(t) = \int_0^\pi (1 - 2t\cos x + t^2)^{-1}\,dx$.

1.3. Find the asymptotics as $A \to +\infty$ of the following integrals:

a) $\int_A^{2A} (\ln x)^{-1}\,dx$;

b) $\int_0^A e^{e^{-x}}\,dx$;

c) $\int_\pi^A |\sin x| e^{a/x}\,dx \quad (a \in \mathbb{R})$;

d) $\int_A^{+\infty} e^{-x^2}\,dx$;

e) $\int_A^{+\infty} x^a e^{-x}\,dx \quad (a \in \mathbb{R})$.

1.4. Find the asymptotics as $A \to +\infty$ of the following integrals:

a) $\int_1^{+\infty} e^{-x^p/A} x^{-1}\,dx \quad (p > 0)$;

b) $\int_0^1 e^{Ax^2}\,dx$;

c) $\int_0^{+\infty} e^{x - x^2/A}\,dx$;

d) $\int_2^{+\infty} e^{-x/A} \ln^p x\,dx \quad (p \in \mathbb{R})$.

1.5. Find the asymptotics as $n \to +\infty$ of the following integrals ($n \in \mathbb{N}$):

a) $\alpha_n = \int_{2\pi n}^{2\pi(n+1)} (\sin x)/x^p\,dx \quad (p \neq 0)$;

b) $\beta_n = \int_{\pi n}^{+\infty} (\cos x)/x^p \, dx \quad (p > 0)$;
c) $\gamma_n = \int_{\pi n}^{+\infty} (\sin x)/x^p \, dx \quad (p > 0)$.

1.6. Prove that as $\varepsilon \to +0$:
a) $\int_1^{+\infty} (\sin^2 x)/x^{1+\varepsilon} \, dx \sim 1/2\varepsilon$;
b) $\int_1^{+\infty} (\cos^2 x)/x^{1+\varepsilon} \, dx \sim 1/2\varepsilon$;
c) $\int_0^{+\infty} (\sin x)/x^{\varepsilon} \, dx \to 1$;
d) $\int_0^{+\infty} (\cos x)/x^{\varepsilon} \, dx \sim \pi\varepsilon/2$.

1.7. Prove that as $p \to +\infty$:
a) $\int_0^{+\infty} \sin(x^p) \, dx \sim \pi/2p$;
b) $\int_0^{+\infty} \cos(x^p) \, dx \to 1$.

1.8. Suppose that the function φ is continuous on $\mathbb{R}$ and has period $T > 0$, and that $c_\varphi = \frac{1}{T}\int_0^T \varphi(t)\,dt \neq 0$. Prove that $\int_1^{+\infty} \varphi(t)/t^{1+\varepsilon} \, dt \sim c_\varphi/\varepsilon$ as $\varepsilon \to +0$.

1.9. Prove that as $A \to +\infty$:

$$\text{a)} \int_0^{\pi/2} \left(\frac{\sin Ax}{\sin x}\right)^2 dx = \frac{\pi}{2}A + O(1);$$

$$\text{b)} \int_0^1 \frac{\cos Ax}{\sqrt{x}} \, dx = \sqrt{\frac{\pi}{2A}} + O\left(\frac{1}{A}\right);$$

$$\text{c)} \int_0^{\pi/2} \frac{\cos^2 x \, dx}{1 + \cos^2 Ax} \to \frac{\pi}{4\sqrt{2}};$$

$$\text{d)} \int_0^{\pi/2} \left(\frac{\sin Ax}{\sin x}\right)^4 dx = \frac{\pi}{3}A^3 + O(A).$$

1.10. Find the asymptotics as $A \to +\infty$ of the following integrals:

$$\text{a)} \int_0^A \frac{\sin^2 x}{x^p} \, dx \quad (p \le 1); \qquad \text{b)} \int_1^A \frac{|\sin x|^p |\cos x|}{x} \, dx \quad (p > -1);$$

$$\text{c)} \int_0^A \frac{|\sin x|}{x^p} \, dx \quad (p \le 1); \qquad \text{d)} \int_A^{+\infty} \frac{|\sin x|}{x^p} \quad (p > 1).$$

1.11. Suppose that the function f decreases to zero on $[a, +\infty)$, the function φ is continuous on $\mathbb{R}$ and has period $T > 0$, and $\int_0^T \varphi(t)\,dt = 0$. Prove that $|\int_A^{+\infty} f(t)\varphi(t)\,dt| \le f(A)\int_0^T |\varphi(t)|\,dt$ for $A \ge a$.

1.12. Suppose that f decreases to zero on $[a, +\infty)$, and φ is continuous on $\mathbb{R}$ and has period $T > 0$. Prove that

$$\int_a^A f(t)\varphi(t)\,dt = c_\varphi \int_a^A f(t)\,dt + I + O(f(A)),$$

where $c_\varphi = \frac{1}{T}\int_0^T \varphi(t)\,dt$ and $I = \int_a^{+\infty} (\varphi(t) - c_\varphi) f(t)\,dt$.

1.13. Suppose that f is nonnegative and monotone on $[a, +\infty)$, φ is continuous on $\mathbb{R}$ and has period $T > 0$, and $c_\varphi = \frac{1}{T}\int_0^T \varphi(t)\,dt \neq 0$. Prove that:

a) if $\int_a^{+\infty} f(t)\,dt = +\infty$ and $\int_A^{A+1} f(t)\,dt = o(\int_a^A f(t)\,dt)$, then $\int_a^A f(t)\varphi(t)\,dt \sim c_\varphi \int_a^A f(t)\,dt$ as $A \to +\infty$;

b) if $\int_a^{+\infty} f(t)\,dt < +\infty$ and $\int_A^{A+1} f(t)\,dt = o(\int_A^{+\infty} f(t)\,dt)$, then $\int_A^{+\infty} f(t)\varphi(t)\,dt \sim c_\varphi \int_A^{+\infty} f(t)\,dt$ as $A \to +\infty$.

1.14. Suppose that $f \in C([a, b])$, and φ is continuous on $\mathbb{R}$ and has period $T > 0$, and let $c_\varphi = \frac{1}{T}\int_0^T \varphi(t)\,dt$. Prove that

$$\lim_{A\to+\infty} \int_a^b f(t)\varphi(At)\,dt = c_\varphi \int_a^b f(t)\,dt.$$

1.15. Let $f \in C([a, b] \times [-1, 1])$. Prove that

$$I(A) = \int_a^b f(x, \sin Ax)\,dx \underset{A\to+\infty}{\longrightarrow} \frac{1}{2\pi}\int_a^b \left(\int_0^{2\pi} f(x, \sin t)\,dt\right) dx.$$

1.16. Find the asymptotics as $\varepsilon \to +0$ of the following integrals:

$$\text{a)} \int_0^{+\infty} e^{-x(1-x^{-\varepsilon})}\,dx; \qquad \text{b)} \int_0^{+\infty} e^{-x(x^{\varepsilon}-1)}\,dx.$$

1.17. Prove that as $\varepsilon \to +0$:

$$\text{a)} \int_0^1 x^{\varepsilon x}\,dx = 1 - \frac{\varepsilon}{4} + \frac{\varepsilon^2}{27} + O(\varepsilon^3);$$

$$\text{b)} \int_0^1 e^{-\varepsilon/\sqrt{x}}\,dx = 1 - 2\varepsilon - \varepsilon^2 \ln \varepsilon + O(\varepsilon^2);$$

$$\text{c)} \int_0^1 e^{-\varepsilon/(x\ln^2(x/e))}\,dx = 1 - \varepsilon - \frac{\varepsilon}{\ln \varepsilon} + o\left(\frac{\varepsilon}{\ln \varepsilon}\right).$$

1.18. Suppose that f is positive and integrable on $(0, 1)$ and that the integral $\int_0^1 |\ln f(x)|^p\,dx$ is finite for some number p in $(1, 2]$. Prove that

$$\int_0^1 f^\varepsilon(x)\,dx = 1 + \varepsilon \int_0^1 \ln f(x)\,dx + O(\varepsilon^p), \qquad \varepsilon \to +0.$$

1.19. Prove that:

$$\text{a)} \int_0^\pi \frac{|\sin nx|}{x}\,dx \sim \frac{2}{\pi} \ln n;$$

$$\text{b)} \int_0^\pi \max_{1\le k\le n} |\sin kx| \frac{dx}{x} \sim \ln n;$$

$$\text{c)} \int_0^\pi \max_{2\le k\le n} \frac{|\sin kx|}{\ln k} \frac{dx}{x} \sim \ln \ln n.$$

§2. The Laplace method

2.1. Let $f \in C([0, \pi/2])$. Find the limit

$$\lim_{n\to\infty} n \int_0^{\pi/2} x f(x) \cos^n x\,dx.$$

2.2. Suppose that $\theta > 0$ and $c_\theta = \int_0^\theta e^{-t^2}\,dt$. Prove that as $A \to +\infty$:

$$\text{a)} \int_0^{\theta/\sqrt{A}} (1-x^2)^A\,dx \sim \frac{c_\theta}{\sqrt{A}}; \qquad \text{b)} \int_0^{\theta/\sqrt{A}} \frac{dx}{(1+x^2)^A} \sim \frac{c_\theta}{\sqrt{A}}.$$

2.3. Let $\theta > 0$. Find the asymptotics as $A \to +\infty$ of the integral

$$\int_0^{\theta/\sqrt{A}} \cos^A x\,dx.$$

2.4. Verify that the asymptotics as $A \to +\infty$ of the following integrals do not depend on the parameter $\theta > 0$:

$$\text{a)} \int_0^\theta (1-x^2)^A\,dx \quad (\theta \le 1); \qquad \text{b)} \int_0^\theta \frac{dx}{(1+x^2)^A};$$

$$\text{c)} \int_0^\theta \cos^A x\,dx \quad \left(\theta \le \frac{\pi}{2}\right); \qquad \text{d)} \int_0^\theta \frac{dx}{(1+x+x^2)^A};$$

$$\text{e)} \int_0^\theta \left(\frac{\ln(1+x)}{x}\right)^A dx; \qquad \text{f)} \int_0^\theta \left(\frac{\sin x}{x}\right)^A dx \quad (\theta \le \pi);$$

$$\text{g)} \int_0^\theta \ln^A(e-x)\,dx \quad (\theta \le e-1).$$

2.5. Find the asymptotics as $A \to +\infty$ of the following integrals:
a) $\int_0^1 (1-x^p)^A\,dx \quad (p > 0)$;
b) $\int_0^{+\infty} (1+x^p)^{-A}\,dx \quad (p > 0)$;
c) $\int_0^{\pi/2} x^p \cos^A x\,dx \quad (p > -1)$.

The characteristic feature of problems 2.2–2.5 consists in the phenomenon of localization—the asymptotic behavior of the integral depends on the behavior of the integrand only in a neighborhood of a single point. This effect also appears clearly in a more general situation in the study of important integrals of the form

$$\Phi(A) = \int_a^b \varphi^A(x)\,dx,$$

where φ is nonnegative and piecewise monotone. For large values of the parameter A the graph of the function φ^A has sharply expressed "humps" in neighborhoods of those points at which φ has a strict local maximum. By breaking up $[a, b]$ into several intervals if necessary it can be assumed that φ is monotone on $[a, b]$. Here it suffices to consider only the case when φ is decreasing. Then the values of φ^A at points far from a are negligibly small in comparison with its values at points near to a, which give the main contribution to the integral $\Phi(A)$. To find the principal part of $\Phi(A)$ it remains to approximate φ in a neighborhood of a by a simpler function and to compute the integral obtained. The Laplace method for investigating integrals of the form $\Phi(A)$ and modifications of them amounts to the realization of the scheme presented.

Most often encountered is the case when the difference $\varphi(a)-\varphi(x)$ is an infinitesimally small quantity of power type, that is,

$$\varphi(a)-\varphi(x) \underset{x\to a+0}{\sim} C\cdot(x-a)^{\sigma} \qquad (C\,,\,\sigma>0)\,.$$

It is investigated in problem 2.6. The result obtained there is called the Laplace asymptotic formula (for brevity of formulation it is assumed in problem 2.6 that $\varphi(a)=1$). The reader can formulate an analogous assertion without difficulty if φ is increasing on $[a\,,\,b]$ and $\varphi(b)-\varphi(x) \underset{x\to b-0}{\sim} C(b-x)^{\sigma}$. This implies at once the Laplace asymptotic formula also for piecewise monotone functions.

2.6. Suppose that $-\infty<a<b\le+\infty$, the function φ is positive and decreasing on $[a\,,\,b)$, $\int_a^b \varphi(x)\,dx<+\infty$, and $1-\varphi(x) \underset{x\to a+0}{\sim} C\cdot(x-a)^{1/p}$ $(C\,,\,p>0)$. Prove that

$$\Phi(A)=\int_a^b \varphi^A(x)\,dx \underset{A\to+\infty}{\sim} \frac{\Gamma(1+p)}{(AC)^p}\,.$$

In particular, $\Phi(A)\sim 1/AC$ for $p=1$, and $\Phi(A)\sim \frac{1}{2}\sqrt{\pi/AC}$ for $p=1/2$.

2.7. Suppose that $-\infty<a<b\le+\infty$, the function φ is positive and decreasing on $[a\,,\,b)$, $\int_a^b \varphi(x)\,dx<+\infty$, and $\lim_{x\to a+0}(1-\varphi(x))/(x-a)^{1/p}=+\infty$ for some $p>0$. Prove that $\int_a^b \varphi^A(x)\,dx=o(A^{-p})$ as $A\to+\infty$.

2.8. Suppose that $-\infty<a<b\le+\infty$, the functions f and g are absolutely integrable on $(a\,,\,b)$, $f>0$, and $f(x)\sim g(x)$ as $x\to a+0$. Assume further that φ is nonnegative and strictly decreasing on $[a\,,\,b)$. Prove that

$$\int_a^b g(x)\varphi^A(x)\,dx \underset{A\to+\infty}{\sim} \int_a^b f(x)\varphi^A(x)\,dx\,.$$

2.9. Find the asymptotics as $A\to+\infty$ of the following integrals:

a) $\displaystyle\int_0^{\pi} \sin^A x\,dx$;

b) $\displaystyle\int_0^{+\infty} \frac{\cos(ax)}{(1+x^2)^A}\,dx \quad (a\in\mathbb{R})$;

c) $\displaystyle\int_0^{A} e^x(A-x)^A\,dx$;

d) $\displaystyle\int_0^{+\infty} \frac{\sin x}{x^p}\frac{dx}{e^{Ax}} \quad (p<2)$;

e) $\displaystyle\int_0^{1} \frac{dx}{((1+x)^{1+x}(1-x)^{1-x})^A}$;

f) $\displaystyle\int_0^{\pi/2} (\cos x)e^{-A\sqrt{\sin 2x}}\,dx$;

g) $\displaystyle\int_0^{+\infty} e^{A(x-x^p)}\,dx \quad (p>1)$;

h) $\displaystyle\int_0^{1} x^{Ax}\,dx$.

2.10. Find the asymptotics as $n\to+\infty$ $(n\in\mathbb{N})$ of the following integrals:

a) $\int_0^{\pi}(a+\cos x)^n\,dx$ $(a>0)$;

b) $\int_0^{+\infty}((\sin x)/x)^n\,dx$;
c) $\int_0^2(1-4x+2x^2)^n\,dx/\sqrt{x}$.

In the following problems we consider integrals of the form $\Phi(A) = \int_a^b f_A(x)\,dx$, where the function f_A, while not now a function of the form considered in problem 2.6, still preserves its characteristic feature: as A increases, the graph of f_A has a sharper and sharper "hump" in a neighborhood of the point x_A at which f_A attains its largest value. Although the result in problem 2.6 is not applicable here, the idea of the solution is preserved: representing f_A in the form $e^{\ln f_A}$, and replacing $\ln f_A$ in some neighborhood of x_A by its Taylor expansion and the integral over (a, b) by the integral over the neighborhood of x_A, we find the principal part of $\Phi(A)$. The choice of the neighborhood is the basic difficulty here. On the one hand, it must not be too large, since otherwise the error due to the use of Taylor's formula becomes influential. On the other hand, in order to neutralize the second error due to the replacement of the integral over (a, b) by the integral over the neighborhood, this neighborhood cannot be too small. The main content of the solution involves a successful choice of the neighborhood, allowing good estimates of both these errors.

We mention also that if $x_A \to x_0$ as $A \to +\infty$, then it can be more convenient to consider the Taylor expansion of the function $\ln f_A$ in a neighborhood of the point x_0.

2.11. Find the asymptotics as $A \to +\infty$ of the following integrals:

$$\text{a)}\ \int_0^{+\infty}\left(\frac{2x}{1+x^2}\right)^A \frac{dx}{e^{x\sqrt{A}}}; \qquad \text{b)}\ \int_0^1 (Ax)^{px}\,dx \quad (p>0);$$

$$\text{c)}\ \int_0^A e^{x^p}(A-x)^A\,dx \quad (0<p<1).$$

2.12. Prove that for any real number p

$$\int_0^1 |\ln x|^p(1-x)^A\,dx \underset{A\to+\infty}{\sim} \frac{\ln^p A}{A}.$$

2.13. Prove that as $A \to +\infty$:

$$\text{a)}\ \int_0^1 \frac{dx}{(Ax)^x} \sim \frac{1}{\ln A}; \qquad \text{b)}\ \int_0^{1/2} x^{Ax}\,dx \sim \frac{1}{A\ln A}.$$

In the following problems we study the asymptotic properties of the Euler Γ-*function* $\Gamma(x) = \int_0^{+\infty} t^{x-1}e^{-t}\,dt$, $x>0$.

2.14. Prove *Stirling's formula*: $\Gamma(1+x) \underset{x\to+\infty}{\sim} \sqrt{2\pi x}(x/e)^x$.

2.15. Prove that $\Gamma(x+c) \underset{x\to+\infty}{\sim} x^c\Gamma(x)$ for any $c\in\mathbb{R}$.

2.16. Prove that $\int_0^x t^x e^{-t}\,dt \underset{x\to+\infty}{\sim} \frac{1}{2}\Gamma(1+x)$.

2.17. Suppose that φ is a positive function defined on $(0, +\infty)$.

Prove the following assertions:

a) $\lim_{x\to+\infty} \frac{\varphi(x)-x}{\sqrt{x}} = 0 \Leftrightarrow \lim_{x\to+\infty} \frac{1}{\Gamma(1+x)} \int_0^{\varphi(x)} t^x e^{-t}\,dt = \frac{1}{2}$;

b) $\lim_{x\to+\infty} \frac{\varphi(x)-x}{\sqrt{x}} = +\infty \Leftrightarrow \lim_{x\to+\infty} \frac{1}{\Gamma(1+x)} \int_0^{\varphi(x)} t^x e^{-t}\,dt = 1$;

c) $\lim_{x\to+\infty} \frac{\varphi(x)-x}{\sqrt{x}} = -\infty \Leftrightarrow \lim_{x\to+\infty} \frac{1}{\Gamma(1+x)} \int_0^{\varphi(x)} t^x e^{-t}\,dt = 0$.

§3. Asymptotics of sums

3.1. Suppose that $a_n > 0$ and $a_n \to 0$. Prove that:
a) if $\sum a_n = +\infty$ and $b_n \sim a_n$, then $\sum_{1\le k\le n} b_k \sim \sum_{1\le k\le n} a_k$;
b) if $\sum a_n = +\infty$ and $b_n = o(a_n)$, then $\sum_{1\le k\le n} b_k = o(\sum_{1\le k\le n} a_k)$;
c) if $\sum a_n < +\infty$ and $b_n \sim a_n$, then $\sum_{k\ge n} b_k \sim \sum_{k\ge n} a_k$;
d) if $\sum a_n < +\infty$ and $b_n = o(a_n)$, then $\sum_{k\ge n} b_k = o(\sum_{k\ge n} a_k)$.

3.2. Suppose that $a_n \downarrow 0$, $\gamma \ge 0$, and $\alpha \in \mathbb{R}$. Prove that:
a) if $\sum_{1\le k\le n} k^\alpha a_k = O(n^\gamma)$, then $a_n = O(n^{\gamma-\alpha-1})$;
b) if $\sum_{1\le k\le n} k^\alpha a_k = o(n^\gamma)$, then $a_n = o(n^{\gamma-\alpha-1})$;
c) if $\sum_{k\ge n} k^\alpha a_k = O(n^{-\gamma})$, then $a_n = O(n^{-\gamma-\alpha-1})$;
d) if $\sum_{k\ge n} k^\alpha a_k = o(n^{-\gamma})$, then $a_n = o(n^{-\gamma-\alpha-1})$.

3.3. Suppose that $\gamma > 0$ and $\sum_{1\le k\le n} k^\alpha x_k \sim n^\gamma$. Prove that:
a) if $\gamma > \alpha$, then $S_n = x_1 + \cdots + x_n \sim \frac{\gamma}{\gamma-\alpha} n^{\gamma-\alpha}$;
b) if $\gamma = \alpha$, then $S_n = x_1 + \cdots + x_n \sim \gamma \ln n$;
c) if $\gamma < \alpha$, then the series $\sum x_n$ converges, and

$$\sigma_n = \sum_{k\ge n} x_k \sim \frac{\gamma}{\alpha-\gamma} n^{\gamma-\alpha}.$$

3.4. Suppose that $\gamma > 0$, $\alpha \in \mathbb{R}$, the series $\sum k^\alpha x_k$ converges, and $\sum_{k\ge n} k^\alpha x_k \sim n^{-\gamma}$. Prove that:
a) if $\gamma < -\alpha$, then $S_n = x_1 + \cdots + x_n \sim \gamma n^{-\alpha-\gamma}/|\alpha+\gamma|$;
b) if $\gamma = -\alpha$, then $S_n = x_1 + \cdots + x_n \sim \gamma \ln n$;
c) if $\gamma > -\alpha$, then the series $\sum x_k$ converges, and $\sigma_n = \sum_{k\ge n} x_k \sim \gamma n^{-\alpha-\gamma}/(\alpha+\gamma)$.

3.5. Suppose that $a_n \downarrow 0$, and $S_n = a_1 + a_2 + \cdots + a_n \sim n^{1-\gamma}/(1-\gamma)$, where $0 < \gamma < 1$.
a) Prove that $a_n \sim n^{-\gamma}$.
b) Can it be asserted that $a_n \sim 1/n$ if $S_n \sim \ln n$?

3.6. Suppose that $a_n \downarrow 0$, $\sum a_n < +\infty$, and $\sum_{k\ge n} a_k \sim n^{-\gamma}$, where $\gamma > 0$. Prove that $a_n \sim \gamma n^{-1-\gamma}$.

3.7. Suppose that $\alpha > 0$ and let

$$x_n = \sum_{1\le k\le n^\alpha} \frac{1}{k}\left(1-\frac{1}{n}\right)^k .$$

Prove that $x_n \sim \min(1, \alpha)\ln n$.

3.8. Let $\tau(k)$ be the number of divisors of a natural number k. Prove that $T(n) = \sum_{1\le k\le n}\tau(k) = n(\ln n + 2\gamma - 1) + O(\sqrt{n})$. Here γ is the Euler constant (see problem II.2.9).

3.9. Suppose that $\alpha_n \downarrow 0$, $\sum \alpha_n < +\infty$, and $f(t) = t\sum \alpha_n/(1+t^2\alpha_n)$ for $t \ge 0$. Prove that the behavior of the function f as $t \to +\infty$ is closely connected with that of the sequences $\{n\sqrt{\alpha_n}\}$; more precisely, there exists a number $C > 0$ such that the following inequalities hold:

a) $\sup_{t\ge 0} f(t) \le C \sup n\sqrt{\alpha_n}$;

b) $\overline{\lim}_{t\to+\infty} f(t) \le C\, \overline{\lim}\, n\sqrt{\alpha_n}$;

c) $\underline{\lim}\, n\sqrt{\alpha_n} \le C\, \underline{\lim}_{t\to+\infty} f(t)$.

3.10. Let $f\colon [0, +\infty) \to \mathbb{R}$ be a nonnegative nonincreasing function, and let $\varphi(h) = h\sum f(kh)$, $h > 0$. Prove that $\lim_{h\to+0}\varphi(h) = \int_0^{+\infty} f(t)\,dt$. Is this true if f is continuous but not monotone?

3.11. Determine the asymptotic behavior of the following sums as $t \to +\infty$:

$$\text{a)}\ \sum_{k\ge 1}\frac{1}{k^2+t^2}; \qquad \text{b)}\ \sum_{k\ge 1} e^{-k^2/t}.$$

3.12. Suppose that $f(p) = \sum_{n\ge 1}(-1)^{n+1}/n^p$ as $p > 0$. Prove that $f(p) = 1/2 + O(p)$.

3.13. Suppose that f is a nonnegative decreasing function on $[1, +\infty)$. Prove that:

a) if $\int_1^{+\infty} f(t)\,dt = +\infty$, then $\sum_{1\le k\le n} f(k) = \int_1^{n+1} f(t)\,dt + C + o(1)$, $0 \le C \le f(1)$, in particular, $\sum_{1\le k\le n} f(k) \sim \int_1^n f(t)\,dt$;

b) if $\int_1^{+\infty} f(t)\,dt < +\infty$, then $\sum_{k\ge n} f(k) = \int_n^{+\infty} f(t)\,dt + O(f(n))$, in particular, if $f(x) = o(\int_x^{+\infty} f(t)\,dt)$ as $x \to +\infty$, then $\sum_{k\ge n} f(k) \sim \int_n^{+\infty} f(t)\,dt$.

3.14. If f is a monotone function on $[m, n]$ $(m, n \in \mathbb{Z})$, then

$$\left|\sum_{m\le k\le n} f(k) - \int_m^n f(t)\,dt\right| \le \max(|f(n)|, |f(m)|).$$

In particular, if f is monotone on $[1, +\infty)$, $\int_1^{+\infty} f(t)\,dt = +\infty$, and $f(x) = o(\int_1^x f(t)\,dt)$ as $x \to +\infty$, then $\sum_{1\le k\le n} f(k) \sim \int_1^n f(t)\,dt$.

3.15. Suppose that $M, N \in \mathbb{Z}$, $M \le N$, and the function f (complex-valued in general) is integrable on the interval $[M - \frac{1}{2}, N + \frac{1}{2}]$. Let $\Delta = \sum_{M\le k\le N} f(k) - \int_{M-1/2}^{N+1/2} f(t)\,dt$.

a) Prove that

$$|\Delta| \le \frac{1}{2} \operatorname*{Var}_{M-1/2}^{N+1/2}(f).$$

In particular, if f is monotone, then $|\Delta| \le \frac{1}{2}|f(N+\frac{1}{2}) - f(M-\frac{1}{2})|$.

b) If $f \in C^2([M-\frac{1}{2}, N+\frac{1}{2}])$, then $\Delta = -\frac{1}{2}\int_{M-1/2}^{N+1/2}(t-[t]-\frac{1}{2})^2 f''(t)\,dt$, and hence $|\Delta| \le \frac{1}{8}\int_{M-1/2}^{N+1/2}|f''(t)|\,dt$. In particular, if f is convex or concave, then $|\Delta| \le \frac{1}{8}|f'(N+\frac{1}{2}) - f'(M-\frac{1}{2})|$.

3.16. Prove that

$$\prod_{1\le k\le n}\left(1+\frac{1}{\sqrt{k}}\right) \sim A\frac{e^{2\sqrt{n}}}{\sqrt{n}} \quad\text{and}\quad \prod_{1\le k\le n}\left(1+\frac{1}{\sqrt[3]{k}}\right) \sim B\sqrt[3]{n}e^{3(n^{2/3}-n^{1/3})/2}$$

for some positive numbers A and B.

3.17. Suppose that $\alpha \in \mathbb{R}\backslash\mathbb{N}$, $n \in \mathbb{N}$, and

$$\binom{n-\alpha}{n} = \frac{(n-\alpha)(n-\alpha-1)\cdots(1-\alpha)}{n!}.$$

Prove that there exists a number $C_\alpha \neq 0$ such that

$$\binom{n-\alpha}{n} \sim \frac{C_\alpha}{n^\alpha}.$$

3.18. Prove that:

a) $\sqrt[n^2]{1\cdot 2^2\cdot 3^3\cdots n^n} \sim \sqrt{n}/\sqrt[4]{e}$;

b) $\sqrt[n^3]{1!(2!)^2\cdots(n!)^n} \sim \sqrt[3]{n}/e^{4/9}$;

c) $\sqrt[n]{1\cdot 2^2\cdot 3^3\cdots n^n} \sim (n/\sqrt{e})^{n/2}\sqrt{n}$.

3.19. Let $\alpha > 0$. Prove that as $n \to +\infty$ $(n \in \mathbb{N})$:

a) $\displaystyle\sum_{1\le k\le n} k^{-\alpha k/n} \sim \frac{n}{\alpha \ln n}$; b) $\displaystyle\sum_{1\le k\le n} k^{\alpha k/n} \sim \frac{n^{1+\alpha}}{\alpha \ln n}$;

c) $\displaystyle\sum_{1\le k\le n} (k!)^{-\alpha/n} \sim \frac{n}{\alpha \ln n}$; d) $\displaystyle\sum_{1\le k\le n} k^{\alpha n} \sim \frac{n^{\alpha n}}{1-e^{-\alpha}}$.

3.20. Prove that $\sum_{1\le k\le n} n^k/k! \sim e^n/2$.

3.21. Prove that

$$\sum_{1\le k\le n}\sqrt{\binom{n}{k}} \sim \sqrt[4]{2\pi n}2^{n/2}.$$

3.22. Suppose that $N \in \mathbb{N}$, $z \in \mathbb{C}$, and $\operatorname{Re} z > 0$. Prove that

$$\sum_{n \geq N} \frac{1}{n} e^{-z\sqrt{n}} = e^{-z\sqrt{N}} \left(\frac{2}{z} + O\left(\frac{1}{N} + \frac{|z|}{\sqrt{N}} \right) \right),$$

where the constant on the O-term is independent of z and N.

3.23. Suppose that $N \in \mathbb{N}$, $z \in \mathbb{C}$, and $\operatorname{Re} z > 0$. Prove that

$$\sum_{n \geq N} e^{-z\sqrt{n}} = \frac{2}{z} e^{-z\sqrt{N-1/2}} \left(\sqrt{N-1/2} + \frac{1}{z} + z^2 O\left(1 + \frac{|z|^2}{\sqrt{N}} \right) \right),$$

where the constant in the O-term is independent of z and N.

3.24. Let $\alpha \in (0, 1)$. Prove that

$$\sum \frac{\cos nx}{n^\alpha} = \frac{A}{x^{1-\alpha}} + O(1) \quad \text{and} \quad \sum \frac{\sin nx}{n^\alpha} = \frac{B}{x^{1-\alpha}} + O(1),$$

as $x \to +0$, where A and B are positive numbers dependent on α.

3.25. Prove that there exists a number $\alpha_0 \in (0, 1)$ such that for any $\alpha \geq \alpha_0$ the partial sums of the series $\sum (\cos nx)/n^\alpha$ are uniformly bounded below, but for any $\alpha \in (0, \alpha_0)$ this is false.

What can be said about boundedness below for the partial sums of the series $\sum (\sin nx)/n^\alpha$, $x \in [0, \pi]$?

3.26. Suppose that $\psi \in C^1([a, +\infty))$, $\psi, \psi' > 0$, and $\psi = O(\psi')$ on $[a, +\infty)$. Prove that $\int_a^{+\infty} e^{-\varepsilon\psi(t)}\, dt = \psi^{-1}(1/\varepsilon) + O(1)$ as $\varepsilon \to +0$.

3.27. Find the asymptotic behavior as $t \to 1-0$ of the sums of the following series([2]):

a) $\sum \frac{1}{n} t^{\ln^p n}$ $(p > 0)$; b) $\sum t^{\ln^p n}$ $(p > 1)$;

c) $\sum t^{n^p}$ $(p > 0)$; d) $\sum t^{a^n}$ $(a > 1)$;

e) $\sum t^{(an)^n}$ $(a > 0)$; f) $\sum t^{n!}$.

3.28. Find the asymptotic behavior as $t \to 1-0$ of the sums of the following series:

a) $\sum n^p t^n$ $(p > -1)$; b) $\sum_{n \geq 2} (\ln n)^p t^n$ $(p \in \mathbb{R})$.

3.29. Find the asymptotic behavior as $t \to 1-0$ of the sums of the

([2]) Recall that $\sum a_n$ stands for the series $\sum_{n \geq 1} a_n$.

following series:

a) $\sum \frac{t^n}{1+t^n}$; b) $\sum \frac{nt^n}{1+t^n}$;

c) $\sum \frac{t^n}{(1+t^n)^2}$; d) $\sum \frac{nt^n}{(1+t^n)^2}$;

e) $\sum \frac{t^n}{1-t^n}$; f) $\sum \frac{nt^n}{1-t^n}$;

g) $\sum \frac{t^n}{(1-t^n)^2}$; h) $\sum \frac{nt^n}{(1-t^n)^2}$;

i) $\sum \frac{(-1)^n t^n}{1-t^n}$; j) $\sum \frac{(-1)^n nt^n}{1-t^n}$;

k) $\sum \frac{(-1)^n t^n}{(1-t^n)^2}$; l) $\sum \frac{(-1)^n nt^n}{(1-t^n)^2}$.

3.30. Find the asymptotic behavior as $t \to 1-0$ of the sums of the following series:

a) $\sum \frac{t^n}{1+t^{2n}}$; b) $\sum \frac{t^n}{(1+t^{2n})^2}$;

c) $\sum \frac{nt^n}{1-t^{2n}}$; d) $\sum \frac{t^n}{1-t^{2n}}$;

e) $\sum \frac{t^n}{(1-t^{2n})^2}$.

3.31. Find the asymptotic behavior as $t \to 1-0$ of the sums of the following series:

a) $T(t) = \sum \tau(n)t^n$; b) $S(t) = \sum \sigma(n)t^n$,

where $\tau(n)$ is the number of divisors and $\sigma(n)$ the sum of the divisors of a number n.

§4. Asymptotics of implicit functions and recursive sequences

4.1. Suppose that $z(t) = \int_t^{+\infty} e^{-u^2/2}\,du$ $(t \in \mathbb{R})$, and let $t(z)$ be the function inverse to $z(t)$. Prove that $t(z) \sim \sqrt{2\ln(1/z)}$ as $z \to +0$.

4.2. Let x_n be the root of the equation $x = \tan x$ lying in interval $(\pi n, \pi(n+1))$, $n \in \mathbb{N}$. Prove that

$$x_n = \pi n + \frac{\pi}{2} - \frac{1}{\pi n} + O\left(\frac{1}{n^2}\right).$$

4.3. Prove that each of the following equations determines an infinitely differentiable implicit function y defined in some neighborhood of the point (a, b), and find the coefficients in the expansion

$$y(x) = b + a_1(x-a) + \cdots + a_n(x-a)^n + \cdots:$$

a) $ye^y - x = 0$, $a = b = 0$, $n = 3$;
b) $y^2 + \ln^2 y - x = 0$, $a = b = 1$, $n = 3$;
c) $xy + e^{xy} - x = 0$, $a = 1$, $b = 0$, $n = 3$;
d) $y \ln y - x = 0$, $a = 0$, $b = 1$, $n = 3$;
e) $e^{xy} + x^2 + y - 1 = 0$, $a = b = 0$, $n = 6$;
f) $\arctan(x + y) - x - 2y = 0$, $a = b = 0$, $n = 6$.

4.4. Prove that *Kepler's equation* $y = M + x\sin y$, where M is a fixed positive parameter, determines an infinitely differentiable function y defined in a neighborhood of the point $(0, M)$, and

$$y(x) = M + x\sin M + x^2\frac{\sin 2M}{2} + O(x^3), \qquad x \to 0.$$

4.5. Verify that the domains of the implicit functions considered in problem 4.3 contain the half-line $[a, +\infty)$, and prove the following as $x \to +\infty$:

$$\text{a)}\quad y(x) = \ln x - \ln\ln x + \frac{\ln\ln x}{\ln x} + o\left(\frac{\ln\ln x}{\ln x}\right);$$

$$\text{b)}\quad y(x) = \sqrt{x} - \frac{\ln^2 x}{8\sqrt{x}} + \frac{\ln^4 x}{128x\sqrt{x}} + o\left(\frac{\ln^4 x}{x\sqrt{x}}\right);$$

$$\text{c)}\quad y(x) = \frac{\ln x}{x} - \frac{\ln x}{x^2} - \frac{1}{2}\frac{\ln^2 x}{x^3} + o\left(\frac{\ln^2 x}{x^3}\right);$$

$$\text{d)}\quad y(x) = \frac{x}{\ln x} + \frac{x\ln\ln x}{\ln^2 x} + \frac{x(\ln\ln x)^2}{\ln^3 x} + O\left(\frac{x\ln\ln x}{\ln^3 x}\right);$$

$$\text{e)}\quad y(x) = -x^2 + 1 - e^{-x^3+x} + O(xe^{2(x-x^3)});$$

$$\text{f)}\quad y(x) = -\frac{x}{2} + \frac{\pi}{4} - \frac{1}{x} + O\left(\frac{1}{x^2}\right).$$

4.6. Let $x_n = h(x_{n-1})$, where $h(x) = x/(1 + ax^p)^{1/p}$ for $x \geq 0$ (a and p are positive constants). Prove that

$$x_n \sim \frac{1}{(an)^{1/p}} \qquad (x_0 > 0).$$

4.7. Suppose that $x_0 = 10^{-3}$, and $x_n = x_{n-1} - x_{n-1}^2$ $(n \in \mathbb{N})$. Prove that $x_{1000} \approx \frac{1}{2}10^{-3}$ with accuracy 10^{-6}.

4.8. Find the asymptotics of the recursive sequences $x_n = f(x_{n-1})$, $x_0 > 0$, in the following cases:

a) $f(x) = x(1-x)$, $x_0 < 1$; b) $f(x) = \sin x$, $x_0 < \pi$; c) $f(x) = \dfrac{x}{1+x^3}$;

d) $f(x) = \arctan x$; e) $f(x) = \ln(1+x)$; f) $f(x) = 1 - e^{-x}$;

g) $f(x) = \dfrac{x}{1+\sqrt{x}}$; h) $f(x) = \dfrac{x}{\sqrt{1+x^4}}$.

4.9. Suppose that $C, p, x_0 > 0$, then $x_n = x_{n-1} + Cx_{n-1}^{1-p}$. Prove that:

a) $x_n \sim (Cpn)^{1/p}$; b) $x_n = (Cpn)^{1/p}\left(1 + \frac{p-1}{2p^2}\frac{\ln n}{n} + O\left(\frac{1}{n}\right)\right)$.

4.10. Construct a positive function f on $(0, +\infty)$ such that the recursive sequence $x_n = f(x_{n-1})$, $x_0 > 0$, satisfies the relation $x_n \sim 1/(\ln n)$.

4.11. Suppose that $x_0 \in (-1, 2)$, $x_n = x_{n-1}(x_{n-1} - 1)$ $(n \in \mathbb{N})$. Find the asymptotics of the sequence $\{x_n\}$.

4.12. Suppose that $a_0 = 1$, and $a_n = a_{n-1} + (\sum_{0\le k<n} a_k)^{-1}$ $(n \in \mathbb{N})$. Prove that $a_n \sim \sqrt{2\ln n}$.

4.13. Suppose that $|p| < 1$, $a_0 = 1$, and $a_n = a_{n-1} + (\sum_{0\le k<n} a_k)^{-p}$ $(n \in \mathbb{N})$. Find the asymptotics of the sequence $\{a_n\}$.

4.14. Suppose that $p \in \mathbb{R}$, $a_0 = 1$, and $a_n = a_{n-1} + (\sum_{0\le k<n} a_k^p)^{-1}$ $(n \in \mathbb{N})$. Find the asymptotics of the sequence $\{a_n\}$.

CHAPTER VII

Functions (Continuation)

§1. Convexity

A function $f: \langle a, b\rangle \to \mathbb{R}$ is said to be *convex* (*strictly convex*) if for all points $x_1, x_2 \in \langle a, b\rangle$ and all numbers $\lambda_1, \lambda_2 > 0$ with $\lambda_1 + \lambda_2 = 1$ the inequality

$$f(\lambda_1 x_1 + \lambda_2 x_2) \le \lambda_1 f(x_1) + \lambda_2 f(x) \ (f(\lambda_1 x_1 + \lambda_2 x_2) < \lambda_1 f(x_1) + \lambda_2 f(x_2))$$

holds for $x_1 \ne x_2$. If $-f$ is convex, then f is said to be *concave.*

1.1. Prove that a convex function f satisfies *Jensen's inequality*

$$f\left(\sum_{1\le j\le n} \lambda_j x_j\right) \le \sum_{1\le j\le n} \lambda_j f(x_j)$$

for any $x_1, \dots, x_n \in \langle a, b\rangle$, $\lambda_1, \dots, \lambda_n \in [0, 1]$, with $\sum_{1\le j\le n} \lambda_j = 1$.

1.2. Suppose that a function f is defined on the interval $\langle a, b\rangle$, and let $x_1 < x_2 < x_3$ be points in $\langle a, b\rangle$. Consider the chords joining the points of the graph of f with abscissae x_1 and x_2, x_1 and x_3, and x_2 and x_3. The slopes of these chords are

$$\frac{f(x_2) - f(x_1)}{x_2 - x_1}, \qquad \frac{f(x_3) - f(x_1)}{x_3 - x_1}, \qquad \frac{f(x_3) - f(x_2)}{x_3 - x_2},$$

respectively. Prove that f is convex if and only if

$$\frac{f(x_2) - f(x_1)}{x_2 - x_1} \le \frac{f(x_3) - f(x_1)}{x_3 - x_1} \le \frac{f(x_3) - f(x_2)}{x_3 - x_2}$$

for any triple of points $x_1 < x_2 < x_3$ in $\langle a, b\rangle$ (the *three chords lemma*). Verify that this pair of inequalities is equivalent to

$$\begin{vmatrix} 1 & x_1 & f(x_1) \\ 1 & x_2 & f(x_2) \\ 1 & x_3 & f(x_3) \end{vmatrix} \ge 0.$$

1.3. Suppose that a function f on an interval Δ satisfies the inequality

$$f\left(\frac{x_1 + x_2}{2}\right) \le \frac{1}{2}(f(x_1) + f(x_2))$$

for all $x_1, x_2 \in \Delta$. Prove that:

a) f satisfies Jensen's inequality

$$f\left(\sum_{1\le j\le n} \lambda_j x_j\right) \le \sum_{1\le j\le n} \lambda_j f(x_j), \qquad x_j \in \Delta\ (j = 1, \dots, n)$$

for any rational $\lambda_j \in [0, 1]$, $\sum_{1\le j\le n} \lambda_j = 1$;

b) if f is continuous, then it is convex.

1.4. Prove that the condition of continuity in problem 1.3b) can be replaced by the following condition: f is bounded above on some nonempty open interval $(p, q) \subset \Delta = \langle a, b\rangle$ (or even on $(p, q)\setminus\mathbb{Q}$).

1.5. Prove that the following three statements are equivalent for a function $f: (a, b) \to \mathbb{R}$:

a) f is convex;

b) the supergraph $\Gamma^+ = \{(x, y) \in \mathbb{R}^2 \mid x \in (a, b),\ y \ge f(x)\}$ of f is a convex set;

c) through each point of the graph of f it is possible to draw a straight line supporting the supergraph (i.e., a line such that all the points of the supergraph lie above it).

1.6. Prove that if f is a convex function on $[0, 1]$, then the function $\varphi(x) = f(x) + f(1 - x)$ is decreasing on $[0, 1/2]$.

1.7. Prove that if a function on an interval is locally convex, then it is convex.

1.8. Prove that a convex function f on $\langle a, b\rangle$ is continuous on (a, b), and has there finite increasing one-sided derivatives $f'_-(x)$ and $f'_+(x)$, and $f'(x)$ exist everywhere except on an at most countable set of points. Prove that $f \in \mathrm{Lip}_1(\Delta)$ for any closed interval $\Delta \subset (a, b)$.

1.9. Prove that a function $f \in C((a, b))$ is convex if and only if for all $x \in (a, b)$

$$L_f(x) = \overline{\lim_{h\to 0}}\, h^{-2}(f(x + h) - 2f(x) + f(x - h)) \ge 0.$$

In particular, if $L_f(x) = 0$ for all $x \in (a, b)$, then f is a linear function; if f is twice differentiable on (a, b), then it is convex if and only if $f'' \ge 0$ on (a, b).

1.10. Prove that a function $f \in C(\langle a, b\rangle)$ is convex if and only if

$$f(x) \le \frac{1}{2h}\int_{x-h}^{x+h} f(t)\, dt$$

for all $x \in (a, b)$ and all $h > 0$ for which $x + h, x - h \in \langle a, b\rangle$.

1.11. Every convex continuous function on $[a, b]$ is the limit of a uniformly convergent decreasing sequence of:

a) piecewise linear convex functions;

b) twice continuously differentiable convex functions.

1.12. Prove that if a function f is convex and strictly monotone, then f^{-1} is either convex or concave.

1.13. Suppose that a function f is convex on $[a, b]$, and a function g is convex and increasing on $[c, d]$. Prove that if the composition $g \circ f$ makes sense, then it is convex.

1.14. Let $f \in C((0, +\infty))$. Prove that the functions $xf(x)$ and $f(1/x)$ are then convex or nonconvex simultaneously.

1.15. Suppose that a function f is convex on $\mathbb{R}$ and $\lim_{x\to+\infty} f(x)/x = \lim_{x\to-\infty} f(x)/x = 0$. Prove that f is constant.

1.16. Suppose that a function f is convex on $[a, +\infty)$. Prove that the limit $\lim_{x\to+\infty} f(x)/x = l \in \overline{\mathbb{R}}$ exists, and $l > -\infty$.

1.17. Suppose that a function f is increasing and concave on $[0, +\infty)$, and $f(0) = 0$. Prove that $f(x+y) \le f(x) + f(y)$ for all $x, y > 0$.

1.18. Prove that if $f\colon [0, \infty) \to \mathbb{R}$ is differentiable, concave, and nonnegative, then $xf'(x) \le f(x)$ for all $x \ge 0$.

1.19. Prove that if f is a convex function on $[a, \infty)$, then the function $\varphi(x) = f(b+x) - f(x)$ is increasing $(b > 0)$.

1.20. Suppose that a function f is convex and increasing on $[0, \infty)$, and $f(0) = 0$. Prove that

$$\sum_{0\le k\le n} (-1)^k f(a_k) \ge f\left(\sum_{0\le k\le n} (-1)^k a_k\right)$$

for any numbers $a_0 \ge a_1 \ge \cdots \ge a_n \ge 0$.

1.21. A sequence $\{a_n\} \subset \mathbb{R}$ is said to be convex down if $\Delta^2 a_n = a_{n+2} - 2a_{n+1} + a_n \ge 0$, i.e., $a_{n+1} \le \frac{1}{2}(a_n + a_{n+2})$ for any $n \in \mathbb{N}$. Prove that if a sequence $\{a_n\}$ is convex, then:

a) $a_{n+p} \le (pa_{n+p+q} + qa_n)/(p+q)$ $(p, q \in \mathbb{N})$;

b) the sequence $\{a_n\}$ is monotone or there is a number $m > 1$ such that $a_1 \ge a_2 \ge \cdots \ge a_m \le a_{m+1} \le a_{m+2} \le \cdots$.

1.22. Let $\{a_n\}_{n\ge0}$ be a bounded convex sequence (see problem 1.21). Prove that:

a) $a_n \downarrow a \in \mathbb{R}$;

b) $a_n - a_{n-1} = o(1/n)$;

c) $\sum_{n\ge0}(n+1)\Delta^2 a_n = a_0 - a$.

1.23. Suppose that $f \in C([0, 1])$, $f \ge 0$, $P = \int_0^1 f(x)\,dx$, and $M = \max f$. Prove the following inequalities:

a) if f is concave, then

$$\text{a}')\ \int_0^1 xf(x)\,dx \le \frac{2}{3}P, \qquad \text{a}'')\ \int_0^1 x^2 f(x)\,dx \le \frac{1}{2}P;$$

b) if f is concave and monotone, then

$$\text{b}')\ \int_0^1 f^2(x)\,dx \le \frac{4}{3}P^2, \qquad \text{b}'')\ \int_0^1 f^3(x)\,dx \le 2P^3;$$

c) if f is convex and $\min f = 0$, then

$$\text{c}')\ \int_0^1 x f(x)\,dx \geq \frac{2}{3}\frac{P^2}{M}, \qquad \text{c}'')\ \int_0^1 x^2 f(x)\,dx \geq \frac{2}{3}\frac{P^3}{M^2};$$

d) if f is convex and monotone, and $\min f = 0$, then

$$\text{d}')\ \int_0^1 f^2(x)\,dx \leq \frac{2}{3}MP; \qquad \text{d}'')\ \int_0^1 f^3(x)\,dx \leq \frac{1}{2}M^2P.$$

All eight inequalities have a simple mechanical meaning. For example, the inequalities b) mean that among equal-area subgraphs of functions of the given class it is the triangles that have the greatest static moment and the greatest moment of inertia with respect to the x-axis.

1.24. Suppose that the function f is nondecreasing on $[0, 1/2]$, and $f(1-x) = f(x)$ on $[0, 1]$. Prove the inequality

$$\int_0^1 f(x)\varphi(x)\,dx \leq \int_0^1 f(x)\,dx \int_0^1 \varphi(x)\,dx$$

for any convex function φ on $[0, 1]$.

1.25. Prove that the integral $\int_0^1 f(x)\varphi(x)\,dx$ is nonnegative for any convex function φ on $[0, 1]$ if and only if the function f satisfies the following conditions:

a) $\int_0^1 f(x)\,dx = 0$;

b) $\int_0^1 x f(x)\,dx = 0$;

c) $\int_0^a (a-x) f(x)\,dx \geq 0$ for any $a \in (0, 1)$.

1.26. Suppose that the function f is nonnegative on $(0, +\infty)$ and $f(x) \to 0$ as $x \to +\infty$. Prove the following inequalities:

a) if f is decreasing and $a > 0$, then $\int_0^{+\infty} f(x)\sin ax\,dx \geq 0$;

b) if f is convex, then $\int_0^{+\infty} f(x)\cos ax\,dx \geq 0$.

In both cases equality is possible only for $f \equiv 0$.

1.27. a) Suppose that $a_n, \lambda_n \geq 0$ $(n \in \mathbb{N})$, and $f(x) = \sum a_n e^{-\lambda_n x}$. Prove that $\ln f$ is a convex function on $\mathbb{R}$.

b) A positive function f is said to be *logarithmically convex* if the function $\ln f$ is convex. Prove that a logarithmically convex function is convex, and that a sum of logarithmically convex functions is logarithmically convex.

1.28. Suppose that $f \in C((0, +\infty))$, $f(1) = 1$.

a) Prove that the following assertions 1) and 2) are equivalent:

1) $f\left(\frac{x+y}{2}\right) \geq \frac{f(x)+f(y)}{2}$, $f(x)f(y) \geq f(xy)$ $(x, y \in (0, \infty))$;

2) there exists a $p \in [0, 1]$ such that $f(x) = x^p$ $(x \in (0, \infty))$.

Verify that condition 1) cannot be replaced by the condition

1′) $f\left(\frac{x+y}{2}\right) \geq \frac{f(x)+f(y)}{2}$, $f(x)f(y) \leq f(xy)$ $(x, y \in (0, \infty))$.

b) Prove that the following assertions 3) and 4) are equivalent:

3) $f\left(\frac{x+y}{2}\right) \le \frac{f(x)+f(y)}{2}$, $f(x)f(y) \le f(xy)$ $(x, y \in (0, \infty))$;

4) there exists a $p \in (-\infty, 0] \cup [1, +\infty)$ such that $f(x) = x^p$ $(x \in (0, \infty))$.

1.29. Suppose that $a_j, b_j \ge 0$, $j = 1, \dots, n$, $p, q > 1$, $\frac{1}{p} + \frac{1}{q} = 1$. Prove the *Hölder inequality*

$$\sum_{1\le j\le n} a_j b_j \le \left(\sum_{1\le j\le n} a_j^p\right)^{1/p} \left(\sum_{1\le j\le n} b_j^q\right)^{1/q}.$$

1.30. Suppose that $x_j > 0$, $j = 1, \dots, n$, and $p \ne 0$. Let

$$\kappa(p) = \left(\frac{1}{n}\sum_{1\le j\le n} x_j^p\right)^{1/p}.$$

Prove that:

a) κ is monotonically increasing;

b)

$$\lim_{p\to 0} \kappa(p) = (x_1 \cdots x_n)^{1/n};$$

c) $\lim_{p\to+\infty} \kappa(p) = \max\{x_1, \dots, x_n\}$;

d) $\lim_{p\to-\infty} \kappa(p) = \min\{x_1, \dots, x_n\}$;

e) the function $\varphi(p) = (\kappa(p))^p$ is logarithmically convex on $(0, +\infty)$.

Formulate analogues of these assertions for nonnegative functions in $C([0, 1])$.

1.31. Suppose that f is a piecewise continuous function, $f\colon [a, b] \to [m, M]$, φ is a convex function on $[m, M]$, p is a continuous and nonnegative function on $[a, b]$, and $\int_a^b p(x)\,dx = 1$.

Prove that in this case:

a)

$$\varphi\left(\frac{1}{b-a}\int_a^b f(x)\,dx\right) \le \frac{1}{b-a}\int_a^b \varphi(f(x))\,dx;$$

b)

$$\varphi\left(\int_a^b f(x)p(x)\,dx\right) \le \int_a^b \varphi(f(x))p(x)\,dx;$$

c) if $f > 0$, then

$$\left(\int_a^b p(x)/f(x)\,dx\right)^{-1} \le \exp\left(\int_a^b p(x)\ln f(x)\,dx\right) \le \int_a^b p(x)f(x)\,dx.$$

1.32. Suppose that $A_0, \dots, A_n$ are the consecutive vertices of a convex polyhedron inscribed in a circle, with A_0 and A_n fixed. How should the points $A_1, \dots, A_{n-1}$ be selected in order to maximize the perimeter and the area of the polyhedron (for a given n)?

1.33. A point source of light located at a point $(0, b)$, $b > 0$, illuminates an area that is the subgraph of a nonnegative convex function $f \in C^1([0, \infty))$, $f(0) > b$, and the light rays are reflected from the graph of f and from the x-axis according to the well-known law. Is the whole area illuminated if $\lim_{x\to\infty} f(x) = 0$?

Let f be a convex function, $f\colon \mathbb{R} \to \overline{\mathbb{R}}$. The *Legendre transform* of f is defined to be the function

$$f^*(x) = \sup_{t\in\mathbb{R}}(xt - f(t)) \ (x \in \mathbb{R}).$$

The quantity $f^*(x)$ shows how much it is necessary to lower a straight line through the origin with slope x in order to make it a supporting line of the supergraph of f (Figure 4).

1.34. Prove that if a function $f\colon \mathbb{R} \to (-\infty, \infty]$ is convex, then:

a) f^* is convex;

b) the set $\mathscr{D}(f) = \{x \in \mathbb{R} \mid f(x) < +\infty\}$ is an interval, and if a point t is not an endpoint of $\mathscr{D}(f)$, then

$$f(t) = \sup\{g(t) \mid g \text{ is a linear function, } g \le f\};$$

c) $(f^*)^*(t) = f(t)$ if t is not an endpoint of $\mathscr{D}(f)$;

d) *Young's inequality* $f(t) + f^*(x) \ge xt$ holds for all $x, t \in \mathbb{R}$;

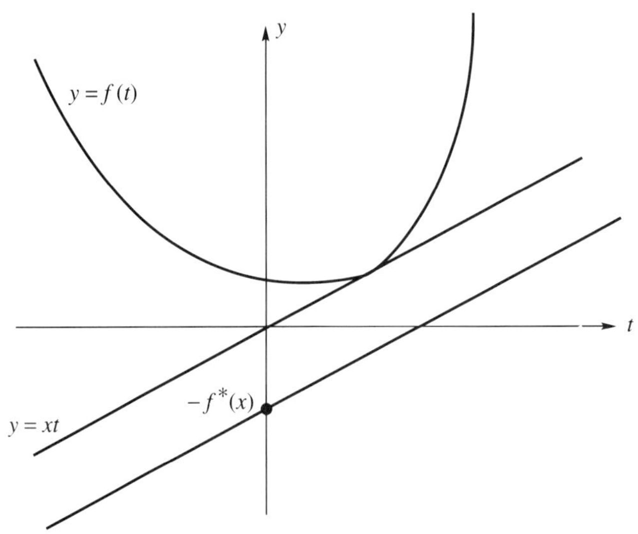

FIGURE 4

e) the function f^* is finite on the interval $\Delta = (\inf f'(t), \sup f'(t))$, where the infimum and the supremum are over those points $t \in \mathscr{D}(f)$ at which $f'(t)$ exists, and for $x \in \Delta$ the supremum in the definition of $f^*(x)$ is attained;

f) if $s_- = f'_-(t) < f'_+(t) = s_+$, then f^* is linear on $[s_-, s_+]$, and if f is linear on $[a, b]$, then f^* is not differentiable at a point $p = f'(t)$, $t \in (a, b)$;

g) if $\mathscr{D}(f) = \langle a, b\rangle$ and f is strictly convex and differentiable on (a, b), then f' is continuous on (a, b), f^* is differentiable on the open interval $\Delta = \{f'(t) \mid t \in (a, b)\}$, and the functions f' and $(f^*)'$ are mutually inverse;

h) if $f \leq g$, then $f^* \geq g^*$.

1.35. Prove that if the function φ is strictly increasing and continuous on $[0, +\infty)$, $\varphi(0) = 0$, and $\psi = \varphi^{-1}$, then for $a \in [0, +\infty)$ and $b \in [0, \sup \varphi)$ the inequality holds:

$$ab \leq \int_0^a \varphi(t)\, dt + \int_0^b \psi(t)\, dt.$$

1.36. Find the Legendre transform and write Young's inequality (see problem 1.34d)) for the following functions:

a) $f(t) = \dfrac{t^2}{2}$;

b) $f(t) = \dfrac{|t|^p}{p}$ $(p \geq 1)$;

c) the graph of f is a convex polygonal curve;

d) $f(t) = \begin{cases} -t^p/p & \text{if } t \geq 0, \\ +\infty & t < 0 \quad (0 < p < 1); \end{cases}$

e) $f(t) = \begin{cases} -\sqrt{a^2 - t^2} & \text{if } |t| \leq a, \\ +\infty & \text{if } |t| > a \quad (a > 0); \end{cases}$

f) $f(t) = \begin{cases} -\frac{1}{2} - \ln t & \text{if } t > 0, \\ +\infty & \text{if } t \leq 0; \end{cases}$

g) $f(t) = e^t$.

1.37. Consider the collection W of pairs (f, g) of convex functions on $\mathbb{R}$ satisfying the condition $xy \leq f(x) + g(y)$ for all $x, y \in \mathbb{R}$. Call a pair $(f_0, g_0) \in W$ extremal if the relations $f \leq f_0$, $g \leq g_0$, and $(f, g) \in W$ imply that $f = f_0$ and $g = g_0$. Prove that the pair (f_0, g_0) is extremal if and only if $f_0^* = g_0$ and $g_0^* = f_0$.

1.38. Give examples of functions satisfying the condition $f^*(x) = f(-x)$ $(x \in \mathbb{R})$. Prove that the only function coinciding with its own Legendre transform is the function $f(t) = t^2/2$.

§2. Smooth functions

2.1. Prove that if f is differentiable m times on the interval (a, b) and $x, x + mh \in (a, b)$, then $\Delta_h^m f(x) = h^m f^{(m)}(x + \theta h)$, where $0 < \theta < m$ (see problem III.3.1 for the definition of $\Delta_h^m f(x)$).

2.2. Suppose that a_k, b_k, $c_k \in \mathbb{R}$, $c_k \neq 0$ for $k = 1, \dots, n$. Prove that if $f \in C(\mathbb{R})$,

$$f(x) = \sum_{1 \le k \le n} a_k f(b_k x + c_k y)$$

for all $x, y \in \mathbb{R}$, then $f \in C^\infty(\mathbb{R})$.

2.3. Prove that for any function $f \in C^\infty([a, b])$ and any numbers $\varepsilon > 0$ and $n \in \mathbb{N}$ there is a polynomial P such that $|f^{(k)}(x) - P^{(k)}(x)| < \varepsilon$ for all $x \in [a, b]$, $k = 0, 1, \dots, n$.

2.4. Suppose that $f \in C^\infty(\mathbb{R})$, $f \not\equiv 0$. Prove that if f vanishes on some nonempty interval, then $\sup_{n,x} |f^{(n)}(x)| = +\infty$.

2.5. Let $f(x) = \sum_{1 \le k \le n} P_k(x) e^{Q_k(x)}$, where P_k, Q_k $(k = 1, \dots, n)$ are real polynomials, $x \in \mathbb{R}$. Prove that f has finitely many zeros.

2.6. Suppose that $f \in C^\infty([0, +\infty))$, $f(0) = \lim_{x \to +\infty} f(x)$. Prove that for each $n \in \mathbb{N}$ the function $f^{(n)}$ has a zero.

2.7. Let $f \in C^2((0, 1])$, and suppose that $f(x) = o(1)$ and $f''(x) = O(x^{-2})$ as $x \to +0$. Prove that $f'(x) = o(x^{-1})$ as $x \to +0$.

2.8. Let f and g be smooth even $2T$-periodic functions that are decreasing on $[0, T]$. Prove that their convolution, that is, the function $h(t) = \int_{-T}^{T} f(x) g(t - x)\, dx$, has the same properties.

2.9. Let f be n times differentiable on $[0, 1]$, and suppose that $f^{(k)}(0) = f^{(k)}(1) = 0$, $k = 1, \dots, n - 1$. Prove that for some x the inequality holds:

$$|f^{(n)}(x)| \ge n! 2^{n-1} |f(0) - f(1)|.$$

2.10. Suppose that Δ is an interval, $f \in C^2(\Delta)$, and $M_k = \sup_\Delta |f^{(k)}(x)|$, $k = 0, 1, 2$.

a) Prove that $M_1 \le \sqrt{2}\sqrt{M_0 M_2}$ in the case $\Delta = \mathbb{R}$, and that the constant $\sqrt{2}$ cannot be decreased.

b) Obtain an analogous inequality in the case when $\Delta = \mathbb{R}_+$.

c) Prove that if $\Delta = [0, 2]$ and M_0, $M_2 \le 1$, then $M_1 \le 2$ (the estimate is sharp).

d) Study the question for an arbitrary interval $\Delta = [a, b]$.

2.11. Prove that if $f \in C^2([0, +\infty))$, then the following inequalities hold:

a) $|f(0)|^2 \le 2 \int_0^\infty |f(x)|\, dx \int_0^\infty |f''(x)|\, dx$;

b) $\sqrt{2} f(0) \le \int_0^\infty (|f(x)| + |f''(x)|)\, dx$.

Are these estimates sharp?

2.12. Find $\min \int_0^1 |f''(x)|^2\, dx$ under the condition that $f \in C^2([0, 1])$, $f(0) = f(1) = 0$, and $f'(0) = a$.

2.13. Let $f \in C^\infty(\mathbb{R})$. Prove that if for each $x \in \mathbb{R}$ there is a number

$n \in \mathbb{N}$ with $f^{(n)}(x) = 0$, then f is a polynomial.

2.14. Let $\{a_n\}$ be an arbitrary sequence of real numbers. Prove that there is a function $f \in C^\infty(\mathbb{R})$ of the form $f(x) = \sum \alpha_n x^n (1 + \beta_n x^{2n})^{-1}$ such that $f^{(n)}(0) = a_n \quad (n \in \mathbb{N})$.

2.15. Suppose that $a = (a_1, \dots, a_n) \in \mathbb{R}^n$ and $f \in C^\infty(\mathbb{R}^n)$. Prove that if $f(a) = 0$, then

$$f(x) = \sum_{1 \le j \le n} (x_j - a_j) g_j(x),$$

where $x = (x_1, \dots, x_n)$, $g_j \in C^\infty(\mathbb{R}^n)$, and $g_j(a) = \frac{\partial f}{\partial x_j}(a)$ $(j = 1, \dots, n)$.

2.16. Let $a \in \mathbb{R}^n$ and suppose that F is a mapping of the set $C^\infty(\mathbb{R}^n)$ into $\mathbb{R}$ satisfying the following conditions $(f, g \in C^\infty(\mathbb{R}^n))$:

a) $F(f + g) = F(f) + F(g)$;

b) $F(\alpha f) = \alpha F(f) \quad (\alpha \in \mathbb{R})$;

c) $F(fg) = F(f)g(a) + f(a)F(g)$.

Prove that

$$F(f) = \sum_{1 \le j \le n} c_j \frac{\partial f}{\partial x_j}(a)$$

for some $c_1, \dots, c_n \in \mathbb{R}$.

2.17. Let f and g be nonnegative continuous functions defined on $[0, \infty)$. Prove that if for some number $C \ge 0$

$$f(t) \le C + \int_0^t f(x) g(x)\, dx, \qquad t \ge 0,$$

then

$$f(t) \le C \exp\left(\int_0^t g(x)\, dx\right)$$

for all $t \ge 0$ (Gronwall's inequality).

Deduce from this that if $h \in C^1([0, \infty))$, $h(0) = 0$, $M \ge 0$ and $|h'(t)| \le M|h(t)|$ for $t \ge 0$, then $h \equiv 0$.

A function $f\colon \mathbb{R}^n \to \mathbb{C}$ is said to be *positive-definite* if

$$\sum_{j,k=1}^{m} f(t_j - t_k) z_j \overline{z}_k \ge 0 \tag{1}$$

for any $t_1, \dots, t_m \in \mathbb{R}^n$, $z_1, \dots, z_m \in \mathbb{C}$, and any $m \in \mathbb{N}$.

A function f is said to be *conditionally positive-definite* if (1) holds under the additional condition that $z_1 + \cdots + z_m = 0$.

2.18. Prove that:

a) the function $f(t) = e^{i(a,t)}$ $(t \in \mathbb{R}^n)$, where $a \in \mathbb{R}^n$, is positive-definite;

b) the function $f(t) = A\cos(a, t) + B$ $(t \in \mathbb{R}^n)$, where $a \in \mathbb{R}^n$, $A, B \in \mathbb{R}$, and $A > 0$, is conditionally positive-definite.

2.19. Prove that if μ is a finite measure on $\mathbb{R}^n$, then the function

$$f(t) = \int_{\mathbb{R}^n} e^{-i(x,t)}\, d\mu(x) \qquad (t \in \mathbb{R}^n)$$

(the Fourier transform of the measure μ) is positive-definite.

2.20. Prove that the functions $f(t) = e^{-|t|}$ and $g(t) = 1/(1+t^2)$ $(t \in \mathbb{R})$ are positive-definite.

2.21. Prove that:

a) the function $f(t) = -\|t\|^2$ $(t \in \mathbb{R}^n)$ is conditionally positive-definite;

b) for $0 < p < 2$ the functions $f(t) = -\|t\|^p$ $(t \in \mathbb{R}^n)$ are conditionally positive-definite.

2.22. Prove that if a function $f\colon \mathbb{R}^n \to \mathbb{C}$ is positive-definite, then:

a) $\overline{f}$ is positive-definite;

b) f is bounded;

c) continuity at zero implies uniform continuity of f.

2.23. Prove that if the function $f(t) = e^{s\varphi(t)}$ is positive-definite for all $s > 0$, then the function $\varphi\colon \mathbb{R}^n \to \mathbb{C}$ is conditionally positive-definite.

Suppose that $\Delta = [a, b]$, $f \in C^3(\Delta)$, and $f'(x) \neq 0$ $(x \in \Delta)$. The *Schwarzian derivative* Sf of f is defined by

$$\begin{aligned} Sf(x) &= \frac{f'''(x)}{f'(x)} - \frac{3}{2}\left(\frac{f''(x)}{f'(x)}\right)^2 \\ &= \frac{6}{(f'(x))^2} \det\begin{pmatrix} f'(x) & \frac{1}{2!}f''(x) \\ \frac{1}{2!}f''(x) & \frac{1}{3!}f'''(x) \end{pmatrix}. \end{aligned}$$

2.24. Verify that the following assertions are valid for real functions satisfying the above conditions:

a) if the composition $f \circ g$ is defined, then

$$S(f \circ g)(x) = Sf(g(x))(g'(x))^2 + Sg(x);$$

b) if $h(x) = (\alpha x + \beta)/(\gamma x + \delta)$ and the composition $g = h \circ f$ is defined, then $Sg(x) = Sf(x)$;

c) if $Sf(x) < 0$ for all $x \in \Delta$, then also $S(f^n)(x) < 0$ $(x \in \Delta)$ for any $n \in \mathbb{N}$;

d) if $Sf(x) < 0$, then $S(f^{-1})(y) > 0$ at the point $y = f(x)$;

e) the inequality $Sf(x) < 0$ $(x \in \Delta)$ holds if and only if the function $g = |f'|^{-1/2}$ is convex;

f) if $Sf(x) < 0$ for all $x \in \Delta$, then the function $|f'|$ does not have a strictly positive local minimum in (a, b) (the "*minimum modulus principle*").

It can be seen from problems X.1.37–40 why it is negativity (and not positivity) of the Schwarzian that is important.

2.25. a) Verify that the functions $f(x) = \sin x$, $f(x) = \arctan x$, $f(x) = xe^{-x}$, $f(x) = e^{-x^2}$, and $f(x) = Ax^2 + Bx + C$ satisfy the condition $Sf(x) <$

0 if $f'(x) \neq 0$, while the functions $f(x) = x + x^3$, $f(x) = \tan x$, and $f(x) = \operatorname{cotan} x$ do not ($x, A, B, C \in \mathbb{R}$).

b) Verify that any polynomial f of degree $n \geq 2$ with all zeros real satisfies the condition $Sf(x) < 0$ if $f'(x) \neq 0$.

2.26. Let

$$R(w, x, y, z) = \frac{(z-w)(y-x)}{(z-y)(x-w)}$$

(w, x, y, z be pairwise distinct numbers). Prove that if $f \in C^1(\Delta)$, f' is piecewise monotone, and

$$R(f(x_1), f(x_2), f(x_3), f(x_4)) > R(x_1, x_2, x_3, x_4)$$

for any points $x_1 < x_2 < x_3 < x_4$ belonging to a single interval of monotonicity of f, then the function $|f'|$ does not have a strictly positive local minimum at interior points.

§3. Bernstein polynomials

3.1. Let $n \in \mathbb{N}$ and $r = 0, 1, \ldots$. For $x \in \mathbb{R}$ let

$$S_{n,r}(x) = \sum_{0 \leq k \leq n} \binom{n}{k} x^k (1-x)^{n-k} (k - nx)^r.$$

Prove that:

a) $S_{n,r+1}(x) = x(1-x)(S'_{n,r}(x) + nrS_{n,r-1}(x))$, $r \in \mathbb{N}$;

b) $S_{n,0}(x) = 1$, $S_{n,1}(x) = 0$, $S_{n,2}(x) = nx(1-x)$, $S_{n,3}(x) = nx(1-x)(1-2x)$, $S_{n,4}(x) = nx(1-x)(1+3(n-2)x(1-x))$.

3.2. Let $n \in \mathbb{N}$ and $\delta > 0$. For $x \in [0, 1]$ let

$$\sigma_{n,\delta}(x) = \sum_{\substack{0 \leq k \leq n \\ |k/n - x| \geq \delta}} \binom{n}{k} x^k (1-x)^{n-k}.$$

Prove that:

a) $\sigma_{n,\delta}(x) \leq \dfrac{1}{4\delta^2 n}$; b) $\sigma_{n,\delta}(x) \leq \dfrac{1}{4\delta^4 n^2}$; c) $\sigma_{n,\delta}(x) \leq \dfrac{17}{\delta\sqrt{n}} e^{-2n\delta^2}$.

Let $n \in \mathbb{N}$. The polynomial

$$B_n(f, x) = \sum_{0 \leq k \leq n} \binom{n}{k} f\left(\frac{k}{n}\right) x^k (1-x)^{n-k}$$

is called the nth Bernstein polynomial of the function $f\colon [0, 1] \to \mathbb{R}$.

3.3. Find the Bernstein polynomials of the following functions:

a) $f(x) = x^m$, $m = 0, 1, 2$; b) $f(x) = a^x$, $a > 0$.

3.4. Suppose that f is defined on $[0, 1]$, and let $n \in \mathbb{N}$. Prove that:

a) $B_n(f, 0) = f(0)$ and $B_n(f, 1) = f(1)$;

b) $B_n(f^2, x) \geq B_n^2(f, x)$ for all $x \in [0, 1]$.

3.5. Prove that:

a) if f is increasing (decreasing) on $[0, 1]$, then the Bernstein polynomial $B_n(f)$ is increasing (decreasing) on $[0, 1]$;

b) if f is convex (concave) on $[0, 1]$, then the polynomial $B_n(f)$ is convex (concave) on $[0, 1]$.

3.6. Suppose that f is defined and bounded above on $[0, 1]$. Prove that:

a) $B_n(f, x) \leq \sup_{[0,1]} f$ for all $n \in \mathbb{N}$ and $x \in [0, 1]$;

b) if the interval Δ is contained in $[0, 1]$, then $\overline{\lim}_{n\to\infty} B_n(f, x) \leq \sup_\Delta f$ for all $x \in \Delta$;

c) if the intervals Δ and Δ' are such that $\overline{\Delta} \subset \Delta' \subset [0, 1]$, and $\sup_{\Delta'\setminus\Delta} f < \sup_\Delta f$, then there is an index $N = N(f, \Delta, \Delta')$ large enough that $B_n(f, x) < \sup_\Delta f$ for all $x \in \Delta$ and $n > N$.

3.7. Suppose that the functions f and g are bounded on $[0, 1]$ and coincide in a neighborhood of a point $x_0 \in [0, 1]$. Prove that the sequences $\{B_n(f, x_0)\}$ and $\{B_n(g, x_0)\}$ are convergent or divergent simultaneously, and in the case of convergence they have a common limit.

3.8. Suppose that the function f is bounded on $[0, 1]$. Prove that:

a) if f is continuous at $x_0 \in [0, 1]$, then $B_n(f, x_0) \to f(x_0)$;

b) if f is continuous at each point of a closed interval $\Delta \subset [0, 1]$, then $B_n(f) \rightrightarrows f$ on Δ;

c) if $x_0 \in (0, 1)$ is a point of discontinuity of f of the first kind, then $B_n(f, x_0) \to \frac{1}{2}(f(x_0 + 0) + f(x_0 - 0))$.

3.9. Suppose that $f \in C([0, 1])$. Prove that:

a) if $\int_0^1 f(x)x^n\,dx = 0$ for $n = 0, 1, \ldots,$ then $f \equiv 0$;

b) if $\int_0^1 f(x)(x^{1+n}(1-x)^2)''\,dx = 0$ for $n \in \mathbb{N}$, then f is a linear function.

3.10. Suppose that $f \in C([0, 1])$ and $\int_0^1 f(x)g''(x)\,dx = 0$ for any function $g \in C^2([0, 1])$ equal to zero in neighborhoods of the points 0 and 1. Prove that f is a linear function.

3.11. Suppose that $f \in C([0, \pi])$ and $\varepsilon > 0$. Prove that there exists a trigonometric polynomial T of the form $T(t) = \sum_{0\leq n\leq N} a_n \cos nt$ such that $|f(t) - T(t)| < \varepsilon$ for all $t \in [0, \pi]$.

3.12. a) Let $f \in C([0, 1] \times [0, 1])$. For $n \in \mathbb{N}$ and $x, y \in \mathbb{R}$ let

$$\mathscr{B}_n(f; x, y) = \sum_{0\leq k\leq n} \sum_{0\leq j\leq n} \binom{n}{k}\binom{n}{j} f\left(\frac{k}{n}, \frac{j}{n}\right) x^k y^j (1-x)^{n-k}(1-y)^{n-j}.$$

Prove that $\mathscr{B}_n(f) \rightrightarrows f$ on $[0, 1] \times [0, 1]$ as $n \to +\infty$.

b) Suppose that $f \in C([0, \pi] \times [0, \pi])$ and $\varepsilon > 0$. Prove that there exists a trigonometric polynomial T of the form

$$T(u, v) = \sum_{0\leq k\leq N} \sum_{0\leq j\leq N} a_{kj} \cos ku \cos jv$$

such that $|f(u, v) - T(u, v)| < \varepsilon$ for all $u, v \in [0, \pi]$.

3.13. Suppose that $f \in C^r([0, 1])$. Prove that $B_n^{(r)}(f) \rightrightarrows f^{(r)}$ on $[0, 1]$.

3.14. a) Let $f \in \mathrm{Lip}_\alpha[0, 1]$, $0 < \alpha \le 1$. Prove that $|B_n(f, x) - f(x)| \le M(x(1-x)/n)^{\alpha/2}$ for all $n \in \mathbb{N}$ and $x \in [0, 1]$, where M is a positive constant depending only on f.

b) Let $\alpha \in [0, 1]$ and let $f_\alpha(x) = |x - \frac{1}{2}|^\alpha$. Prove that $|B_n(f_\alpha, \frac{1}{2}) - f_2(\frac{1}{2})| \ge \frac{1}{4} n^{-\alpha/2}$ for all $n \in \mathbb{N}$.

3.15. a) If f is convex on $[0, 1]$, then $B_n(f, x) \ge f(x)$ for all $n \in \mathbb{N}$ and $x \in [0, 1]$.

b) If a continuous function f on $[0, 1]$ is such that $B_n(f, x) \ge f(x)$ for all $n \in \mathbb{N}$ and $x \in [0, 1]$, then f is convex on $[0, 1]$.

3.16. Let $f \in C^2([0, 1])$. Prove that:

$$B_n(f, x) - f(x) = \frac{x(1-x)}{2n}(f''(x) + \varepsilon_n(x)),$$

where the sequence $\{\varepsilon_n\}$ of functions converges uniformly on $[0, 1]$ to zero (the Voronovskaja formula).

3.17. Suppose that $f \in C([0, 1])$, $g \in C^2([0, 1])$, and g is equal to zero outside (a, b), where $0 < a < b < 1$. Prove that:

$$\sum_{0 \le k \le n} \left(B_n\left(f, \frac{k}{n}\right) - f\left(\frac{k}{n}\right)\right) g\left(\frac{k}{n}\right) \to \frac{1}{2}\int_0^1 f(x)(x(1-x)g(x))''\,dx.$$

3.18. Let the function $f\colon [0, 1] \to \mathbb{R}$ be such that

$$\sup_{0 \le x \le 1} |B_n(f, x) - f(x)| = o(1/n).$$

Prove that f is linear.

3.19. Let $f \in C([0, 1])$. Prove that the condition $f(0), f(1) \in \mathbb{Z}$ is necessary and sufficient for f to be the limit of a uniformly convergent sequence of algebraic polynomials with integer coefficients on $[0, 1]$.

3.20. Let f be defined on $[0, 1]$. Prove that:

$$B_n(f, x) = f(0) + \sum_{1 \le k \le n} x^k \binom{n}{k} \Delta_k^n f(0),$$

where $\Delta_1^n f(y) = f(y + \frac{1}{n}) - f(y)$, $\Delta_{k+1}^n f(y) = \Delta_1^n(\Delta_k^n f(y))$, $k \in \mathbb{N}$.

3.21. Let $f(x) = x^m$ $(x \in \mathbb{R})$, $m = 0, 1, 2, \ldots$. Prove that:

a) $B_n(f) \rightrightarrows f$ on any finite interval;

b) $|B_n(f, x)| \le \max(1, |x|^m)$ for any $x \in \mathbb{R}$.

3.22. Let $f(x) = \sum_{m \ge 0} c_m x^m$ for $x \in (-R, R)$, where $R > 1$. Prove that:

a) $B_n(f) \rightrightarrows f$ on any interval $[-r, r]$ if $0 < r < R$;

b) if $c_m \ge 0$ for any m, then $B_n(f) \ge f$ on $[0, 1]$, and $B_n(f) \le f$ on $[1, R)$.

3.23. a) Prove that for a fixed $n \in \mathbb{N}$ the polynomials $x^k(1-x)^{n-k}$, $k = 0, 1, \ldots, n$, form a basis in the space of polynomials of degree at most n.

b) Prove that the coefficients a_n^k in the expansion of an mth-degree polynomial P with respect to such a base with $n \geq m$, which form the table

$$\begin{array}{ccccccccccccc}
 & & & a_m^0 & a_m^1 & \cdots & \cdots & \cdots & a_m^{m-1} & a_m^m & & & \\
 & & a_{m+1}^0 & a_{m+1}^1 & \cdots & \cdots & \cdots & \cdots & \cdots & a_{m+1}^m & a_{m+1}^{m+1} & & \\
 & & & \cdots & \cdots & \cdots & \cdots & \cdots & \cdots & \cdots & & & \\
a_n^0 & a_n^1 & \cdots & \cdots & \cdots & \cdots & \cdots & \cdots & \cdots & \cdots & \cdots & a_n^{n-1} & a_n^n \\
 & & \cdots & \cdots & \cdots & \cdots & \cdots & \cdots & \cdots & \cdots & \cdots & &
\end{array}$$

satisfy the equalities

$$a_{n+1}^0 = a_n^0, \quad a_{n+1}^{n+1} = a_n^n, \quad a_{n+1}^k = a_n^{k-1} + a_n^k \qquad (k = 1, 2, \ldots, n)$$

(cf. the equalities for the binomial coefficients $\binom{n}{k}$).

c) Prove that $P(x) > 0$ for $x \in (0, 1)$ if and only if for sufficiently large n the numbers a_n^k $(k = 0, \ldots, n)$ are not all zero and are nonnegative.

§4. Almost periodic functions and sequences

We write $x \overset{\varepsilon}{=} y$ if $|x - y| < \varepsilon$, and $x \overset{\varepsilon}{=} y \pmod 1$ if there exists an integer k such that $|x - y - k| < \varepsilon$.

A set $X \subset \mathbb{R}$ is said to be relatively dense if there exists an $L > 0$ such that each interval of length L contains at least one number in X.

4.1. Prove that if λ is an irrational number, then for any $a \in [0, 1)$ and any $\varepsilon > 0$ there is an $n \in \mathbb{Z}$ such that $n\lambda \overset{\varepsilon}{=} a \pmod 1$. Verify that the set of such n is relatively dense on the line (cf. problem I.3.2).

4.2. a) Prove that if $\lambda_1, \lambda_2 \neq 0$ and the ratio λ_1/λ_2 is irrational, then for any $a_1, a_2 \in [0, 1)$ and any $\varepsilon > 0$ there is a $t \in \mathbb{R}$ such that

$$\lambda_1 t \overset{\varepsilon}{=} a_1 \pmod 1, \qquad \lambda_2 t \overset{\varepsilon}{=} a_2 \pmod 1. \tag{1}$$

Conversely, if the system (1) is solvable with respect to t for any a_1, a_2, and ε, then λ_1/λ_2 is irrational.

b) Prove that if λ_1, λ_2, and λ_1/λ_2 are irrational, then the set of integer solutions of the system (1) is relatively dense on the line.

4.3. What condition must the numbers $\lambda_1, \ldots, \lambda_k \in \mathbb{R}$ satisfy in order that the system of equations

$$\lambda_i t \overset{\varepsilon}{=} 0 \pmod 1, \qquad i = 1, 2, \ldots, k,$$

have for any $\varepsilon > 0$ a solution t with $|t| > |\lambda_i|^{-1}$ for some i?

4.4. Find a condition which the numbers $\lambda_1, \ldots, \lambda_k \in \mathbb{R}$ must satisfy in order that the system of equations

$$\lambda_i t \overset{\varepsilon}{=} a_i \pmod 1, \qquad i = 1, \ldots, k,$$

have for any $a_1, \ldots, a_k \in [0, 1]$ and any $\varepsilon > 0$:

a) a real solution t;

b) a relatively dense set of solutions.

4.5. Find a condition which the numbers $\lambda_1, \dots, \lambda_k \in \mathbb{R}$ must satisfy in order that the system of equations

$$\lambda_i m \overset{\varepsilon}{=} a_i \pmod 1, \qquad i = 1, \dots, k,$$

have for any $a_1, \dots, a_k \in [0, 1)$ and any $\varepsilon > 0$:

a) an integer solution m;

b) a relatively dense set of integer solutions.

4.6. Let $f(x) = \sin(ax+b) + \sin(cx+d)$, $x \in \mathbb{R}$, where $ac \neq 0$. Prove that if f is not identically equal to zero, then it takes values of different signs.

A complex-valued function f on $\mathbb{R}$ is said to be uniformly almost periodic (UAP) if for any $\varepsilon > 0$ there is a number $L = L(\varepsilon) > 0$ such that each interval of length L contains at least one number τ (an ε-almost period) for which

$$|f(x+\tau) - f(x)| < \varepsilon \quad \text{for all } x \in \mathbb{R}.$$

4.7. Prove that each trigonometric quasipolynomial

$$f(x) = \sum_{1 \le k \le n} a_k e^{i\lambda_k x} \qquad (\lambda_k \in \mathbb{R},\ a_k \in \mathbb{C})$$

has the property of uniform almost periodicity. Verify that f is periodic if and only if the factors $\lambda_1, \dots, \lambda_n$ are pairwise commensurable (that is, their ratios are rational).

4.8. Let $f \in C(\mathbb{R}^n)$ be a function having period 1 with respect to each argument, let $\alpha_1, \dots, \alpha_n \in \mathbb{R}$ be rationally independent numbers, and let $\beta_1, \dots, \beta_n \in \mathbb{R}$. Prove that the function

$$\varphi(t) = f(\alpha_1 t - \beta_1, \dots, \alpha_n t - \beta_n), \qquad t \in \mathbb{R}$$

is uniformly almost periodic.

4.9. Prove that if f is a UAP function, and the numbers $L(\varepsilon)$ are bounded for small ε, then f is periodic.

4.10. Prove that a continuous function on $\mathbb{R}$ that is UAP is uniformly continuous on the line and is bounded. Does there exist an unbounded non-periodic UAP function? Does continuity follow from boundedness?

4.11. Prove that the limit of a uniformly convergent sequence of UAP functions on $\mathbb{R}$ is a UAP function.

4.12. Prove that if a UAP function has a derivative that is uniformly continuous, then the derivative is also a UAP function.

4.13. Prove that if a primitive F of a continuous UAP function f is bounded, then it is also a UAP function.

4.14. Suppose that $f \in C(\mathbb{R})$, $h \in \mathbb{R}$, and $f_h(x) = f(x+h)$. Prove that f is UAP if and only if for each numerical sequence $\{h_n\}$ the sequence $\{f_{h_n}\}$ has a subsequence uniformly convergent on $\mathbb{R}$ (compactness of the family of translates).

4.15. Prove that a sum or a product of continuous UAP functions is again a UAP function.

4.16. Prove that the mean value

$$M(f) = \lim_{T\to+\infty} \frac{1}{2T} \int_{-T}^{T} f(x)\,dx$$

exists for each continuous UAP function.

4.17. Let $\langle f, g\rangle = M(f\overline{g})$, where f and g are continuous UAP functions, and the functional M is defined in the preceding problem. Prove that:

a) $\langle f, g\rangle$ has the properties of an inner product (in particular, $\langle f, f\rangle = 0$ if and only if $f \equiv 0$).

b) the functions $\varphi_\lambda(x) = e^{i\lambda x}$, $\lambda \in \mathbb{R}$, form an uncountable orthogonal family with respect to this inner product;

c) the quantities $c_{\lambda_k} = M(f\varphi_{\lambda_k})$ ($\lambda_k \in \mathbb{R}$, $k = 1, \dots, n$) satisfy the inequality

$$\sum_{1\le k\le n} |c_{\lambda_k}|^2 \le M(|f|^2)$$

(Bessel's inequality for the system $\{\varphi_\lambda\}$). Conclude from this that $c_\lambda \neq 0$ for at most countably many values λ. Can this set be an arbitrary countable subset of $\mathbb{R}$?

4.18. Let $\{a_n\}$, $a_n \in \{0, 1, 2\}$, be a nonperiodic sequence. Define the sequences $\{b_n\}$ and $\{c_n\}$ as follows:

$$b_n = \begin{cases} 0 & \text{if } a_n = 0, \\ 1 & \text{if } a_n \neq 0; \end{cases} \qquad c_n = \begin{cases} 1 & \text{if } a_n = 2, \\ 0 & \text{if } a_n \neq 2. \end{cases}$$

Prove that at least one of the sequences $\{b_n\}$ and $\{c_n\}$ is nonperiodic.

A binary sequence $\varepsilon = \{\varepsilon_k\}_{k\in\mathbb{N}}$ is said to be *almost periodic* if for any "word" $(\varepsilon_{k+1}, \dots, \varepsilon_{k+l})$ (we also write $\varepsilon_{k+1}\cdots\varepsilon_{k+l}$) in ε there exists a number $L \in \mathbb{N}$ such that the set $\{n \in \mathbb{N} \mid \varepsilon_{n+j} = \varepsilon_{k+j},\ j = 1, \dots, l\}$ intersects every segment of the natural numbers of length L. Almost periodicity for two-sided sequences $\{\varepsilon_k\}_{k\in\mathbb{Z}}$ is defined similarly.

4.19. Let $\{\varepsilon_k\}_{k\in\mathbb{Z}}$ be an almost periodic sequence, and let $f(x) = \varepsilon_{[x]}$ ($x \in \mathbb{R}$, $[x]$ the integer part of a number x). Is f a UAP function?

4.20. Is it true that for every almost periodic sequence $\{\varepsilon_k\}_{k\in\mathbb{Z}}$ the Cesàro means have a limit

$$\lim_n \frac{1}{2n+1} \sum_{-n\le k\le n} \varepsilon_k?$$

4.21. Let $S = \{(x, y) \mid kx \le y \le k(x+1)+1\}$, a strip in $\mathbb{R}^2$ with k an irrational number, and let $\Sigma = S \cap \mathbb{Z}^2$ be the set of integer points in S. Prove that the orthogonal projections of the points in Σ on a side of the strip (see Figure 5) partition it into segments with lengths taking exactly two values a and b. After numbering the segments by the integers in order of their location on the line, let $\varepsilon_i = 0$ if the ith segment has length a, and $\varepsilon_i = 1$ otherwise. Prove that the sequence $\{\varepsilon_i\}_{k\in\mathbb{Z}}$ is almost periodic.

4.22. Consider the sequence $\varepsilon = \{\varepsilon_k\} = 0110100110010110\ldots$ (the *Morse sequence*) formally defined by the equalities $\varepsilon_1 = 0$, $\varepsilon_2 = 1$, $\varepsilon_k = 1 - \varepsilon_{k-2^n}$ for $2^n < k \le 2^{n+1}$ $(n \in \mathbb{N})$. Prove that it is almost periodic.

4.23. For a finite sequence ("word") $A = \varepsilon_1\varepsilon_2\cdots\varepsilon_n$ of 0's and 1's let A^0 denote A, and $A^1 = \overline{\varepsilon}_1\overline{\varepsilon}_2\cdots\overline{\varepsilon}_n$, where $\overline{\varepsilon}_k = 1 - \varepsilon_k$. Define a sequence ε by determining its initial segments A_p recursively: A_0 is an arbitrary word, $A_0 = \varepsilon_1\cdots\varepsilon_n$; $A_1 = A_0A_0^{\xi_1}A_0^{\xi_2}\cdots A_0^{\xi_l}$, where $\xi_1\cdots\xi_l$ is some word of 0's and 1's, $A_2 = A_1A_1^{\eta_1}A_1^{\eta_2}\cdots A_1^{\eta_m}$, $\eta_k \in \{0, 1\}$, and so on. Thus, A_0 serves as the initial part of A_1, A_1 as the initial part of A_2, and so on. The infinite sequence $\{A_p\}$ determines some sequence ε. Prove that ε is periodic or almost periodic.

4.24. A *sequence of folds* is defined to be a sequence $s = \{s_n\}_{n\in\mathbb{N}}$ of 0's and 1's whose initial segments A_n of length $2k-1$ $(k = 2^n,\ n \in \mathbb{N})$ have the form $s_1s_2\cdots s_{k-1}s_{2^n}\overline{s}_{k-1}\cdots\overline{s}_1$ (where $\overline{s}_j = 1 - s_j$). The name "sequence of folds" is due to the fact that with the word A_n it is possible to associate a sequence of folds of a paper strip that are formed on the strip as a result of bending it $n+1$ times (first in halves, then once more in halves, and so

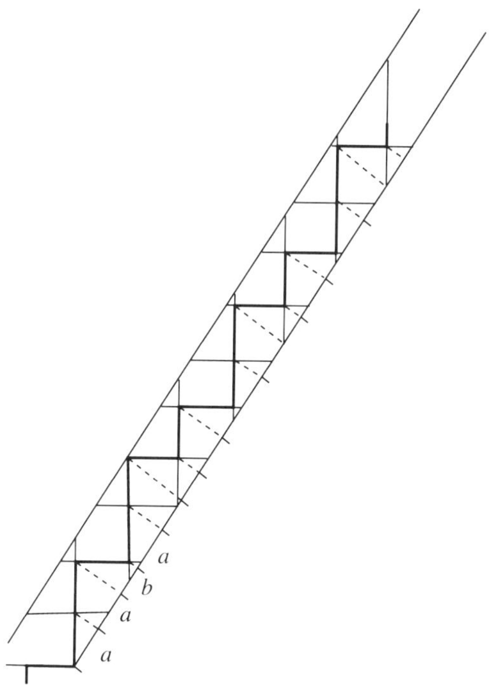

FIGURE 5

on). If upon moving along the strip a 1 is associated with each fold leading to a bend to the left, while a 0 is associated with a "right bend" fold, then A_n is obtained (see Figure 6). The sequence s is uniquely determined by specifying the numbers s_{2^n}, $n = 0, 1, 2, \ldots$.

Prove that every sequence of folds is almost periodic, but not periodic, not even if we confine ourselves to terms with sufficiently large indices.

4.25. Prove that sequences of folds (see problem 4.24) have the following property (Besicovitch almost periodicity): for each $\varepsilon > 0$ there exist a number $L > 0$ and numbers $\tau_m \in \mathbb{N}$ $(m \in \mathbb{N})$ such that:

a) if $\nu(\Delta)$ is the number of τ_m falling in the interval $\Delta \subset \mathbb{R}^+$, then $\nu(\Delta_1) < 2\nu(\Delta_2)$ for all intervals of length L;

b) for each $m \in \mathbb{N}$

$$\overline{\lim} \frac{1}{n} \sum_{1 \le j \le n} |s_{j+\tau_m} - s_j| < \varepsilon;$$

c)

$$\overline{\lim_n}\, \overline{\lim_p} \frac{1}{np} \sum_{1 \le j \le n} \sum_{1 \le m \le p} |s_{j+\tau_m} - s_j| < \varepsilon.$$

4.26. Prove that if $x_n = (-1)^{s_n}$, $n \in \mathbb{N}$, where $\{s_n\}$ is a sequence of folds, then for any $t \in \mathbb{R}$ the limit

$$c_t = \lim \frac{1}{n} \sum_{1 \le k \le n} x_k e^{-2\pi k t i}$$

(Fourier coefficient) exists; moreover,

$$c_t = -i(-1)^l 2^{-(m+1)} x_{2^m} \text{ if } t = (2l+1)2^{-(m+2)}, \quad m \ge 0,$$

and $c_t = 0$ otherwise. Verify "Parseval's equality"

$$\lim \frac{1}{n} \sum_{1 \le k \le n} |x_k|^2 = \sum_{0 \le t < 1} |c_t|^2.$$

4.27. Consider the binary sequence

$$r = \{r_n\}_{n \in \mathbb{N}} = 0001001000011101000100101110001 0\ldots$$

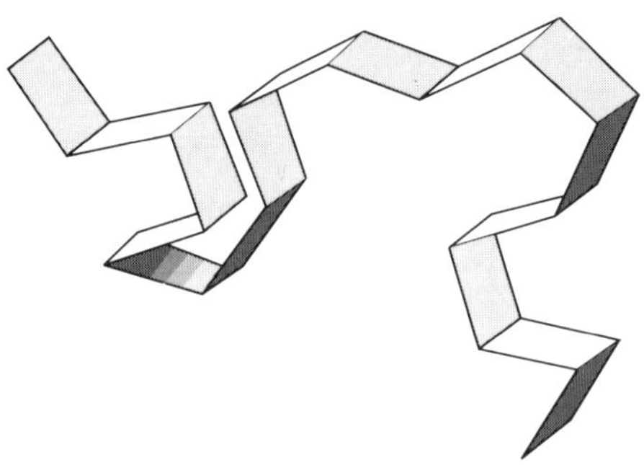

FIGURE 6

(the *Rudin-Shapiro sequence*), whose initial segments A_n of length 2^n have the form $A_n = C_n D_n$, where C_n and D_n are words of length 2^{n-1}, and $A_1 = C_1 D_1 = 00$, while A_n, C_n, and D_n are defined recursively for $n > 1$: $C_n = A_{n-1}$, $D_n = C_{n-1} D^1_{n-1}$, $A_n = C_n D_n$ (D^1_{n-1} is obtained from D_{n-1} by the substitution $0 \leftrightarrow 1$).

Prove the almost periodicity of this sequence. We note that the sequence $\{(-1)^{r_n}\}$ is the sequence of coefficients of the Rudin-Shapiro polynomials (see problem IV.6.29).

CHAPTER VIII

Lebesgue Measure and the Lebesgue Integral

In this chapter the word "measure" means Lebesgue measure on $\mathbb{R}^m$ or on the sphere $S^{m-1} = \{x \in \mathbb{R}^m \mid \|x\| = 1\}$. Lebesgue measure on $\mathbb{R}^m$ is denoted by the symbol λ_m, and for $m = 1$ by the letter λ.

§1. Lebesgue measure

1.1. Suppose that $E_k \subset (0, 1)$, $(k = 1, 2, \dots, N)$, and $\sum_{1\le k\le N} \lambda(E_k) > N - 1$. Prove that $\lambda(\bigcap_{1\le k\le N} E_k) > 0$.

1.2. Suppose that $E \subset S^1$, $z_1, \dots, z_N \in S^1$, and μ is Lebesgue measure on S^1 ("arclength"). Prove that if $\mu(E) > 2\pi(1 - 1/N)$, then the set E, suitably rotated, contains all the points $z_1, \dots, z_N$, that is, there is a point $z_0 \in S^1$ such that $z_0 z_k \in E$ for $k = 1, \dots, N$.

1.3. a) Prove that if $E \subset \mathbb{R}^m$ and $\lambda_m(E) > 1$, then there are two (distinct) points $x, x^1 \in E$ such that the differences between the corresponding coordinates are integers.

b) Let $V \subset \mathbb{R}^m$ be a convex set centrally symmetric with respect to zero, and let $\lambda_m(V) > 2^m$. Prove that V contains a point $a \neq 0$ with integer coordinates.

c) Let $V \subset \mathbb{R}^m$ be a convex set centrally symmetric with respect to zero, and let $\lambda_m(V) > N2^m$, where N is some positive integer. Prove that V contains at least $2N$ nonzero points with integer coordinates.

1.4. Find the measure of the set of points in $(0, 1)$ whose decimal fraction expansion:

a) contains a 4 at a given place;

b) contains a prime number at a given place;

c) contains given numbers at two given places;

d) contains numbers of different parity at two given places.

1.5. What is the measure of the set of numbers in $(0, 1)$ whose decimal expansions contain a 0?

1.6. Let $E_k \subset (0, 1)$ $(k \in \mathbb{N})$. Determine whether there is a subsequence $\{E_{k_j}\}$ such that $\lambda(\bigcap_{j=1}^{\infty} E_{k_j}) > 0$ if one of the following conditions holds:

a) $\overline{\lim}\, \lambda(E_k) = 1$; b) $\underline{\lim}\, \lambda(E_k) > \frac{3}{4}$?

1.7. Suppose that $\alpha_k > 0$, and $\sum \alpha_k < \infty$. Consider the set

$$A = \{x \in (0, 1) | \quad \text{the inequality} \\ |x - p/q| < \alpha_q/q \ (p \in \mathbb{Z},\, q \in \mathbb{N}) \text{ has infinitely many solutions}\}.$$

Prove that $\lambda(A) = 0$.

1.8. Suppose that $0 < \theta < 1$, $E \subset \mathbb{R}^m$, and $0 < \lambda(E) < \infty$. Prove that there is a cube Δ such that $\theta\lambda(\Delta) < \lambda(E \cap \Delta)$.

1.9. Let $E, E_0 \subset \mathbb{R}$ be arbitrary measurable sets of positive measure. Prove that:

a) 0 is an interior point of the set $E - E = \{x - y \mid x, y \in E\}$;

b) E contains two (distinct) points separated by a rational distance;

c) the sets $E + E_0 = \{x + y \mid x \in E,\ y \in E_0\}$ and $E \cdot E_0 = \{xy \mid x \in E,\ y \in E_0\}$ have interior points.

1.10. Suppose that $E \subset \mathbb{R}$ and $\lambda(E) > 0$. Prove that if $(x + y)/2 \in E$ for any points $x, y \in E$, then E is an interval (cf. problem I.1.19).

1.11. Suppose that $\nu = \{n_k\}$ is a strictly increasing sequence of positive integers, and let

$$E_\nu = \left\{\sum \varepsilon_k 2^{-n_k} \mid \varepsilon_k = 0 \text{ or } 1\right\}$$

(see problem I.1.37).

a) When is $\lambda E_\nu = 0$?

b) Prove that if $\overline{\lim}(n_{k+1} - n_k) \geq 2 + \log_2 m \quad (m \in \mathbb{N})$, then the set $E^{(m)} = E_\nu + E_\nu + \cdots + E_\nu$ (m terms) has measure zero.

c) Let $H = \{\sum_{1 \leq k \leq n} r_k x_k \mid n \in \mathbb{N},\ r_k \in \mathbb{Q},\ x_k \in E_\nu,\ k = 1, 2, \ldots, n\}$. Prove that if $\overline{\lim}(n_{k+1} - n_k) = \infty$, then $\lambda(H) = 0$.

1.12. Let $E = \{\sum_{k \geq 1} \varepsilon_k / k! \mid \varepsilon_k = 0 \text{ or } 1\}$. Prove that E is a closed set without interior points, and $\lambda E = 0$. Is it true that the measure of the set $E + E + \cdots + E$ (m terms) is equal to zero for any $m \in \mathbb{N}$?

Let Δ be a closed interval of length l_0. Let $\{l_n\}_{n \geq 0}$ be a sequence of positive numbers satisfying the condition $2l_n < l_{n-1}$ $(n \in \mathbb{N})$. Consider a set $E_1 \subset \Delta$ containing the endpoints of Δ and consisting of two closed intervals Δ_0 and Δ_1 of length l_1 each. The intervals Δ_0 and Δ_1 are called the intervals of the first rank. Replacing l_1 by l_2 and repeating the same operation with Δ_0 and Δ_1 instead of Δ, we get four intervals $\Delta_{00}, \Delta_{01}, \Delta_{10}, \Delta_{11}$ of second rank ($\Delta_{00} \cup \Delta_{01} \subset \Delta_0$, $\Delta_{10} \cup \Delta_{11} \subset \Delta_1$), whose union is denoted by E_2. A similar construction applies for the intervals of the third, fourth, etc., ranks. The union of the 2^n intervals of rank n is denoted by E_n. The set $E = \bigcap_{n=1}^{\infty} E_n$ is called the *set of Cantor type* constructed with the help of the *defining sequence* $\{l_n\}$ on the interval Δ (cf. the construction of the Cantor set and of generalized Cantor sets (problems I.1.27 and I.1.31)).

1.13. a) What is the defining sequence for the Cantor set?

b) What is the defining sequence for the set in problem 1.12?

c) When is the set in problem 1.11 a set of Cantor type? In this case what is the defining sequence for it?

1.14. Construct a set $E \subset [0, 1]$ of Cantor type having positive measure. Show that the measure of such a set can be arbitrarily close to 1.

1.15. Prove that there exists a set $A \subset [0, 1]$ such that for any (nonempty) interval $\Delta \subset (0, 1)$

$$\lambda(A \cap \Delta) > 0 \quad \text{and} \quad \lambda(\Delta \backslash A) > 0.$$

1.16. Let $A \subset \mathbb{R}^2$ be a compact convex set, and let A_ε be its ε-neighborhood, that is, the set $\bigcup_{x \in A} B(x, \varepsilon)$. Prove that $\lambda_2(A) = a + b\varepsilon + c\varepsilon^2$, and find the coefficients a, b, and c.

1.17. Prove that every (nonempty) open subset of $\mathbb{R}^n$ can be represented as the union of a sequence of disjoint balls and a set of measure zero.

1.18. Let $\{E_n\}$ be a sequence of pairwise congruent subsets of S^1 satisfying the following conditions (see problem I.1.11):

$S^1 = \bigcup_{n \geq 1} E_n$, $E_n \cap E_m \neq \varnothing$ for $n \neq m$ $(n, m \in \mathbb{N})$. Prove that the sets E_n are not measurable.

1.19. The mapping $x \mapsto \{x + \theta\}$ where $\{y\}$ is the fractional part of a number y and $x \in [0, 1)$, is called *translation by θ modulo* 1. Replacing rotation by the angle $2\pi\theta$ by translation by θ modulo 1, construct a nonmeasurable subset of $[0, 1)$ by analogy with problems I.1.11 and 1.18.

1.20. Using the result of problem 1.19, show that every set $E \subset \mathbb{R}^1$ having positive measure contains a nonmeasurable subset.

1.21. Let $\mathfrak{A}$ be a system of subsets of $\mathbb{N}$ satisfying the following conditions:

1) if $\operatorname{card}(A) = \operatorname{card}(\mathbb{N} \backslash A) = \infty$, then $A \in \mathfrak{A}$ if and only if $\mathbb{N} \backslash A \notin \mathfrak{A}$;

2) if $A \in \mathfrak{A}$, $C \subset \mathbb{N}$, and $\operatorname{card}(C) < \infty$, then $A \cup C \in \mathfrak{A}$ and $A \backslash C \in \mathfrak{A}$.

(The existence of such a system of sets can be proved with the help of Zorn's lemma (see, for example, [13], p. 33)).

Prove that the set $E = \{\sum_{k \in A} 2^{-k} \mid A \in \mathfrak{A}\}$ is nonmeasurable.

1.22. A set $E \subset \mathbb{R}^m$ is said to generate a tesselation if its translates by all possible vectors with integer coordinates cover $\mathbb{R}^m$ without overlapping, that is, $\mathbb{R}^m = \bigcup_{l \in \mathbb{Z}^m} (l + E)$, and $(l + E) \cap (l' + E) = \varnothing$ for $l \neq l'$ $(l, l' \in \mathbb{Z}^m)$.

Prove that the measure of a measurable set generating a tesselation is equal to 1.

1.23. Suppose that $E \subset \mathbb{R}^m$, $\lambda_m(E) > 0$, and A is a countable dense subset of $\mathbb{R}^m$. Prove that $\lambda_m(\mathbb{R}^m \backslash \bigcup_{a \in A} (a + E)) = 0$.

§2. Measurable functions

The symbol $\mathscr{L}^0(E)$ denotes the set of all Lebesgue-measurable functions on a (measurable) set $E \subset \mathbb{R}^n$ that are finite almost everywhere. Two functions in $\mathscr{L}^0(E)$ are said to be equivalent on a set $E_0 \subset E$ if they coincide almost everywhere on E_0. The symbol χ_A denotes the characteristic function of a set $A \subset \mathbb{R}^n$.

2.1. Suppose that $f \in \mathscr{L}^0(E)$, and $|f| \leq 1$ almost everywhere (a.e.) on $E \subset \mathbb{R}^n$. Prove that if $|f| < 1$ on a set of positive measure, then there exist mutually nonequivalent functions $f_1, f_2 \in \mathscr{L}^0(E)$ such that

$$|f_1| \leq 1 \text{ and } |f_2| \leq 1 \text{ a.e. on } E, \text{ and } f = (f_1 + f_2)/2.$$

2.2. Give an example of a monotone function on $\mathbb{R}$ that is not equivalent to a continuous function, not on any (nonempty) open interval.

2.3. The Dirichlet function (equal to 1 at rational points and to 0 at irrational points of the real line) is discontinuous at each point, but equivalent to a continuous function. Does there exist a function in $\mathscr{L}^0([0, 1])$ such that any function equivalent to it on $[0, 1]$ is discontinuous at each point of this interval?

2.4. Let K be the Cantor set, f the Cantor function (see problem III.3.17)), and $g(x) = \frac{1}{2}(x + f(x))$ $(x \in [0, 1])$. Prove that:

a) g is strictly increasing;

b) $\lambda(g(K)) > 0$;

c) there is a (measurable!) set $H \subset K$ such that $g(H)$ is nonmeasurable.

Let $E \subset \mathbb{R}^n$ and $f \in \mathscr{L}^0(E)$, and suppose that for any $t \in \mathbb{R}$ the measure of the set $E(f < t) = \{x \in E \mid f(x) < t\}$ is finite. The function

$$F(t) = \lambda_n E(f < t) \ (t \in \mathbb{R})$$

is called the *distribution function* of f (on the set E). Functions having the same distribution functions are said to be *equimeasurable.*

2.5. Prove that a distribution function is left-continuous. When is it continuous at a point $t_0 \in \mathbb{R}$?

2.6. Find the distribution functions on $[0, 2\pi]$ of the following functions:

$$\text{a) } \sin x, \quad \text{b) } \cos x, \quad \text{c) } \sin \tfrac{3x}{2}, \quad \text{d) } \sin(2x - \tfrac{\pi}{4}).$$

2.7. Determine whether the functions $\sin x$ and $\sin(nx+\alpha)$ $(n \in \mathbb{N}, \alpha \in \mathbb{R})$ are equimeasurable on the following sets: a) $[0, 2\pi]$; b) $[0, \pi]$.

2.8. Let F be the distribution function of a function $f \in \mathscr{L}^0([0, T])$.

a) Find the distribution functions of the functions $f + c$, cf $(c > 0)$, and f^3.

b) Let $\tilde{f}$ be the extension of f to $\mathbb{R}$ with period T, and let $g(x) = \tilde{f}(x - c)$ $(x \in (0, T))$. Find the distribution function of g.

2.9. Suppose that $E \subset \mathbb{R}^n$, $\lambda_n(E) = 1$, and $f \in \mathscr{L}^0(E)$.

a) Prove that there exists a unique number t_0 such that $\lambda_n E(f \geq t_0) \geq \frac{1}{2}$ and $\lambda_n E(f \geq t_0 + \varepsilon) < \frac{1}{2}$ for any $\varepsilon > 0$.

b) Prove that $\lambda_n E(f \leq t_0) \geq \frac{1}{2}$.

c) A number M satisfying the inequalities

$$\lambda_n E(f \geq M) \geq \tfrac{1}{2}, \qquad \lambda_n E(f \leq M) \geq \tfrac{1}{2}$$

is called a *median* of f. Is a median unique?

Let $E \subset \mathbb{R}^n$, $\lambda_n(E) < +\infty$, and $f \in \mathscr{L}^0(E)$. Define

$$f^*(\tau) = \inf\{t \mid \lambda_n E(f > t) \leq \tau\}, \quad \tau \in [0, \lambda_n(E)].$$

The function f^* is called the *nonincreasing rearrangement* of f.

2.10. Prove that the functions f and f^* are equimeasurable. Assuming that the distribution function F of a function f is continuous and strictly increasing, determine how F and f^* are connected.

2.11. Find the nonincreasing rearrangements of the following functions: a) $\sin x$ on $[0, 2\pi]$; b) $\sin x/2$ on $[0, 2\pi]$; c) $\tan x$ on $(0, \pi)$.

2.12. Find the nonincreasing rearrangement of the function

$$f(x, y) = x^2 + y^2 \qquad (x^2 + y^2 \leq 1).$$

2.13. Let $f(x) = \sum_{1\leq k\leq n} a_k/(x - c_k)$, where $a_1 > 0, \ldots, a_n > 0$, and $c_1, \ldots, c_n \in \mathbb{R}$. Prove that for any $t > 0$

$$\lambda\{x \in \mathbb{R} \mid f(x) > t\} = \frac{A}{t}, \qquad \lambda\{x \in \mathbb{R} \mid f(x) < -t\} = \frac{A}{t},$$

where $A = \sum_{1\leq k\leq n} a_k$.

§3. Integrable functions

In this section all the sets and functions under consideration are assumed to be measurable. The set of functions integrable with respect to Lebesgue measure on a set $E \subset \mathbb{R}^m$ is denoted by $\mathscr{L}(E)$; as usual, $\int_E f(x)\,dx$ denotes the integral of a function f over a set E with respect to Lebesgue measure. The characteristic function of a set A is denoted by χ_A.

3.1. Suppose that the union of any three sets in the family $E_1, \ldots, E_N$ coincides with $[0, 1]$. Let

$$S_1 = \sum_{1\leq k\leq N} \lambda(E_k) \quad \text{and} \quad S_2 = \sum_{1\leq j<k\leq N} \lambda(E_j \cap E_k).$$

Express the integral

$$\mathscr{I} = \int_0^1 \left(\sum_{1\leq k\leq N} \chi_{E_k}(x) \right)^3 dx$$

in terms of S_1 and S_2.

3.2. Suppose that each point of $[0, 1]$ belongs to at least k of the sets in the family $E_1, \ldots, E_N$. Prove that $\lambda(E_n) \geq k/N$ for some n.

3.3. Suppose that $E \subset \mathbb{R}^m$, $\lambda_m(E) < \infty$, and $f \in \mathscr{L}^0(E)$. Prove that the following conditions are equivalent:

a) $f \in \mathscr{L}(E)$,
b) $\sum_{k\geq 1} \lambda_m(E(|f| \geq k)) < \infty$,
c) $\sum_{k\geq 1} k\lambda_m(E(k \leq |f| < k+1)) < \infty$.

3.4. Suppose that the sets $E_k \subset \mathbb{R}^m$ $(k \in \mathbb{N})$ are distinct and that $\sum_{k\geq 1} \lambda_m(E_k) < \infty$, and let

$$A_n = \{x \in \mathbb{R}^m \mid x \in E_k \text{ for precisely } n \text{ values of } k\},$$
$$B_n = \{x \in \mathbb{R}^m \mid x \in E_k \text{ for at least } n \text{ values of } k\}.$$

Prove that the sets A_n and B_n are measurable, and that

$$n\lambda_m(B_n) \leq \sum_{k\geq n} \lambda_m(A_k),$$
$$\sum_{n\geq 1} n\lambda_m(A_n) = \sum_{n\geq 1} \lambda_m(B_n) = \sum_{n\geq 1} \lambda_m(E_n).$$

3.5. a) Prove that the series $\sum 2^{-n}|x-r_n|^{-1/2}$, where $\{r_n\}$ is an arbitrary numerical sequence, converges almost everywhere on $\mathbb{R}$.

b) Construct a function in $\mathscr{L}^0(\mathbb{R})$ that is nonintegrable on every (nonempty) interval.

3.6. Can an everywhere differentiable function on $[0, 1]$ have a derivative that is not integrable on $[0, 1]$?

3.7. Determine for what $p \in \mathbb{R}$ the function f is integrable on the given set in the following cases:

a) $f(x) = x^{-p}(-1)^{[x]}$, $x \in (+1, +\infty)$;
b) $f(x) = x^{-p}\sin x$, $x \in (0, +\infty)$;
c) $f(x) = x^{-p}(-1)^{[1/x]}$, $x \in (0, 1)$;
d) $f(x) = e^{-x^p \sin^2 x}$, $x \in (0, +\infty)$;
e) $f(x) = x^p/(1 + x^6 \sin^2 x)$, $x \in (0, +\infty)$.

3.8. Are the functions f below integrable on the set $[-1, 1] \times [-1, 1]$? Find their iterated integrals

$$\int_{-1}^{+1}\left(\int_{-1}^{+1} f(x, y)\,dy\right)dx \quad \text{and} \quad \int_{-1}^{+1}\left(\int_{-1}^{+1} f(x, y)\,dx\right)dy$$

for:

$$\text{a) } f(x, y) = \frac{xy}{x^2+y^2}, \qquad \text{b) } f(x, y) = \frac{x^2-y^2}{x^2+y^2},$$
$$\text{c) } f(x, y) = \frac{xy}{(x^2+y^2)^2}, \qquad \text{d) } f(x, y) = \frac{x^2-y^2}{(x^2+y^2)^2},$$

(in all these cases the value $f(0, 0)$ can be defined arbitrarily).

3.9. Determine for what $p \in \mathbb{R}$ the function $f(x, y) = |1 - xy|^{-p}$ $((x, y) \in \mathbb{R}^2)$ is integrable on the set E in the following cases:

a) $E = [0, 1] \times [0, 1]$,
b) $E = [0, 2] \times [0, 2]$,
c) $E = \{(x, y) \in [0, 1] \times [0, 1] \mid (x-2)^2 + y^2 \geq 2,\ x^2 + (y-2)^2 \geq 2\}$.

3.10. a) Let $p > 0$ and $E \subset \mathbb{R}^m$. Prove that if $\lambda_m(E) = \lambda_m(B(0, r))$, then $\int_E dy/\|x-y\|^p \leq \int_{B(0,r)} dz/\|z\|^p$ for any $x \in \mathbb{R}^m$.

b) Prove that $|\int_E e^{ix}\,dx| \leq 2\sin(\lambda(E)/2)$ for any set $E \subset [0, 2\pi]$.

c) Prove that

$$\left|\iint_E \frac{dx\,dy}{x+iy}\right| \leq \sqrt{\pi\lambda_2(E)}$$

for any set E $(E \subset \mathbb{R}^2)$ of finite measure.

3.11. Let $p, t > 0$, $E \subset R^m$, and $f \in \mathscr{L}^0(E)$. Prove that

$$\lambda_m E(|f| \geq t) \leq t^{-p} \int_F |f(x)|^p\,dx.$$

In particular, for $p = 2$ we get Chebyshev's inequality:

$$\lambda_m E(|f| \geq t) \leq t^{-2} \int_E |f(x)|^2\,dx.$$

3.12. Prove that if $|f|^p \in \mathscr{L}(E)$ $(E \subset \mathbb{R}^m)$, then

$$\lambda_m E(|f| \geq t) = o(t^{-p}) \qquad (t \to +\infty).$$

Given an example showing that the converse is false.

3.13. Suppose that $p > 0$, $E \subset \mathbb{R}^m$, $f \in \mathscr{L}^0(E)$, and $\lambda_m E(|f| \geq t) = O(t^{-p})$ $(t \to +\infty)$. Prove that if $\lambda_m(E) < \infty$, then $|f|^{p-\varepsilon} \in \mathscr{L}(E)$ for any $\varepsilon \in (0, p)$.

3.14. Let $E \subset \mathbb{R}^m$ and $f \in \mathscr{L}(E)$. Prove that

$$\int_E |f(x)|^p\,dx \underset{p \to +0}{\longrightarrow} \lambda_m E \qquad (|f| \neq 0).$$

3.15. Let $f, g \in \mathscr{L}^0(E)$, $f, g \geq 0$. Prove that if $\int_E g(x)\,dx = 1$ and $\int_E g(x)f(x)\,dx < \infty$ then

$$\left(\int_E f^p(x)g(x)\,dx\right)^{1/p} \underset{p \to +0}{\longrightarrow} e^{\int_E g(x)\ln f(x)\,dx}.$$

Can the condition $\int_E g(x)f(x)\,dx < \infty$ be dropped?

3.16. Let $x \in \mathbb{R}$, and let $x = [x] + \sum \varepsilon_k 2^{-k}$, where $\varepsilon_k = \varepsilon_k(x) = 0$ or 1, be the binary representation of x.

a) Find the integrals $\int_0^1 \varepsilon_k(x)\,dx$ and $\int_0^1 \varepsilon_k(x)\varepsilon_m(x)\,dx$ $(k, m \in \mathbb{N},\ k \neq m)$.

b) Find the integral $\int_0^1 |\sigma_n(x) - \frac{1}{2}|^2\,dx$, where $\sigma_n(x) = \frac{1}{n}\sum_{1\le k\le n}\varepsilon_k(x)$.

c) Prove that $\int_0^1 |\sigma_n(x) - \frac{1}{2}|^4\,dx = O(n^{-2})$.

3.17. Let $x \in \mathbb{R}$, and let $x = [x] + \sum \varepsilon_k 10^{-k}$, where $\varepsilon_k = \varepsilon_k(x) = 0, 1, \ldots, 9$, be the decimal representation of x. Let i be one of the numbers $0, 1, \ldots, 9$, and let

$$a_j(x) = \begin{cases} 1 & \text{if } \varepsilon_1(x) = j, \\ 0 & \text{if } \varepsilon_1(x) \neq j. \end{cases}$$

Find the integrals $(k, m \in \mathbb{N},\ k \neq m)$

$$\text{a) } \int_0^1 a_j(10^{k-1}x)\,dx, \qquad \text{b) } \int_0^1 a_j(10^{k-1}x)a_j(10^{m-1}x)\,dx.$$

3.18. Suppose that $a \in \mathbb{R}$, $a \neq 0, 1$, and $x = \sum_{k\ge1}\varepsilon_k(x)\cdot 10^{-k}$ is the decimal representation of $x \in (0, 1)$. Let

$$f(x) = \begin{cases} a & \text{if } \varepsilon_n(x) \neq 0 \text{ for any } n; \\ 1 & \text{if the first } \varepsilon_n(x) \neq 0 \text{ has even index}; \\ 0 & \text{otherwise}. \end{cases}$$

Prove that f is measurable, and find $\int_0^1 f(x)\,dx$.

3.19. Open intervals of length 1/3, 1/9, etc., were removed successively from $[0, 1]$ in the construction of the Cantor set (see the discussion preceding problem I.1.27). Let $f(x) = n$ on the open intervals of length 3^{-n}, and $f(x) = 0$ on the Cantor set.

a) Prove that f is measurable, and find $\int_0^1 f(x)\,dx$.

b) Find the distribution function F of f and the nonincreasing rearrangement f^* of f (see definitions before problems 2.5 and 2.10).

3.20. Let K be a closed subset of $[a, b]$, and let $\rho(x) = \inf\{|x-t| \mid t \in K\}$ be the distance from a point $x \in \mathbb{R}$ to the set K. Prove that for any $s > 0$ the function $\mathcal{I}_s(y) = \int_a^b \rho^s(x)|x-y|^{-s-1}\,dx$ $(y \in K)$ is integrable on K.

3.21. Prove the equalities $(0 < a \le \infty)$

a) $\int_{[-a,a]^m} f(\max_{1\le k\le m}|x_k|)\,dx = m\cdot 2^m \int_0^a t^{m-1}\cdot f(t)\,dt$,

b) $\int_{[-a,a]^m} f(x)\,dx = \int_0^a t^{m-1}\left(\int_{\max|y_k|=1} f(ty)\,d\lambda_{m-1}(y)\right)dt$, where f is assumed to be nonnegative, and λ_{m-1} is $(m-1)$-dimensional Lebesgue measure on the boundary of the cube $[-1, 1]^m$.

Below, the symbol μ_{m-1} denotes Lebesgue measure (surface area) on the sphere $S^{m-1} = \{x \in \mathbb{R}^m \mid \|x\| = 1\}$. The area of the sphere, i.e., the quantity $\mu_{m-1}(S^{m-1})$, is denoted by σ_{m-1}.

3.22. Prove by induction that

a) $$\int_{S^{n-1}} f(x_1, \dots, x_{n-1}, x_n)\, d\mu_{n-1}(x)$$
$$= \int_0^\pi \sin^{n-2}\theta \left\{ \int_{S^{n-2}} f(u_1 \sin\theta, \dots, u_{n-1}\sin\theta, \cos\theta)\, d\mu_{n-2}(u) \right\} d\theta \qquad (n \geq 3),$$

b) $$\int_{\mathbb{R}^n} f(x)\, dx = \int_0^\infty t^{n-1} \left\{ \int_{S^{n-1}} f(tw)\, d\mu_{n-1}(w) \right\} dt \qquad (n \geq 2),$$

where f is assumed to be integrable on S^{n-1} or $\mathbb{R}^n$, respectively. Note that the formula a) remains true also for $n = 2$ if S^0 is understood to be the boundary of $[-1, 1]$, that is, the two-point set $\{-1, 1\}$, and the measure μ_0 is assumed to consist of a unit load at the points $+1$ and -1.

3.23. a) Let r be a nonnegative continuous function on S^{n-1}, and let $T = \{x \in \mathbb{R}^n \mid 0 < \|x\| \leq r(x/\|x\|)\}$. Prove that

$$\lambda_n(T) = \frac{1}{n} \int_{S^{n-1}} r^n(w)\, d\mu_{n-1}(w).$$

b) Let A be a bounded neighborhood of zero in $\mathbb{R}^n$, and let $\rho(x) = \sup\{|(x, y)| \mid y \in A\}$ and $V = \{x \in \mathbb{R}^n \mid \rho(x) \leq 1\}$. Prove that

$$\lambda_n(V) = \tfrac{1}{n} \int_{S^{n-1}} \rho^{-n}(x) d\mu_{n-1}(x).$$

In the solution of problems 3.24–3.26 it is useful to employ the formula (see problem 3.22)

$$\int_{a \leq \|x\| \leq b} f(\|x\|)\, dx = n\alpha_n \int_a^b t^{n-1} f(t)\, dt,$$

where f is a (measurable) nonnegative function on the interval $\langle a, b \rangle$ $(0 \leq a < b \leq \infty)$, and $\alpha_n = \lambda_n(B(0, 1))$.

3.24. Let E be a (measurable) subset of $(0, +\infty)$, and let

$$A_E = \{x \in \mathbb{R}^n \mid \|x\| \in E\}.$$

Prove that $\lambda_n(A_E) = n\alpha_n \int_E t^{n-1}\, dt$.

In problems 3.25, 3.26, and others the symbol γ_n denotes the *standard Gaussian measure* on $\mathbb{R}^n$, that is, the measure with density $(2\pi)^{-n/2} e^{-\|x\|^2/2}$.

3.25. Compute $I_n(r) = \gamma_n(B(0, r))$ $(r > 0)$.

3.26. For $a > 0$ let K_a be the cone in $\mathbb{R}^4$ given by

$$K_a = \{(x_1, x_2, x_3, x_4) \in \mathbb{R}^4 \mid x_2^2 + x_3^2 + x_4^2 \leq a^2 x_1^2\}.$$

Find $\gamma_4(K_a)$.

3.27. Compute the volume of the generalized torus $T \subset \mathbb{R}^4$ given by

$$T = \{(x_1, x_2, x_3, x_4) \in \mathbb{R}^4 \mid (x_1, \sqrt{x_2^2 + x_3^2 + x_4^2}) \in B \subset \mathbb{R}^2\},$$

where B is the disk of radius $R \leq a$ about the point $(0, a)$.

3.28. Suppose that A is a compact subset in the upper half-plane of the space $\mathbb{R}^2$. If T_0 is the body obtained by rotating A about the x_1-axis in $\mathbb{R}^3$, then by Guldin's theorem, $\lambda_3(T_0) = \lambda_2(A)2\pi y_c$, where y_c is the ordinate of the center of gravity of A. Regarding $\mathbb{R}^2$ as canonically imbedded in $\mathbb{R}^4$, consider the body T "obtained by rotating A about the x_1-axis in $\mathbb{R}^4$," that is,

$$T = \{(x_1, x_2, x_3, x_4) \in \mathbb{R}^4 \mid (x_1, \sqrt{x_2^2 + x_3^2 + x_4^2}) \in A\}.$$

Prove the following modification of Guldin's theorem: $\lambda_4(T) = \lambda_2(A)\cdot 4\pi\rho^2$, where ρ is the radius of inertia of A with respect to the x_1-axis, that is, the positive number determined by the equality

$$\rho^2 = (\lambda_2(A))^{-1} \iint_A y^2\, dx\, dy.$$

3.29. Let A be a compact set in the half-plane $x_2 \geq 0$ in $\mathbb{R}^2$. Regarding $\mathbb{R}^2$ as canonically imbedded in $\mathbb{R}^n$, consider the "body T obtained by rotating A about the x_1-axis in $\mathbb{R}^n$," that is,

$$T = \{(x_1, x_2, \dots, x_n) \in \mathbb{R}^n \mid (x_1, \sqrt{x_2^2 + \cdots + x_n^2}) \in A\}.$$

Prove that

$$\text{a) } \lambda_n(T) = \sigma_{n-2} \iint_A x_2^{n-2}\, dx_1\, dx_2,$$

$$\text{b) } \gamma_n(T) = \frac{\sigma_{n-2}}{(2\pi)^{n/2}} \iint_A e^{-(x_1^2+x_2^2)/2} x_2^{n-2}\, dx_1\, dx_2,$$

where $\gamma_n(T)$ is the "Gaussian volume of T" (γ_n is the measure on $\mathbb{R}^n$ with density $(2\pi)^{-n/2} e^{-\|x\|^2/2}$).

3.30. Find the measure of the set

$$A_n(p) = \{(x_1, \dots, x_n) \in \mathbb{R}^n \mid \sum_{1 \leq k \leq n} |x_k|^p \leq 1\} \qquad (p > 0).$$

3.31. a) Compute the integral $\mathfrak{J} = \sigma_{n-1}^{-1} \int_{S^{n-1}} |(x, a)|^p\, d\mu_{n-1}(x)$.

b) Prove that

$$\sigma_{n-1}^{-1} \int_{S^{n-1}} (x_1^2 + \cdots + x_m^2)^{q/2}\, d\mu_{n-1}(x) = \frac{\Gamma((m+q)/2)}{\Gamma(n+q)/2)} \cdot \frac{\Gamma(n/2)}{\Gamma(m/2)} \qquad \left(\underset{m\to\infty}{\sim} \left(\frac{m}{n}\right)^{q/2} \right)$$

for $m, n \in \mathbb{N}$, $1 \leq m \leq n$, and $m + q > 0$.

3.32. Let $a, b \in S^{n-1}$, and let θ be the angle between the vectors a and b $(0 \leq \theta \leq \pi)$. Prove that

$$\int_{S^{n-1}} \operatorname{sgn}(a, x) \operatorname{sgn}(b, x)\, d\mu_{n-1}(x) = (1 - 2\theta/\pi)\sigma_{n-1}.$$

3.33. Let $I_n = \int_0^\infty \gamma_n(B(0, r))(1 - \gamma_n(B(0, r)))\, dr$. Prove that:

a) $I_n = \frac{1}{2} \int_{\mathbb{R}^n} \int_{\mathbb{R}^n} |\|x\| - \|y\||\, d\gamma_n(x)\, d\gamma_n(y)$;

b) $I_n \underset{n\to\infty}{\longrightarrow} (2\pi)^{-1/2}$.

3.34. Let $A \subset \mathbb{R}^n$ be an open set star-like with respect to zero, let $q_A(x) = \inf\{t > 0 \mid t^{-1}x \in A\}$ $(x \in \mathbb{R}^n)$ be the gauge function of the set A, and let μ be a probability measure on $\mathbb{R}^n$ such that $\int_{\mathbb{R}^n} |(x, y)|\, d\mu(x) < +\infty$ for any $y \in \mathbb{R}^n$. Prove that

$$\int_0^{+\infty} \mu(tA)(1 - \mu(tA))\, dt = \frac{1}{2} \int_{\mathbb{R}^n} \int_{\mathbb{R}^n} |q_A(x) - q_A(y)|\, d\mu(x)\, d\mu(y) < +\infty.$$

3.35. Suppose that f is an integrable function on the cube $Q = [0, 1)^m$ and has period 1 in each variable. Prove that:

a) the integral $\int_Q f(x)\, dx$ does not change under a translation of Q by any vector, that is, $\int_Q f(x)\, dx = \int_{a+Q} f(x)\, dx$ for any $a \in \mathbb{R}^m$, where $a + Q = \{a + y \mid y \in Q\}$;

b) if A is an arbitrary $m \times m$ integer matrix with determinant 1, then

$$\int_Q f(x)\, dx = \int_{A(Q)} f(x)\, dx.$$

Is this true if the determinant of A is equal to 1, but its elements are not necessarily integers?

3.36. Let T be an arbitrary compact convex body in $\mathbb{R}^n$. Prove that if the $(n-1)$-dimensional volume of the projection of T on any hyperplane is not less than S, then $\operatorname{diam} T \leq nV/S$, where $V = \lambda_n(T)$.

3.37. Suppose that $T \subset \mathbb{R}^n$ is a compact convex centrally symmetric body with respect to zero, and let Э be an ellipsoid of maximal volume contained in T:

Э $\subset T$, $\lambda_n($Э$) = \sup\{\lambda_n(\mathscr{E}) \mid \mathscr{E} \subset T,\ \mathscr{E}$ an ellipsoid$\}$.

Prove that $T \subset \sqrt{n}\,$Э.

3.38. Let $K \subset \mathbb{C}$ be a closed bounded set of positive (planar) measure. Define the function φ by

$$\varphi(z) = \iint_K \frac{d\xi\, d\eta}{z - (\xi + i\eta)} \qquad (z \in \mathbb{C}).$$

Prove that:

a) $\varphi \in C(\mathbb{C})$, $z\varphi(z) \underset{z\to\infty}{\longrightarrow} \lambda_2(K)$;

b) φ is holomorphic outside K.

3.39. Identifying $\mathbb{C}^2$ in the natural way with $\mathbb{R}^4$, consider on the sphere S^3 the functions $\varphi_{m,n}$ $(m, n = 0, 1, 2, \ldots)$, defined by the equalities $\varphi_{m,n}(\zeta_1, \zeta_2) = \zeta_1^m \zeta_2^n$.

a) Compute the integrals

$$\int_{S^3} |\varphi_{m,n}(\zeta_1, \zeta_2)|^2 \, d\mu_3(\zeta_1, \zeta_2) \qquad (m, n = 0, 1, 2, \dots);$$

b) Prove that the system $\{\varphi_{m,n}\}_{m,n=0}^{\infty}$ is orthogonal.

c) Prove that if the series $\sum_{m,n\geq 0} a_{mn}\varphi_{mn}$ is uniformly convergent on $\overline{B^4}$ and its sum is equal to f, then the following generalization of the Cauchy integral formula is valid:

$$f(\zeta_1, \zeta_2) = \frac{1}{2\pi^2} \int_{S^3} \frac{f(z_1, z_2)}{(1 - \zeta_1\overline{z}_1 - \zeta_2\overline{z}_2)^2} \, d\mu_3(z_1, z_2),$$

where $(\zeta_1, \zeta_2) \in B^4$, $B^4 = \{(\zeta_1, \zeta_2) \in \mathbb{C}^2 \mid |\zeta_1|^2 + |\zeta_2|^2 < 1\}$.

§4. The Stieltjes integral

In problems 1–4 the function h is defined on the interval over which the integral on the right-hand side is taken, and it is measurable and nonnegative.

4.1. a) For $0 \leq a < b \leq \infty$ let $E = \{(x_1, \dots, x_m) \in \mathbb{R}^m \mid a \leq \max_{1\leq k\leq m} |x_k| \leq b\}$. Prove that

$$\int_E h\left(\max_{1\leq k\leq m} |x_k|\right) dx = m2^m \int_a^b t^{m-1} h(t)\, dt.$$

b) Prove that

$$\int_{[1,+\infty)^m} h(\max_{1\leq k\leq m}|x_k|)\, dx = m \int_1^{\infty} (t-1)^{m-1} h(t)\, dt.$$

4.2. a) Let $O_m(r) = \{(x_1, \dots, x_m) \in \mathbb{R}^m \mid \sum_{1\leq k\leq m} |x_k| \leq r\}$. Prove that

$$\int_{O_m(r)} h\left(\sum_{1\leq k\leq m} |x_k|\right) dx = \frac{2^m}{(m-1)!} \int_0^r t^{m-1} h(t)\, dt.$$

b) For $a_k > 0$ $(k = 1, 2, \dots, m)$, let

$$E = \left\{(x_1, \dots, x_m) \in \mathbb{R}^m \mid \sum_{1\leq k\leq m} a_k x_k \geq 1,\ x_1 \geq 0, \dots, x_m \geq 0\right\}.$$

Compute the integral $\int_E e^{-\sum_{1\leq k\leq m} a_k x_k}\, dx$.

4.3. a) For $0 \leq a < b \leq \infty$ let $E = \{x \in \mathbb{R}^m \mid a \leq \|x\| \leq b\}$. Prove that

$$\int_E h(\|x\|)\, dx = m\alpha_m \int_a^b t^{m-1} h(t)\, dt.$$

b) For $p > 0$ let $A(p) = \{(x_1, \dots, x_m) \in \mathbb{R}^m \mid \sum_{1\leq k\leq m} |x_k|^p \leq 1\}$. Prove that

$$\int_{\mathbb{R}^m} h\left(\left(\sum_{1\leq k\leq m} |x_k|^p\right)^{1/p}\right) dx = mC_m(p) \int_0^{\infty} t^{m-1} h(t)\, dt,$$

where $C_m(p) = \lambda_m(A(p))$ (see problem 3.30 about the value of $C_m(p)$).

4.4. For $E \subset \mathbb{R}^m$ let $g_E(t) = \lambda_m(E \cap B(0, t))$ $(t > 0)$. Prove that:

a) $\int_E h(\|x\|)\,dx = \int_0^\infty h(t)\,dg_E(t)$;

b) if $g_E \in C^1[0, +\infty)$, then $\int_E h(\|x\|)\,dx = \int_0^\infty h(t)g_E'(t)\,dt$.

4.5. Suppose that $A \subset \mathbb{R}^2$ is a convex bounded set, and let $\rho(x) = \inf\{\|x - y\| \mid y \in A\}$ be the distance from a point $x \in \mathbb{R}^2$ to A. Find the integral $\int_{\mathbb{R}^2} e^{-\rho(x)}\,dx$.

4.6. Let σ be the Cantor function (see problem III.3.17). Compute the following integrals:

$$\text{a)} \int_0^1 x\,d\sigma(x), \qquad \text{b)} \int_0^1 x^2\,d\sigma(x), \qquad \text{c)} \int_0^1 \sigma(x)\,dx,$$

$$\text{d)} \int_0^1 \sigma(x)\,d\sigma(x), \qquad \text{e)} \int_0^1 \sigma(1-x)\,d\sigma(x).$$

4.7. Let K be the Cantor set, and σ the Cantor function. Determine the values of $p \in \mathbb{R}$ for which the following integrals are finite:

$$\text{a)} \int_0^1 \int_0^1 \frac{d\sigma(x)\,d\sigma(y)}{(x^2 + y^2)^{p/2}};$$

$$\text{b)} \int_0^1 \int_0^1 \frac{d\sigma(x)\,d\sigma(y)}{((x-u)^2 + (y-v)^2)^{p/2}}, \qquad \text{where } (u, v) \in K \times K;$$

4.8. Let K be the Cantor set, and σ the Cantor function. Define the function φ by

$$\varphi(z) = \int_0^1 \int_0^1 \frac{d\sigma(\xi)\,d\sigma(\eta)}{z - (\xi + i\eta)} \qquad (z \in \mathbb{C}).$$

Prove that (cf. problem 3.38):

a) $\varphi \in C(\mathbb{C})$, $z\varphi(z) \underset{z\to\infty}{\longrightarrow} 1$;

b) φ is holomorphic outside $K \times K$.

§5. ε-entropy and Hausdorff measures

Let A be a subset of $\mathbb{R}^n$ and ε a positive number. A set $C \subset \mathbb{R}^n$ is called an *ε-net* for A if $A \subset \bigcup_{x\in C} B(x, \varepsilon)$. Let

$$N(A, \varepsilon) = \min\{\text{card}(C) \mid C \text{ is an } \varepsilon\text{-net for } A\}$$

be the minimal cardinality of an ε-net for A. If the set A is bounded, then $N(A, \varepsilon) < \infty$ for any $\varepsilon > 0$. The function $H(A, \varepsilon) = \log_2 N(A, \varepsilon)$ is called the *ε-entropy* of A.

5.1. A set $E \subset \mathbb{R}^n$ is said to be *ε-distinguishable* if the distance between any two points in it is at least ε. Let

$$M(A, \varepsilon) = \max\{\text{card}(E) \mid E \subset A, \text{ and } E \text{ is } \varepsilon\text{-distinguishable}\}$$

be the maximal cardinality of an ε-distinguishable subset of A. Prove that $N(A, \varepsilon) \le M(A, \varepsilon) \le N(A, \varepsilon/2)$.

5.2. Let A be a bounded subset of $\mathbb{R}^n$.

a) Prove that if A has an interior point, then

$$H(A, \varepsilon) \sim n \log_2(1/\varepsilon) \qquad (\varepsilon \to +0).$$

b) Recall that the *outer measure* $\lambda_n^*(A)$ of a set $A \subset \mathbb{R}^n$ is defined to be the quantity $\lambda_n^*(A) = \inf\{\lambda_n(G) \mid A \subset G, G \text{ an open subset of } \mathbb{R}^n\}$. Prove that if $\lambda_n^*(A) > 0$, then $N(A, \varepsilon) \ge C\varepsilon^{-n}$, where C is some positive constant.

5.3. Find the asymptotic behavior (as $\varepsilon \to +0$) of the ε-entropy of the following sets:

a) $\{2^{-n} \mid n \in \mathbb{N}\}$;
b) $\{n^{-n} \mid n \in \mathbb{N}\}$;
c) $\{2^{-n^2} \mid n \in \mathbb{N}\}$;
d) $\{n^{-\alpha} \mid n \in \mathbb{N}\}$ $(\alpha > 0)$;
e) $\{1/\ln(n+1) \mid n \in \mathbb{N}\}$.

The *metric dimension* (*metric order*) of a bounded subset A of $\mathbb{R}^n$ is defined to be

$$M\text{-}\dim A = \overline{\lim_{\varepsilon\to+0}} \frac{H(A, \varepsilon)}{\log_2(1/\varepsilon)} \quad \left(\text{respectively, } r(A) = \underline{\lim_{\varepsilon\to+0}} \frac{H(A, \varepsilon)}{\log_2(1/\varepsilon)}\right).$$

5.4. Find the metric dimensions and metric orders of the graphs of the following functions:

a) $f(x) = \sin \pi x$, $0 \le x \le 1$;
b) $f(x) = \sin(\pi/x)$, $0 < x \le 1$.

5.5. Find the metric dimensions and metric orders of the following sets:

a) $A \times [-1, 1] \subset \mathbb{R}^2$, where $A = \{n^{-1/2} \mid n \in \mathbb{N}\}$;
b) the graph of the function $f(x) = \sin(\pi/x^2)$, $0 < x \le 1$;
c) $\bigcup_{n=1}^{\infty} S(n^{-1/2})$, where $S(r)$ is the circle about zero with radius r.

5.6. Find the metric dimensions and metric orders of the following sets;

a) $A \times [-1, 1] \subset \mathbb{R}^2$, where $A = \{1/\ln(n+1) \mid n \in \mathbb{N}\}$;
b) the graph of the function $f(x) = \sin(\pi e^{1/x})$, $0 < x \le 1$;
c) $\bigcup_{n=1}^{\infty} S(1/\ln(n+1))$.

5.7. Find the metric dimension and metric order of the Cantor set.

5.8. Find the asymptotic behavior of the ε-entropy of the set

$$A = \left\{ \sum_{k\ge0} \varepsilon_k/k! \mid \varepsilon_k = 0 \text{ or } 1 \right\}.$$

5.9. Define an increasing sequence $\{n_k\}$ of positive integers by setting $n_k = k + m!$ if $m! \le k < (m+1)!$, and consider the set $A = \{\sum \varepsilon_k 2^{-n_k} \mid$

$\varepsilon_k = 0$ or $1\}$. Further, let $A_l = A + A + \cdots + A$ (the arithmetic sum, with l terms). Prove that:

a) M-dim$(A) = 1$, $r(A) = 1/2$;

b) $\lambda(A_l) = 0$ for any $l \in \mathbb{N}$.

5.10. Let $\psi\colon [0, +\infty) \to [0, +\infty)$ be a strictly increasing function such that $\psi(\mathbb{N}) \subset \mathbb{N}$. Consider the set $A = \{\sum \varepsilon_k 2^{-\psi(k)} \mid \varepsilon_k = 0 \text{ or } 1\}$. Prove that

$$H(A, \varepsilon) \sim \psi^{-1}(\log_2(1/\varepsilon)) \qquad (\varepsilon \to +0).$$

5.11. Find the metric dimension and metric order of the set

$$A = \left\{ (x, y) \in \mathbb{R}^2 \;\middle|\; \begin{array}{l} x = \sum \varepsilon_k 2^{-k}, \ \varepsilon_k = 0, 1, \\ y = \sum \eta_k 3^{-k}, \ \eta_k = 0, 1, 2, \\ \varepsilon_k \text{ and } \eta_k \text{ have the same parity for any } k, \end{array} \right\}$$

(the *Hironaka curve*).

5.12. Let $A \subset \mathbb{R}^1$, and let $A_l = A + \cdots + A$ (l terms).

Prove that:

a) if $r(A) = 0$, then $\lambda(A_l) = 0$ for any $l \in \mathbb{N}$;

b) if $r(A) < l^{-1}$, then $\lambda(A_l) = 0$.

c) Is it true that if $r(A) > 1/2$, then $\lambda(A + A) > 0$? (Consider the set in problem 5.3e).)

Hausdorff measures. L. et $A \subset \mathbb{R}^n$, and let $p, \varepsilon > 0$. Define

$$\mu_p(A, \varepsilon) = \inf\left\{ \sum_{k\geq 1} r_k^p \mid A \subset \bigcup_{k\geq 1} B(x_k, r_k),\ r_k < \varepsilon \right\}.$$

The infimum in the definition of $\mu_p(A, \varepsilon)$ is over all possible countable coverings of A by balls, including balls with centers not in A. The function μ_p defined on the system of all subsets of $\mathbb{R}^n$ by

$$\mu_p(A) = \sup\left\{\mu_p(A, \varepsilon) \mid \varepsilon > 0\right\} = \lim_{\varepsilon\to+0} \mu_p(A, \varepsilon),$$

is called the p-dimensional (outer) Hausdorff measure.

5.13. Let $p > 0$. Prove that:

a) if A is countable, then $\mu_p(A) = 0$;

b) if $A \subset A'$, then $\mu_p(A) \leq \mu_p(A')$;

c) if $A \subset \bigcup_{k\geq 1} A_k$, then $\mu_p(A) \leq \sum_{k\geq 1} \mu_p(A_k)$;

d) if A and B are separated by a positive distance, that is, if $d = \inf\{\|x - y\| \mid x \in A,\ y \in B\} > 0$, then $\mu_p(A \cup B) = \mu_p(A) + \mu_p(B)$, and $\mu_p(A \cup B, \varepsilon) = \mu_p(A, \varepsilon) + \mu_p(B, \varepsilon)$ for $\varepsilon < d/2$.

5.14. Let $A \subset \mathbb{R}^n$. Prove that:

a) if $0 < p < n$ and $\mu_p(A) < \infty$, then $\lambda_n(A) = 0$;

b) $\mu_n(A) = 0$ if and only if $\lambda_n(A) = 0$.

5.15. Let $A \subset S^{n-1}$ and let ν_{n-1} be Lebesgue measure on S^{n-1} ("surface area"). Prove that:

a) if $0 < p < n-1$ and $\mu_p(A) = 0$, then $\nu_{n-1}(A) = 0$;
b) $\mu_{n-1}(A) = 0$ if and only if $\nu_{n-1}(A) = 0$.

5.16. Prove that:
a) if $0 < p < q$, then $\mu_p(A) \geq \mu_q(A)$;
b) if $0 < \mu_p(A) < \infty$, then $\mu_q(A) = 0$ for $q > p$, and $\mu_q(A) = \infty$ for $q < p$.

The number $p = \inf\{q \mid \mu_q(A) = 0\} = \sup\{q \mid \mu_q(A) = \infty\}$ is called the *Hausdorff dimension* of the set A. We denote it by the symbol $\dim_H(A)$.

5.17. a) Prove that

$$\dim_H(A) \leq r(A) = \varlimsup_{\varepsilon \to +0} \frac{H(A, \varepsilon)}{\log_2 1/\varepsilon}$$

for a bounded set $A \subset \mathbb{R}^n$.

b) Let $A_k \subset \mathbb{R}^n$ $(k = 1, 2, \dots)$. Prove that

$$\dim_H\left(\bigcup_{k\geq 1} A_k\right) = \sup_k \dim_H(A_k).$$

5.18. The *diameter* of a set $C \subset \mathbb{R}^n$ is defined to be the number $\operatorname{diam}(C) = \sup\{\|x-y\| \mid x, y \in C\}$. Let $p > 0$ and $A \subset \mathbb{R}^n$. Define the *modified p-dimensional Hausdorff measure* $\tilde{\mu}_p(A)$ by

$$\tilde{\mu}_p(A) = \sup_{\varepsilon>0}\left\{\inf\left(\sum_{k\geq 1}(\operatorname{diam}(C_k))^p \mid A \subset \bigcup_{k\geq 1} C_k,\ \operatorname{diam}(C_k) < \varepsilon\right)\right\}.$$

Prove that $\mu_p(A) \leq \tilde{\mu}_p(A) \leq 2^p \mu_p(A)$, and hence $\tilde{\mu}_p$ can be used instead of μ_p in the definition of the Hausdorff dimension.

5.19. Find the Hausdorff dimension p and compute $\mu_p(A)$ for the following sets:
a) $A = [a, b]$;
b) $A = S^1$;
c) $A = [0, 1] \times [0, 1]$;
d) $A = [0, 1]^n$.

5.20. Find the Hausdorff dimension p and compute $\mu_p(K)$ for the Cantor set K. Prove that the function φ defined by $\varphi(x) = 2^{-p}\mu_p(K \cap [0, x])$ $(0 \leq x \leq 1)$ coincides with the Cantor function.

5.21. Find $\dim_H(A)$ and M-$\dim(A)$ for the following sets:
a) $A = K$, K the Cantor set;
b) $A = A_0$, where $A_0 = \{1 + 1/\ln(n+1) \mid n \in \mathbb{N}\}$;
c) $A = K \cup A_0$.

Let us recall now the concept of a set of Cantor type, which is needed in the following problems (cf. the definitions of the Cantor set and of a generalized Cantor set in §1 of Chapter I). Let $\{l_n\}_{n\geq 0}$ be a sequence of positive numbers such that $2l_n < l_{n-1}$ $(n \in \mathbb{N})$. A *set of Cantor type* with *defining sequence* $\{l_n\}_{n\geq 0}$ is defined to be an intersection of sets K_n constructed as follows. Let Δ be an arbitrary closed interval of length l_0. Remove from Δ an open interval δ symmetric with respect to its midpoint in such a way that each of the closed intervals making up $\Delta\backslash\delta$ has length l_1. The left-hand one of these closed intervals is denoted by Δ_0, and the right-hand one by Δ_1. The closed intervals Δ_0 and Δ_1 are called the intervals of first rank. Let $K_1 = \Delta_0 \cup \Delta_1$. Suppose that the set K_n, which is the union of the 2^n intervals of nth rank with length l_n, has already been constructed. From each closed interval $\Delta_{\varepsilon_1\cdots\varepsilon_n}$ of nth rank remove an open interval $\delta_{\varepsilon_1\cdots\varepsilon_n}$ symmetric with respect to its midpoint in such a way that each of the closed intervals making up the set $\Delta_{\varepsilon_1\cdots\varepsilon_n}\backslash\delta_{\varepsilon_1\cdots\varepsilon_n}$ has length l_{n+1}. The left-hand one of these closed intervals is denoted by $\Delta_{\varepsilon_1\cdots\varepsilon_n 0}$, and the right-hand one by $\Delta_{\varepsilon_1\cdots\varepsilon_n 1}$. The closed intervals $\Delta_{\varepsilon_1\cdots\varepsilon_n\varepsilon_{n+1}}$ are called the intervals of $(n+1)$st rank; let $K_{n+1} = \bigcup_{\varepsilon_1,\dots,\varepsilon_{n+1}\in\{0,1\}} \Delta_{\varepsilon_1\cdots\varepsilon_n\varepsilon_{n+1}}$.

The set of Cantor type generated on $[0, 1]$ by the defining sequence $\{2^{-np}\}_{n\geq 0}$, where $p > 1$, will be denoted by $K(p)$. Note that for $p = \log_2 3$ the set $K(p)$ is the classical Cantor set.

5.22. Suppose that $E \subset [0, 1]$ is a set of Cantor type, $0 < \mu_p(E) < \infty$, and σ_E is the Cantor function corresponding to E. Prove that

$$\mu_p(E \cap [0, x]) = \mu_p(E)\sigma_E(x) \qquad (0 \leq x \leq 1).$$

5.23. Suppose that $0 < p \leq 1$, $E \subset \mathbb{R}$ is a set of Cantor type, and l_n is the length of the intervals of nth rank. Prove that:

a) if $\underline{\lim}\, 2^n l_n^p = 0$, then $\mu_p(E) = 0$;

b) if $\underline{\lim}\, 2^n l_n^p > 0$, then $\mu_p(E) > 0$;

c) $\mu_p(E) = \infty$ if and only if $2^n l_n^p \to \infty$.

d) $\dim_H(E \times E) = 2\dim_H(E)$.

5.24. Suppose that A is the set in problem 5.9.

a) Verify that A is a set of Cantor type, and find its defining sequence $\{l_n\}_{n\geq 0}$.

b) Find the Hausdorff dimension p of A.

c) Find $\overline{\lim}\, 2^n l_n^p$.

5.26. Let A and B be sets of Cantor type with defining sequences $\{\frac{1}{n+1}3^{-n}\}_{n\geq 0}$ and $\{\sqrt{n+1}3^{-n}\}_{n\geq 0}$, respectively, and let $p = \dim_H(A)$ and $q = \dim_H(B)$. Prove that:

a) $p = q = \log_3 2$ (recall that $\log_3 2$ is the Hausdorff dimension of the Cantor set);

b) $\mu_p(A) = 0$, $\mu_p(B) = +\infty$, and hence for any $t > 0$ the quantities $\mu_t(A)$ and $\mu_t(B)$ can take only the values 0 and $+\infty$.

5.26. Suppose that $n \in \mathbb{N}$, $E \subset \{0, \dots, n-1\}$, $m = \mathrm{card}(E)$, and

$1 \le m < n$. Find the Hausdorff dimension of the set

$$A = \left\{ \sum_{k \ge 1} a_k n^{-k} \mid a_k \in E \text{ for any } k \in \mathbb{N} \right\}.$$

5.27. Let $A = \{\sum_{k\ge1} a_k 10^{-k} \mid a_k \in \{0, 1, 2, 3, 4\} \text{ for any } k \in \mathbb{N}\}$. Find $\dim_H(A)$ and $\dim_H(A+A)$ and verify that $\dim_H(A) > 1/2$, but $\dim_H(A+A) < 1$.

5.28. Let K be a set of Cantor type, with $\dim_H(K) > 1/2$. Prove that there exists a continuous function $\varphi: \mathbb{C} \to \mathbb{C}$ such that $\lim_{z\to\infty} z\varphi(z) = 1$ and φ is holomorphic outside $K \times K$ (cf. problem 4.8).

§6. Asymptotics of integrals of higher multiplicity

In this section the symbols $B^n(r)$ and $S^{n-1}(r)$ denote the n-dimensional (open) ball and the $(n-1)$-dimensional sphere about zero with radius r. If the radius is equal to 1, then we write simply B^n and S^{n-1}. The Lebesgue measures in $\mathbb{R}^n$ and S^{n-1} are denoted by λ_n (volume) and μ_{n-1} (surface area), respectively. The volume of the ball B^n is denoted by α_n, and the area of the sphere S^{n-1} by σ_{n-1}. The symbol γ_n denotes the standard Gaussian measure on $\mathbb{R}^n$, that is, the measure with density $(2\pi)^{-n/2}e^{-\|x\|^2/2}$. If $x = (x_1, \dots, x_n)$ is an n-dimensional vector and $0 < p < +\infty$, then $\|x\|_p$ is defined to be $(\sum_{1\le k\le n} |x_k|^p)^{1/p}$, while $\|x\|_\infty = \max_{1\le k\le n} |x_k|$. As before, the notation $\|x\|$ is used without an index to denote the Euclidean norm of a vector x, that is, the quantity $\|x\|_2$. Finally, as usual, the mean value of a function f integrable with respect to a measure μ on a set E $(0 < \mu E < +\infty)$ is understood to be the quantity $(1/\mu E)\int_E f\,d\mu$.

In the solution of a number of problems in this section it is convenient to use the formula (see problem VIII.3.22b))

$$\int_{\mathbb{R}^n} f(x)\,dx = \int_0^\infty r^{n-1}\left(\int_{S^{n-1}} f(rw)\,d\mu_{n-1}(w)\right) dr, \qquad f \in \mathscr{L}(\mathbb{R}^n). \quad (*)$$

6.1. Prove that:

a) $\displaystyle \frac{1}{\alpha_n}\int_{B^n(r)} x_1^2\,dx = \frac{r^{n+2}}{n+2}$;

b) $\displaystyle \frac{1}{\alpha_n}\int_{B^n(r)} |x_n|\,dx \sim_{n\to\infty} \sqrt{\frac{2}{\pi n}}\, r^{n+1}$;

c) $\displaystyle \frac{1}{\alpha_n}\int_{B^n(r)} |x_1|^p\,dx \sim_{n\to\infty} \frac{\Gamma((1+p)/2)}{\sqrt{\pi}}\left(\frac{2}{n}\right)^{p/2} r^{n+p} \qquad (p > -1).$

6.2. The thickness of the rind of a watermelon is $1/20$ of the radius of the watermelon (assumed to be a ball). What percentage of the volume of the watermelon is filled by its rind: a) in 3-dimensional space? b) in 100-dimensional space?

6.3. After falling into 1000-dimensional space, Alice was asked how to make thin glass hoops (spherical zones) in such a way that their width is equal to 1/10 of the diameter. "It couldn't be simpler," replied Alice. "You need to blow up spheres, and then cut off the excess" (see Figure 7). What percentage of glass is wasted with this technique?

6.4. A point $x = (x_1, \dots, x_n)$ is chosen at random in the ball B^n. What is more probable for large n: $|x_1| < 10^{-3}$ or $|x_1| > 10^{-3}$? (It is assumed that the probability of the point falling into a given set is proportional to the measure of the set.)

6.5. Two points are chosen at random on the sphere S^{n-1}. Let L_n be the mean value of the distance between them. Find $\lim_{n\to\infty} L_n$.

6.6. Let P_n be normalized Lebesgue measure on the sphere $S^{n-1}(\sqrt{n})$. Prove that the coordinates on this sphere "are asymptotically distributed according to a normal law," that is, that for any $a, b \in \mathbb{R}$ with $a < b$

$$\lim_{n\to\infty} P_n\{(x_1, \dots, x_n) \in S^{n-1}(\sqrt{n}) \mid a < x_n < b\} = \frac{1}{\sqrt{2\pi}} \int_a^b e^{-t^2/2}\, dt$$

6.7. Suppose that the function f is continuous on the closed ball $\overline{B^n}$, $M = \max_{\overline{B^n}} |f|$, and $|f(r, w) - f(w)| \le h(1-r)$ for any $w \in S^{n-1}$ and $r \in [0, 1]$, where h is an increasing continuous function on $[0, 1]$ that vanishes at zero. Prove that the mean values of f on the ball B^n and on the sphere S^{n-1} are close in the following sense: for any $\varepsilon \in (0, 1/2)$

$$\left| \alpha_n^{-1} \int_{B^n} f(x)\, dx - \sigma_n^{-1} \int_{S^{n-1}} f(w)\, d\mu_{n-1}(w) \right| \le 2(M(1-\varepsilon)^n + h(\varepsilon)).$$

6.8. Prove that for any $q \in \mathbb{R}$ the integrals $\int_{\mathbb{R}^n} |x_1 \cdots x_n|^{q/n}\, d\gamma_n(x)$ have a finite positive limit as $n \to \infty$.

6.9. Prove that for any $q \in \mathbb{R}$ and $0 < p < \infty$

$$\alpha_n^{-1} \int_{B^n} \|x\|_p^q\, dx \underset{n\to\infty}{\sim} \sigma_{n-1}^{-1} \int_{S^{n-1}} \|x\|_p^q\, d\mu_{n-1}(x).$$

6.10. Let $\tilde{\gamma}_n$ be the measure on $\mathbb{R}^n$ with density $(n/2\pi)^{n/2} \cdot e^{-n\|x\|^2/2}$.

a) Verify that $\tilde{\gamma}_n(\mathbb{R}^n) = 1$.

b) Find the limit $\lim_{n\to\infty} \tilde{\gamma}_n(B^n)$.

c) Prove that $\int_{|\|x\|-1|>\varepsilon} d\tilde{\gamma}_n(x) < 2\sqrt{n}e^{-n\varepsilon^2/2}$ for $0 < \varepsilon < 1$.

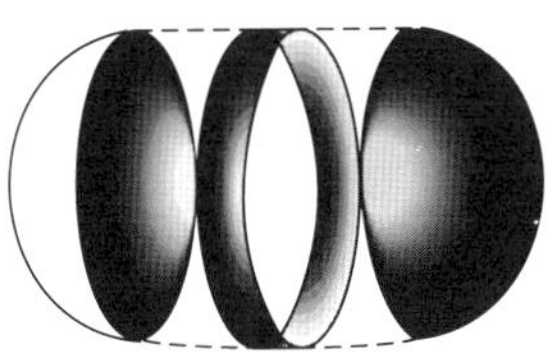

FIGURE 7

6.11. Prove that for any $q \in \mathbb{R}$ and $0 < p \le \infty$

$$\int_{\mathbb{R}^n} \|x\|_p^q \, d\tilde{\gamma}_n(x) \underset{n\to\infty}{\sim} \sigma_{n-1}^{-1} \int_{S^{n-1}} \|x\|_p^q \, d\mu_{n-1}(x)$$

(see problem 6.10 for the definition of the measure $\tilde{\gamma}_n$).

6.12. Prove that $\int_{\mathbb{R}^n} \|x\|^q \, d\gamma_n(x) \underset{n\to\infty}{\sim} n^{q/2}$ $(q \in \mathbb{R})$.

6.13. Prove that for any $q \in \mathbb{R}$

$$\int_{\mathbb{R}^n} \|x\|_\infty^q \, d\gamma_n(x) \underset{n\to\infty}{\sim} (2\ln n)^{q/2}.$$

If $\{a_n\}$ and $\{b_n\}$ are positive numerial sequences (used in problems 6.14 and 6.15 and some other problems), then the expression $a_n \asymp b_n$ means that

$$0 < \inf_n \frac{a_n}{b_n} \le \sup_n \frac{a_n}{b_n} < \infty.$$

We use this notation in an analogous sense not only for sequences but also for arbitrary families of positive numbers.

6.14. Suppose that $q \in \mathbb{R}$, $m, n \in \mathbb{N}$, $2 \le m \le n$, $m + q > 0$, and $x = (x_1, \dots, x_n) \in \mathbb{R}^n$, and let

$$\mathscr{I}_{m,n} = \sigma_{n-1}^{-1} \int_{S^{n-1}} \max_{1\le k\le m} |x_k|^q \, d\mu_{n-1}(x).$$

Prove that:

a) $\mathscr{I}_{m,n} \asymp ((\ln m)/n)^{q/2}$; b) $\mathscr{I}_{m,n} \underset{m\to\infty}{\sim} ((2\ln m)/n)^{q/2}$.

6.15. Prove that for any $q \in \mathbb{R}$ and $0 < p < \infty$

$$\mathscr{I}_n(p, q) = \int_{\mathbb{R}^n} \|x\|_p^q \, d\gamma_n(x) \asymp n^{q/p}.$$

Consider successively the following cases:

a) $q > 0$, $0 < p \le 2$;
b) $0 < q \le p$, $p \ge 2$;
c) $2 \le p \le q$;
d) $q < 0$.

6.16. Prove that for any $q \in \mathbb{R}$ and $0 < p < \infty$

$$\sigma_{n-1}^{-1} \int_{S^{n-1}} \|x\|_p^q \, d\mu_{n-1}(x) \asymp n^{q(1/p-1/2)}.$$

6.17. Prove that the radius r_n of the ball $B^n(r_n)$ with volume 1 is expressed by the formula

$$r_n = \sqrt{\frac{n}{2\pi e}} \left(1 + a\frac{\ln n}{n} + \frac{b}{n} + o\left(\frac{1}{n}\right)\right),$$

and find a and b.

6.18. a) The ball $B^n(r)$ is tangent to all the edges of the n-dimensional cube $[-1, 1]^n$. Compare for large n the volume of the cube and the volume of the ball.

b) The cube $[-1, 1]^n$ is partitioned by the coordinate planes into 2^n cubes. In each of them we inscribe a ball, and then we consider the ball $B^n(\rho)$ tangent from the outside to all the balls inscribed in the cubes. For what n is $B^n(\rho) \subset [-1, 1]^n$? For large n which is larger: the volume of the cube $[-1, 1]^n$ or the volume of the ball $B^n(\rho)$?

6.19. Find the asymptotics of the radius ρ_n of the ball $B^n(\rho_n)$ whose Gaussian volume is equal to $1/2$.

6.20. Let a_n, b_n, and c_n be the mean values of the function $1/\|x\|_1$ on the respective sets O_n, B^n, and $[-1, 1]^n$, where O_n is the n-dimensional octahedron: $O_n = \{(x_1, \dots, x_n) \mid \sum_{1\le k\le n} |x_k| \le 1\}$. Prove that $a_n = n/(n-1)$, $b_n \asymp 1/\sqrt{n}$, and $c_n \asymp 1/n$.

6.21. Let I_n, J_n, and K_n be the mean values of the function $\|x\|^2/\|x\|_1$ on the respective sets O_n, B_n, and $[-1, 1]^n$. Prove that:

$$\text{a) } I_n \sim \frac{2}{n}; \qquad \text{b) } J_n \asymp \frac{1}{\sqrt{n}}; \qquad \text{c) } K_n \asymp 1.$$

6.22. Let $O_n + \varepsilon B^n$ be the ε-neighborhood of the octahedron O_n.

a) Find the limit

$$\lim_{n\to+\infty} \left(\frac{\lambda_n(O_n + \varepsilon B^n)}{\alpha_n} \right)^{1/n}.$$

b) Find the asymptotics of the ratio $\lambda_n(O_n + \varepsilon B^n)/\alpha_n$ as $n \to \infty$.

6.23. Prove that $\lambda_n(A_n) \sim 2(2/\pi)^{n+1}$, where

$$A_n = \left\{ (x_1, \dots, x_n) \in \mathbb{R}^n \,\middle|\, \begin{matrix} x_1 + x_2 \le 1, & x_2 + x_3 \le 1, \dots, x_{n-1} + x_n \le 1 \\ x_1 \ge 0, & x_2 \ge 0, \dots, x_n \ge 0 \end{matrix} \right\}.$$

6.24. Let V_n be a convex set centrally symmetric with respect to zero that is contained in the cube $(-1, 1)^n$. Prove that if $\lambda_n(V_n) > (3/2)^n$, then for sufficiently large n the set $2V_n$ contains a point $a = (a_1, \dots, a_n)$ with integer coordinates such that $\sum_{1\le k\le n} |a_k| > n/10$.

6.25. Prove that:

$$\text{a) } \frac{1}{\sqrt{n}} \int_{[0,1]^n} \|x\|\, dx \underset{n\to\infty}{\to} \frac{1}{\sqrt{3}};$$

$$\text{b) } \int_{[0,1]^n} f\left(\frac{x_1 + \dots + x_n}{n} \right) dx \underset{n\to\infty}{\to} f\left(\frac{1}{2}\right) \qquad (f \in C([0, 1]));$$

$$\text{c) } \int_{[0,1]^n} f\left(\sqrt[n]{x_1 \cdots x_n} \right) dx \underset{n\to\infty}{\to} f(e^{-1}) \qquad (f \in C([0, 1]));$$

$$\text{d) } \int_{\mathbb{R}^n} f\left(\frac{|x_1| + \dots + |x_n|}{n} \right) d\gamma_n(x) \underset{n\to\infty}{\to} f\left(\sqrt{\frac{2}{\pi}} \right)$$

$$(f \in C(\mathbb{R}^n),\ \sup|f| < \infty).$$

6.26. Suppose that $f \ge 0$, $\int_{-\infty}^{+\infty} f(x)\, dx = 1$, and $0 < p < \infty$. Prove that:

$$\text{a)} \int \cdots \int_{\|x\|_p \le c} f(x_1) \cdot f(x_2) \cdots f(x_n)\, dx_1 \cdots dx_n \underset{n\to\infty}{\to} 0$$

for any $c > 0$;

$$\text{b)} \int \cdots \int_{\|x\|_p \le n^\theta} f(x_1) \cdot f(x_2) \cdots f(x_n)\, dx_1 \cdots dx_n \underset{n\to\infty}{\to} 0$$

for any $\theta \in (0, 1/p)$.

c) Can n^θ in b) be replaced by $(nw_n)^{1/p}$, where $w_n > 0$, $w_n \to 0$?

6.27. Prove that for any n and any vector $a \in \mathbb{R}^n$

$$A_p\|a\| \le \left(2^{-n} \int_{Q_n} |(x, a)|^p\, dx\right)^{1/p} \le B_p\|a\|,$$

where $Q_n = [-1, 1]^n$, and A_p and B_p are positive constants dependent only on p (cf. problem 3.31a)).

CHAPTER IX

Sequences of Measurable Functions

In this chapter, as in the preceding one, the symbol λ_m stands for Lebesgue measure on the space $\mathbb{R}^m$ $(\lambda = \lambda_1)$, and the symbol $\mathscr{L}^0(E)$ stands for the set of all measurable functions defined on a set E and almost everywhere finite.

We assume that the reader is familiar with the basic facts in the theory of functions (the theorems of Lebesgue and Riesz about the connection between convergence in measure and convergence almost everywhere, Lebesgue's theorem on taking the limit under the integral sign, the theorem on termwise integration of positive function series, Bessel's inequality, the Riesz-Fischer theorem, the completeness of the trigonometric system of functions). Some of these theorems are constantly used in the solutions (for example, the theorem on termwise integration of positive series), and it seems to us that without others (for example, the theorems of Lebesgue and Riesz) the very formulation of a number of problems cannot be sufficiently understood, even if these theorems are not formally used in the solutions.

§1. Convergence in measure and almost everywhere

1.1. Determine whether the function sequences $\{f_n\}_{n\geq 1}$ below converge in measure and almost everywhere on the intervals Δ, and whether they have integrable majorants:

a) $f_n(x) = nx/(1+n^2x^2)$, $\Delta = (0, 1)$;

b) $f_n(x) = nx/(1+n^2x^2)$, $\Delta = (1, +\infty)$;

c) $f_n(x) = nxe^{-nx^2}$, $\Delta = (0, 1)$;

d) $f_n(x) = n^2xe^{-nx^2}$, $\Delta = (1, +\infty)$;

e) $f_n(x) = n^2x^2/(n^4+x^2)$, $\Delta = (1, +\infty)$;

f) $f_n(x) = \dfrac{1}{\ln(n+1)} \cdot \dfrac{nx}{1+n^2x^4}$, $\qquad \Delta = (0, 1)$;

g) $f_n(x) = |x - 1/n|^{-1/2}$, $\Delta = (0, 1)$.

1.2. For $a \in \mathbb{R}$ let $\{a\} = a - [a]$ be the fractional part of a, and let f_n be the characteristic function of the interval with endpoints $\{\ln n\}$, $\{\ln(n+1)\}$

($n \in \mathbb{N}$) (if $\{\ln(n+1)\} < \{\ln n\}$, then we take f_n to be identically zero). Prove that:

a) $\overline{\lim}_{n\to\infty} f_n(x) = 1$ for any $x \in (0, 1)$;

b) $f_{[c^k]}(x) \underset{k\to\infty}{\longrightarrow} 0$ almost everywhere on $(0, 1)$ for any $c > 1$. Compare this result with the result of problem II.1.25.

1.3. Construct a sequence $\{g_n\} \subset \mathscr{L}^0(0, 1)$ such that $g_n \to 0$ in measure as $n \to \infty$, $\overline{\lim}_{n\to\infty} g_n(x) = +\infty$ a.e. on $(0, 1)$, and $\underline{\lim}_{n\to\infty} g_n(x) = -\infty$ a.e. on $(0, 1)$.

1.4. Suppose that f_n ($n \in \mathbb{N}$) is monotone on $(0, 1)$, and $f_n \to f$ in measure as $n \to \infty$. Prove that f coincides almost everywhere with a monotone function f_0, and $f_n(x) \underset{n\to\infty}{\longrightarrow} f_0(x)$ at the points of continuity of f_0.

1.5. Let $f_n(x) = |\sin(A_n x + \varphi_n)|^{p_n}$, where $x \in (0, 2\pi)$, and $\{A_n\}$, $\{\varphi_n\}$, and $\{p_n\}$ are real numerical sequences such that $\underline{\lim}|A_n| > 0$ and $p_n \to +\infty$. Prove that $f_n \to 0$ in measure on $(0, 2\pi)$ as $n \to \infty$. It is necessary that $f_n(x) \underset{n\to\infty}{\longrightarrow} 0$ almost everywhere on $(0, 2\pi)$? Consider the example $f_n(x) = |\cos(\sqrt{m}(x - \frac{k\pi}{m}))|^m$, where $m = [\sqrt{n}]$ and $k = n - m^2$.

1.6. Let $\{A_n\}$ and $\{\varphi_n\}$ be real numerical sequences with $A_n \to \infty$ as $n \to \infty$. Prove that $\overline{\lim}_{n\to\infty} \sin(A_n x + \varphi_n) = 1$ a.e. on $\mathbb{R}$ and $\underline{\lim}_{n\to\infty} \sin(A_n x + \varphi_n) = -1$ a.e. on $\mathbb{R}$.

1.7. Prove that $n^{1/3} \int_0^\infty e^{-x} \sin^n(nx + \varphi_n)\, dx \underset{n\to\infty}{\longrightarrow} 0$ for any sequence $\{\varphi_n\}_{n\geq 1} \subset \mathbb{R}$.

1.8. Suppose that $\{f_n\}_{n\geq 1} \subset \mathscr{L}^0(\mathbb{R})$, $a > 0$. Prove that if $\sum_{n\geq 1} \lambda\{x \in \mathbb{R} | f_n(x) > a\} < \infty$, then $\overline{\lim}_{n\to\infty} f_n(x) \leq a$ a.e. on $\mathbb{R}$.

1.9. Suppose that $\{f_n\}_{n\geq 1} \subset \mathscr{L}^0(\mathbb{R})$ and $\varepsilon_n \underset{n\to\infty}{\longrightarrow} 0$. Prove that if $\sum_{n\geq 1} \lambda\{x \in \mathbb{R} | |f_n(x)| \geq \varepsilon_n\} < \infty$, then:

a) $f_n(x) \underset{n\to\infty}{\longrightarrow} 0$ a.e. on $\mathbb{R}$;

b) for any number $\varepsilon > 0$ there is a set $e \subset \mathbb{R}$ such that $\lambda(e) < \varepsilon$ and $f_n \rightrightarrows 0$ on $\mathbb{R}\backslash e$ as $n \to \infty$.

1.10. Prove that convergence almost everywhere is "stable" in the sense that if $\{f_n\}_{n\geq 1} \subset \mathscr{L}^0(0, 1)$ and $f_n(x) \underset{n\to\infty}{\longrightarrow} 0$ a.e. on $(0, 1)$, then there is a sequence $\{A_n\}_{n\geq 1} \subset \mathbb{R}$ such that $A_n \underset{n\to\infty}{\longrightarrow} \infty$ and $A_n f_n(x) \underset{n\to\infty}{\longrightarrow} 0$ a.e. on $(0, 1)$. Does this assertion remain true if the interval $(0, 1)$ is replaced by $\mathbb{R}$?

1.11. Suppose that $\{f_n\}_{n\geq 1} \subset \mathscr{L}^0(0, 1)$, and $f_n(x) \underset{n\to\infty}{\longrightarrow} f(x)$ a.e. on $(0, 1)$. Prove that there exist a function $g \in \mathscr{L}^0(0, 1)$ (a "convergence regulator") and a sequence $\{\varepsilon_n\}_{n\geq 1} \subset \mathbb{R}$ such that $\varepsilon_n \to 0$ as $n \to \infty$, and $|f_n(x) - f(x)| \leq \varepsilon_n g(x)$ a.e. on $(0, 1)$ for any $n \in \mathbb{N}$. Does this assertion remain true if the interval $(0, 1)$ is replaced by $\mathbb{R}$?

1.12. Using the result of the preceding problem, prove Egorov's theorem:

if $\{f_n\}_{n\geq 1} \subset \mathscr{L}^0(0, 1)$, and $f_n(x) \underset{n\to\infty}{\longrightarrow} f(x)$ a.e. on $(0, 1)$, then for any $\varepsilon > 0$ there is a set $e \subset (0, 1)$ such that $\lambda(e) < \varepsilon$ and $f_n \rightrightarrows f$ on $(0, 1)\backslash e$ as $n \to \infty$. Is this theorem still valid if the interval $(0, 1)$ is replaced by $\mathbb{R}$?

1.13. Let $\{f_n\}_{n\geq 1} \subset \mathscr{L}^0(0, 1)$. Prove that $f_n \to 0$ in measure as $n \to \infty$ if and only if every subsequence $\{f_{n_k}\}$ has, in turn, a subsequence $\{f_{n_{k_j}}\}$ converging to zero almost everywhere on $(0, 1)$.

1.14. Suppose that $\{f_n\}_{n\geq 1} \subset \mathscr{L}^0(\mathbb{R})$, $g \in \mathscr{L}^0(\mathbb{R})$, and $|f_n(x)| \leq g(x)$ a.e. on $\mathbb{R}$ for any $n \in \mathbb{N}$. Prove that if g satisfies the condition

$$\lambda\{x \in \mathbb{R} | g(x) > \varepsilon\} < \infty \quad \text{for any } \varepsilon > 0, \tag{1}$$

then convergence of the sequence $\{f_n\}_{n\geq 1}$ to zero almost everywhere on $\mathbb{R}$ implies that $f_n \to 0$ in measure as $n \to \infty$. Is this true if the condition (1) fails?

1.15. Let $\{f_{n,k}\}_{n,k\geq 1} \subset \mathscr{L}^0(0, 1)$, $\{f_n\}_{n\geq 1} \subset \mathscr{L}^0(0, 1)$, and $f_0 \in \mathscr{L}^0(0, 1)$, and suppose that $f_{n,k} \underset{k\to\infty}{\longrightarrow} f_n$ in measure, and $f_n \underset{n\to\infty}{\longrightarrow} f_0$ in measure. Prove that there is a strictly increasing sequence $\{k_n\}_{n\geq 1}$ of indices such that $f_{n,k_n} \underset{n\to\infty}{\longrightarrow} f_0$ in measure ("diagonal sequence theorem "). Determine whether the analogous assertion is true if convergence in measure is replaced by:

a) uniform convergence on $(0, 1)$;
b) convergence almost everywhere on $(0, 1)$;
c) pointwise convergence on $(0, 1)$?

§2. Convergence in the mean. The law of large numbers

In this section the letter E denotes a measurable subset of $\mathbb{R}^m$ with finite (Lebesgue) measure. The symbol $\mathscr{L}^r(E)$ $(\mathscr{L}^\infty(E))$, where r is a positive number, denotes the set of functions in $\mathscr{L}^0(E)$ satisfying the condition $\int_E |f(x)|^r\,dx < \infty$ $(\|f\|_\infty \equiv \operatorname{esssup}_{x\in E}|f(x)| < \infty)$. The symbol $\|f\|_r$ denotes $(\int_E |f(x)|^r\,dx)^{1/r}$. We say that $f_n \to f_0$ in the space $\mathscr{L}^r(E)$ (convergence in the mean with exponent r) if $\|f_n - f_0\|_r \underset{n\to\infty}{\longrightarrow} 0$.

2.1. Let $f \in \mathscr{L}^r(E)$. Prove that:
a) if $0 < s < r$, then $f \in \mathscr{L}^s(E)$;
b) if $\lambda_m(E) = 1$, then the function $s \to \|f\|_s$ is increasing on $(0, r]$.

2.2. Suppose that $\{f_n\}_{n\geq 1}$, $\{g_n\}_{n\geq 1} \subset \mathscr{L}^r(E)$, and $f_0, g_0 \in \mathscr{L}^r(E)$. Prove that:
a) if $\|f_n - f_0\|_r \underset{n\to\infty}{\longrightarrow} 0$, then $f_n \to f_0$ in measure as $n \to \infty$;
b) if $\|f_n - f_0\|_r \underset{n\to\infty}{\longrightarrow} 0$ and $\|g_n - g_0\|_r \underset{n\to\infty}{\longrightarrow} 0$, then $\|f_n g_n - f_0 g_0\|_{r/2} \underset{n\to\infty}{\longrightarrow} 0$.

2.3. Suppose that $\{f_n\}_{n\geq 1} \subset \mathscr{L}^1(E)$. Prove that if $f_n \to 0$ in measure as $n \to \infty$ and $\sup_n \|f_n\|_1 < \infty$, then $\int_E \sqrt{|f_n(x)g(x)|}\,dx \underset{n\to\infty}{\longrightarrow} 0$ for any function $g \in \mathscr{L}^1(E)$.

2.4. Suppose that $\{f_n\}_{n\geq 1} \subset \mathscr{L}^2(E)$ and $f_0 \in \mathscr{L}^2(E)$. Prove that if $f_n \to f_0$ in measure as $n \to \infty$ and $\|f_n\|_2 \underset{n\to\infty}{\longrightarrow} \|f_0\|_2$, then $\|f_n - f_0\|_2 \underset{n\to\infty}{\longrightarrow} 0$.

2.5. Suppose that $f, g \in \mathscr{L}^0(E)$ and $0 < \alpha < \lambda_m(E)$. Let

$$\mathscr{J}_\alpha(f) = \inf\{t | \lambda_m\{x \in E \,|\, |f(x)| > t\} < \alpha\}.$$

The quantity $\mathscr{J}_\alpha(f)$ is called the *quantile* (compare the definition of $\mathscr{J}_\alpha(f)$ with the definition of the nonincreasing rearrangement of f before problem VIII.2.10; the quantile $\mathscr{J}_{1/2}(f)$ is the median of the function $|f|$—see problem VIII.2.9). Prove that:

a) $\mathscr{J}_\alpha(f) \leq \mathscr{J}_\beta(f)$ for $0 < \beta < \alpha$;

b) $\mathscr{J}_\alpha(af) = |a|\mathscr{J}_\alpha(f)$ for any $a \in \mathbb{R}$;

c) $\mathscr{J}_\alpha(f+g) \leq \mathscr{J}_{\alpha/2}(f) + \mathscr{J}_{\alpha/2}(g)$;

d) $f_n \underset{n\to\infty}{\longrightarrow} f_0$ in measure ($f_n, f_0 \in \mathscr{L}^0(E)$) if and only if $\mathscr{J}_\alpha(f_n - f_0) \underset{n\to\infty}{\longrightarrow} 0$ for any α with $0 < \alpha < \lambda_m(E)$.

2.6. Let $M \subset \mathscr{L}^2(E)$, and suppose that $\|f\|_2 \leq C\|f\|_1$ for any function $f \in M$. Prove that:

a) if $0 < r < 1$, then there is a number $\theta = \theta(r)$ such that $\|f\|_1 \leq C^\theta \|f\|_r$ for any function $f \in M$;

b) there is a number $\alpha = \alpha(C) > 0$ such that $\|f\|_1 \leq 2\lambda_m(E)\mathscr{J}_\alpha(f)$ for any $f \in M$;

c) if $\{f_n\}_{n\geq 1} \subset M$, then $\|f_n\|_2 \underset{n\to\infty}{\to} 0$ if and only if $f_n \underset{n\to\infty}{\to} 0$ in measure.

2.7. Let $\{f_n\} \subset \mathscr{L}^2(E)$ and suppose that $\|f_n\|_2 \leq C$ and $\|f_n\|_1 \geq \delta$ for some $C, \delta > 0$ and any $n \in \mathbb{N}$. Prove that if the series $\sum_{n\geq 1} a_n f_n(x)$ converges absolutely almost everywhere on E, then $\sum_{n\geq 1} |a_n| < \infty$.

2.8. Prove that if $\sum_{n\geq 1} |a_n \sin(nx + \varphi_n)|$ converges on a set of positive measure, then $\sum_{n\geq 1} |a_n| < \infty$.

2.9. Let $\{f_n\}_{n\geq 1}$ be an orthogonal system in $\mathscr{L}^2(E)$, and let $\sigma_n = \frac{1}{n}\sum_{1\leq k\leq n} f_k$. Prove that if $\frac{1}{n^2}\sum_{1\leq k\leq n} \|f_k\|_2^2 \underset{n\to\infty}{\longrightarrow} 0$ then $\sigma_n \to 0$ in measure. (Diverse variants of this assertion bear the name of the *law of large numbers*.)

2.10. Prove that the following sequences $\{f_n\}_{n\geq 1}$ converge in measure on the interval $(-\pi, \pi)$:

a) $f_n(x) = \frac{1}{n}\sum_{1\leq k\leq n} \sin^2 kx$,

b) $f_n(x) = \frac{1}{n}\sum_{1\leq k\leq n} \sqrt[3]{k} \sin kx$,

c) $f_n(x) = \frac{1}{n}\sum_{1\leq k\leq n} (1 + 1/k)^k \sin^2 kx$,

d) $f_n(x) = \frac{1}{n}\sum_{1\leq k\leq n} (1 - k/n) \sin^2 kx$.

2.11. Find the limit

$$\lim_{n\to\infty} \int_{-\pi}^{\pi} \arctan\left(\frac{1}{n}\sum_{1\leq k\leq n} k \sin^2 kx\right) dx.$$

2.12. Let $\{f_n\}_{n\geq 1}$ be an orthogonal system in $\mathscr{L}^2(E)$ with $|f_n(x)| \leq M$ almost everywhere on E for any $n \in \mathbb{N}$, and let $\sigma_n = \frac{1}{n}\sum_{1\leq k\leq n} f_k$. Prove that:

a) $\sum_{k\geq 1} \|\sigma_{k^2}\|_2^2 < \infty$;

b) $\sigma_n(x) \to 0$ almost everywhere on E.

(The last statement is often formulated by asserting that the *strong law of large numbers* holds for the sequence $\{f_n\}_{n\geq 1}$.)

2.13. Let $\{f_n\}_{n\geq 1}$ be an orthogonal system in $\mathscr{L}^2(E)$, and let $\sigma_n = \frac{1}{n}\sum_{1\leq k\leq n} f_k$. Prove that if $\sum_{n\geq 1} n^{-3/2}\|f_n\|_2^2 < \infty$, then $\sigma_n(x) \to 0$ a.e. on E, that is, the strong law of large numbers holds for the sequence $\{f_n\}_{n\geq 1}$. Prove that, moreover, the function $h = \sup_n |\sigma_n|$ belongs to $\mathscr{L}^2(E)$.

2.14. Let $\varepsilon_n(x)$ be the nth digit in the binary expansion of a number $x \in (0, 1]$, and let $\sigma_n(x) = \frac{1}{n}\sum_{1\leq k\leq n} \varepsilon_k(x)$. Prove that:

a) $\sigma_n \underset{n\to\infty}{\longrightarrow} 1/2$ in measure;

b) $\sigma_n(x) \underset{n\to\infty}{\longrightarrow} 1/2$ a.e. on $(0, 1)$.

The latter means that, for almost all numbers in $(0, 1)$, 0's and 1's occur equally often in their binary expansions.

2.15. Fix one of the digits $0, \dots, 9$, say m, and let $c_m(x) = 1$ if the decimal expansion of the number $x-[x]$ has the form $0.m\dots$, while $c_m(x) = 0$ otherwise. Let $\sigma_n(x) = \frac{1}{n}\sum_{0\leq k<n} c_m(10^k x)$. Using the same method as in the preceding problem, prove that:

a) $\sigma_n \underset{n\to\infty}{\longrightarrow} 1/10$ in measure;

b) $\sigma_n(x) \underset{n\to\infty}{\longrightarrow} 1/10$ a.e. on $(0, 1)$.

A number $x \in (0, 1)$ is said to be *normal* if for any $p \in \mathbb{N}$ its p-ary expansion contains all "digits" (the numbers $0, 1, \dots, p-1$) equally often. It is known, for example (see [12]) that the number

$$x = 0.1234567891011121314151617181920 2122\cdots$$

is normal (the decimal expansion of x consists of all the natural numbers written one after the other (in decimal notation)).

2.16. Generalizing the problems 2.14 and 2.15, prove that almost all the numbers in the interval $(0, 1)$ are normal.

2.17. Suppose that $f \in \mathscr{L}^2(0, 1)$ and $\int_0^1 f(x)\,dx = 0$. For arbitrary $n \in \mathbb{N}$ and $\varepsilon > 0$ let

$$E_n(\varepsilon) = \left\{(x_1, \dots, x_n) \in (0, 1)^n \,\middle|\, \frac{1}{n}\left|\sum_{1\leq x\leq n} f(x_k)\right| > \varepsilon\right\}.$$

Prove that $\lambda_n(E_n(\varepsilon)) \underset{n\to\infty}{\longrightarrow} 0$.

2.18. Suppose that $\{f_n\}_{n\geq 1} \subset \mathscr{L}^2(0, 1)$, and $\int_0^1 f_n(x)\,dx = 0$ for any

$n \in \mathbb{N}$. For arbitrary $n \in \mathbb{N}$ and $\varepsilon > 0$ let

$$E_n(\varepsilon) = \left\{ (x_1, \ldots, x_n) \in (0, 1)^n \,\middle|\, \left|\frac{1}{n} \sum_{1 \le k \le n} f_k(x_k)\right| > \varepsilon \right\}.$$

Prove that $\lambda_n(E_n(\varepsilon)) \underset{n\to\infty}{\longrightarrow} 0$ if $\frac{1}{n^2}\sum_{1\le k\le n} \|f_n\|_2^2 \underset{n\to\infty}{\longrightarrow} 0$.

2.19. Suppose that $\varphi \in C(\mathbb{R})$ and $\sup_{\mathbb{R}} |\varphi| < \infty$. Find the limit

$$\lim_{n\to\infty} \int_{\mathbb{R}^n} \varphi\left(\frac{1}{n}\|x\|^2\right) e^{-\pi\|x\|^2}\, dx.$$

2.20. Suppose that $\{a_n\}_{n\ge1} \subset \mathbb{R}$ and $\{f_n\}_{n\ge1}$ is an orthogonal sequence in $\mathscr{L}^2(E)$, and let $\mathscr{L}^2(E)$, $S_n = \sum_{1\le k\le n} a_k f_k$. Prove that if $\sum_{n\ge1} \sqrt{n} a_n^2 < \infty$, then:

a) $\sum_{k\ge1} \|S - S_{k^2}\|_2^2 < \infty$, where $S = \sum_{k\ge1} a_k f_k$;

b) the series $\sum_{n\ge1} a_n f_n(x)$ converges almost everywhere on E;

c) $\sup_n |S_n| \in \mathscr{L}^2(E)$.

§3. The Rademacher functions. Khintchine's inequality

Let $n \in \mathbb{N}$, $k \in \mathbb{Z}$, and $\Delta_{n,k} = (k/2^n, (k+1)/2^n)$. Define the function r_n on $\mathbb{R}$ by $r_n(x) = (-1)^k$ for $x \in \Delta_{n,k}$, and $r_n(k/2^n) = 0$. The functions r_n are called the *Rademacher functions.*

3.1. a) Prove that for any $n \in \mathbb{N}$ and any collection $\{\varepsilon_k\}_{1\le k\le n}$, where the numbers ε_k take the values $+1$ and -1, there is a point $t \in (0, 1)$ such that $r_k(t) = \varepsilon_k$ for $k = 1, \ldots, n$. What is the measure of the set of such points?

b) Prove that $r_1(t+1) = r_1(t)$, and $r_n(t) = r_1(2^{n-1}t)$ $(t \in \mathbb{R}, n \in \mathbb{N})$.

c) Let $a_n(x)$ be the nth digit in the binary expansion of a number $x \in (0, 1)$. Prove that $a_n(x) = (1 - r_n(x))/2$ almost everywhere on $(0, 1)$.

d) Find the sum of the series $\sum_{n\ge1} 2^{-n} r_n(x)$ $(x \in (0, 1))$.

3.2. Prove that:

a)

$$\int_0^1 \varphi(r_1(t), \ldots, r_n(t))\, dt = 2^{-n} \sum_{\varepsilon_1=0,1;\ldots;\varepsilon_n=0,1} \varphi(\varepsilon_1, \ldots, \varepsilon_n)$$

b)

$$\int_0^1 \varphi_1(r_1(t))\varphi_2(r_2(t))\cdots\varphi_n(r_n(t))\, dt = \prod_{1\le k\le n} \int_0^1 \varphi_k(r_1(t))\, dt.$$

3.3. Let $m_1, \ldots, m_n$ be distinct positive integers. Prove that:

a) $\int_0^1 \varphi(r_{m_1}(t), r_{m_2}(t), \ldots, r_{m_n}(t))\, dt = \int_0^1 \varphi(r_1(t), r_2(t), \ldots, r_n(t))\, dt$;

b) $\int_0^1 r_{m_1}(t) r_{m_2}(t) \cdots r_{m_n}(t)\, dt = 0$,

in particular, $\{r_n\}_{n\ge1}$ is an orthonormal system in $\mathscr{L}^2(0, 1)$.

3.4. Compute the following integrals:

$$\text{a)}\quad \int_0^1 e^{it\sum_{k\ge 1} 2^{-k} r_k(x)}\,dx,$$

$$\text{b)}\quad \int_0^1 e^{it\sum_{k\ge 1} 3^{-k} r_k(x)}\,dx,$$

$$\text{c)}\quad \int_0^1 e^{z\sum_{1\le k\le n} c_k r_k(x)}\,dx \qquad (z, c_k \in \mathbb{C}).$$

d) Prove that if $t, a_n \in \mathbb{R}$, $a_n \to 0$, and $\sum_{n\ge 1} a_n^2 = \infty$, then

$$\int_0^1 e^{it\sum_{1=k\le n} a_k r_k(x)}\,dx \underset{n\to\infty}{\longrightarrow} 0 \quad \text{for } t \ne 0.$$

3.5. Prove that for any numbers $c_1, \dots, c_n$

a) $\int_0^1 |\sum_{1\le k\le n} c_k r_k(x)|^4\,dx \le A(\sum_{1\le k\le n} |c_k|^2)^2$,

b) $(\sum_{1\le k\le n} |c_k|^2)^{1/2} \le B\int_0^1 |\sum_{1\le k\le n} c_k r_k(x)|\,dx$, where A and B are absolute constants.

3.6. Prove that for any numbers c_{kj} $(k, j \in \mathbb{N})$

a)

$$\int_0^1\int_0^1 \left|\sum_{1\le k,j\le n} c_{kj} r_k(t) r_j(s)\right|^4 dt\,ds \le A\left(\sum_{1\le k,j\le n} |c_{k,j}|^2\right)^2,$$

b)

$$\left(\sum_{1\le k,j\le n} |c_{k,j}|^2\right)^{1/2} \le B\int_0^1\int_0^1 \left|\sum_{1\le k,j\le n} c_{kj} r_k(t) r_j(s)\right| dt\,ds,$$

where A and B are absolute constants.

3.7. Prove that:

a) $\{r_j r_k\}_{1\le j<k}$ is an orthonormal system in $\mathscr{L}^2(0, 1)$;

b) for any numbers c_{jk} $(1 \le j < k)$

$$\int_0^1 \left|\sum_{1\le j<k\le n} c_{jk} r_j(x) r_k(x)\right|^4 dx \le A\left(\sum_{1\le j<k\le n} |c_{jk}|^2\right)^2,$$

$$\left(\sum_{1\le j<k\le n} |c_{jk}|^2\right)^{1/2} \le B\int_0^1 \left|\sum_{1\le j<k<n} c_{jk} r_j(x) r_k(x)\right| dx,$$

where A and B are absolute constants.

3.8. Prove that for any numbers $c_1, \dots, c_n$

$$\text{a)} \quad \int_0^{2\pi} \left| \sum_{1\le k\le n} c_k \sin 2^k x \right|^4 dx \le A \left(\sum_{1\le k\le n} |c_k|^2 \right)^2,$$

$$\text{b)} \quad \left(\sum_{1\le k\le n} |c_k|^2 \right)^{1/2} \le B \int_0^{2\pi} \left| \sum_{1\le k\le n} c_k \sin 2^k x \right| dx,$$

where A and B are absolute constants.

3.9. Using the results of problems 3.5–3.8, prove that for any $t > 0$ and any numbers c_k and c_{jk}

$$\text{a)} \quad \lambda \left\{ x \in (0, 1) \,\middle|\, \left| \sum_{1\le k\le n} c_k r_k(x) \right| > t \right\} \le A_1 t^{-4} \left(\sum_{1\le k\le n} |c_k|^2 \right)^2,$$

$$\text{b)} \quad \lambda \left\{ x \in (0, 1) \,\middle|\, \left| \sum_{1\le j<k\le n} c_{jk} r_j(x) r_k(x) \right| > t \right\}$$

$$\le A_2 t^{-4} \left(\sum_{1\le j<k\le n} |c_{jk}|^2 \right)^2,$$

$$\text{c)} \quad \lambda \left\{ x \in (0, 2\pi) \,\middle|\, \left| \sum_{1\le k\le n} c_k \sin 2^k x \right| > t \right\} \le A_3 t^{-4} \left(\sum_{1\le k\le n} |c_k|^2 \right)^2,$$

where A_1, A_2, and A_3 are absolute constants.

3.10. Let $\sigma_n(x) = \frac{1}{n} \sum_{1\le k\le n} r_k(x)$ $(x \in (0, 1))$. Prove that:

a) $\sum_{n\ge1} \lambda\{x \in (0, 1) | \sigma_n(x) \ge n^{-1/5}\} < \infty$;

b) $\sum_{n\ge1} |\sigma_n(x)|^4 < \infty$ a.e. on $(0, 1)$.

Each of the assertions a) and b) implies the equality $\lim_{n\to\infty} \sigma_n(x) = 0$ almost everywhere on $(0, 1)$, which is known under the name of the strong law of large numbers. Using the connection between the Rademacher functions and the digits of the binary expansion (see problem 3.1c)), deduce from this that the binary expansion of almost everywhere number contains "equally many" 0's and 1's (see problems 2.14 and X.2.30).

3.11. Let C be an arbitrary real $n \times n$ matrix, and A the set of all possible vectors in $\mathbb{R}^n$ with coordinates equal to ±1. Prove that

a) $\sum_{x,y\in A} (Cx, y)^4 \le K_1 \cdot 2^{-2n} (\sum_{x,y\in A} (Cx, y)^2)^2$,

b) $(\sum_{x,y\in A} (Cx, y)^2)^{1/2} \le K_2 \cdot 2^{-n} \sum_{x,y\in A} |(Cx, y)|$, where K_1 and K_2 are absolute constants.

3.12. Using the uniqueness theorem for Fourier transforms of finite Borel measures, find the distribution functions of the following functions:

a) $f(x) = \frac{1}{2} + \sum_{k\ge1} 3^{-k} r_k(x)$ $(x \in (0, 1))$;

b) $f(x) = \sum_{k \geq 1} 2^{-k} r_{m_k}(x) \quad (x \in (0, 1))$.
where $\{m_k\}_{k \geq 1}$ is an arbitrary (not necessarily increasing) sequence of distinct positive integers.

3.13. Let Δ be one of the intervals $\Delta_{m,j}$ (see the definition of the Rademacher functions). Prove that for $m \leq n$ $(m, n \in \mathbb{N},\ t > 0,\ a_k \in \mathbb{R})$ the following inequalities hold:

$$\text{a)} \quad \int_\Delta \left| \sum_{1 \leq k \leq m} a_k r_k(x) \right| dx \leq \int_\Delta \left| \sum_{1 \leq k \leq n} a_k r_k(x) \right| dx,$$

$$\text{b)} \quad \int_\Delta e^{t \sum_{1 \leq k \leq m} a_k r_k(x)} dx \leq \int_\Delta e^{t \sum_{1 \leq k \leq n} a_k r_k(x)} dx.$$

Moreover, if $E = \{x \in (0, 1) \mid |\sum_{1 \leq k \leq m} a_k r_k(x)| > t\}$, then:

$$\text{c)} \quad \int_E \left| \sum_{1 \leq k \leq m} a_k r_k(x) \right| dx \leq \int_E \left| \sum_{1 \leq k \leq n} a_k r_k(x) \right| dx;$$

$$\text{d)} \quad \int_E e^{t \sum_{1 \leq k \leq m} a_k r_k(x)} dx \leq \int_E e^{t \sum_{1 \leq k \leq n} a_k r_k(x)} dx.$$

3.14. Using the inequality $\cosh(s) \leq e^{s^2/2}$ $(s \in \mathbb{R})$, prove that the following inequalities hold for $a_1, a_2, \ldots \in \mathbb{R}$ and $c_1, c_2, \ldots \in \mathbb{C}$:

a) $\lambda\{x \in (0, 1) \mid |\sum_{1 \leq k \leq n} a_k r_k(x)| > t\} \leq 2 \cdot e^{-t^2/2A^2}$, where $A^2 = \sum_{1 \leq k \leq n} a_k^2$;

b) $\lambda\{x \in (0, 1) \mid |\sum_{1 \leq k \leq n} c_k r_k(x)| > t\} \leq 4e^{-t^2/4c^2}$, where $c^2 = \sum_{1 \leq k \leq n} |c_k|^2$.

Moreover:

c) if

$$E = \left\{ x \in (0, 1) \;\middle|\; \sup_{1 \leq m \leq n} \left| \sum_{1 \leq k \leq m} a_k r_k(x) \right| > t \right\},$$

then $\lambda(E) \leq A/t$ and $\lambda(E) \leq 2e^{-t^2/2A^2}$, where $A^2 = \sum_{1 \leq k \leq n} a_k^2$;

d) if $\sum_{k \geq 1} a_k^2 < \infty$ and

$$g(x) = \sup_{n \geq 1} \left| \sum_{1 \leq k \leq n} a_k r_k(x) \right|,$$

then $\int_0^1 e^{sg^2(x)} dx < \infty$ for any number $s > 0$.

3.15. Prove that if $\sum_{k \geq 1} |a_k|^2 < \infty$, then the series $\sum_{k \geq 1} a_k r_k(x)$ converges almost everywhere.

3.16. Prove that if the series $\sum_{k \geq 1} a_k r_k$ converges in measure on some set E of positive measure, then $\sum_{k \geq 1} |a_k|^2 < \infty$.

3.17. Prove that the following statements are equivalent:

a) $\sum_{k\geq 1}|a_k|^2<\infty$;

b) the series $\sum_{k\geq 1}a_kr_k$ converges in $\mathscr{L}^p(0,1)$ for some $p\in(0,2)$;

c) the series $\sum_{k\geq 1}a_kr_k$ converge in measure on the set $(0,1)$;

d) the series $\sum_{k\geq 1}a_kr_k$ converges almost everywhere on $(0,1)$.

3.18. Prove that if the series $\sum_{k\geq 1}a_k\sin 2^kx$ converges in measure on some set of positive measure, then $\sum_{k\geq 1}|a_k|^2<\infty$.

3.19. Let $\sum_{k\geq 1}|a_k|^2<\infty$. Prove that:

a) if

$$\operatorname*{esssup}_{x\in\Delta}\left|\sum_{k\geq 1}a_kr_k(x)\right|<\infty$$

for some (nonempty) interval Δ, then $\sum_{k\geq 1}|a_k|<\infty$;

b) if

$$\operatorname*{esssup}_{x\in\Delta}\left|\sum_{k\geq 1}a_k\sin 2^kx\right|<\infty$$

for some (nonempty) interval, then $\sum_{k\geq 1}|a_k|<\infty$.

In problems 3.22 and 3.21 we consider the *Haar system* $\{h_m\}_{m\geq 0}$. This system is defined as follows. Each positive integer m can be represented uniquely in the form $m=2^n+k$, where n and k are nonnegative integers, and $0\leq k<2^n$. By definition, let $h_0(x)=1$ for any $x\in(0,1)$ and

$$h_m(x)=\begin{cases}2^{n/2} & \text{for } \frac{k}{2^n}<x<\frac{k+1/2}{2^n},\\ -2^{n/2} & \text{for } \frac{k+1/2}{2^n}<x<\frac{k+1}{2^n},\\ 0 & \text{for other values of } x\in(0,1)\end{cases}$$

for $m\in\mathbb{N}$.

3.20. Prove that:

a) $\{h_m\}_{m\geq 0}$ is an orthonormal system in $\mathscr{L}^2(0,1)$;

b) if $E=\{x\in(0,1)\,|\,|\sum_{0\leq k\leq m}a_kh_k(x)|>t\}$, then for $n\geq m$

$$\int_E\left|\sum_{0\leq k\leq m}a_kh_k(x)\right|dx\leq\int_E\left|\sum_{0\leq k\leq n}a_kh_k(x)\right|dx;$$

c) if

$$E=\left\{x\in(0,1)\,\middle|\,\max_{0\leq m\leq n}\left|\sum_{0\leq k\leq m}a_kh_k(x)\right|>t\right\},$$

then $\lambda(E)=At^{-1}$, where $A=(\sum_{0\leq k\leq n}|a_k|^2)^{1/2}$;

d) if $\sum_{k\geq 1}|a_k|^2<\infty$, then the series $\sum_{k\geq 0}a_kh_k(x)$ converges almost everywhere on $(0,1)$.

3.21. Using the fact that for any function f integrable on $(0, 1)$ the Fourier series in the Haar system converges to f in $\mathscr{L}^1(0, 1)$, prove that:

a) the series converges almost everywhere on $(0, 1)$;

b) if $E = \{x \in (0, 1) \mid \sup_n |S_n(x)| > t\}$, where S_n is the nth partial sum of the Fourier series for f in the Haar system, then $\lambda(E) \leq \|f\|_1 \cdot t^{-1}$.

3.22. Using the result of problem 3.14, prove that Khintchine's inequality holds for any $p > 0$ and any $a_1, \dots, a_n \in \mathbb{R}$:

$$A_p\Big(\sum_{1\leq k\leq n} a_k^2\Big)^{1/2} \leq \Big\|\sum_{1\leq k\leq n} a_k r_k\Big\|_p \leq B_p\Big(\sum_{1\leq k\leq n} a_k^2\Big)^{1/2},$$

where A_p and B_p are positive constants dependent only on p, and $\overline{\lim}_{p\to+\infty} B_p/\sqrt{p} \leq e^{-1/2}$. The precise values of the quantities A_p and B_p are found in [37].

3.23. Prove that the assertions a)–d) of problem 3.17 are equivalent to the assertion

e) the series $\sum_{k\geq1} a_k r_k$ converges in $\mathscr{L}^p(0, 1)$ for any p in $(0, +\infty)$.

3.24. Prove the following for any trigonometric polynomial $\sum_{|k|\leq n} c_k e^{ikx}$ $(n \geq 2)$:

a) there exist numbers $\varepsilon_k = \pm 1$ $(k = 0, \pm1, \pm2, \dots, \pm n)$, such that

$$\int_0^{2\pi}\Big|\sum_{|k|\leq n} \varepsilon_k c_k e^{ikx}\Big|\,dx \geq c\Big(\sum_{|k|\leq n} |c_k|^2\Big)^{1/2},$$

where $c > 0$ is an absolute constant;

b) there exist numbers $\varepsilon_k = \pm 1$ $(k = 0, \pm1, \pm2, \dots, \pm n)$ such that

$$\sup_{x\in\mathbb{R}}\Big|\sum_{|k|\leq n} \varepsilon_k c_k e^{ikx}\Big| \leq 10\sqrt{\ln n}\Big(\sum_{|k|\leq n} |c_k|^2\Big)^{1/2}.$$

3.25. Prove that for any $a > 2$ there are numbers $p_n > 0$ such that the series

$$\sum_{n\geq2}\left(\frac{|r_1(x) + \cdots + r_n(x)|}{\sqrt{an\ln n}}\right)^{p_n}$$

converges almost everywhere. Deduce from this that

$$\overline{\lim_{n\to\infty}} \frac{|r_1(x) + \cdots + r_n(x)|}{\sqrt{2n\ln n}} \leq 1$$

almost everywhere.

3.26. Let $\{n_j\}_{j\geq1}$ be an arbitrary increasing sequence of positive integers. Prove that

$$\overline{\lim_{j\to+\infty}} \frac{|r_1(x) + r_2(x) + \cdots + r_{n_j}(x)|}{\sqrt{2n_j\ln j}} \leq 1$$

almost everywhere on $(0, 1)$. If $j = O(\ln n_j)$, then we get from this a very weak variant of the *law of the iterated logarithm:*

$$\varlimsup_{j\to+\infty} \frac{|r_1(x) + r_2(x) + \cdots + r_{n_j}(x)|}{\sqrt{2n_j \ln\ln n_j}} \leq 1$$

almost everywhere on $(0, 1)$.

§4. Fourier series and the Fourier transform

4.1. Compute the convolution $f_{c_1} * f_{c_2}$ in the following cases $(c > 0)$:

a) $f_c(x) = \dfrac{1}{c\sqrt{2\pi}} e^{-x^2/2c^2} \qquad (x \in \mathbb{R});$

b) $f_c(x) = (c\sqrt{2\pi})^{-n} e^{-\|x\|^2/2c^2} \quad (x \in \mathbb{R}^n);$

c) $f_c(x) = \dfrac{1}{\pi} \cdot \dfrac{c}{c^2 + x^2} \qquad (x \in \mathbb{R}).$

4.2. Suppose that $0 < \alpha < 1$ and f and g_α are 2π-periodic (measurable) functions with $\operatorname{esssup}_{\mathbb{R}} |f(x)| < \infty$, $g_\alpha(x) = |x|^{-\alpha}$ for $|x| \leq \pi$.

Prove that the function $\varphi(x) = \int_0^{2\pi} f(y) g_\alpha(x - y)\, dy$ $(x \in \mathbb{R})$ is in the class $\mathrm{Lip}_{1-\alpha}$.

In problems 4.3–4.15 the symbol $\hat{f}(n)$ denotes the Fourier coefficient of a function f integrable in $(-\pi, \pi)$ with respect to the orthogonal system $\{e^{inx}\}_{n\in\mathbb{Z}}$:

$$\hat{f}(n) = \frac{1}{2\pi} \int_{-\pi}^{\pi} f(x) e^{-inx}\, dx \qquad (n \in \mathbb{Z}).$$

4.3. a) Let $f, g \in \mathscr{L}^2(-\pi, \pi)$, $h = fg$. Prove that if $\hat{f}(n) = \hat{g}(n) = 0$ for $n < 0$, then $\hat{h}(n) = 0$ for $n < 0$.

b) Let $f \in \mathscr{L}^\infty(-\pi, \pi)$ and let $\Phi = e^f$. Prove that if $\hat{f}(n) = 0$ for $n < 0$, then also $\hat{\Phi}(n) = 0$ for $n < 0$.

4.4. Let $f \in \mathscr{L}^1(-\pi, \pi)$. Prove that $\sum_{n\in\mathbb{Z}} |\hat{f}(n)| < \infty$ if and only if $f(x) = \int_{-\pi}^{\pi} g(y) h(x - y)\, dy$ almost everywhere on $(-\pi, \pi)$, where g and h are 2π-periodic functions in the class $\mathscr{L}^2(-\pi, \pi)$.

4.5. Let $f \in \mathscr{L}^\infty(-\pi, \pi)$. Prove that the numbers $\hat{f}(n)/\sqrt{|n|}$ $(n \in \mathbb{Z}$, $n \neq 0)$ are the Fourier coefficients of a function in the class $\mathrm{Lip}_{1/2}$.

4.6. Prove that if $0 < \alpha \leq 1$ and f is a 2π-periodic function in the class Lip_α, then $\hat{f}(n) = O(1/|n|^\alpha)$.

4.7. Prove that if f is a function of bounded variation on $[-\pi, \pi]$, then $\hat{f}(n) = O(1/n)$.

4.8. Let $f(x) = x \cdot \cos 1/x$ for $0 < |x| \leq \pi$, and $f(0) = 0$. Prove that, although f is not a function of bounded variation on $[-\pi, \pi]$ (see problem III.3.13 a)), $\hat{f}(n) = O(1/n)$.

4.9. Let $0 < \alpha < 1$ and $a_n = O(n^{-(1+\alpha)})$. Prove that the function $f(x) = \sum_{n\geq 1} a_n e^{inx}$ $(x \in \mathbb{R})$ is in the class Lip_α. Is this true for $\alpha = 1$?

4.10. Let $f(x) = \sum_{n\geq 1} \frac{1}{n^2} \sin(2^{n!}|x|)$ $(x \in \mathbb{R})$. Prove that the Fourier series of f diverges at the point $x = 0$.

Suppose that $f \in \mathscr{L}^1(-\pi, \pi)$, and $S_n(x) = \sum_{|k|\leq n} \hat{f}(k)e^{ikx}$ $(|x| \in \pi,$ $n = 0, 1, 2, \dots)$ is the corresponding partial sum of the Fourier series of f. Let

$$\sigma_n(x) = \frac{S_0(x) + \cdots + S_{n-1}(x)}{n}.$$

The sums σ_n are called the *Cesàro-Fejér sums.* It is easy to see that

$$\sigma_n(x) = \frac{1}{2\pi n} \int_{-\pi}^{\pi} f(y) \frac{\sin^2(n(x-y)/2)}{\sin^2((x-y)/2)}\, dy.$$

4.11. Prove that the sequence of *Cesàro-Fejér sums* corresponding to a 2π-periodic continuous function converges to the function uniformly on $\mathbb{R}$.

4.12. Suppose that f is a 2π-periodic function in Lip_α, where $0 < \alpha < 1$. Prove that:

a) $|\sigma_n(x) - f(x)| \leq C/n^\alpha$ for any $x \in \mathbb{R}$, where C is a constant depending on α and f, but not on x;

b) $\|f - S_n\|_2 = O(n^{-\alpha})$, where S_n is the nth partial sum of the Fourier series of f.

4.13. Let f be a 2π-periodic function in Lip_α. Prove that if $\alpha > 1/2$, then $\sum_{n\in\mathbb{Z}} |\hat{f}(n)| < \infty$.

For $\alpha = 1/2$ the assertion becomes false (see problem IV.6.34).

4.14. Let $f \in \mathscr{L}^1(-\pi, \pi)$. Prove that if $\|f - \sigma_n\|_1 = o(1/n)$, then $f \equiv \mathrm{const}$ almost everywhere.

4.15. Let $f \in \mathscr{L}^2(-\pi, \pi)$. Consider the function $\tilde{f}$ defined by

$$\tilde{f}(x) = -i \sum_{n\in\mathbb{Z}} \hat{f}(n)\operatorname{sgn}(n) e^{inx} \qquad (x \in (-\pi, \pi)).$$

Prove that $\tilde{f}$ is real if f is real.

The function $\tilde{f}$ is called the *conjugate* of f. The reader can become acquainted with one of the numerous applications of this important concept in the following two problems, borrowed from [41].

4.16. Let $1 = n_0 < \cdots < n_m$ be positive integers, let $a_0, \dots, a_m$ be arbitrary complex numbers, and let $f_k(x) = \sum_{2^{k-1}\leq j < 2^k} a_j e^{in_j x}$ (for $j > m$ the numbers a_j are taken to be zero). Further, let $g_k = |f_k|$, $h_k = g_k - i\tilde{g}_k$, and $F_\varepsilon = \sum_{k\geq 1} f_k e^{-\varepsilon \sum_{j\geq k} h_j}$, where ε is an arbitrary positive number, and $\tilde{g}_k$ is the function conjugate to g_k (see the definition in problem 4.15). Prove that:

a) $|F_\varepsilon(x)| \le \varepsilon^{-1}$ for any $x \in (-\pi, \pi)$;
b) if $2^{p-1} \le q < 2^p$, then

$$|\widehat{F}_\varepsilon(n_q) - a_q| \le \frac{\varepsilon}{2} \sum_{k \ge p} \|f_k\|_2 \sum_{j \ge k} \|f_j\|_2.$$

4.17. Let $1 \le n_1 < n_2 < \cdots < n_m$ $(n_j \in \mathbb{N})$. Prove that for any complex numbers $c_1, \ldots, c_m$

$$\int_{-\pi}^{\pi} \left| \sum_{1 \le j \le m} c_j e^{ik_j x} \right| dx \ge \gamma \sum_{1 \le j \le m} \frac{|c_j|}{j},$$

where γ is a positive absolute constant $(\gamma \ge 10^{-2})$.

The *Fourier transform* of a function $f \in \mathscr{L}^1(\mathbb{R}^n)$ is defined to be the function $\hat{f}$ given by

$$\hat{f}(s) = \int_{\mathbb{R}^n} f(t) e^{-i(s,t)}\, dt \qquad (s \in \mathbb{R}^n),$$

where (s, t) is the inner product of vectors $s, t \in \mathbb{R}^n$.

4.18. Compute the Fourier transform of the functions in problem 4.1.

4.19. Find the Fourier transform of the Hermite functions $h_n(x) = e^{x^2/2}(e^{-x^2})^{(n)}$ $(x \in \mathbb{R},\ n = 0, 1, 2, \ldots)$.

The *Fourier transform* of a finite *measure* μ defined at least on the Borel subsets of $\mathbb{R}^n$ is defined to be the function $\hat{\mu}$ with

$$\hat{\mu}(s) = \int_{\mathbb{R}^n} e^{-i(s,t)}\, d\mu(t) \qquad (s \in \mathbb{R}^n),$$

where (s, t) is the inner product of the vectors $s, t \in \mathbb{R}^n$.

The Fourier transform of the measure on $\mathbb{R}^1$ generated by a nondecreasing bounded function g is denoted by the symbol $\hat{g}$.

4.20. Let F be the distribution function of a function $f \in \mathscr{L}^0(E)$, where $E \subset \mathbb{R}^m$ and $\lambda_m(E) < \infty$. Prove that

$$\widehat{F}(s) = \int_E e^{-isf(x)}\, dx \qquad (s \in \mathbb{R}).$$

4.21. Let $\hat{\sigma}(s) = \int_0^1 e^{-ist}\, d\sigma(t)$ $(s \in \mathbb{R})$, where σ is the Cantor function (see problem III.3.17). Prove that $\hat{\sigma}(s) \not\to 0$ as $s \to \infty$.

4.22. Let $\sigma_1(x) = \int_0^1 \sigma(x - y)\, d\sigma(y)$ $(x \in \mathbb{R})$, where σ is the Cantor function ($\sigma(x) = 0$ for $x < 0$, and $\sigma(x) = 1$ for $x > 1$). Prove that:
a) σ_1 is strictly increasing on $[0, 2]$;
b) $\hat{\sigma}_1(s) \not\to 0$ as $s \to \infty$.

The solutions of problems 4.23–4.27 are based on the following uniqueness theorems:

If $f \in \mathscr{L}^1(\mathbb{R}^n)$ and $\hat{f} \equiv 0$, then $f(x) = 0$ a.e. on $\mathbb{R}^n$.

If μ is a finite Borel measure on $\mathbb{R}^n$ and $\hat{\mu} \equiv 0$, then $\mu = 0$ (see, for example, [20], p. 302).

4.23. Prove that if the convolution of an integrable function f on $\mathbb{R}^n$ with the function $e^{-\|x\|^2}$ is identically equal to zero, then $f(x) = 0$ a.e.

4.24. Let $g(t) = \sup\{\sum_{k\geq 1} \varepsilon_k 2^{-k} \mid \sum_{k\geq 1} \varepsilon_k 2^{-(2k-1)} \leq t,\ \varepsilon_k = 0 \text{ or } 1\}$ for $t > 0$, and let $g(t) = 0$ for $t < 0$. Prove that:

a) $g \in C(\mathbb{R})$;

b) $g'(t) = 0$ almost everywhere on $\mathbb{R}$;

c) $\hat{g}(s) = e^{-is/3} \prod_{n\geq 1} \cos(s4^{-n})$;

d)

$$\int_{-\infty}^{+\infty} g(x-t)\, dg(2t) = \begin{cases} 0 & \text{for } x < 0, \\ x & \text{for } 0 \leq x \leq 1, \\ 1 & \text{for } x > 1. \end{cases}$$

4.25. Let $\nu = \{n_k\}_{k\geq 1}$ be an arbitrary strictly increasing sequence of positive integers, and let $\nu_1 = \{n_{2k-1}\}_{k\geq 1}$ and $\mu_2 = \{n_{2k}\}_{k\geq 1}$. Further, let

$$g(t) = \sup\left\{ \sum_{k\geq 1} \varepsilon_k 2^{-k} \,\middle|\, \sum_{k\geq 1} \varepsilon_k 2^{-n_k} \leq t,\ \varepsilon_k = 0 \text{ or } 1 \right\}$$

for $t > 0$ and $g(t) = 0$ for $t < 0$, and let the functions g_1 and g_2 be constructed analogously with the help of the sequences ν_1 and ν_2, respectively. Prove that:

a) $g \in C(\mathbb{R})$;

b) $g_1'(t) = 0$ almost everywhere on $\mathbb{R}$, and $g_2'(t) = 0$ almost everywhere on $\mathbb{R}$.

c) Verify that the function g is "divisible" in the sense that $g(x) = \int_{-\infty}^{+\infty} g_1(x-t)\, dg_2(t)$ for any $x \in \mathbb{R}$.

4.26. Prove that if two finite Borel measures on $\mathbb{R}^n$ take the same values on all possible half-spaces (that is, on sets of the form $\{x \in \mathbb{R}^n \mid (x, a) \geq c\}$, where $a \in \mathbb{R}^n$ and $c \in \mathbb{R}$, then they coincide.

4.27. Find the general form of a Borel probability measure on $\mathbb{R}^n$ that is invariant with respect to orthogonal transformations and is at the same time a product of two measures concentrated on nontrivial orthogonal subspaces.

CHAPTER X

Iterates of Transformations of an Interval

§1. Topological dynamics

1.1. Suppose that the function $f\colon \mathbb{N} \to \mathbb{N}$ satisfies the inequality $f(n+1) > f(f(n))$ for any $n \in \mathbb{N}$. Prove that $f(n) = n$ for all $n \in \mathbb{N}$.

1.2. Determine whether there exists a continuous function $f\colon \mathbb{R} \to \mathbb{R}$ satisfying the equation $f \circ f = F$ in the following cases:

$$\text{a) } F(x) = -x; \quad \text{b) } F(x) = e^x; \quad \text{c) } F(x) = x^2 - 2.$$

Denote by Id_X (or simply Id) the identity mapping of X into itself (i.e., $\mathrm{Id}(x) = x$ for any x in X). A mapping $h\colon X \to X$ is called an *involution* if $h \circ h = \mathrm{Id}_X$.

1.3. Let $f(x) = x+1$ and $g(x) = x-1$ $(x \in \mathbb{R})$. Choose two involutions h and k such that $h \circ k = f$ and $k \circ h = g$.

1.4. Prove that each bijection is a composition of two involutions.

A *fixed point* of a mapping $f\colon X \to X$ is defined to be a point $x \in X$ such that $x = f(x)$.

1.5. Prove that a continuous mapping $f\colon \Delta \to \mathbb{R}$ of an interval Δ has a fixed point in each of the following cases:

a) $\Delta = [a, b]$, $f(\Delta) \subset \Delta$;

b) $\Delta = [a, b]$, $f(\Delta) \supset \Delta$;

c) $\Delta = \langle a, b\rangle$, f is an involution.

1.6. Suppose that a function $f \in C([a, b])$ is monotonically increasing, and $f([a, b]) \subset [a, b]$. Prove that for each $x_0 \in [a, b]$ the sequence $x_n = f(x_{n-1})$ $(n \in \mathbb{N})$ converges to a fixed point. Is this true if f is a decreasing function?

1.7. Let $f \in C(\mathbb{R})$. For $x_0 \in \mathbb{R}$ let $x_1 = f(x_0)$ and $x_k = f(x_{k-1})$ $(k \in \mathbb{N})$. Prove that if the sequence $\{\frac{1}{n}\sum_{0\le k<n} f(x_k)\}$ is bounded for some $x_0 \in R$, then f has a fixed point. Is this true for a continuous mapping $f\colon \mathbb{C} \to \mathbb{C}$?

1.8. Prove that a finite group of affine transformations of the plane has a fixed point (common for all transformations in the group).

Mappings $f\colon \Delta_1 \to \Delta_1$ and $g\colon \Delta_2 \to \Delta_2$ (Δ_1, Δ_2 are open intervals on the line) are said to be *topologically conjugate* if there exists a continuous bijection (homeomorphism) $h\colon \Delta_2 \to \Delta_1$ such that $f = h \circ g \circ h^{-1}$ (that is, $h \circ g = f \circ h$). In this case we use the notation $f \sim g$, and the mapping h will be called a *conjugating mapping*.

In problems 1.9–1.12 it is assumed that the mappings are continuous, are defined on open intervals, and map them into themselves.

1.9. Prove that the relation of topological conjugacy is an equivalence relation, i.e.: a) $f \sim f$; b) $f \sim g$ if and only if $g \sim f$; c) if $f_1 \sim f_2$ and $f_2 \sim f_3$, then $f_1 \sim f_3$.

1.10. Prove that topologically conjugate mappings have or do not have the following properties simultaneously:

a) being one-to-one;

b) being monotone of a definite type;

c) being bounded above or below (without specification of which);

d) boundedness.

1.11. Prove that topologically conjugate mappings have:

a) the same number of intervals of monotonicity;

b) the same number of fixed points.

1.12. Prove that if $f \sim g$, then intervals on which $f(x) > x$ are carried by a conjugating mapping h into intervals on which the same or the opposite strict inequality holds for g (depending on the character of monotonicity of h). Prove that if f and g are defined on $\mathbb{R}$ and one of the mappings f and g is odd, then the bijection h can be assumed to be increasing.

1.13. Determine whether the mappings f and g on $\mathbb{R}$ are topological conjugates in the following cases:

a) $f(x) = x^2$, $g(x) = x^{2n}$ $(n = 2, 3, \dots)$;

b) $f(x) = \cos x$, $g(x) = \sin x$;

c) $f(x) = \cos x$, $g(x) = -\cos x$;

d) $f(x) = \sin x$, $g(x) = -\sin x$;

e) $f(x) = x + \sin x$, $g(x) = x + \cos x$.

1.14. a) Let $f(x) = x^2$ and $g(x) = x^2 + ax + b$ $(x \in \mathbb{R})$. Describe the set of pairs (a, b) such that $f \sim g$.

b) For which $a \in \mathbb{R}$ is there a $b \in \mathbb{R}$ such that the mappings $f(x) = 1 - ax^2$ and $g(x) = bx(1-x)$ $(x \in \mathbb{R})$ are topological conjugates? Describe the set of all numbers b arising in this way.

c) Prove that the mappings g and $\tilde{g}\colon [0, 1] \to [0, 1]$, $g(x) = 2\tilde{g}(x) = 3x(1-x)$, are not topological conjugates.

d) For which $a \in \mathbb{R}$ and which intervals $\Delta = [p, q] \subset \mathbb{R}$ is the mapping $f\colon \Delta \to \Delta$, $f(x) = 1 - ax^2$, topologically conjugate to a mapping $g\colon [0, 1] \to [0, 1]$ of the form $g(x) = bx(1-x)$?

1.15. Prove the topological conjugacy of the continuous mappings f, g:

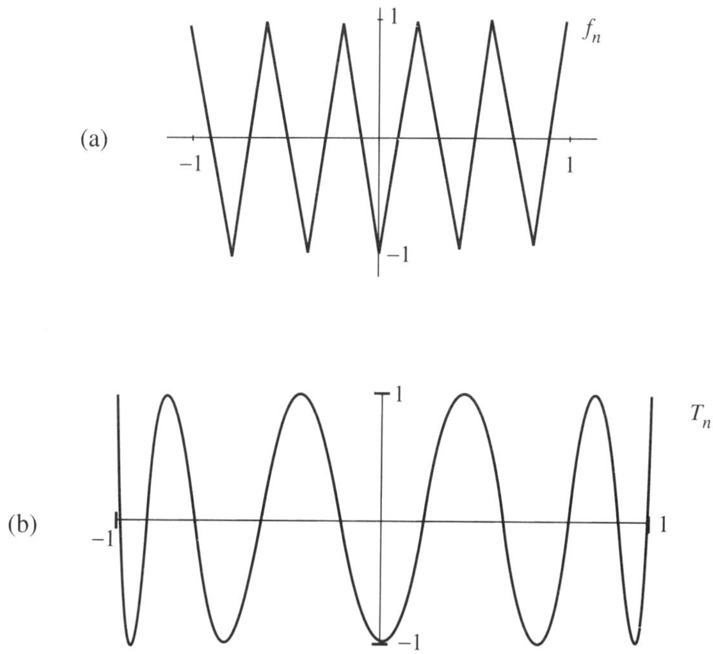

FIGURE 8

$[-1, 1] \to [-1, 1]$ given as follows:

a) $f(x) = 1 - 2|x|$, $g(x) = 1 - 2x^2$ $(x \in [-1, 1])$;

b) $f = f_n$ $(n \in \mathbb{N})$ is a "saw-tooth" mapping that takes the values $(-1)^{k+n}$ at the points $-1 + 2k/n$, $0 \le k \le n$, and is linear in the intervals between them, and $g = T_n$ is the corresponding *Chebyshev polynomial* determined by the relation $T_n(\cos t) = \cos nt$ (see Figure 8);

c) Prove that the functions in a) are conjugate as mappings from $\mathbb{R}$ to $\mathbb{R}$.

1.16. Suppose that $f: [a_1, b_1] \to [a_1, b_1]$ and $g: [a_2, b_2] \to [a_2, b_2]$ are homeomorphisms of intervals that do not have interior points as fixed points. Prove that $f \sim g$. Verify that the assertion remains in force if the closed intervals are replaced by open intervals.

1.17. Classify, up to topological conjugacy, the homeomorphisms of a closed interval having a finite set of fixed points that includes the endpoints of the interval.

Suppose that f is a mapping of an interval $\Delta \subset \mathbb{R}$ into itself, and let $f^0(x) = x$, $f^1(x) = f(x)$, and $f^n(x) = f(f^{n-1}(x))$ for $n > 1$. We say that the *method of successive approximations with initial point* $x_0 \in \Delta$ *converges* if the limit $\lim x_n$ exists, where $x_n = f^n(x_0)$.

1.18. Suppose that the function $f \in C([a, b])$ is increasing. Prove that the method of successive approximations converges for any initial point $x_0 \in [a, b]$ if:

a) $f(x) < x$ for $x > a$, $f(a) = a$;

b) $f(x) > x$ for $x < b$, $f(b) = b$.

1.19. Prove convergence of the method of successive approximations for any initial point $x_0 \in \Delta$ and find $\lim x_n$ for the following mappings f:

a) $f(x) = \arctan x$, $\Delta = \mathbb{R}$;

b) $f(x) = \sqrt{x+2}$, $\Delta = [-2, +\infty)$;

c) $f(x) = (x-2)/2$, $\Delta = \mathbb{R}$;

d) $f(x) = x(x^2+3)/(3x^2+1)$, $\Delta = \mathbb{R}$;

e) $f(x) = (x + 1/x)/2$, $\Delta = (0, +\infty)$;

f) $f(x) = \frac{2}{\pi} \arcsin(2x/(x^2+1))$, $\Delta = (0, +\infty)$;

g) $f(x) = 4/(2+x)$, $\Delta = [0, +\infty)$.

1.20. In the following cases determine for which initial points x_0 in the domain of a mapping f the method of successive approximations converges, and find $\lim x_n$:

a) $f(x) = x + \sin x$;

b) $f(x) = x(x-1)$;

c) $f(x) = 1 + ax^2$ $(0 < a \leq 1/4)$;

d) $f(x) = 1 - ax^2$ $(0 < a \leq 3/4)$;

e) $f(x) = x^2 + c$ $(|c| \leq 1/4)$;

f) $f(x) = x^2 + (1-2c)x + c^2$ $(c \in \mathbb{R})$;

g)

$$f(x) = \begin{cases} 1/(4-3x) & \text{for } x \neq 4/3, \\ 1/3 & \text{for } x = 4/3; \end{cases}$$

h)

$$f(x) = \begin{cases} 2x^3/(3x^2-1) & \text{if } |x| \neq 1/\sqrt{3}, \\ \pm 1 & \text{if } x = \pm 1/\sqrt{3}. \end{cases}$$

A fixed point x^* of a mapping f is said to be *attracting* if for x_0 in some neighborhood of x^* the sequence $x_n = f^n(x_0)$ converges to x^*. If for some neighborhood U of the fixed point x^* and an arbitrary point $x_0 \in U$, $x_0 \neq x^*$, there is an index n_0 such that $x_{n_0} \notin U$, then x^* is called a *repelling* fixed point.

1.21. Let x^* be a fixed point of a mapping f, and suppose that f is differentiable at x^*. Prove that:

a) if $|f'(x^*)| < 1$, then x^* is attracting;

b) if $|f'(x^*)| < 1$, then x^* is repelling;

c) show by examples that a fixed point can be neither attracting nor repelling.

1.22. For what values of the parameter a and for what $x_0 \in \mathbb{R}$ does the sequence of iterates $x_n = f^n(x_0)$ converge if:

a) $f(x) = 1 + ax$;

b) $f(x) = 1 - a|x|$ $(a > 0)$;
c) $f(x) = 1 - ax^2$ $(a > 0)$;
d) $f(x) = a^x$ $(a > 0)$.

Let φ be a differentiable function on an interval Δ, and let N_φ be the mapping defined at the points $x \in \Delta$ with $\varphi'(x) \neq 0$ by the formula

$$N_\varphi(x) = x - \varphi(x)/\varphi'(x).$$

It is easy to see that the fixed points of N_φ coincide with the zeros of the function φ. We say that the *Newton iteration process with initial point* x_0 *converges to* c if $x_k = N_\varphi^k(x_0) \to c$.

1.23. Suppose that $\varphi \in C^2(\Delta)$, $c \in \Delta$, $\varphi(c) = 0$, and $|\varphi'(x)| > m > 0$ for $x \in \Delta$. Prove that if x_0 is sufficiently close to c, then:
a) the Newton iteration process starting at x_0 converges to c;
b) "superconvergence" holds, namely, $|x_k - c| \leq Mq^{2^k}$, where $M > 0$, $0 < q < 1$, and $k \in \mathbb{N}$.

1.24. Prove that if $\varphi \in C^{(n)}(\Delta)$, $c \in \Delta$, $\varphi(c) = \varphi'(c) = \cdots = \varphi^{(n-1)}(c) = 0$, and $\varphi^{(n)}(c) \neq 0$, then for initial points x_0 sufficiently close to c the Newton iteration process converges to c, and $|x_k - c| \leq Mq^k$ $(M > 0$, $0 < q < 1)$.

1.25. Let $\varphi \in C^2([a, b])$ be a function such that $\varphi\varphi'' > 0$ in (a, b) and $\varphi(a) = \varphi''(b) = 0$ (or $\varphi(b) = \varphi''(a) = 0$). Verify that for any initial point in $[a, b]$ the Newton iteration process converges to a zero of φ.

1.26. Let P be a polynomial having only simple real roots. Prove that if x_0 is an inflection point of P (that is, $P''(x_0) = 0$), then the Newton iteration process starting at x_0 converges to a root of P, and all roots except for the minimal and maximal roots can be found in this way.

The set $\{x_n = f^n(x_0) | n \in \mathbb{N}\}$ is called the *trajectory of a point* x_0. If the trajectory is finite, then there is obviously an $n \in \mathbb{N}$ such that the points $x_0, \ldots, x_{n-1}$ are distinct, and $x_n = x_0$, $x_{n+1} = x_1, \ldots$. In this case x_0 is called a *periodic point*, the number n is called its *period*, and the tuple $(x_0, \ldots, x_{n-1})$ is called an *n-cycle*. We say that an n-cycle is *attracting* if x_0 is an attracting fixed point for f^n, and *repelling* if x_0 is a repelling fixed point of f^n.

1.27. Let $(x_0^*, x_1^*, \ldots, x_{n-1}^*)$ be an n-cycle of a mapping f. Prove that:
a) if $|(f^n)'(x_0^*)| < 1$, then for x_0 in some neighborhood of x_0^*

$$f^{mn+k}(x_0) \xrightarrow[m \to \infty]{} x_k^*, \qquad k = 0, \ldots, n-1;$$

b) if $|(f^n)'(x_0^*)| > 1$, then for some $\varepsilon > 0$ and any $x_0 \in (x_0^* - \varepsilon, x_0^* + \varepsilon)$, $x_0 \neq x_0^*$, there is an m such that $|f^m(x_0) - x_k^*| \geq \varepsilon$ for all $k = 0, \ldots, n-1$.

1.28. Let f be a continuous mapping of a closed interval Δ into itself, and let $I_0, I_1, \dots$ be a finite or infinite sequence of closed intervals such that $I_0 \subset \Delta$, and $I_{n+1} \subset f(I_n)$ for any $n \geq 0$. Prove that:

a) there exists a sequence of closed intervals $K_0 \supset K_1 \supset \dots$, $K_0 \subset I_0$, such that $f^n(K_n) = I_n$ $(n \geq 0)$;

b) there exists a point $x \in \Delta$ such that $x_n = f^n(x) \in I_n$ for any n.

1.29. Prove that if a continuous mapping of a closed interval into itself has a 4-cycle, then it also has a 2-cycle.

1.30. Let $f: \Delta \to \Delta$ be a continuous mapping of a closed interval into itself. Prove that the following statements are equivalent:

a) there exists a point $a \in \Delta$ such that the points $b = f(a)$, $c = f^2(a)$, and $d = f^3(a)$ satisfy the inequalities $d \leq a < b < c$ or $c < b < a \leq d$;

b) Δ contains a point having period three;

c) for any $m \in \mathbb{N}$, Δ contains a point having period m.

The implication b) $\to$ c) and (problem 1.29) is a special case of the following theorem of Sharkovskiĭ [30]. Let $f: \Delta \to \Delta$ be a continuous mapping $(\Delta = [a, b])$. We consider the series of positive integers, ordered as follows:

$$1, 2, 4, 8, \dots, 2^n, \dots, \dots, 2^m \cdot 7, 2^m \cdot 5, 2^m \cdot 3, \dots,$$
$$2 \cdot 7, 2 \cdot 5, 2 \cdot 3, \dots, 7, 5, 3.$$

If p is to the left of q in this series and f has a q-cycle, then f also has a p-cycle.

1.31. Prove that for $n \geq 2$ the Chebyshev polynomial T_n, defined on $[-1, 1]$ (see problem 1.15), has cycles of arbitrary order.

1.32. Find the values of the parameter a such that the mapping $f(x) = 1 - a|x|$ $(x \in \mathbb{R})$ has: a) a 2-cycle; b) a 4-cycle; c) a 3-cycle. Determine whether these cycles are attracting.

1.33. Let $f(x) = 1 - ax^2$ $(x \in \mathbb{R})$. Find the infimum of the values of the parameter a such that f has: a) an attracting 2-cycle; b) a repelling 2-cycle; c) a 4-cycle; d) a 3-cycle.

1.34. Let $f(x) = bx(1-x)$ $(x \in [0, 1])$, $0 < b \leq 4$. Find the infimum of the values of the parameter b such that f has: a) an attracting 2-cycle; b) a repelling 2-cycle; c) a 4-cycle; d) a 3-cycle.

1.35. Let $x_0 \in (0, 1)$ be a fixed point of the continuous mapping f: $[0, 1] \to [0, 1]$. Prove that if the one-sided derivatives $f'_\pm(x_0)$ at x_0 are negative and $f'_+(x_0) f'_-(x_0) > 1$, then f has a 2-cycle. In particular, a 2-cycle exists if $f'(x_0) < -1$.

1.36. Let g be a piecewise smooth mapping of $[0, 1]$ into itself. Prove that if $|g'(x)| \geq a > 1$ at all points where the derivative exists, then g has a 2^n-cycle for any $n \in \mathbb{N}$. Show by an example that this is false if differentiability of the mapping fails at countably many point.

In problems 1.37–1.39 we consider mappings of $\Delta = [a, b]$ into itself that belong to the class $C^3(\Delta)$ and are such that the Schwarzian derivative Sf (see problems VII.2.24 and VII.2.25 and the definition preceding them) is negative at the points x that are not critical points of f (that is, where $f'(x) \neq 0$). The class of such mappings is denoted by $S(\Delta)$.

1.37. Let $f \in S(\Delta)$, and suppose that the set of critical points of f is finite. Prove that for each $n \in \mathbb{N}$ the transformation f can have only finitely many n-cycles.

1.38. Let $f \in S(\Delta)$, let $g = f^n$ $(n \geq 1)$, and let $x < y < z$ be successive fixed points of g. Prove that if $[x, y]$ does not contain critical points of g, then $g'(y) > 1$.

1.39. Suppose that $f \in S(\Delta)$ has a unique critical point $c \in (a, b)$. Prove that each attracting cycle of f attracts the trajectory of one of the points a, c, or b, and hence f can have at most three attracting cycles.

1.40. Prove that for $a \leq 2$ the mapping $f(x) = 1 - ax^2$ $(x \in \mathbb{R})$ cannot have more than one attracting cycle, and for $a = 2$ it has an n-cycle for any $n \in \mathbb{N}$, but none of them are attracting.

1.41. Let φ be a continuous mapping of the circle into itself having a fixed point and a 3-cycle.

a) Prove that φ also has n-cycles for all $n > 3$.

b) Does φ necessarily have a 2-cycle?

1.42. Let f be a continuous increasing function on the line that satisfies the condition

$$f(x+1) = f(x) + 1.$$

Prove that:

a) the mapping $z = e^{2\pi i x} \mapsto \varphi(z) = e^{2\pi i f(x)}$ gives a homeomorphism of the circle $S^1 = \{z \mid |z| = 1\}$ that preserves the order of succession of any three points, and an arbitrary orientation-preserving homeomorphism of the circle can be given in this way;

b) if the limit

$$\lim \frac{f^n(x)}{n} = \alpha$$

exists for some $x_0 \in \mathbb{R}$, then it exists also for any $x \in \mathbb{R}$;

c) if the mapping φ^m has a fixed point for some $m \in \mathbb{N}$, then the limit α exists and is a rational number;

d) if no power of φ has a fixed point, then the limit α exists and is an irrational number;

e) if two distinct functions f_1 and f_2 determine a single homeomorphism φ, then the corresponding limits α_1 and α_2 differ by an integer.

The number $\alpha \bmod 1$ determined in this way for an arbitrary orientation-preserving homeomorphism $\varphi: S^1 \to S^1$ is called the *rotation number* of this homeomorphism. This concept was introduced by Poincaré.

§2. Transformations with an invariant measure

Let X be a measurable subset of $\mathbb{R}^n$, and μ a σ-finite measure on the (Lebesgue-) measurable subsets of X. A nonnegative measurable function f is said to be a *density* of μ if $\mu(A) = \int_A f(x)\,d\lambda_n(x)$ for an arbitrary measurable set $A \subset X$ (in this case the notation used is $d\mu = f\,d\lambda_n = f\,dx$). Two densities give rise to the same measure if and only if they coincide almost everywhere on X. By the well-known Radon-Nikodým theorem, μ has a density if and only if it is *absolutely continuous* (with respect to Lebesgue measure), that is, if and only if $\mu(A) = 0$ when $\lambda_n(A) = 0$ (see [14]). A measure μ is said to be *equivalent* to Lebesgue measure if $\mu(A) = 0$ if and only if $\lambda_n(A) = 0$. It will be said that *a mapping* $T\colon X \to X$ *preserves the measure* μ, or that the measure μ is *invariant* with respect to T, if $\mu(A) = \mu(T^{-1}(A))$ for any measurable set $A \subset X$. If ν is a measure on the measurable subsets of a set $Y \subset \mathbb{R}^m$, then ν is said to be the *image* of μ under a mapping $T\colon X \to Y$ if $\nu(B) = \mu(T^{-1}(B))$ for any measurable set $B \subset Y$ (in this case we write $\nu = T\mu$).

2.1. Verify that the following transformations preserve Lebesgue measure:

a) $T(x) = (2x) \bmod 1$ $(x \in [0, 1])$;
b) $T(z) = z^n$ $(z \in \mathbb{C},\ |z| = 1)$, $n \in \mathbb{N}$;
c) $T(x) = x - 1/x$ $(x \in \mathbb{R}\backslash\{0\})$.

2.2. Verify that the transformation

$$T(x, y) = ((2x) \bmod 1,\ \tfrac{1}{2}(y + [2x])), \qquad (x, y) \in [0, 1) \times [0, 1)$$

(the "baker's transformation") preserves the Lebesgue measure λ_2.

2.3. Suppose that $X = [0, 1) \times [0, 1)$, $A = \left(\begin{smallmatrix} a_{11} & a_{12} \\ a_{21} & a_{22} \end{smallmatrix}\right)$ is an integer matrix, and $\det A = \pm 1$. Define a transformation $T_A\colon X \to X$ by the formulas

$$y_1 = (a_{11}x_1 + a_{12}x_2) \bmod 1, \qquad y_2 = (a_{21}x_1 + a_{22}x_2) \bmod 1,$$

where $(x_1, x_2) \in X$. Prove that:

a) T_A is one-to-one;
b) T_A preserves Lebesgue measure.

The mapping $(x_1, x_2) \mapsto (e^{2\pi i x_1}, e^{2\pi i x_2})$ allows us to identify X with the torus $T^2 = S^1 \times S^1$ and T_A with an automorphism of T^2 as a group. This lets us call the transformation T_A an automorphism of the torus.

We recall a fundamental result: on every compact group G there is a nontrivial finite measure μ that is translation invariant: $\mu(gA) = \mu(Ag) = \mu(A)$, where $A \subset G$, $g \in G$, $gA = \{gx | x \in A\}$, and $Ag = \{xg | x \in A\}$. This measure is uniquely determined (up to a constant factor) and is called *Haar measure*. Its uniqueness can be employed in the solution of the following problem.

2.4. Let the mapping $T_A\colon X \to X$ be the same as in the preceding problem, but let $\det A \in \mathbb{Z}$, $|\det A| > 1$. Such a mapping (now not one-to-one) is called an *endomorphism of the torus*. Prove that it preserves Lebesgue measure.

2.5. Verify that the mapping $T(x) = \frac{1}{x} \bmod 1$ $(x \in (0, 1])$ preserves the measure $d\mu = (C/(1+x))\,dx$.

2.6. Find an absolutely continuous invariant measure:

a) for the mapping $T_n\colon [-1, 1] \to [-1, 1]$, where T_n is the corresponding Chebyshev polynomial (see problem 1.15b));

b) for the mapping $f\colon \mathbb{R} \to \mathbb{R}$ given by $f(x) = 1 - 2x^2$.

2.7. Prove that a mapping T has an invariant (respectively, finite invariant, absolutely continuous invariant) measure if and only if some power T^n $(n \in \mathbb{N})$ of it has this property.

2.8. Let $T\colon [a, b] \to [a, b]$ be a strictly monotone continuous transformation. Prove that it has an absolutely continuous invariant measure.

2.9. Suppose that the mapping $T\colon [a, b] \to [a, b]$ is piecewise monotone and smooth. Prove that if it has an attracting cycle (see §1), then T does not preserve any finite measure equivalent to Lebesgue measure.

2.10. Determine whether there is a nontrivial measure on $\mathbb{R}$ that is invariant with respect to all:

a) homotheties $H_a\colon x \mapsto ax$ $(a \neq 0)$;

b) affine transformations $A_{a,b}\colon x \mapsto ax + b$ $(a, b \in \mathbb{R},\ a \neq 0)$.

If the answer is positive, give an example.

2.11. Determine whether there is a nontrivial measure on $(0, +\infty)$ that is invariant with respect to all transformations of the form:

a) $T_p\colon x \to x^p$ $(p \neq 0)$;

b) $S_{b,p}\colon x \to bx^p$ $(p \neq 0,\ b > 0)$.

If the answer is positive, give an example.

A measure μ on $\mathbb{R}^n$ is said to be *quasi-invariant with respect to translations if* for each $a \in \mathbb{R}^n$ the condition $\mu(A) = 0$ is equivalent to the condition $\mu(A + a) = 0$.

2.12. Prove that:

a) if μ is equivalent to Lebesgue measure λ_n on $\mathbb{R}^n$, then μ is quasi-invariant with respect to translations;

b) if μ is quasi-invariant with respect to translations, and the measure of any bounded set is finite, then μ is equivalent to λ_n.

2.13. Let T be a transformation of a set $X \subset \mathbb{R}^n$ into itself such that the image $T\lambda_n$ of λ_n is absolutely continuous with respect to λ_n, and let f be an arbitrary integrable function on X.

a) Prove that there exists a function $f^* \in \mathscr{L}_1(X)$ such that

$$\int_{T^{-1}(A)} f(x)\,dx = \int_A f^*(x)\,dx \qquad (A \subset X) \tag{1}$$

and f^* is uniquely determined up to values on a set of measure zero.

The correspondence $P\colon f \mapsto f^*$ is called the *Perron-Frobenius operator* associated with T.

b) If $X \subset \mathbb{R}^n$ is a domain, T is a smooth one-to-one mapping, and $\det T'(x) \neq 0$ $(x \in X)$, then

$$Pf(Tx) = \frac{f(x)}{|\det T'(x)|} \qquad (x \in X).$$

c) If $X = [0, 1]$, T is piecewise differentiable, and the set $T^{-1}(x)$ is finite $(x \in X)$, then

$$Pf(y) = \sum_{x \in T^{-1}(y)} \frac{f(x)}{|T'(x)|}$$

at all points for which the right-hand side is defined.

d) If $X = [0, 1]$, $f \in C([0, 1])$, and T is piecewise monotone, then

$$Pf(x) = \frac{d}{dx} \int_{T^{-1}([0,x])} f(t)\,dt.$$

e) The measure $d\mu = f\,dx$ is invariant with respect to T if and only if $Pf(x) = f(x)$ for almost all $x \in X$.

Let $n \in \mathbb{N}$, and

$$\Delta_i = \left[\frac{i-1}{n}, \frac{i}{n}\right] \qquad (i = 1, \dots, n).$$

Consider the nonnegative functions on $[0, 1)$ that are constant on the intervals Δ_i. Each such function is determined by the vector $c = (c_1, \dots, c_n)$ of values it takes on $\Delta_1, \dots, \Delta_n$. Let us denote it by f_c. The set of nonnegative vectors in $\mathbb{R}^n$ is denoted by $\mathbb{R}^n_+$.

2.14. Let $T\colon [0, 1) \to [0, 1)$ be an arbitrary transformation.

a) Prove that for each function of the form f_c $(c \in \mathbb{R}^n)$ there is a unique function f_d $(d \in \mathbb{R}^n_+)$ satisfying the equalities

$$\int_{T^{-1}(\Delta_i)} f_c(x)\,dx = \int_{\Delta_i} f_d(x)\,dx, \qquad i = 1, \dots, n,$$

and the vector d has the form $d = \pi c$, where $\pi = (p_{ij})_{i,j=1}^n$ is some matrix determined by T.

The mapping $c \mapsto \pi c$ is a discrete (finite-dimensional) variant of a Perron-Frobenius operator.

b) Find the matrix π and prove that $\sum_{1 \le i \le n} p_{ij} = 1$ $(j = 1, \dots, n)$.

c) Prove that there exists a vector $v \in \mathbb{R}_+^n$, $v \neq 0$, such that $\pi v = v$ (with this goal consider the sequence of vectors $v^{(m)} = (1/m)\sum_{0\leq k<m} \pi^k u$, $u \in \mathbb{R}_+^n$).

2.15. a) Let $T\colon [0, 1] \to [0, 1]$ be the mapping defined by the formula

$$T(x) = \begin{cases} 2x & \text{for } x \leq \frac{1}{2}, \\ \frac{3}{2} - x & \text{for } x > \frac{1}{2}. \end{cases}$$

Find the general form of an absolutely continuous measure invariant with respect to T.

b) Let $T\colon [0, 1] \to [0, 1]$ be the mapping defined by

$$T(x) = \begin{cases} x + \frac{1}{3} & \text{for } x \leq \frac{2}{3}, \\ 2 - 2x & \text{for } x > \frac{2}{3}. \end{cases}$$

Find an absolutely continuous measure that is invariant with respect to T.

c) Let $T\colon [0, 1] \to [0, 1]$ be the mapping given by $T(x) = x/2$. Find a fixed point v_n for the discrete Perron-Frobenius operator (for even n) and prove that there is no invariant finite absolutely continuous measure for T. What is the limit of measures $d\mu_n = \int_{v_n} dx$ as $n \to \infty$ if μ_n is normalized by the condition $\mu_n([0, 1)) = 1$?

2.16. Let $\Omega = \{z \in \mathbb{C} \mid \operatorname{Im} z > 0\}$. Find an absolutely continuous measures that is invariant with respect to all the conformal automorphisms of Ω, that is, the mappings $T(z) = (az + b)/(cz + d)$, where $a, b, c, d \in \mathbb{R}$ and $\det\left(\begin{smallmatrix} a & b \\ c & d \end{smallmatrix}\right) > 0$.

2.17. Let $P = \{(x, y) \in \mathbb{R}^2 | x > 0\}$ and suppose that for each point $(a, b) \in P$ the transformations $L_{a,b}$ and $R_{a,b}$ of the half-plane P into itself are defined by

$$L_{a,b}(x, y) = (ax, ay + b), \qquad R_{a,b}(x, y) = (ax, bx + y).$$

Find absolutely continuous measures μ_L and μ_R on P that are invariant with respect to all the transformations $L_{a,b}$ and $R_{a,b}$, respectively.

We remark that the transformations under discussion correspond to left and right shift transformations on the group (with respect to multiplication) of matrices of the form

$$M_{x,y} = \begin{pmatrix} x & y \\ 0 & 1 \end{pmatrix} \quad ((x, y) \in P),$$

that is, to the transformations

$$M_{x,y} \mapsto M_{a,b}M_{x,y} \quad \text{and} \quad M_{x,y} \mapsto M_{x,y}M_{a,b}.$$

The measures μ_L and μ_R are called *Haar measures* (*left* and *right*) for this group.

2.18. Prove that if a transformation $T\colon [a, b] \to [a, b]$ preserves some finite absolutely continuous measure, then T is topologically conjugate to some transformation of the same interval that preserves Lebesgue measure.

The existence of an invariant measure (not necessarily absolutely continuous) is often established with the help of the following *theorem of Bogolyubov and Krylov*:

For every continuous mapping of a compact set into itself there exists a finite invariant measure.

2.19. Prove that if T is a homeomorphism of the circle S^1 with an irrational rotation number α (see problem 1.42), then T is conjugate to the rotation T_α of the circle through the angle α $(T(e^{2\pi ix}) = e^{2\pi i(x+\alpha)})$ in the sense that $h \circ T = T_\alpha \circ h$ for some continuous mapping $h: S^1 \to S^1$.

In what follows, it is assumed that the measure μ is finite and satisfies the normalization condition $\mu(X) = 1$ (such measures are called *probability measures*).

Let T be a transformation of the set X. A set $A \subset X$ is said to be *invariant* if $T^{-1}(A) = A$. A transformation T preserving a measure μ is said to be *ergodic* if for each *invariant set* $A \subset X$ either $\mu(A) = 0$ or $\mu(X \backslash A) = 0$.

2.20. A measurable function f defined on a set X with measure μ is said to be *invariant* with respect to a transformation T if $f(Tx) = f(x)$ for almost all x. Prove that T is ergodic if and only if every invariant measurable function coincides with a constant function almost everywhere.

2.21. Determine whether the following transformation are ergodic:

a) $T(x) = (x + \alpha) \bmod 1$, $X = [0, 1)$, $\mu = \lambda$;

b) $T(x) = (2x) \bmod 1$, $X = [0, 1)$, $\mu = \lambda$;

c) $T(z) = z^n (n \in \mathbb{N})$, $z \in X = S^1$, μ is Lebesgue measure on the circle;

d) T is an automorphism of the torus (see problem 2.3);

e) T is an endomorphism of the torus (see problem 2.4).

2.22. Suppose that X is a subset of $\mathbb{R}^n$ that is the closure of the set of its interior points, and T is a transformation of X preserving Lebesgue measure. Prove that if T is ergodic, then for almost all $x \in X$ the trajectory of x is dense.

2.23. Let $X = [0, 1)^n$ be the n-dimensional torus $(n \in N)$, and let S be the shift $x \mapsto y = S(x)$ defined by the formula $y_k = (x_k + \alpha_k) \bmod 1$, $k = 1, \dots, n$. For what values of $\alpha_1, \dots, \alpha_n$ is the shift ergodic?

2.24. Prove that the mappings of $[0, 1]$ into itself given by the formulas

$$T(x) = (2x) \bmod 1, \qquad S(x) = 1 - |2x - 1|$$

are *metrically conjugate*. (Transformations $T: X \to X$ and $S: Y \to Y$ preserving the *respective measures* μ and ν are said to be *metrically conjugate* if there exists a measurable mapping $h: X \to Y$ such that $h\mu = \nu$ and $S \circ h(x) = h \circ T(x)$ for almost all $x \in X$.)

2.25. Prove that a mapping T of $[-1, 1]$ into itself preserving a measure μ is ergodic if:

a) $T(x) = 1 - 2|x|$, $d\mu = dx/2$;

b) $T(x) = 1 - 2x^2$, $d\mu = dx/\pi\sqrt{1-x^2}$;

c) $T = T_n$ is a Chebyshev polynomial (see problem 1.15b)), $d\mu = dx/\pi\sqrt{1-x^2}$.

2.26. Prove that if $T\colon X \to X$ is a transformation preserving a probability measure μ, and f is a positive function, then $\sum f(T^k(x)) = +\infty$ for almost all $x \in X$ (with respect to μ).

A fundamental theorem in the theory of transformations with an invariant measure (or "ergodic theory") is the *Birkhoff-Khintchine ergodic theorem* (see [3], [15]):

Let T be a transformation of a set X preserving a probability measure μ, and let f be a measurable function with $\int_X |f(x)|\, d\mu(x) < +\infty$. Then for almost every $x \in X$ (in the sense of μ) the limit

$$\lim \frac{1}{n} \sum_{0 \le k < n} f(T^k(x)) = \overline{f}(x) \qquad (*)$$

exists. Furthermore, the function $\overline{f}$ is invariant, and

$$\int_X \overline{f}(x)\, d\mu(x) = \int_X f(x)\, d\mu(x).$$

If T is ergodic, then $\overline{f}$ turns out to be constant (see problem 2.20) and equal to $\int_X f(x)\, d\mu(x)$ (that is, to the mean value of f on X). The limit on the left-hand side of $(*)$ is called the *mean value of f along the trajectory* of the point x.

2.27. Let $T\colon X \to X$ be an ergodic transformation preserving the probability measure μ, and let A be a measurable subset of X. Let $\varepsilon_n = 0$ if $T^n(x) \notin A$, and $\varepsilon_n = 1$ if $T^n(x) \in A$ $(n \ge 0)$. Prove that the limit $\lim \frac{1}{n} \sum_{0 \le k < n} \varepsilon_k$ exists (the mean number of times the trajectory hits the set A) for almost all $x \in X$, and find it.

Probability measures μ_1 and μ_2 on X are said to be *mutually singular* if there exists a set $A \subset X$ such that $\mu_1(A) = 1$ and $\mu_2(A) = 0$.

2.28. Let μ_1 and μ_2 be two probability measures on X, and let T be a transformation that is ergodic with respect to both μ_1 and μ_2. Prove that the measures μ_1 and μ_2 either coincide or are mutually singular.

2.29. Consider an irrational shift

$$S(x) = (x + \alpha) \bmod 1 (x \in [0, 1) = X, \ \alpha \in \mathbb{R}\backslash\mathbb{Q}).$$

It is ergodic, according to 2.21a). By the ergodic theorem, the equality

$$\lim \frac{1}{n} \sum_{0\le k<n} f(S^k(x)) = \int_0^1 f(x)\,dx \tag{1}$$

holds for any integrable function f on X and for almost all $x \in X$.

a) Prove that if $f \in C([0, 1])$, then for the shift S it is possible to assert more, namely, that (1) is valid for any $x \in [0, 1)$ (assume first that $f(0) = f(1)$, and then consider the general case).

b) Prove that if $\Delta = \langle a, b\rangle \subset [0, 1)$, $\chi_\Delta(x) = 1$ for $x \in \Delta$, and $\chi_\Delta(x) = 0$ for $x \notin \Delta$, then

$$\lim \frac{1}{n} \sum_{0\le k<n} \chi_\Delta(S^k(x)) = b - a \tag{2}$$

for all $x \in [0, 1)$.

If $\{S^k(x)\}$ is replaced by an arbitrary sequence $\{x_k\}$, then the property in b) is referred to by saying that the sequence is *uniformly distributed*, and the equality (1) is called Weyl's criterion for uniform distributivity.

2.30. Let $x = \sum x_k/2^k$, $x_k \in \{0, 1\}$, be the binary expansion of a number $x \in [0, 1)$. Prove that the limit $\lim \frac{1}{n} \sum_{0\le k<n} x_k$ (the frequency of 1's) exists for almost all $x \in [0, 1)$, and find it. Solve the analogous problem for p-ary expansions.

2.31. Prove that if a transformation $T: X \to X$ preserving a probability measure μ has the property

$$\lim \mu(A \cap T^{-n}(B)) = \mu(A)\mu(B) \qquad (A, B \subset X) \tag{$*$}$$

(the *mixing property with respect to* μ), then T is ergodic.

2.32. Prove that the transformation $T(x) = (2x) \bmod 1$ $(x \in [0, 1))$ is mixing (with respect to Lebesgue measure).

2.33. Suppose that $p \in (0, 1)$, $q = 1 - p$, and μ_p is the measure on $[0, 1]$ given on closed intervals with dyadic rational endpoints as follows: if $\varepsilon_1, \dots, \varepsilon_n \in \{0, 1\}$ and

$$\Delta_{\varepsilon_1\cdots\varepsilon_n} = \{x \in [0, 1] | x_k = \varepsilon_k \ (k = 1, \dots, n)\}$$

(here $x = \sum x_k/2^k$ is the binary expansion of a number x), then

$$\mu_p(\Delta_{\varepsilon_1\cdots\varepsilon_n}) = \prod_{1\le k\le n} p^{1-\varepsilon_k} q^{\varepsilon_k},$$

and on the remaining closed intervals the measure is extended by additivity. Verify that:

a) the transformation $T(x) = (2x) \bmod 1$ is mixing with respect to the measure μ_p;

b) $\mu_{1/2} = \lambda$;

c) the measures μ_p and λ are mutually singular for $p \neq 1/2$;

d) μ_{p_1} and μ_{p_2} are mutually singular when $p_1 \neq p_2$.

2.34. Let a_k be the first digit in the decimal expansion of the number 2^k $(k = 0, 1, \ldots)$, that is, $a_0 = 1$, $a_1 = 2$, $a_2 = 4$, $a_3 = 8$, $a_4 = 1, \ldots$. What is the frequency of the digit $p \in \{1, \ldots, 9\}$ in this sequence?

As known, every number x in $(0, 1]$ can be represented uniquely as a *continued fraction*

$$x = \cfrac{1}{a_1 + \cfrac{1}{a_2 + \cdots}},$$

which is also written in the form $x = [a_1, a_2, \ldots]$, where $a_1 = a_1(x)$, $a_2 = a_2(x), \ldots$ are positive numbers. Such a continued fraction has finitely many terms if and only if x is rational. A finite continued fraction $[a_1, \ldots, a_n]$ can be written uniquely in the form of an irreducible fraction p_n/q_n. An infinite continued fraction is a limit of finite continued fractions:

$$[a_1, a_2, \ldots] = \lim[a_1, \ldots, a_n] = \lim(p_n/q_n),$$

and the number $p_n/q_n = p_n(x)/q_n(x)$ is called a *convergent*. It is not hard to verify that

$$\left| x - \frac{p_n(x)}{q_n(x)} \right| \le \frac{1}{q_n(x)q_{n+1}(x)}.$$

The *Gauss transformation*

$$T(x) = \frac{1}{x} \bmod 1 \ (x \in (0, 1])$$

is connected with expansion of numbers in continued fractions. Indeed, let $x = [a_1, a_2, \ldots]$ and $y = [a_2, a_3, \ldots]$. Then $x = 1/(a_1 + y)$, which implies that $y = T(x)$. With the help of T the numbers a_n can be written as follows:

$$a_1(x) = [1/x], \ a_2(x) = a_1(T(x)), \ldots, a_n(x) = a_1(T^{n-1}(x)).$$

The Gauss transformation preserves the Gauss measure $d\mu = A\,dx/(1+x)$ (problem 2.5). The measure μ is a probability measure when $A = 1/(\ln 2)$.

2.35. a) For the Gauss transformation T prove that for some $C > 0$ the inequalities

$$\frac{1}{C}\mu(A)\mu(B) \le \mu(T^{-n}(A) \cap B) \le C\mu(A)\mu(B)$$

are valid for $A, B \subset (0, 1]$ and $n \in \mathbb{N}$. Deduce from this the ergodicity of T.

b) Prove the following for almost all $x = [a_1, a_2, \ldots]$:

1) the relative frequency of a number $k \in \mathbb{N}$ among the numbers a_1, $a_2, \ldots$ is equal to

$$\frac{1}{\ln 2} \ln \frac{(k+1)^2}{k(k+2)};$$

2)
$$\lim \frac{a_1 + \cdots + a_n}{n} = +\infty\,;$$

3)
$$\lim \sqrt[n]{a_1 \cdots a_n} = \prod_{k \in \mathbb{N}} \left(1 + \frac{1}{k^2 + 2k}\right)^{\ln k / \ln 2}\,;$$

4)
$$\lim \sqrt[n]{q_n} = \exp\left(\frac{\pi^2}{12 \ln 2}\right),$$

where q_n is the denominator of the nth convergent.

Solutions

CHAPTER I

Introduction

§1. Sets

1.1. b) Consider the mapping $\varepsilon \mapsto (\varepsilon', \varepsilon'')$, where $\varepsilon = \{\varepsilon_k\}$, $\varepsilon' = \{\varepsilon_{2k-1}\}$, and $\varepsilon'' = \{\varepsilon_{2k}\}$, and show that it is bijective.

1.2. Use the binary expansions of numbers in $[0, 1)$.

1.3. Use the results of problems 1.2 and 1.1b).

1.4. Replace the set $\mathbb{R}$ by the set Ξ of binary sequences, and prove that it has the same cardinality as the set $\Xi^{\mathbb{N}} = \Xi \times \Xi \times \Xi \times \cdots$. For this consider the mapping $\varepsilon = \{\varepsilon_n\} \to \{\eta^{(1)}, \eta^{(2)}, \dots\} \in \Xi^{\mathbb{N}}$, where $\eta^{(k)} = \{\eta_1^{(k)}, \eta_2^{(k)}, \dots\}$, $\eta_j^{(k)} = \varepsilon_{2^{k-1}(2j-1)}$ $(k, j \in \mathbb{N})$, and show that it is bijective (compare with the solution of problem 1.1b)).

1.5. Use the result of problem 1.4, with the fact that a continuous function on $[a, b]$ is determined by its values at the rational points.

1.6. Replace the set $\mathbb{N}$ be the set $\mathbb{Q}$ and consider a system of convergent sequences of rational numbers with distinct limits.

1.7. Use the result of problem 1.6, replacing the set $\mathbb{N}$ by the set $\{2, 2^2, \dots\}$.

1.8. Replace the set $\mathbb{N}$ by the set $\mathbb{Q}$ and consider the system of sets of the form $(-\infty, \alpha) \cap \mathbb{Q}$, where $\alpha \in \mathbb{R}$.

1.9. b) Consider the set $\mathscr{F}_n = \{A \in \mathscr{F}(\mathbb{N}) | \max A = n\}$ $(n \in \mathbb{N})$, and prove that $\operatorname{card}(\mathscr{F}_n) = 2^{n-1}$. Setting $\varphi(\varnothing) = 0$, define φ by induction with the help of the equality $\varphi(A) = 2^{n+1} + \varphi(A \setminus \{n\})$ $(A \in \mathscr{F}_n,\ n \in \mathbb{N})$ and show that the function $\varphi + 1$ has all the required properties.

1.11. Using the Axiom of Choice, form a set E_1 by choosing one point from each equivalence class described in problem 1.10b). Consider all possible rotations of E_1 through angles $2\pi\theta$, where $\theta \in \mathbb{Q}$ and $0 < \theta < 1$.

1.12. Associate with each figure eight a pair of rational coordinates, taking one point inside each loop.

1.13. Consider first the bird tracks with lengths of segments bounded off zero. Use the same idea as in problem 1.12, taking points with rational coordinates in all three angles corresponding to a bird track and sufficiently close to its vertex.

1.14. Associate with each T-shaped figure a half-disk of radius $1/4$

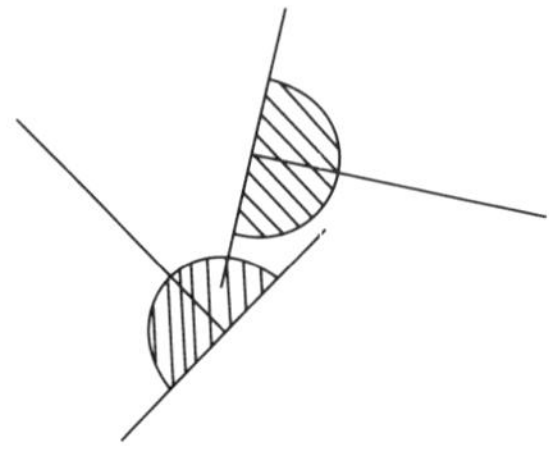

FIGURE 9

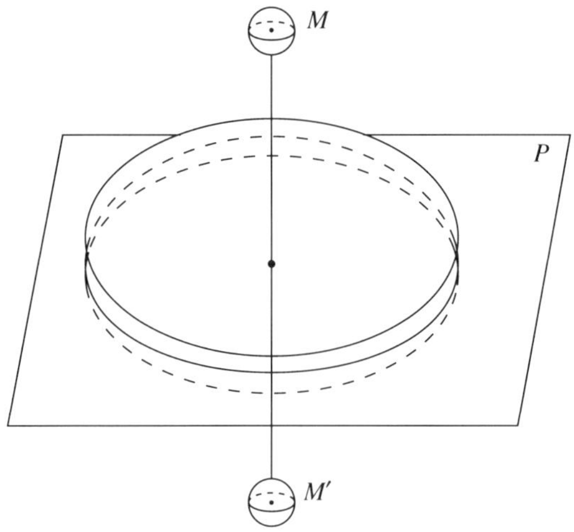

FIGURE 10

constructed as shown in Figure 9. Show that these half-disks are disjoint. Adding their areas, we get that $\frac{1}{2}\pi(\frac{1}{4})^2 N \le a^2$, that is, $N \le \frac{32}{\pi}a^2 < 11a^2$.

1.15. Let P be the plane of symmetry of a hoop O perpendicular to its axis. On this axis fix points M and M' symmetric with respect to P and separated by a distance equal to the diameter of the hoop. Consider the balls about M and M' with radius δ (see Figure 10). We say that these balls accompany the hoop O. Consider hoops of width at least $\varepsilon > 0$. Prove that for any $\varepsilon > 0$ there is a number $\delta > 0$ such that if both balls accompanying a hoop O intersect the balls accompanying a hoop O', then O and O' themselves intersect.

1.16. See the solution of problem 1.17a).

1.17. a) Let $\Delta_m = \{n \in \mathbb{N} | A_n \ni m\}$. Arguing by contradiction, consider an infinite set $M = \{m_1, m_2, \dots\} \subset \mathbb{N}$ such that $\operatorname{card}(\Delta_{m_j}) < +\infty$ for any $j \in \mathbb{N}$. Show that in some infinite subset of M there cannot be two points in the same set A_n.

b) Let $\Delta_k = \{n | k \in A_n\}$. By a), there is an index m_1 such that the set $E_1 = \bigcup_{n \in \Delta_{m_1}} A_n$ is infinite. Replacing $\mathbb{N}$ by E_1 and A_n by $A_n' = A_n \cap E_1$

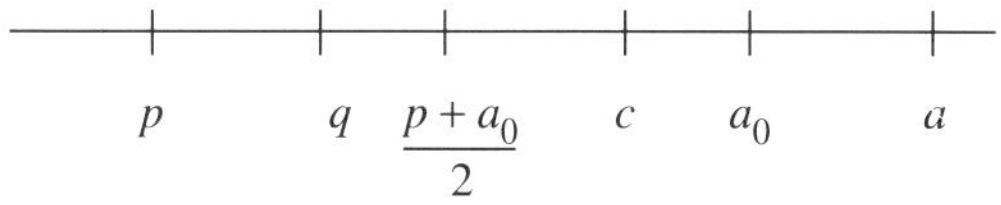

FIGURE 11

and again using a), we see that there is an index $m_2 > m_1$ such that the set $E_2 = \bigcup_{n\in\Delta_{m_2}} A_n'$ is infinite. Using induction, we construct a set $E = \{m_1, m_2, \dots\}$ which is the desired object.

1.18. Consider the set E in b) of problem 1.17 and show that if $\operatorname{card}(E \cap B_q) \geq 2$, then there is an index p such that $\operatorname{card}(A_p \cap B_q) \geq 2$.

1.19. b) First prove by induction that if x, $y \in E$ and $k = 0, \dots, 2^n$ then $kx/2^n + (1 - k/2^n)y \in E$ for any $n \in \mathbb{N}$.

Let $[p, q] \subset E$ and $a \in E$, and assume for definiteness that $a > q$. We prove that $[p, q] \subset E$. Let $c = \sup\{x > p \mid [p, x] \subset E\}$. It is clear that $[p, c) \subset E$. Assume that $c < a$. Consider a point $a_0 \in E$ such that $c < a_0$ and $(p + a_0)/2 < c$ (see Figure 11). Then

$$E \supset \left\{ \frac{x + a_0}{2} \,\middle|\, x \in (p, c) \right\} = \left(\frac{p + a_0}{2}, \frac{c + a_0}{2} \right),$$

and hence $E \supset [p, (c + a_0)/2)$, which contradicts the definition of c, since $(c + a_0)/2 > c$.

1.21. Show that along with each of its points a the set E contains all points of the form $a\sqrt{n}$ $(n \in \mathbb{N})$, and E also contains arbitrarily small numbers.

1.22. a) Let $\mathscr{E}$ be the system of all possible disks with rational center and radius. It is clear that if $x \in B_\alpha$, then there is a disk $B' \in \mathscr{E}$ such that $x \in B' \subset B_\alpha$. For each disk in $\mathscr{E}$ fix arbitrarily a disk B_α containing it (if such a disk exists). Show that the system of disks chosen in this way is the desired object.

b) For each $n \in \mathbb{N}$ find an at most countable set $E_n \subset E$ such that $E \subset \bigcup_{x\in E_n} B(x, 1/n)$. Consider $\bigcup_{n=1}^{\infty} E_n$.

1.23. a) Prove the countability of the set

$$E_n = \{x \in E \mid E \cap B(x, 1/n) \setminus \{x\} = \varnothing\}.$$

b) Prove the countability of each of the sets

$$E_n^+ = \{x \in E \mid E \cap (x, x + \tfrac{1}{n}) = \varnothing\}, \qquad E_n^- = \{x \in E \mid E \cap (x - \tfrac{1}{n}, x) = \varnothing\}.$$

To do this verify that for distinct points x, $x' \in E_n^+$, (or x, $x' \in E_n^-$) the intervals $(x, x + 1/n)$ and $(x', x' + 1/n)$ (respectively, $(x - 1/n, x)$ and $(x' - 1/n, x')$) are disjoint.

1.24. Apply the result of problem 1.25 to the set

$$G = \bigcup_{n\in E} (\ln n, \ln(n + 1)).$$

1.25. Note first that if $0 < p < q$, then there is a number C such that $[C, +\infty) \subset \bigcup_{n=1}^{\infty}(np, nq)$.

Let $0 < p < q$ and $m \in \mathbb{N}$. Using the remark above, we see that there exist numbers $x_1 \in (0, q)$ and $n_1 \in \mathbb{N}$ such that $n_1 > m$ and $n_1 x_1 \in G$. Consequently, there is a nondegenerate closed interval $[p_1, q_1] \subset [p, q]$ such that $n_1 x \in G$ for any $x \in [p_1, q_1]$. Replacing $[p, q]$ by $[p_1, q_1]$ and m by n_1 and repeating the arguments, we construct a nondegenerate closed interval $[p_2, q_2] \subset [p_1, q_1]$ and a positive integer $n_2 > n_1$ such that $n_2 x \in G$ for any $x \in [p_2, q_{12}]$. Continuing this process, we get a sequence of nested intervals whose common point is the desired one.

1.26. Argue as in the solution of problem 1.25, replacing the set G at each step by a set G_k in such a way that each of them is used infinitely many times (for examples, in the sequence $G_1, G_2, G_1, G_2, G_3, G_1, G_2, G_3, G_4, \dots$).

1.27. a) Show that the Cantor set has the same cardinality as the set of binary sequences (see problem 1.2).

1.28. b) Use the result in part a).

1.29. Consider, for example, the set $E = K \setminus \mathbb{Q}$, where K is the Cantor set.

1.32. Show that A contains a set with the cardinality of the continuum. To do this consider the sets

$$Q_n = \bigcup_{\varepsilon_1, \dots, \varepsilon_n \in \{0, 1\}} \overline{\Delta}_{\varepsilon_1 0 \varepsilon_2 0 \cdots \varepsilon_{n-1} 0 \varepsilon_n} \qquad (n \in \mathbb{N}),$$

where $\overline{\Delta}$ denotes the closure of Δ, and prove as in problem 1.27a) that the set $Q = \bigcap_{n=1}^{\infty} Q_n$ has the cardinality of the set of binary sequences. Verify that $Q \setminus A$ is at most countable.

1.33. Assume that (a, b) is a bounded interval, and $\{F_n\}$ is a sequence of disjoint nonempty closed sets contained in (a, b). Prove that $(a, b) \setminus \bigcup_{n=1}^{\infty} F_n \neq \varnothing$. Let $p = \inf F_1$ and $q = \sup F_1$. Then the intervals $\Delta_0 = (a, p)$ and $\Delta_1 = (q, b)$ are nonempty, and the sets $F_n^0 = F_n \cap \Delta_0$ and $F_n^1 = F_n \cap \Delta_1$ are closed. Replacing (a, b) by Δ_0 and the sets F_n by the nonempty sets F_n^0, we construct the intervals Δ_{00} and Δ_{01}. Carrying out this construction for Δ_1, we get the intervals Δ_{10} and Δ_{11}. The intervals Δ_{00}, Δ_{01}, Δ_{10}, and Δ_{11} are disjoint from the set $F_1 \cup F_2$ and from each other. Continuing this process by induction, we construct a family $\{\Delta_{\varepsilon_1 \cdots \varepsilon_n}\}$ of open intervals satisfying the conditions of problem 1.32, and also the equality $\Delta_{\varepsilon_1 \cdots \varepsilon_n} \cap \bigcup_{k=1}^{n} F_k = \varnothing$ for any n and any $\varepsilon_1, \dots, \varepsilon_n$. It now remains to refer to problem 1.32.

1.34. Find a straight line not containing points of tangency of the given disks. Use the result of problem 1.33.

1.35. Let $\overline{E}_n$ be the closure of E_n. Arguing by contradiction, construct a sequence of nested closed intervals $\Delta_n \subset (a, b)$ such that $\Delta_n \cap \overline{E}_n = \varnothing$ for any $n \in \mathbb{N}$.

1.36. Use the result of problem 1.35.

§2. Inequalities

2.1. Let $S'_m = \sum_{1\le k\le m} a_k - \sum_{m<k\le n} a_k$. It is clear that $S_0 = -S_n$. Using the fact that the sequence $\{S_m\}_0^n$ changes sign, choose an index p such that S_p and S_{p-1} have different signs, and show that at least one of the numbers $|S_p|$ and $|S_{p-1}|$ does not exceed $\max_{1\le k\le n}|a_k|$.

2.2. a) Use the inequalities

$$2\sqrt{\frac{m}{M}} \le \frac{a_k}{M} + \frac{m}{a_k} \le 1 + \frac{m}{M}.$$

b) The right-hand inequality follows at once from the right-hand inequality of a). The left-hand inequality is proved by induction. Note also that the left-hand inequality in b) is a special case of the inequality in 2.7 for $b_k = 1/a_k$.

2.3. The right-hand inequalities are easy to prove by induction. To prove the left-hand inequalities it suffices to observe that $\prod_{1\le k\le n}(1+a_k) \times \prod_{1\le k\le n}(1-a_k) \le 1$ and to use the right-hand inequalities.

2.4. a) Use the Cauchy inequality about the arithmetic mean and the geometric mean, along with the binomial formula.

b) Use the inequalities $1 + a \ge e^{a/1+a}$ and $e^\sigma \ge 1 + \frac{\sigma}{1!} + \cdots + \frac{\sigma^n}{n!}$ for $a > -1$ and $\sigma \ge 0$.

2.5. a) Represent each factor on the left-hand side as the sum of a geometric progression, and multiply the series obtained.

b) Using the inequalities a) and $\ln(1+t) \le t$, we have that

$$\sum_{1\le n\le p_m} \ln\left(1+\frac{1}{n}\right) \le \sum_{1\le n\le p_m} \frac{1}{n} \le \prod_{1\le k\le m}\left(1-\frac{1}{p_k}\right)^{-1} = \prod_{1\le k\le m} \ln\left(1+\frac{1}{p_k-1}\right).$$

Taking logarithms, we get that

$$\ln\ln(1+p_m) = \ln\left(\sum_{1\le n\le p_m}\ln\left(1+\frac{1}{n}\right)\right) \le \sum_{1\le k\le m}\ln\left(1+\frac{1}{p_k-1}\right) \le \sum_{1\le k\le m}\frac{1}{p_k-1} \le 1 + \sum_{1\le k\le m}\frac{1}{p_k}.$$

2.6. It suffices to prove only the right-hand equation (the left-hand inequality follows from the right-hand inequality, applied to the sequences $\{a_k\}_1^n$ and $\{-b_k\}_1^n$). Here it can be assumed that the sequence $\{b_k\}_1^n$ is nondecreasing—this can be achieved by changing the numbering. Assuming that $a_j > a_i$ for some indices $j < i$, prove that transposing the numbers a_j and a_i in the sequence $\{a_k\}_1^n$ increases the sum $\sum_{1\le k\le n} a_k b_k$.

These inequalities have a simple mechanical interpretation. Suppose that n weights of magnitude $a_1, \dots, a_n$ and fixed on a lever rotating in the vertical plane above a fixed point O at distances $b_1, \dots, b_n$ to one side of O. Then $\sum_{1\le k\le n} a_k b_k$ is the static moment of the system. It is maximal if the magnitudes of the weights increase in the direction away from O. But if the magnitudes of the weights decrease, then the static moment is minimal.

2.7. To prove the right-hand inequality consider the double sum

$$\sum_{1\le k\le n}\sum_{1\le j\le n}(\hat{a}_k-\hat{a}_j)(\hat{b}_k-\hat{b}_j)$$

and use the fact that its terms are nonnegative.

The left-hand inequality can be proved similarly or can be derived from the right-hand inequality, applied to the sequences $\{a_k\}_1^n$ and $\{-b_k\}_1^n$.

2.8. To prove the inequalities a) and c) use the Abel transformation. The inequality b) follows from a).

The factor M cannot be decreased: fix $j\in\{1,\dots,n\}$ and consider the sequence $a_k=1$ for $1\le k\le j$, $a_k=0$ for $j<k\le n$. Then

$$\sum_{1\le k\le m} a_k b_k = b_1+\cdots+b_j\le Mj = M\sum_{1\le k\le n} a_k .$$

Consequently, $M\ge (b_1+\cdots+b_j)/j$ for any $j\in\{1,\dots,n\}$.

2.9. To prove the first inequality use the Abel transformation and the inequality $b_1+\cdots+b_k\le(\max_{1\le j\le n} b_j)\min(k, m)$. The second inequality follows from the first, applied to the sequences $\{a_1-a_{n-k+1}\}_1^n$ and $\{b_{n-k+1}\}_1^n$.

2.10. Use the identity

$$\frac{1}{n}\sum_{1\le k\le n} a_k^2-\left(\frac{1}{n}\sum_{1\le k\le n} a_k\right)^2=\frac{1}{2n^2}\sum_{1\le k\le n}\sum_{1\le j\le n}(a_k-a_j)^2 .$$

To get an upper estimate of the double sum use the inequality $|a_k-a_j|\le \Delta|k-j|$, and to get a lower estimate use the inequality $|a_k-a_j|\ge\delta|k-j|$ (it follows from the monotonicity of the sequence $\{a_k\}_1^n$). Note that the inequality b) is valid also for a nonomonotone sequence if we set

$$\delta=\min_{j\ne k}\left|\frac{a_k-a_j}{k-j}\right| .$$

2.11. The solution is analogous to that of the preceding problem, but the inequalities $(j-k)(a_{k+1}-a_k)\le a_j-a_k\le(j-k)(a_j-a_{j-1})$ for $1\le k<j\le n$ are used instead of the inequalities $\delta|k-j|\le|a_k-a_j|\le\Delta|k-j|$.

2.12. a) Using the Cauchy inequality about the arithmetic mean and the geometric mean, prove that the derivative of the function $\varphi(x) = \sqrt[n]{(a_1+x)\cdots(a_n+x)}$ is not less than 1.

b) Using the inequality a), prove first the particular case when $a_k = 1$ for $k=1,\dots,n$. The general case reduces to the particular one if we set $b_k=c_k a_k$ for $k=1,\dots,n$.

2.13. To determine the sign of the derivatives use the inequalities $t/(1+t) < \ln(1+t) < t$ for $t > -1$, $t \neq 0$.

2.14. The inequalities can be proved by induction. To prove the induction step use the inequalities $(1+1/n)^n < e < (1+1/n)^{n+1}$, which follow from the result of the preceding problem. To prove b) for $n = 11$ use the fact that $e < 11/4$.

2.15. The inequality c) is stronger than the inequality b), because

$$\theta(x) = 2^{p-2}(1+x^p) - (1+x^{p/(p-1)})^{p-1} \geq 0.$$

This follows from the easily verified relations $\theta(1) = 0$ and $\theta'(x) \leq 0$.

c) For $p = 2$ the inequality is obvious. For $p > 2$ let $q = p/(p-1)$, $q \in (1, 2)$, and prove that the function

$$f(x) = \frac{(1+x)^p + (1-x)^p}{(1+x^q)^{p/q}}$$

takes its largest value on $[0, 1]$ at the endpoints of this interval. To do this, show that f first decreases and then increases, that is, $f'(x) < 0$ for $x \in (0, c)$ and $f'(x) > 0$ for $x \in (c, 1)$, where c is some point in $(0, 1)$. It is clear that

$$f'(x) = \frac{p(1-x^2)^{p-1}}{(1+x^q)^{1+p/q}}\left(\frac{1-x^{q-1}}{(1-x)^{p-1}} - \frac{1+x^{q-1}}{(1+x)^{p-1}}\right).$$

Since we are interested in the sign of $f'(x)$, we study the function

$$g(x) = \frac{1-x^{q-1}}{(1-x)^{p-1}} - \frac{1+x^{q-1}}{(1+x)^{p-1}}.$$

Let $r = p-1$ and $t = 1/x$, and note that $q - 1 = 1/r$ and $r > 1$. Then $g(x) = -t^{q-p}\varphi(t)/(t+1)^{p-1}$, where

$$\varphi(t) = t^{1/r} + 1 - \left(\frac{t+1}{t-1}\right)^r (t^{1/r} - 1), \qquad t > 1.$$

It is clear that $\varphi(t) \to -\infty$ as $t \to 1+0$, and $\varphi(t) \to 2$ as $t \to +\infty$. It now suffices for us to verify that φ is increasing. Uncomplicated transformations give us that

$$\varphi'(t) = \frac{(t+1)^{r-1}}{(t-1)^r} t^{-\sigma}(u(t) - v(t)),$$

where $\sigma = 1 - 1/r$,

$$u(t) = 2r\frac{t - t^\sigma}{t-1} \quad \text{and} \quad v(t) = \frac{t+1}{r}\left(1 - \left(\frac{t-1}{t+1}\right)^r\right).$$

Note that $\lim_{t\to 1+0} u(t) = \lim_{t\to+\infty} v(t) = 2$. Therefore, to prove the inequality $\varphi'(t) > 0$ it suffices to see that both the functions u and v are strictly increasing (in this case $v(t) < 2 < u(t)$ for $t > 1$). Since

$$u'(t) = \frac{2rt^\sigma}{(t-1)^2}\left(\frac{1}{r} + \frac{\sigma}{t} - t^{-\sigma}\right)$$

and the function $\alpha(t) = 1/r + \sigma/t - t^{-\sigma}$ is positive for $t > 1$ $(\alpha(1) = 0$ and $\alpha'(t) = \sigma(t^{-1-\sigma} - t^{-2}) > 0)$, it follows that $u'(t) > 0$ for $t > 1$. Now consider the function v. Since

$$rv'(t) = 1 - \left(1 + \frac{2r}{t-1}\right)\left(\frac{t-1}{t+1}\right)^r,$$

it remains for us to see that the function

$$\beta(t) = \left(\frac{t+1}{t-1}\right)^r - 1 - \frac{2r}{t-1}$$

is positive for $t > 1$. This follows from the obvious relations $\beta(t) \to 0$ as $t \to +\infty$ and

$$\beta'(t) = \frac{2r}{(t-1)^2}\left(1 - \left(\frac{t+1}{t-1}\right)^{r-1}\right) < 0.$$

d) Prove that the derivative of the function $\varphi(t) = 4^{1-1/p}(1+x^p)^{2/p} - (1+x)^2 - (p-1)(1-x)^2$ is negative on $(0, 1)$. To do this, it is useful to represent $\varphi'(x)$ in the form $x\psi(1/x)$ and investigate the sign of the function ψ on $(1, +\infty)$.

2.16. The left-hand inequality follows from the relation $\ln(1+x) \le x$, $x > -1$. To prove the right-hand inequality, verify that the function $\varphi(t) = t^2/n + e^t(1-t/n)^n$ is increasing on $[0, n]$. To do this, represent $\varphi'(t)$ in the form $\varphi'(t) = t\psi(t)$ and show that $\psi(t) \ge \psi(1) > 0$.

b) It must be verified that the function $\varphi(t) = t^2/n^2 + e^t(1-t/n)^n$ does not exceed 1 on $[0, n]$. Since the inequality holds at the endpoints of the interval, is suffices to establish that $\varphi(t) \le 1$ if $\varphi'(t) = 0$, that is, if $e^t(1-t/n)^{n-1} = 2/n$. But in this case

$$\varphi(t) = \frac{t^2}{n^2} + \frac{2}{n}\left(1 - \frac{t}{n}\right) \le 1, \qquad t \in [0, n],\ n \ge 2.$$

c) It can be assumed that $t \in [0, 2\sqrt{n})$. Let $t = \theta\sqrt{n}$, where $\theta \in [0, 2)$. Then the inequality to be proved is equivalent to the inequality $1 - e^{\theta\sqrt{n}}(1-\theta/\sqrt{n})^n \le \theta/2$, that is, $0 \le \varphi_n(\theta) = \theta\sqrt{n} + n\ln(1-\theta/\sqrt{n}) - \ln(1-\theta/2)$. Differentiating with respect to n, show that $\varphi_n \le \varphi_{n+1}$:

$$\frac{d\varphi_n}{dn} = \frac{\theta}{2\sqrt{n}} + \ln\left(1 - \frac{\theta}{\sqrt{n}}\right) + \frac{\theta}{2(\sqrt{n}-\theta)} = \frac{1}{2}\psi\left(\frac{\theta}{\sqrt{n}}\right),$$

where $\psi(u) = u + 2\ln(1-u) + u/(1-u)$. Since $\psi(0) = 0$ and $\psi'(u) = (u/(1-u))^2 \ge 0$, it follows that $\psi(u) \ge 0$, that is, $\varphi_n \le \varphi_{n+1}$. It remains to prove that $\varphi_{36}(\theta) \ge 0$ on $[0, 2)$. Since

$$\varphi_{36}'(\theta) = \frac{6\theta^2 - 13\theta + 6}{(6-\theta)(2-\theta)},$$

it follows that

$$\min_{[0,2)} \varphi_{36} = \varphi_{36}(\tfrac{3}{2}) = 9 + 36\ln\tfrac{3}{4} + \ln 4 = 0.0297\ldots.$$

Note that the inequality c) holds for $n \geq 30$, but does not hold for $n \leq 29$.

2.17. The left-hand inequality follows from the relation $\ln(1-x) \leq -x-x^2/2$ for $0 < x < 1$. To prove the right-hand inequality we consider the function $\varphi(t) = e^{-t^2/2n} - e^t(1-t/n)^n$. Since $\varphi(0) \leq 1/\sqrt{en}$ and $\varphi(n) \leq 1/\sqrt{en}$, it suffices to prove that $\varphi(t) \leq 1/\sqrt{en}$ if $\varphi'(t) = 0$, that is, $e^{-t^2/2n} = e^t(1-t/n)^{n-1}$. But in this case $\varphi(t) = te^{-t^2/2n}/n \leq 1/\sqrt{en}$.

2.18. a) In proving the inequality $\varphi(x) = x^2/(1+x) - \ln^2(1+x) \geq 0$, represent the function φ' in the form $\varphi'(x) = (1+x)\psi(x)$ and verify that $\operatorname{sgn}\psi(x) = \operatorname{sgn} x$.

b) Study the character of monotonicity of the function $\varphi(x) = \arctan x - x/(1+2x/\pi)$ and use the fact that $0 = \varphi(0) = \lim_{x\to+\infty}\varphi(x)$.

d) Consider the function $\varphi(x) = 2\sin(\pi x/2) + x^3 - 3x$ for $x \in [0, 1]$. Starting from the graph of φ''', construct successively the graphs of the functions φ'', φ', and φ.

e) Consider the function $\varphi(x) = \sin^2 x \tan x - x^3$, and prove successively that $\varphi''' > 0$, $\varphi'' > 0$, $\varphi' > 0$, $\varphi > 0$ $(\varphi'''(x) = (6+8\cos^2 x)\tan^4 x)$.

f) Compare the Taylor expansions of the functions $\cos x$ and $((\sin x)/x)^{3+\varepsilon}$.

2.19. These inequalities are special cases of the inequality

$$f(x)f^{-1}(y) \geq xy, \tag{$*$}$$

where x, $y > 0$, $y \leq f(x)$, and f is such that $f(x)/x \uparrow$. The inequality $(*)$ follows from the inequalities $f(x)/x \geq f(z)/z$ for $z = f^{-1}(y) \leq x$. The sign should be changed in the inequality c) for $p < 0$. For the proof it suffices to observe that for $p < 0$

$$|(1+x)^p - 1| \leq |p|x \quad \text{and} \quad |(1+x)^{1/p} - 1| < x/|p|.$$

For $p \in (-1, 0)$ the inequality d) is valid for x close to 1, but it does not hold in a neighborhood of zero—for a proof it suffices to consider the Taylor expansion of the function on the left-hand side of the inequality.

2.20. The right-hand inequality is a special case of the inequality $(*)$. The left-hand inequality is a special case of the inequality opposite to $(*)$:

$$f(x)f^{-1}(y) \leq xy, \tag{$**$}$$

where $x > 0$, $y \geq f(x)$, and the positive function f is such that $f(x)/x \uparrow$. The inequality $(**)$ follows from the inequality $f(x)/x \leq f(z)/z$ for $z = f^{-1}(y) \geq x$.

2.28. See [34]. It suffices to prove the inequality for step functions in K. If f_1, $f_2 \in K$, then $f = \alpha f_1 + (1-\alpha)f_2 \in K$ for any number α in $(0, 1)$. It is easy to see that for any function g

$$\max(f, g) \leq \alpha\max(f_1, g) + (1-\alpha)\max(f_2, g).$$

Therefore,

$$\int_a^b \max(f(x), g(x))\,dx \le \max\left(\int_a^b \max(f_1(x), g(x))\,dx, \int_a^b \max(f_2(x), g(x))\,dx\right).$$

If the number of "steps" corresponding to the function $f \in K$ is greater than two, then the function is easy to represent as half the sum of two different functions in K. Indeed, let $c_1 \ge c_2 \ge \dots \ge c_n > 0$ be all the values of f, $n \ge 3$. Then there are numbers ε, $\delta > 0$ such that the functions $f_\pm$ having the same intervals of constancy and taking on them the values $c_1(1\pm\varepsilon)$, $c_2(1\mp\delta)$, $c_3, \dots, c_{n-1}, c_n(1\pm\varepsilon)$ are in K. This leads to the integral $\int_a^b \max(f(x), g(x))\,dx$ attaining its maximal value on functions consisting of two steps. Each such function has the form $f_s = c_1\chi_{[a,s)} + c_2\chi_{[s,b]}$, where $s \in (a, b)$, and the coefficients c_1 and c_2 are uniquely determined from the condition $f_s \in K$: $c_1 = b/(s(b-a))$ and $c_2 = a/(s(b-a))$. If $a < t < s < b$, then

$$\int_a^b \max(f_t(x), f_s(x))\,dx = \frac{1}{b-a}\left(2b - b\frac{t}{s} - a\frac{s}{t}\right).$$

The right-hand side is maximal for $s/t = \sqrt{b/a}$ and is equal to $2\sqrt{b}/(\sqrt{a}+\sqrt{b})$ in this case.

§3. Irrationality

3.1. a) $\Rightarrow$ b). It suffices to prove that for any number $Q \in \mathbb{N}$ there exist integers p and q such that $1 \le q \le Q$ and $|x - p/q| < 1/(qQ)$ (since the right-hand side of the inequality tends to zero as $Q \to +\infty$, while the left-hand side is positive by condition a), this implies that there are solutions of the inequality $|x - p/q| < 1/q^2$ with arbitrarily large denominators q).

Fixing a number $Q \in \mathbb{N}$, we partition the interval $[0, 1)$ into Q congruent intervals $[0, 1/Q), [1/Q, 2/Q), \dots, [1 - 1/Q, 1)$ and consider the points $\delta_n = nx - [nx]$ for $n = 0, \dots, Q$. It is clear that $\delta_n \in [0, 1)$. Since the number of these points is greater than the number of intervals, at least one of them contains two points, say δ_k and δ_j, $0 \le k < j \le Q$. Consequently, $1/Q > |\delta_j - \delta_k| = |jx - [jx] - kx + [kx]|$. Taking $q = j - k$ and $p = [jx] - [kx]$, we get that $1/Q > |qx - p|$.

c) $\Rightarrow$ a). Assuming that $x = p/q$, where $p \in \mathbb{Z}$ and $q \in \mathbb{N}$, we get that $pq_k - qp_k \to 0$ as $k \to +\infty$. Since $pq_k - qp_k \in \mathbb{Z}$, this implies that $pq_k - qp_k = 0$, that is, $p/q = p_k/q_k$ for all sufficiently large indices k, and this contradicts the assumption.

3.2. Use the result of problem 3.1.

3.3. a) It suffices to prove that the sequence $\{\sin 4^n\}$ does not have a limit. Assume that this is not so: $\sin 4^n \to l$. If $l = 0$, then $4^n = \pi k_n + \varepsilon_n$ and $4^{n+1} = \pi k_{n+1} + \varepsilon_{n+1}$, where $k_n, k_{n+1} \in \mathbb{Z}$ and $\varepsilon_n, \varepsilon_{n+1} \to 0$.

Consequently, $0 = \pi(4k_n - k_{n+1}) + 4\varepsilon_n - \varepsilon_{n+1}$. Therefore, for sufficiently large n we have $4k_n = k_{n+1}$, that is, $k_n = A4^n$, where $A \in \mathbb{Q}$. But then the equality $4^n = \pi k_n + \varepsilon_n$ implies that $1 = \pi A$. This is impossible, because the number π is irrational (see problem 3.13).

If $l \neq 0$, then we get from the equality

$$\sin 4^{n+1} = 4 \sin 4^n \cos 4^n (1 - 2\sin^2 4^n)$$

that the sequence $\{\cos 4^n\}$ has a limit. Consequently, $4^n = 2\pi k_n + \varphi + \varepsilon_n$, where $0 < |\varphi| < \pi$, $k_n \in \mathbb{Z}$, and $\varepsilon_n \to 0$. Since $4^{n+1} = 2\pi k_{n+1} + \varphi + \varepsilon_{n+1}$, it follows that $0 = 2\pi(4k_n - k_{n+1}) + 3\varphi + 4\varepsilon_n - \varepsilon_{n+1}$, and this is possible only for $\varphi = \pm 2\pi/3$. Therefore, for sufficiently large n the difference $4k_n - k_{n+1}$ is constant and equal either to 1 (if $\varphi = -2\pi/3$) or to -1 (if $\varphi = 2\pi/3$). In the first case it is easy to show by induction that $k_n = A \cdot 4^n + 1/3$, where $A \in \mathbb{Q}$. But then it follows from the equality $4^n = 2\pi k_n - 2\pi/3 + \varepsilon_n$ that $\pi \in \mathbb{Q}$. The case $\varphi = 2\pi/3$ is handled analogously.

b) Represent the number θ in the form $\theta = \pi m + \pi \sum \varepsilon_k 2^{-k}$, where $m = [\theta/\pi]$ and $\varepsilon_k = 0$ or 1.

3.5. Consider the numbers $0.101001000100001000001\ldots$ (the nth 1 is followed by n 0's) and $0.1234567891011121314151617181920\ldots$ (all the natural numbers are written out in the decimal expansion).

3.6. In the case a), b), and d) it is possible to use the result in problem II.1.21. In the case c) consider the numbers $n = (2k)^4 + j$, $j = 1, \ldots, 4k$. Assuming k to be large, show by means of the Taylor formula that

$$\sin(\pi n^{3/2}) = \sin\left(\frac{3\pi}{32}\frac{j^2}{k^2}\right) + O\left(\frac{1}{k^3}\right).$$

3.7. b) Use the result of problem 3.2.

c) Use the result of problem II.1.21.

d) Use the same device as in the solution of problem 3.6c).

3.8. Prove that for any interval $[p, q]$, $0 \le p < q \le 1$, there exists a number $x \in \mathbb{R}$ such that $\delta(x^n) \in [p, q]$ for $n \in \mathbb{N}$. To do this choose a number $m \in \mathbb{N}$ such that the numbers $a = m + p$ and $b = m + q$ satisfy the inequality

$$a(q - p) \ge 2. \qquad (*)$$

It is clear that $\delta(y) \in [p, q]$ for all $y \in [a, b]$. It follows from the inequality $(*)$ that $b^2 - a^2 \ge 2$. Therefore, the interval $[a^2, b^2]$ contains an interval of the form $[m_1 + p, m_1 + q]$, where $m_1 \in \mathbb{N}$. Let $[a_1, b_1] \subset [a, b]$ be an interval such that $a_1^2 = m_1 + p$ and $b_1^2 = m_1 + q$. It is clear that $\delta(y^2) \in [p, q]$ for all $y \in [a_1, b_1]$. It follows from the inequality $(*)$ that the length of $[a_1^3, b_1^3]$ is at least two:

$$b_1^3 - a_1^3 \ge a_1(b_1^2 - a_1^2) = a_1(q - p) \ge a(q - p) \ge 2.$$

Therefore, $[a_1^3, b_1^3]$ contains an interval of the form $[m_2+p, m_2+q]$, where $m_2 \in \mathbb{N}$. Suppose that $[a_2, b_2] \subset [a_1, b_1]$ is an interval such that $a_2^3 = m_2+p$ and $b_2^3 = m_2+q$. It is clear that $\delta(y^3) \in [p, q]$ for all $y \in [a_2, b_2]$. By $(*)$, the length of the interval $[a_2^4, b_2^4]$ is no less than two: $b_2^4 - a_2^4 \geq a_2(b_2^3 - a_2^3) \geq a_2(q-p) \geq a(p-q) \geq 2$. Continuing the construction by induction, we get a sequence $\{[a_k, b_k]\}$ of nested intervals. The desired point is taken from their intersection.

In conclusion note that the numbers x satisfying the conditions of the problem form a set having the cardinality of the continuum. For the proof it suffices to make a slight change in the solution, replacing the inequality $(*)$ by the inequality $a(q-p) \geq 3$. This enables us to double at each step the number of intervals having the required properties. As a result we get a generalized Cantor set with all points satisfying the condition of the problem.

3.9. Direct computations prove the assertion for functions of the form $f(t) = \cos(2\pi m t)$ $(m = 0, 1, 2, \dots)$. This implies that the assertion is valid for all functions f of the form $f(t) = \sum_{0 \leq m \leq M} a_m \cos(2nmt)$. It remains to use the Weierstrass theorem (see VII.3.11).

3.10. Use the same idea as in the proof of the assertion a) $\Rightarrow$ b) in problem 3.1: for any number $Q \in \mathbb{N}$ there exist integers $q, p_1, \dots, p_m$ such that $1 \leq q \leq Q^m$ and $|x_j - p_j/q| < 1/qQ$ for $j = 1, \dots, m$. For a proof, partition the cube $[0, 1)^m$ into Q^m congruent disjoint cubes and consider the points $\delta_n = (\delta(nx_1), \dots, \delta(nx_m)) \in [0, 1]^m$ for $n = 0, \dots, Q^m$.

3.11. a) The right-hand inequality is obvious. Assuming that $|\sqrt{2} - m/n| \leq 1/4n^2$ for some m, $n \in \mathbb{N}$, we get that

$$\left|2 - \frac{m^2}{n^2}\right| \leq \frac{\sqrt{2} + m/n}{4n^2} \leq \frac{2\sqrt{2} + 1/4}{4n^2} < \frac{1}{n^2}.$$

Consequently, $|2n^2 - m^2| < 1$, which is impossible.

b) It follows from the inequalities

$$\alpha_{n_j} = |\sqrt{2} - m_j/n_j| \leq C/n_j^2 \quad \text{and} \quad \alpha_{n_{j+1}} = |\sqrt{2} - m_{j+1}/n_{j+1}| \leq C/n_{j+1}^2$$

that $|\sqrt{2}(n_{j+1} - n_j) - (m_{j+1} - m_j)| \leq 2c/n_j$, that is,

$$\alpha_{n_{j+1} - n_j} \leq \frac{2C}{n_j(n_{j+1} - n_j)}.$$

Since $\alpha_n > 1/4n^2$ for any $n \in \mathbb{N}$, it follows that

$$\frac{1}{4(n_{j+1} - n_j)^2} < \alpha_{n_{j+1} - n_j} \leq \frac{2C}{n_j(n_{j+1} - n_j)},$$

which implies the inequality to be proved.

3.12.–3.14. After verifying that the remainder of the series is majorized by the first term of this remainder, show that the assertion c) of problem 3.1 holds.

3.15. Let P be an algebraic polynomial of degree n with integer coefficients such that $P(\alpha) = 0$. It suffices to consider fractions p/q close enough to α that $|\alpha - p/q| < 1$ and $P(p/q) \neq 0$. Since $q^n P(p/q) \in \mathbb{Z} \setminus \{0\}$, it follows that $|P(p/q)| \geq q^{-n}$. On the other hand, expanding the polynomial $P(t)$ in powers of $t - \alpha$, we get that

$$\left| P\left(\frac{p}{q}\right) \right| \leq \left| \sum_{1 \leq k \leq n} c_k \left(\frac{p}{q} - \alpha\right)^k \right| \leq \left| \frac{p}{q} - \alpha \right| \sum_{1 \leq k \leq n} |c_k| .$$

Thus, $q^{-n} \leq |P(p/q)| \leq \mathrm{const}\,|p/q - \alpha|$. The assertion proved is called Liouville's theorem.

3.16. b) Let $\alpha = \sum \varepsilon_k / n_k$ and prove that for any $\varepsilon > 0$ there exist solutions of the inequality $|\alpha - p/q| < \varepsilon/q^2$ with arbitrarily large $q \in \mathbb{N}$. Since $\alpha \notin \mathbb{Q}$ (see problem 3.14), this gives us by Liouville's theorem that the number α is not a quadratic irrational.

Since $|\alpha - \sum_{1 \leq k \leq m} (\varepsilon_k / n_k)| \leq C_Q / n_{m+1}$, it suffices to prove that $\underline{\lim}((n_1^2 \cdots n_m^2)/n_{m+1}) = 0$. Assuming the opposite, we get that $n_{m+1} \leq C n_1^2 \cdots n_m^2$. Let $L_m = n_1 n_2 \cdots n_m$. Then the last inequality takes the form $L_{m+1} \leq C L_m^3$, and hence

$$n_{m+1} \leq L_{m+1} \leq C (C L_{m-1})^3 \leq \cdots \leq C^{1+3+\cdots+3^m} L_1^{3^m} \leq (C L_1)^{3^{m+1}} .$$

Consequently, $\sqrt[3^{m+1}]{n_{m+1}} \leq C L_1$, which contradicts the condition.

3.17. See the solution of the next problem.

3.18. e) Arguments analogous to the solution of problem 3.16 show that the equality $\overline{\lim}_{k \to \infty} \sqrt[N^k]{n_k} = +\infty$ for some $N \in \mathbb{N}$, $N > 1$, implies that the sum of the series $\sum \varepsilon_k / n_k$ is not an algebraic number of degree N. It remains to see that the condition $\overline{\lim} \sqrt[k!]{n_k} > 1$ implies that $\overline{\lim}_{k \to \infty} \sqrt[N^k]{n_k} = +\infty$ for all $N \in \mathbb{N}$.

3.19. a) It is clear that $e - \sum_{1 \leq k \leq n} (1/k!) = \theta_n / (n+1)!$, where $0 < \theta_n < 3$. Assuming that $e = p/q$, where p, $q \in \mathbb{N}$, and multiplying by $qn!$, we get that

$$pn! - qn! \sum_{0 \leq k \leq n} \frac{1}{k!} = \frac{q\theta_n}{n+1} > 0 .$$

The left-hand side of this inequality is a positive integer, while the right-hand side tends to zero with increasing n, which leads to a contradiction.

b) If $Ae^2 + Be + C = 0$, then $Ae + Ce^{-1} \in \mathbb{Z}$. This is possible only for $A = C = 0$; the proof is analogous to the solution of problem a).

3.20. a) Integrating by parts, verify that

$$I_{n+1} = (4n+2) I_n - \pi^2 I_{n-1} \qquad (n \in \mathbb{N}) .$$

b) Assume the opposite: $\pi^2 = p/q$, where p, $q \in \mathbb{N}$. Then

$$q^n P_n(p/q) = q^n I_n > 0 .$$

This equality leads to a contradiction, since its left-hand side is a positive integer, while the right-hand side tends to zero as n increases.

The result obtained, like the results in the next two problems, was established (by another method) by Lambert. In the book [32] the reader can find a history and expanded exposition of this theme.

3.21. The solution is analogous to that of the preceding problem.

3.22. a) The solution is analogous to that of problem 3.20a).

b) Assume the opposite: $\tan r = p/q$, where $p \in \mathbb{Z}$ and $q \in \mathbb{N}$. Let $r = a/b$ where a, $b \in \mathbb{N}$. Then

$$C_n\left(\frac{a}{b}\right)\cos r + S_n\left(\frac{a}{b}\right)\sin r = I_n(r).$$

From this,

$$qb^n C_n\left(\frac{a}{b}\right) + b^n S_n\left(\frac{a}{b}\right) p = \frac{qb^n}{\cos r} I_n(r).$$

The left-hand side of this equality is an integer, while the right-hand side tends to zero as n increases. To reach a contradiction it remains to see that $I_n(r) \neq 0$ for sufficiently large n. If $r \leq \pi/2$, then this is obvious. Suppose that $r > \pi/2$. Then

$$\begin{aligned} n! I_n(r) &= 2\int_0^{\pi/2} (r^2 - t^2)^n \cos t\, dt + O\left(\left(r^2 - \frac{\pi^2}{4}\right)^n\right) \\ &\geq \int_0^{\pi/3} r^n (r-t)^n\, dt + O\left(\left(r^2 - \frac{\pi^2}{4}\right)^n\right) \sim \frac{r^{2n+1}}{n+1}. \end{aligned}$$

CHAPTER II

Sequences

§1. Computation of limits

1.2. Verify that the sequences $\{x_{2k}\}$ and $\{x_{2k-1}\}$ increase. The Wallis formula can be used to compute their limits.

1.3. Show that the sums under consideration differ little from a geometric progression.

1.4. Write x_n in the form $x_n = \prod_{1\le k\le n} \frac{\sin(k/n^{3/2})}{k/n^{3/2}}$ and study $\ln x_n$ using Taylor's formula.

1.5. The first half of the terms gives an infinitesimally small contribution. Therefore,

$$x_n = \sum_{n/2<k\le n} \left(\frac{a+k}{n}\right)^n + o(1) = \sum_{0\le j<n/2} \left(1+\frac{a-j}{n}\right)^n + o(1)$$
$$= \sum_{0\le j<n/2} e^{a-j}\left(1+O\left(\frac{(a-j)^2}{n}\right)\right) + o(1) = \sum_{j\ge 0} e^{a-j} + o(1).$$

1.6. a) The main contribution is given by the last terms:

$$\sum_{n-\sqrt[3]{n}\le k\le n} \left(\frac{k}{n}\right)^k = \sum_{0\le j\le \sqrt[3]{n}} \left(1-\frac{j}{n}\right)^{n-j}$$
$$= \sum_{0\le j\le \sqrt[3]{n}} e^{-j}\left(1+O\left(\frac{j^2}{n}\right)\right) = \sum_{j\ge 0} e^{-j} + o(1).$$

The sum of the remaining terms is infinitesimally small, since $\max_{1<k<n-\sqrt[3]{n}} (k/n)^k = 4/n^2$.

b) Represent the difference $S_n - S_{n+1}$ in the form of a sum $u_n + v_n$, where

$$u_n = \left(\frac{1}{n}+\frac{4}{n^2}+\frac{27}{n^3}\right) - \left(\frac{1}{n+1}+\frac{4}{(n+1)^2}+\frac{27}{(n+1)^3}+\frac{256}{(n+1)^4}\right),$$
$$v_n = \sum_{4\le k\le n} \left(\left(\frac{k}{n}\right)^k - \left(\frac{k+1}{n+1}\right)^{k+1}\right).$$

Obtain the inequality $v_n \geq 0$ by using the fact that the function $f(x) = (x/(a+x))^x$ is decreasing on $(0, +\infty)$ for any $a > 0$. Verify that $u_n > 0$ for $n \geq 8$. Establish directly the inequalities between the S_n for $1 \leq n \leq 8$.

1.7. To investigate the quantities $(n+1)(n+2)\cdots(n+n) = (2n)!/n!$ use the result in problem I.2.14. Another solution is obtained by noting that $\ln(\sqrt[n]{(n+1)\cdots(n+n)}/n)$ is an integral sum corresponding to the integral $\int_0^1 \ln(1+x)\,dx$.

1.8. Use the result of problem I. 2.14.

1.9. In problems a)–d) use the fact that the sums under consideration differ little from integral sums. In problem e) show that the sum lies between the integrals $\int_{1/n}^{1-1/n}(t(1-t))^{-1/2}\,dt$ and $\int_0^1 (t(1-t))^{-1/2}\,dt$.

1.10. a)–c) Use the equality $e = \sum_{k\geq 0} 1/k!$.

d) Compare $(\sqrt{2}+1)^n$ with the integer $(\sqrt{2}+1)^n + (1-\sqrt{2})^n$.

1.11. The convergence of the sequence $\{\frac{x_1}{a}\cdot\frac{x_2}{a}\cdots\frac{x_n}{a}\}_{n\in\mathbb{N}}$ to a positive limit is equivalent to the convergence of the series

$$\sum \ln\frac{x_n}{a} = \sum \ln\left(1 - \frac{a-x_n}{a}\right).$$

1.12. a) The convergence of the sequence follows from its monotonicity and the inequality $x_n \leq 1+\sqrt{a}$, which is proved by induction. To compute the limit use the identity $x_n = \sqrt{a + x_{n-1}}$.

b) Since $l_a^2 = a + l_a$, it follows that

$$l_a - x_n = \frac{l_a^2 - x_n^2}{l_a + x_n} = \frac{l_a - x_{n-1}}{l_a + x_n},$$

and hence $l_a - x_n = (l_a - x_1)\prod_{1\leq k\leq n}(l_a + x_k)^{-1}$. Therefore,

$$\sum (l_a - x_n) \leq (l_a - x_1)\sum \frac{1}{l_a^n} < +\infty,$$

and it is possible to use the result of problem 1.11:

$$l_a - x_n = (l_a - x_1)\prod_{1\leq k\leq n}(l_a + x_k)^{-1} \sim c_a/(2l_a)^n.$$

For $a = 2$ it is not hard to see that $x_n = 2\cos(\pi/2^{n+1})$, and hence $2 - x_n \sim \pi^2/4^{n+1}$.

1.13. Since

$$2 - x_n = \frac{8 - x_n^3}{4 + 2x_n + x_n^2} = \frac{2 - x_{n-1}}{4 + 2x_n + x_n^2}$$
$$= \cdots = (2 - x_1)\prod_{1\leq k\leq n}(4 + 2x_k + x_k^2)^{-1},$$

it follows that $2 - x_n = O(4^{-n})$. Using the result of problem 1.11, we get that $\prod_{1\leq k\leq n}(4 + 2x_k + x_k^2) \sim \text{const}(12)^n$.

1.14. The monotonicity of the sequence $\{x_n\}$ is obvious. Boundedness follows from the inequality $x_n \leq \sqrt[p]{1+x_n} \leq \sqrt[p]{2x_n}$, that is, $x_n \leq 2^{1/(p-1)}$.

1.15. If $x_n \to c < +\infty$, then $x_n = \sqrt[p]{c_1 + \cdots + \sqrt[p]{c_n}} \leq c$, and hence $\sqrt[p]{\sqrt[p]{\cdots \sqrt[p]{c_n}}} \leq c$, that is, $c_n \leq c^{p^n}$. To prove the converse assertion it suffices to observe that the inequality $c_n \leq c^{p^n}$ implies that

$$x_n \leq \sqrt[p]{c^p + \sqrt[p]{c^{p^2} + \cdots + \sqrt[p]{c^{p^n}}}} = c\sqrt[p]{1 + \sqrt[p]{1 + \cdots + \sqrt[p]{1}}}.$$

Using the result of the preceding problem, we get that $\{x_n\}$ is bounded, and this suffices for convergence, since the sequence is increasing.

1.16. It follows from the result of the preceding problem that the convergence of the sequence

$$x_n = \sqrt{a_1 + b_1\sqrt{a_2 + \cdots + b_{n-1}\sqrt{a_n}}}, \tag{$*$}$$

where $a_k, b_k \geq 0$, is equivalent to boundedness above of the sequence $\{\frac{1}{2^n}\ln a_n + \sum_{1\leq k<n} \frac{1}{2^k}\ln b_k\}$. Therefore, the expressions on the right-hand sides of equalities a)–e) are the limits of convergent sequences (use the inequality $u_n \leq 2^n$ in example d), and the inequality $T_n \leq (2x)^n$ for $x \geq 1$ in example e); these inequalities are easy to prove by induction). To find the limit of the sequence defined by $(*)$ consider a sequence $\{\theta_n\}$ with

$$\theta_n^2 = a_n + b_n\theta_{n+1} \qquad (n \in \mathbb{N}). \tag{$**$}$$

Replacing a_n by θ_n^2 in $(*)$, we get an upper estimate for x_n:

$$x_n \leq y_n = \sqrt{a_1 + b_1\sqrt{a_2 + b_2\sqrt{\cdots + b_{n-2}\sqrt{a_{n-1} + b_{n-1}\theta_n}}}} = \theta_1$$

(the last equality follows from $(**)$). To get a lower estimate for x_n use the inequality $\sqrt{a+cb} \leq \sqrt{c}\sqrt{a+b}$ (for $a, b \geq 0$ and $c \geq 1$) $n-1$ times. This gives us that $y_n \leq x_n \sqrt[2^n]{\theta_n/\sqrt{a_n}}$. Thus, $\theta_1 \sqrt[2^n]{\sqrt{a_n}/\theta_n} \leq x_n \leq \theta_1$, and hence $x_n \to \theta_1$ if $\frac{1}{2^n}\ln(\sqrt{a_n}/\theta_n) \to 0$.

Verify that in examples a)–e) the relation $(**)$ is satisfied by the following sequences $\{\theta_n\}$:

a) $n+2$; b) $n+2$; c) $\cos(x/2^{n-1})$;
d) $3u_{2n+2}$; e) $2xT_n$.

In the solution of examples d) and e) use the following representations of the

Fibonacci numbers and the Chebyshev polynomials:

$$u_n = \frac{1}{\sqrt{5}}\left(\left(\frac{1+\sqrt{5}}{2}\right)^n - \left(\frac{1-\sqrt{5}}{2}\right)^n\right),$$

$$T_n = \frac{1}{2}\left(\left(x+\sqrt{x^2-1}\right)^n + \left(x-\sqrt{x^2-1}\right)^n\right).$$

These equalities can be proved by induction.

1.17. The convergence of the sequence $\{x_n\}$ follows from the result of problem 1.15. To estimate the difference $a - x_N$, write it in the form $(a - x_N) = \sum_{n\geq N}(x_{n+1} - x_n)$, and consider the quantity $x_{n+1} - x_n$:

$$\begin{aligned} x_{n+1} - x_n &= \sqrt{1+\sqrt{2+\sqrt{3+\cdots+\sqrt{n+\sqrt{n+1}}}}} \\ &\quad - \sqrt{1+\sqrt{2+\sqrt{3+\cdots+\sqrt{n}}}} \\ &= \frac{\sqrt{2+\sqrt{3+\cdots+\sqrt{n+\sqrt{n+1}}}} - \sqrt{2+\sqrt{3+\cdots+\sqrt{n}}}}{\sqrt{1+\sqrt{2+\cdots+\sqrt{n+1}}} + \sqrt{1+\sqrt{2+\cdots+\sqrt{n}}}} \\ &= \cdots = \frac{\sqrt{n+1}}{\prod_{1\leq k\leq n}(x_{n+1}^{(k)} + x_n^{(k)})}, \end{aligned}$$

where $x_n^{(k)} = \sqrt{k+\sqrt{(k+1)+\cdots+\sqrt{n}}}$. Using the recursion formula $x_n^{(k)} = \sqrt{k + x_n^{(k+1)}}$, prove successively that:

a) $\sqrt{k} \leq x_n^{(k)} \leq \sqrt{k} + 1$;

b) $x_n^{(k)} = \sqrt{k} + 1/2 + O(1/\sqrt{k})$ for $1 \leq k < n$;

c) $x_n^{(k)} = \sqrt{k}\exp(1/2\sqrt{k} + O(1/k^{3/2}))$ for $1 \leq k < n$.

(Here the constants in the O-terms are independent not only of k but also of n.) With the help of the equality c) and the result of problem 2.12a) show that

$$x_{n+1} - x_n = \frac{1+o(1)}{2^n\sqrt{(n-1)!}}\exp\left(-\sum_{1\leq k<n}\left(\frac{1}{2\sqrt{k}} + O\left(\frac{1}{k^{3/2}}\right)\right)\right),$$

that is, $x_{n+1} - x_n \sim \mathrm{const}/(2^n e^{\sqrt{n}}\sqrt{(n-1)!})$. This implies immediately that $a - x_N \sim x_{N+1} - x_N \sim \mathrm{const}/(2^N e^{\sqrt{N}}\sqrt{(N-1)!})$. Therefore, using the result of problem I.2.14, we get that $\sqrt[N]{a - x_N} \sim \sqrt{e/N}/2$.

1.18. Let $\tilde{x}_n = x_n - x_{n-1}/2$. Then

$$x_n - \frac{x_1}{2^{n-1}} = \sum_{1<k\leq n}\frac{\tilde{x}_k}{2^{n-k}} = \sum_{1<k\leq m} + \sum_{m<k\leq n}.$$

Since $\tilde{x}_k \to 0$, for large m the second sum is small for all $n \geq m$. For fixed m the first sum is arbitrarily small if n is sufficiently large.

1.19. a) Assuming the opposite, we get for sufficiently large n that

$$\left(\frac{a_1 + a_{n+1}}{a_n}\right)^n < \left(\frac{1+n}{n}\right)^n,$$

which implies that

$$\frac{a_1}{n+1} < \frac{a_n}{n} - \frac{a_{n+1}}{n+1}.$$

This is impossible, because the partial sums of the series

$$\sum \left(\frac{a_n}{n} - \frac{a_{n+1}}{n+1}\right)$$

do not exceed a_1.

b) Consider the sequences $a_n = n^c$ (for $c > 1$) and $a_n = n\ln(n+1)$.

1.20. Let $l = \inf(a_n/n) \in [-\infty, +\infty)$ and $L > l$. Fix an index m such that $a_m/m < L$. Representing an arbitrary index $n > m$ in the form $n = km + j$, where $k \in \{0, 1, \dots\}$ and $j \in \{1, \dots, m\}$, we get that $a_n = a_{km+j} \leq a_{km} + a_j \leq ka_m + a_j$, and hence

$$l \leq \frac{a_n}{n} \leq \frac{k}{n}a_m + \frac{a_j}{n} < \frac{km}{n}L + \frac{a_j}{n} \leq L + O\left(\frac{1}{n}\right).$$

Thus, $l \leq \underline{\lim}(a_n/n) \leq \overline{\lim}(a_n/n) \leq L$ for any number $L > l$, which implies that $\underline{\lim}(a_n/n) = \overline{\lim}(a_n/n) = l$, that is, $a_n/n \to l$.

1.22. Apply the result of the preceding problem to the sequences $\{x_{2k}\}$ and $\{x_{2k-1}\}$.

1.23. Apply the theorem of Stoltz (see problem 2.6) to the sequences $\{x_{2k}/k\}$ and $\{x_{2k+1}/k\}$.

1.24. b) To construct such a sequence proceed as follows. Consider a sequence $\{\Delta_n\}$ of closed intervals such that Δ_n is located to the left of Δ_{n+1}. Define the function $f \in C(\mathbb{R})$ by the equality $f(t) = (-1)^n$ if $t \in \Delta_n$, and let f be linear between the intervals Δ_n and Δ_{n+1}. Let $x_n = f(n)$. It is clear that if the lengths of the open intervals adjacent to the Δ_n increase without bound, then $x_{n+1} - x_n \to 0$. On the other hand, by choosing successively the lengths of the Δ_n to be sufficiently large it is possible to ensure that there are numbers arbitrarily close to $+1$ and arbitrarily close to -1 among the numbers $\frac{1}{n}(x_1 + x_2 + \cdots + x_n)$.

We get another example by considering the sequence $\{\sin(\ln n)\}$.

1.25. Let $\underline{\lim}\, x_n < a < b < \overline{\lim}\, x_n$, and consider the sets $E_1 = \{n \in \mathbb{N} | x_n < a\}$ and $E_2 = \{n \in \mathbb{N} | x_n > b\}$. As established in problem I.1.24–I.1.26, there is a $c > 1$ such that each of the sets E_1 and E_2 contain infinitely many numbers of the form $[c^k]$, but this is incompatible with the assumption that the limit $\lim_{k\to+\infty} x_{[c^k]}$ exists.

§2. Averaging of sequences

2.2. The left-hand inequality follows from the right-hand inequality, applied to the sequence $\{-x_n\}$. To prove the right-hand inequality it suffices to consider the case when $\overline{\lim}\, x_n = l < +\infty$. Let $L > l$ and let $m = m_L$ be an index such that $x_n < L$ for $n > m$. Then

$$\begin{aligned}\tilde{x}_n &= \frac{1}{A_n}\sum_{1\le k\le m} a_k x_k + \frac{1}{A_n}\sum_{m<k\le n} a_k x_k \\ &\le o(1) + \frac{1}{A_n}\sum_{m<k\le n} La_k = L + o(1).\end{aligned}$$

Consequently, $\overline{\lim}\, \tilde{x}_n \le L$ for any $L > l$, that is, $\overline{\lim}\, \tilde{x}_n \le l$.

2.3. a) If $n_k \le n < n_{k+1}$, then $k/n_{k+1} \le p_n/n \le k/n_k$.

b) If $0 < \alpha \le 1$, then, for example, $A = \{n \in \mathbb{N} | [\alpha n] > [\alpha(n-1)]\}$. In this case $\alpha - 1/n < p_n/n \le \alpha$.

2.4. a) Apply the result of problem 2.1 to the sequence $\{|x_n - a|\}$.

b) Let $\varepsilon > 0$, and let $A_\varepsilon = \{n \in \mathbb{N} | |x_n - a| \ge \varepsilon\}$. Verify that $\theta(A_\varepsilon) = 0$. Let $p_k(n) = \operatorname{card}\{A_{1/k} \cap [1, n]\}$, and let the numbers $1 = n_0 < n_1 \cdots$ be such that $\frac{1}{n}p_{k+1}(n) < \frac{1}{k}$ for $n > n_k$. Define $A = \bigcup_{k\ge1} A_{1/k} \cap [n_{k-1}, n_k]$ and $B = \mathbb{N}\backslash A = \{m_1, m_2, \dots\}$. Then $\theta(B) = 1$ and $x_{m_k} \to a$.

2.5. To construct a sequence $\{a_n\}$ having the limit $\lim((a_1 + \cdots + a_n)/n)$ but not a limit $\lim((a_1^2 + \cdots + a_n^2)/n)$ it suffices to change somewhat the solution of problem 1.24b): let $\{b_k\}$ form an alternating sequence of $+1$ and -1 on each closed interval Δ_n, and on the adjacent open intervals let $b_k = 0$. For a suitable choice of the lengths of the closed intervals and the lengths of the adjacent open intervals the sequence $\{(2 + b_n)/4\}$ is the required sequence. To construct a sequence $\{a_n\}$ having the limit $\lim((a_1^2 + \cdots + a_n^2)/n)$ but not having a limit $\lim((a_1 + \cdots + a_n)/n)$, consider the sequence $\{\sqrt{(2 + b_n)/4}\}$.

Another example is obtained by considering the sequence $\{c_n\}$ with

$$c_n = \begin{cases} \frac{3+(-1)^n}{6} \text{ if } [\log_2 n] & \text{is even}, \\ \frac{3+2(-1)^n}{6} \text{ if } [\log_2 n] & \text{is odd}. \end{cases}$$

Of the three limits $\lim \frac{1}{n}(c_1^2 + \cdots + c_n^2)$, $\lim \frac{1}{n}(c_1 + \cdots + c_n)$, and $\lim \frac{1}{n}(\sqrt{c_1} + \cdots + \sqrt{c_n})$, only the second exists.

b) The inequality $a^2 \le b$ follows from the Cauchy inequality. To construct a sequence $\{a_n\}$ with given limits $a = \lim((a_1 + \cdots + a_n)/n)$ and $b = \lim((a_1^2 + \cdots + a_n^2)/n)$, where $0 \le a^2 \le b \le a \le 1$, consider the sequence taking the value α on a set $A \subset \mathbb{N}$ of density a (see problem 2.3) and the value β on the set $\mathbb{N}\backslash A$. Prove that the required equalities can be ensured by suitably choosing the parameters α and β.

2.6. It can be assumed that $l \geq 0$. Here it suffices to consider only the case $l = 0$: if $0 < l < +\infty$, then the sequence $\{x_n - ly_n\}$ should be considered instead of $\{x_n\}$, and if $l + \infty$, then the sequences $\{x_n\}$ and $\{y_n\}$ should be interchanged. (The conditions $x_{n+1} > x_n$ and $x_n \to +\infty$ hold because $x_{n+1} - x_n > y_{n+1} - y_n$ for sufficiently large n.) Assuming that

$$\varepsilon_n = \frac{x_n - x_{n-1}}{y_n - y_{n-1}} \to 0,$$

prove that $x_n/y_n \to 0$. Since

$$x_n = x_1 + \sum_{1<k\leq n} (x_k - x_{k-1}) = x_1 + \sum_{1<k\leq n} \varepsilon_k(y_k - y_{k-1}),$$

it follows that

$$\frac{x_n}{y_n} = \frac{x_1}{y_n} + \sum_{1<k\leq m} \frac{y_k - y_{k-1}}{y_n}\varepsilon_k + \sum_{m<k\leq n} \frac{y_k - y_{k-1}}{y_n}\varepsilon_k .$$

For large m the second sum is small for all $n > m$, because $\varepsilon_k \to 0$. For fixed m the first sum is small, because $y_n \to +\infty$.

We get another solution if we apply the result of problem 2.2 to the sequences $a_n = y_n - y_{n-1}$ and $b_n = (x_n - x_{n-1})/(y_n - y_{n-1})$ (instead of $\{x_n\}$).

2.7. Use the theorem of Stoltz.

2.8. Use the theorem of Stoltz. In examples d) and e) use the result of problem I.2.14.

2.9. To prove convergence of the sequence $\{\ln n - \sum_{1\leq k\leq n} 1/k\}$ use the fact that convergence of the sequence $\{x_n\}$ is equivalent to convergence of the series $\sum(x_n - x_{n-1})$. The series arising here converges, as follows from the easily verified inequality $0 < 1/k - \ln(1 + 1/k) < 1/2k^2$.

Another solution is obtained by using the inequality

$$0 < \frac{1}{[x]} - \frac{1}{x} < \frac{1}{x-1} - \frac{1}{x} < \frac{1}{(x-1)^2}$$

for $x > 1$ and observing that the improper integral

$$\int_2^\infty \left(\frac{1}{[x]} - \frac{1}{x}\right) dx$$

converges; therefore, the sequence

$$I_n = \int_2^n \left(\frac{1}{[x]} - \frac{1}{x}\right) dx = \frac{1}{2} + \frac{1}{3} + \cdots + \frac{1}{n-1} - \ln n + \ln 2$$

has a finite limit.

2.10. The solution is analogous to the solution of problem 2.9.

2.11. It follows from Stirling's formula that to prove the right-hand inequality it suffices to see that the sequence $x_n = n!/(\sqrt{2\pi n}(n/e)^n e^{1/12n})$ is increasing. Since

$$\ln x_{n+1} - \ln x_n = 1 + \frac{1}{12n(n+1)} - \left(n + \frac{1}{2}\right) \ln \frac{n+1}{n}$$
$$= \left(n + \frac{1}{2}\right) \varphi\left(\frac{1}{n}\right),$$

where

$$\varphi(t)=\frac{2t}{2+t}+\frac{t^3}{6(1+t)(2+t)}-\ln(1+t),$$

it suffices to verify that $\varphi(t)>0$ for $t>0$, but this follows from the equalities $\varphi(0)=0$ and $\varphi'(t)=t^4/(6(1+t)^2(2+t)^2)$. The left-hand inequality can be proved analogously.

2.12–2.14. The solution is analogous to that of problem 2.9.

§3. Recursive sequences

3.1. Prove by induction that the sequence $\{x_n\}$ has the form $\{Aq^n+B\}$, and determine the parameters A, B, and q.

3.2. a) Since

$$x_n-x_{n-1}=-\frac{1}{n}(x_{n-1}-x_{n-2})=\cdots=\frac{x_0-x_1}{n!}(-1)^n,$$

it follows that

$$\begin{aligned}\lim x_n&=x_0+\sum_{n\ge1}(x_n-x_{n-1})\\&=x_0+(x_0-x_1)\sum_{n\ge1}\frac{(-1)^n}{n!}\\&=x_0+(x_0-x_1)\left(\frac{1}{e}-1\right).\end{aligned}$$

3.3. Since at each step the term in the sequence $\{x_n\}$ almost doubles, it is natural to consider the sequence $\{y_n\}=\{x_n/2^n\}$. It is obvious that

$$y_{n+1}=\frac{n+1}{n}y_n-\frac{1}{2^{n+1}},$$

that is,

$$\frac{y_{n+1}}{n+1}-\frac{y_n}{n}=-\frac{2^{-(n+1)}}{n+1}.$$

Consequently, $y_n=n(y_1-\sum_{1<k\le n}2^{-k}/k)$. From this,

$$x_n=n2^n\left(\frac{x_1}{2}+\frac{1}{2}-\ln 2+\sum_{k>n}\frac{1}{k2^k}\right).$$

Therefore, $x_n\sim n2^{n-1}(x_1-\ln(e/4))$ for $x_1\ne\ln(e/4)$, and otherwise

$$x_n=n2^n\sum_{k>n}\frac{1}{k2^k}=\sum_{k\ge1}\frac{1}{(1+k/n)\,2^k}\to1.$$

3.4. Assuming that $b\ne0$, we consider the sequence $y_n=x_n/b^n$. Then $y_{n+1}=y_n+\alpha_n$, where $\alpha_n=a_n/b^{n+1}$. It is clear that $y_{n+1}=y_1+\sum_{1\le k\le n}\alpha_k$.

If $|b| > 1$ and $S \neq 0$, then $x_{n+1}/b^{n+1} = y_{n+1} \to y_1 + \sum \alpha_k = S/b$, that is, $x_{n+1} \sim S \cdot b^n$. But if $|b| > 1$ and $S = 0$, then

$$y_{n+1} = y_1 + \sum_{1 \le k \le n} \alpha_k - \frac{S}{b} = -\sum_{k>n} \frac{a_k}{b^{k+1}}$$
$$= -\sum_{k \ge 1} \frac{a_{n+k}}{b^{n+k+1}} \sim -\frac{a}{b^{n+1}} \sum \frac{1}{b^k} = \frac{1}{b^{n+1}} \frac{a}{1-b}.$$

This yields b). It remains to consider the case when $0 < |b| < 1$. We have that

$$y_{n+1} = y_1 + \sum_{1 \le k \le n} \frac{a}{b^{k+1}} + \sum_{1 \le k \le n} \frac{a_k - a}{b^{k+1}} = \frac{a + o(1)}{b^{n+1}(1-b)}.$$

Therefore, $x_{n+1} = b^{n+1} y_{n+1} \to a/(1-b)$.

3.5. If $p \le 0$, then the sequence $x_n = (p-1)^n$ does not have a limit, although $x_{n+1} = px_n + (1-p)x_{n-1}$. If $p \ge 1$, then either $x_{n+1} \ge x_n$ for all $n \in \mathbb{N}$, or $x_{n+1} \le x_n$ for $n \ge n_0$ (since the inequality $x_{n_0+1} \le x_{n_0}$ implies that $x_{n+1} \le x_n$ for $n \ge n_0$).

Let $0 < p < 1$, and define $l = \underline{\lim}\, x_n$ and $L = \overline{\lim}\, x_n$. Since the sequence $\{x_n\}$ is bounded above (by the number $\max\{x_1, x_2\}$), $L < +\infty$. Assuming that $l < L$, fix two numbers l' and L' with $l < l' < L < L'$; they will be made more precise later. Then $x_{n-1} < L'$ for all sufficiently large n. Further, n can be chosen so that $x_n < l'$, and hence $x_{n+1} < pl' + (1-p)L' = \lambda$. Since $x_n < l' < \lambda$, it follows that $x_{n+2} < \lambda$, $x_{n+3} < \lambda$, ... Therefore, $L = \overline{\lim}\, x_n \le \lambda$, that is, $L \le pl' + (1-p)L'$. But this is not true if the number l' is chosen sufficiently close to l, and L' sufficiently close to L.

3.6. Since the sequence $\{\ln x_n\}$ has a limit for $p \in (0, 2)$ (see the preceding problem), it suffices to prove that for such values of p the sequence $\{x_n\}$ is bounded. This is obvious if $p \in (0, 1]$, since in this case $x_n \le \max\{x_1, x_2\}$. If $p \in (1, 2)$, by multiplying the inequalities $x_{n+1} \le x_n^p x_{n-1}^{1-p}$ for $n = 2, \dots, N$ we get that $x_{N+1} \le x_1^{1-p} x_2 x_N^{p-1} = Kx_N^{p-1}$. Consequently,

$$x_{N+1} \le K(Kx_{N-1}^{p-1})^{p-1} \le \dots \le K^{1+(p-1)+(p-1)^2+\dots+(p-1)^N} \cdot x_1^{(p-1)^N}.$$

This implies that the sequence $\{x_n\}$ is bounded for $p \in (1, 2)$.

If $p \notin (0, 2]$, then the sequence $\{\exp(p-1)^n\}$ diverges, although $x_{n+1} = x_n^p x_{n-1}^{1-p}$. For $p = 2$ it is possible to take the sequence $\{2^n\}$.

3.7. Prove that the inequalities $x_{n-1} \le c/(n-1)!$ and $x_{n-2} \le c/(n-2)!$ imply the inequality $x_n \le c/n!$. The constant c is chosen so that the last inequality holds for $n = 1, 2$.

3.8. The proofs of these assertions are similar. We confine ourselves to the cases c). Fix a $\beta \in (0, \alpha/k)$. By assumption, there exists a number $c = c_\beta > 0$ such that

$$x_n \le C\frac{x_{n-1} + \cdots + x_{n-k}}{e^{\alpha n}}$$

for $n = k+1, k+2, \ldots$. Let $M > 0$ be a number such that $x_j \le Me^{-\beta j^2/2}$ for $j = 1, \ldots, n-1$. Then $x_n \le kCMe^{-\beta(n-k)^2/2}/e^{\alpha n}$. Therefore, $x_n \le Me^{-\beta n^2/2}$ if $kC \le \exp(n(\alpha - \beta k) + \beta k^2/2)$. Now fix an index n_β large enough that $kC \le \exp(n(\alpha - \beta k) + \beta k^2/2)$ for $n \ge n_\beta$. It can be assumed that $n_\beta \ge k$. The inequality $x_n \le Me^{-\beta n^2/2}$ holds for $n = 1, \ldots, n_\beta$ and for sufficiently large $M = M_\beta$. Its validity for all subsequent indices n is ensured by the choice of n_β.

3.9. Arguing just as in the solution of the preceding problem, we get that it suffices to prove that $\beta(n-k)\ln A_{n-k} - (\beta n - 1)\ln A_n \to +\infty$. For brevity let $a_n = \ln A_n$. Fix a number $\gamma > \theta$; its choice will be made more precise below. For sufficiently large indices j we have that $j(1 - a_{j-1}/a_j) < \gamma$, that is, $a_{j-1}/a_j > 1 - \gamma/j$. Consequently,

$$\frac{a_{n-k}}{a_n} = \prod_{n-k\le j\le n} \frac{a_{j-1}}{a_j} > \prod_{n-k\le j\le n}\left(1 - \frac{\gamma}{j}\right) = 1 - \gamma\frac{k}{n} + O\left(\frac{1}{n^2}\right).$$

Thus,

$$a_{n-k} > a_n\left(1 - \gamma\frac{k}{n} + O\left(\frac{1}{n^2}\right)\right),$$

and hence

$$\begin{aligned}&\beta(n-k)a_{n-k} - (\beta n - 1)a_n \\ &\quad \ge a_n\left(1 - \beta k(1+\gamma) + O\left(\frac{1}{n}\right)\right) \underset{n\to+\infty}{\to} +\infty\end{aligned}$$

if γ is chosen sufficiently close to the number θ.

3.10. Let $C > 0$ be a number such that $x_m \le Ce^{s_m}/\sqrt[k]{A_1\cdots A_m}$ for $1 \le m < n$, $n > k$. Then

$$\begin{aligned} x_n &\le \frac{C}{A_n}\left(\frac{e^{s_{n-1}}}{\sqrt[k]{A_1\cdots A_{n-1}}} + \cdots + \frac{e^{s_{n-k}}}{\sqrt[k]{A_1\cdots A_{n-k}}}\right) \\ &\le \frac{Ce^{s_{n-1}}}{\sqrt[k]{A_1\cdots A_n}}(A_n^{(1-n)/k} + A_n^{(2-n)/k} + \cdots + A_n^{-1/k} + 1).\end{aligned}$$

Therefore,

$$x_n \le \frac{Ce^{s_n}}{\sqrt[k]{A_1\cdots A_n}}\frac{1}{1 - A_n^{-1/k}}e^{-1/(A_n^{1/k}-1)} \le \frac{Ce^{s_n}}{\sqrt[k]{A_1\cdots A_n}}$$

(the last inequality follows from the inequality $e^t \ge 1 + t$). Choosing the factor C large enough that $x_n \le Ce^{s_n}/\sqrt[k]{A_1\cdots A_n}$ for $n = 1, \ldots, k$, we get from this that the inequality holds for all indices n.

3.11. Use the result of the preceding problem. In example d) use the result of problem 2.14b).

CHAPTER III

Functions

§1. Continuity and discontinuities of functions

1.2. g) Consider, for example, a discontinuous strictly increasing function.

1.3. Let $\omega_f(x)$ be the oscillation of f at a point x, that is,

$$\omega_f(x) = \lim_{\delta\to 0}\Big(\sup_{|y-x|<\delta} f(y) - \inf_{|y-x|<\delta} f(y)\Big),$$

and let A be the set of points of discontinuity of the first kind for f. It is clear that $A = \bigcup_{n\geq 1} A_n$, where $A_n = \{x \in A | \omega_f(x) \geq 1/n\}$. If A is assumed to be uncountable, then at least one of the sets A_n is uncountable, say A_m. Let x_0 be a nonisolated point of A_m (see problem I.1.23a)), and let $\{x_k\} \subset A_m$, $x_k \to x_0$, $x_k \neq x_0$ $(k \in \mathbb{N})$. It can be assumed without loss of generality that $x_k > x_0$ for any $k \in \mathbb{N}$. We prove that f does not have a limit from the right at the point x_0. Indeed, since $x_k \in A_m$, there are points x_k' and x_k'' such that $|x_k' - x_k| < |x_k - x_0|$, $|x_k'' - x_k| < |x_k - x_0|$, and $|f(x_k') - f(x_k'')| > 1/2m$. It is clear that $x_k' \to x_0$ and $x_k'' \to x_0$. Thus,

$$\varlimsup_{x\to x_0+0} f(x) - \varliminf_{x\to x_0+0} f(x) \geq \varlimsup |f(x_k') - f(x_k'')| \geq \frac{1}{2m} > 0.$$

Another solution can be obtained by proving countability of the sets

$$\begin{aligned} A_1 = \{x \in E \mid & \text{ there exists the finite right-side} \\ & \text{limit } f(x+0) \text{ and} f(x+0) > f(x)\}, \\ A_2 = \{x \in E \mid & \text{ there exists the finite right-side} \\ & \text{limit } f(x+0) \text{ and } f(x+0) < f(x)\}. \end{aligned}$$

The sets A_3, A_4 are defined in the similar way using left-side limits. To prove countability of A_1, associate with each point $x \in A_1$ a rectangle $P_x = (x, x+\varepsilon) \times (l, L)$, where $f(x) < l < L < f(x+0)$, and $\varepsilon = \varepsilon_x > 0$ is small enough that $f(t) > L$ for all $t \in (x, x+\varepsilon)$. It is easy to see that $P_x \cap P_y = \varnothing$ if $x, y \in A_1$ and $x \neq y$. Choosing in each rectangle P_x a point with rational coordinates, we get a one-to-one mapping of A_1 into $\mathbb{Q}\times\mathbb{Q}$.

The proofs that A_2, A_3, and A_4 are countable are analogous.

1.4. Consider the following functions φ and ψ:

$$\varphi(x)=\begin{cases}0 & \text{for } x\le 0,\\ 1 & \text{for } x>0;\end{cases}\qquad \psi(x)=\begin{cases}0 & \text{for } x<0,\\ 1 & \text{for } x\ge 0.\end{cases}$$

The desired function can be obtained as the sum of a series with terms of the form $\lambda(\varphi(x-t)+\psi(x-s))$, where $t\in E_L$ and $s\in E_R$.

1.5. Assume that the set $D_R(f)$ is at most countable, and prove that the set $D_L(f)$ is at most countable. Let $\omega_f(x)$ be the oscillation of f at a point x (see the solution of problem 1.3). It is clear that $\mathscr{D}_L(f)=\bigcup_{k=1}^{\infty}A_k$, where $A_k=\{x\in\mathscr{D}_L(f)\mid\omega_f(x)\ge 1/k\}$. If we assume that the set $D_L(f)$ is uncountable, then at least one of the sets A_k, say A_m, must be uncountable, and with it also the set $A=A_m\setminus D_R(f)$. Let x_0 be a point in A such that $(x_0, x_0+\varepsilon)\cap A\neq\varnothing$ for any $\varepsilon>0$ (see problem I.1.23b)). Fix a sequence $\{x_n\}\subset A$ such that $x_n\to x_0$, $x_n>x_0$ $(n\in\mathbb{N})$. Since $x_n\in A$, there are points x_n' and x_n'' such that $x_0<x_n'<x_n$, $x_0<x_n''<x_n$, and $|f(x_n')-f(x_n'')|\ge 1/2m$. Using the continuity of f from the right at x_0 and passing to the limit in the last inequality, we arrive at a contradiction.

1.6 b) Assume that a sequence $\{f_n\}$ with the indicated properties exists. We find a point $c\in\mathbb{R}$ such that the sequence $\{f_n(c)\}$ does not converge, and thereby come to a contradiction.

Let Δ_0 be an arbitrary closed interval, and x_1 an arbitrary rational number interior to Δ_0. Take an index n_1 such that $|f_0(x_1)-f_{n_1}(x_1)|<1/3$, that is, $|1-f_{n_1}(x_1)|<1/3$. Using the continuity of f_{n_1}, fix a closed interval Δ_1 centered at x_1 such that $\Delta_1\subset\Delta_0$ and $|1-f_{n_1}(x)|<1/3$ for any $x\in\Delta_1$. Let x_2 be an irrational number interior to Δ_1. Take an index $n_2>n_1$ such that $|f_0(x_2)-f_{n_2}(x_2)|<1/3$, that is, $|f_{n_2}(x_2)|<1/3$. Fix a closed interval Δ_2 centered at x_2 such that $\Delta_2\subset\Delta_1$ and $|f_{n_2}(x)|<1/3$ for any $x\in\Delta_2$. Now take an arbitrary rational number x_3 interior to Δ_2 and repeat the constructions above, replacing Δ_0 by Δ_2 and x_1 by x_3, etc. By induction we construct a sequence $\{n_k\}$ of indices and a sequence $\{\Delta_k\}$ of closed nested intervals such that $|1-f_{n_{2l-1}}(x)|<1/3$ for $x\in\Delta_{2l-1}$ and $|f_{n_{2l}}(x)|<1/3$ for $x\in\Delta_{2l}$. It is clear that $|f_{n_{2l-1}}(c)-f_{n_{2l}}(c)|\ge 1/3$ at a point $c\in\bigcap_{k=1}^{\infty}\Delta_k$, and hence the sequence $\{f_n(c)\}$ cannot have a limit.

1.8. a) Consider the function equal to 1 on an at most countable subset of E with closure E (see problem I.1.22b)) and equal to 0 at the remaining points.

b) Consider the series $\sum 3^{-n}\varphi_n$, where φ_n is the function constructed as described in the hint to part a), for the set $E=E_n$.

1.9. Study the function

$$f(x) = \begin{cases} 1/(n+1) & \text{if the ternary expansion of } x \in (0\,,\,1) \\ & \text{has } n \text{ 1's } (n = 0\,,\,1\,,\,\dots), \\ 0 & \text{if the ternary expansion of } x \in (0\,,\,1) \\ & \text{has infinitely many 1's.} \end{cases}$$

1.10. Prove that if $\underline{\lim}_{x\to x_0} f(x) < c < \overline{\lim}_{x\to x_0} f(x)$ and $c \neq f(x_0)$, then the set $f^{-1}(\{c\})$ is not closed.

1.11. Use the result of the preceding problem.

1.12. Use the fact that it can be assumed without loss of generality that f has a maximum at each point of E. Consider the set $E_\varepsilon = \{x \in E | f(y) \leq f(x) \text{ for } |x-y| < \varepsilon\,,\ y \in E\}$ (ε is a fixed positive number) and prove that f is locally constant on E_ε. Use the result in problem I.1.22a). Another solution can be obtained by modifying the arguments in the second solution of problem 1.3.

The arguments remain in force for a separable metric space. The result is not true in the case of a nonseparable space. To get a counterexample consider the space $X \times S$ without isolated points, where X is a closed interval with the discrete metric and $S = \{z \in \mathbb{C} \mid |z| = 1\}$, along with the function $f(x\,,\,z) = x$.

1.13. Use the result of problem 1.12.

1.14. a) Consider, for example, the function

$$f(x) = \begin{cases} 1 - x & \text{for } -1 < x \leq 0\,, \\ \dfrac{x^{-1} + [x^{-1}]}{1 + x^{-1} + [x^{-1}]} & \text{for } 0 < x < 1\,, \\ \dfrac{1 + (2-x)^{-1} + [(2-x)^{-1}]}{2 + (2-x)^{-1} + [(2-x)^{-1}]} & \text{for } 1 < x < 2 \end{cases}$$

and the point $c = 0$.

1.15. Arguing by contradiction, use the Bolzano-Cauchy theorem to prove the absence of bijectivity.

1.16. a) Use the mean value theorem on the interval $[x\,,\,(1+\delta)x]$.

b) Assuming that $\varepsilon < 1$ and arguing by contradiction, find a sequence $\{z_n\}$ such that $z_n \to +\infty$ and $f^{-1}(\varepsilon z_n) \leq \frac{1}{1+\delta} f^{-1}(z_n)$, where δ is some positive number. Use a).

1.17. Assume that the limit $\lim_{x\to+\infty} f(x)$ does not exist. Then there are numbers a and b such that $a < b$ and the sets $G_a = \{x > 0 | f(x) < a\}$ and $G_b = \{x > 0 | f(x) > b\}$ are not bounded. Let x_0 be a point such that each of the sets G_a and G_b contains infinitely many points of the form nx_0 ($n \in \mathbb{N}$) (see problem I.1.26). Since the sequence $\{f(nx_0)\}$ cannot have a limit, we arrive at a contradiction.

1.18. Consider the function

$$f(x) = \begin{cases} x_n & \text{if } x = \sqrt{p_n}, \\ 0 & \text{if } x \neq \sqrt{p_n}, \text{ for any } n \in \mathbb{N}, \end{cases}$$

where $\{p_n\}$ is the sequence of prime numbers, and $\{x_n\}$ is a sequence whose set of limits of convergent subsequences fills $\mathbb{R}$ (for example, enumerate the set $\mathbb{Q}$ arbitrarily).

1.19. Consider a partition of $\mathbb{R}_+ = (0, +\infty)$ into equivalence classes, with $x \sim y$ if $x/y \in \mathbb{Q}$. After establishing arbitrarily a one-to-one correspondence between the set of equivalence classes constructed in this way and the set $\mathbb{R}$, define the function f on $(0, +\infty)$ by $f(x) = t/(1+[x])$, where $t \in \mathbb{R}$ corresponds to the equivalence class containing x.

1.20. Prove that the assertion $f(x+h) - f(x) \rightrightarrows 0$ as $x \to +\infty$ on any finite interval is equivalent to the assertion that for all $\varepsilon > 0$ there exist a_ε and b_ε such that $a_\varepsilon < b_\varepsilon$ and $\overline{\lim}_{x\to+\infty} \sup_{a_\varepsilon \le h \le b_\varepsilon} |f(x+h) - f(x)| \le \varepsilon$. Prove the last assertion by arguing by contradiction and constructing a sequence $\{\Delta_n\}$ of nested closed intervals and a numerical sequence $\{x_n\}$ such that $x_n \to +\infty$ and $|f(x_n + h) - f(x_n)| \ge \varepsilon$ for $h \in \Delta_n$.

1.21. Prove that the function $\varphi(x) = f(x + 1/n) - f(x)$ takes a zero value on $[0, 1 - 1/n]$. If $\delta \neq 1/n$ and $\delta > 0$, then consider the function $f(x) = \varphi(x) - Ax$, where $\varphi \in C(\mathbb{R})$, $\varphi(0) = 0$, $\varphi(1) = A \neq 0$, and φ has period δ.

1.22. Prove that if $f(x)$ has the property indicated in the condition of the problem, then so does $f(-x)$, and reduce the problem to the cases when f is either even or odd. In the first case show that $f(x) \to f(0)$ as $x \to +\infty$, and conclude from this that $f \equiv \text{const}$. In the case when f is odd prove that $\frac{1}{n} f(nx) \to f(x)$ for any $x \in \mathbb{R}$, and using this verify that $f(\lambda x) = \lambda f(x)$ for any $\lambda \in \mathbb{R}$.

1.23. As in the preceding problem, reduce the question to the case when f is either even or odd. If f is even, then assuming that $f(x_0) \neq f(0)$, consider the sequence $\{(-1)^n x_0\}$. If f is odd, then prove that $f(\lambda x) = \lambda f(x)$, considering successively the cases $\lambda \in \mathbb{N}$, $\lambda \in \mathbb{Z}$, $\lambda \in \mathbb{Q}$, and $\lambda \in \mathbb{R}$.

1.24. a) It follows from the equality $(1+x)(1+x^2)\cdots(1+x^{2^n})\cdots = (1-x)^{-1}$ $(|x| < 1)$ that $f(x)f(x^2) = (1-x)^{-1}$. Therefore, $f^2(x^2) \le (1-x)^{-1} \le f^2(x)$, that is, $1 \le \sqrt{1-x}\, f(x) \le \sqrt{1+\sqrt{x}} \le \sqrt{2}$.

b) Computing the logarithmic derivative of the function

$$F(x) = (1-x)f^2(x) = \prod_{k\ge 0} \frac{1+x^{4^k}}{1+x^{2\cdot 4^k}},$$

we get that

$$x\frac{F'(x)}{F(x)} = \sum_{n\ge 0} (-1)^n \frac{2^n x^{2^n}}{1+x^{2^n}} = S(x) + \sum_{n\ge 7} (-1)^n \frac{2^n x^{2^n}}{1+x^{2^n}}.$$

Since $t/(1+t) < 2t^2/(1+t^2)$ for $t \in (0, \sqrt{2}-1)$, it follows that the series $\sum_{n\geq 7}(-1)^n 2^n x^{2^n}/(1+x^{2^n})$ satisfies the Leibnitz test when $x^{2^7} \in (0, \sqrt{2}-1)$, and its sum is negative. Direct computations show that $S(x) < 0$ for $0.93 < x < 0.945$.

c) By what was proved in b), there are points s and t in $(\sqrt{2}-1, 1)$ such that $s < t$ and $F(s) > F(t)$. Let $s_n = s^{4^{-n}}$, $t_n = t^{4^{-n}}$ $(n = 0, 1, 2, \dots)$. Using the identity

$$F(x) = \frac{1+x}{1+x^2} F(x^4)$$

and the fact that the function $(1+x)/(1+x^2)$ is decreasing on $(\sqrt{2}-1, 1)$, we get the inequality

$$\begin{aligned} F(s_n) - F(t_n) &= \frac{1+s_n}{1+s_n^2} F(s_{n-1}) - \frac{1+t_n}{1+t_n^2} F(t_{n-1}) \\ &\geq \frac{1+s_n}{1+s_n^2}(F(s_{n-1}) - F(t_{n-1})) \geq F(s_{n-1}) - F(t_{n-1}). \end{aligned}$$

Thus, $F(s_n) - F(t_n) \geq F(s) - f(t) > 0$, although s_n, $t_n \to 1$.

1.25. Consider the set $Q = f^{-1}(K) \subset [0, 1]$, where $f = (\varphi, \psi)$ is a Peano curve (see problem 4.5), and $K = \{(x, y) | 0 \leq x \leq 1, y = 0, 1, \frac{1}{2}, \frac{1}{3}, \dots\}$. Verify that the restriction $\varphi|_Q$ has the required property (use the result of problem 4.4c)). Extend the function $\varphi|_Q$ in a suitable way from Q to $[0, 1]$.

1.26. For each $n \in \mathbb{N}$ and $x \in [0, 1]$ define $f_n(x, y)$ by linear interpolation, with values $f(x, k/n)$ at the nodes $y_k = k/n$ $(k = 0, 1, 2, \dots, n)$.

1.27. Arguing by contradiction, we see that $|f(a, y)| \geq c$ for $|y - b| < \varepsilon$ for some $a, b \in \mathbb{R}$, $c > 0$, and $\varepsilon > 0$. Let $E_k = \{y \in (b-\varepsilon, b+\varepsilon) | f(x, y) \geq c/2 \text{ for } |x - a| \leq 1/k\}$ $(k \in \mathbb{N})$. Show that $\bigcup_{k=1}^{\infty} E_k = (b - \varepsilon, b + \varepsilon)$ and use the result of problem I.1.33. Prove that if the closure of E_m contains an interval (α, β), then $f(x, y) \neq 0$ on $(a - 1/m, a + 1/m) \times (\alpha, \beta)$.

§3. Continuous and differentiable functions

3.1. Show that the mth-degree interpolation polynomial P for f constructed for the nodes $0, h, 2h, \dots, mh$ coincides with f at all points kh $(k \in \mathbb{Z})$. Conclude from this that $P(t) = f(t)$ for $t = kh/2^n$ $(k \in \mathbb{Z}, n \in \mathbb{N})$.

3.2. Show that $f'(x) \to 0$ as $x \to +\infty$, assuming that the limit $\lim_{x\to+\infty} f(x)$ exists. Prove the existence of the last limit by arguing by contradiction: if $\alpha = \underline{\lim}_{x\to+\infty} f(x) < \beta = \overline{\lim}_{x\to\infty} f(x)$, then there is a sequence $x_n \to +\infty$ such that $f'(x_n) = 0$, $f(x_{2n-1}) \to \alpha$, and $f(x_{2n}) \to \beta$.

Another proof of the equality $\lim_{x\to+\infty} f(x) = l$, extremely short but not as instructive, can be obtianed by applying L'Hôspital's rule to the pair of functions $e^x f(x)$ and e^x.

b) To prove the converse, show that if $f'(x)$ does not tend to 0 as $x \to +\infty$, then (since f' is uniformly continuous) the function does not satisfy the Bolzano-Cauchy criterion at infinity.

3.3. Assuming that $f'(a) < 0 < f'(b)$, prove that the smallest value of f is attained in the interior of $[a, b]$.

3.4. Use the results of problems 3.3 and 1.10.

3.5. In checking the necessity use the uniform continuity of f' on $[a, b]$.

3.6. a) Assume first that $g(x) > 0$ for $x \in (a, b)$, and suppose that there are points $x_1, x_2 \in (a, b)$ such that $x_1 < x_2$ and $f(x_1) > f(x_2)$. It can be assumed without loss of generality that $f(x_1) = 1$ and $f(x_2) = -1$. Let $\overline{x} = \sup\{x \in (x_1, x_2) | f(y) \geq 0$ for $y \in (x_1, x)\}$. There exists a sequence $\{h_n\}$ such that $h_n > 0$, $h_n \to 0$, and $f(\overline{x} + h_n) < 0$. Therefore,

$$g(x) = \lim(f(\overline{x} + h_n) - f(\overline{x} - h_n))/2h_n \leq 0,$$

which is impossible. Consequently, f is increasing. In the general case consider the function $f_\varepsilon(x) = f(x) + \varepsilon x$, where ε is an arbitrary positive number. If $g \equiv 0$, then the result already obtained should be applied to the functions f and $-f$.

b) Apply part a) to the function $f - G$, where G is a primitive of g.

3.7. If f is differentiable, then $g = f'$, and hence $f \in C^\infty((a, b))$. Differentiating the identity $f(x+h) - f(x-h) = 2hg(x)$ twice with respect to h, show by using 3.6a) that $f'' \equiv$ const. Using 3.6b), we see that this result is preserved if f and g are continuous on (a, b). If only the continuity of f is assumed, then the result becomes false ($f(x) = |x|$, $g(x) = \operatorname{sgn} x$).

3.8. Consider a primitive of the function φ defined by $\varphi(x) = \inf\{|x - y| \mid y \in F\}$ $(x \in \mathbb{R})$.

3.9. To prove the assertion b) $\Rightarrow$ a), study the set $E_\varepsilon = \{x \in [0, 1] | \, |f(x) - f(0)| \leq \varepsilon x\}$, where ε is an arbitrary positive number. Prove that $\sup(E_\varepsilon \cap (0, c)) = c$ for any $c \in (0, 1)$. The statement c) does not imply a). A corresponding counterexample can be obtained by considering one of the coordinate functions of the Peano curve constructed in problem 4.4 and using the fact that a set of constancy of this function does not have isolated points (see assertion d) of this problem).

3.12. Estimate the increment of the function f on a region of monotonicity. For $\alpha = 1$ it is useful to employ the result in the preceding problem. For $\alpha = 1$ we have $g'(0) = 2/\pi$.

3.13. a) See the hint for the preceding problem.

b) Estimate $f'(x)$.

c) Show that $|f(x+h)-f(x)| = O(|f(x_0+h)-f(x_0)|)$, where x_0 is an endpoint of the smallest region of monotonicity of f whose length is not less than h.

3.14. Consider the series with terms of the form $\lambda_k f(x-a_k)$, where f is the function in the preceding problem with suitably chosen parameters.

3.15. Consider the functions $f(x) = (\ln(2/x))^{-1}$ and $g(x) = e^{-1/x}\cos e^{1/x}$ for $0 < x \leq 1$, $f(0) = g(0) = 0$.

3.16. Show that

$$|\varphi(x+h)-\varphi(x)| \leq \max_{x\in\mathbb{R}} |f(x)| \int_{-\pi}^{\pi} |\omega(z+h) - |z|^{-1/2}|\, dz,$$

where ω is a 2π-periodic function coinciding with $|z|^{-1/2}$ on the interval $[-\pi, \pi]$, and estimate the integral on the right-hand side of the inequality.

3.17. To prove the continuity of f use its monotonicity.

3.18. Fixing an $n \in \mathbb{N}$, get a lower estimate of the length of the polygonal curve with vertices at the points $M_j = (j3^{-n}, f(j3^{-n}))$ $(j = 0, 1, \ldots, 3^n)$ by considering the sums $\sum_{j3^{-n}\in K} \rho(M_j, M_{j+1})$ and $\sum_{j3^{-n}\notin K} \rho(M_j, M_{j+1})$, where K is the Cantor set and ρ is the metric in $\mathbb{R}^2$.

3.19. Compute the increment of the Cantor function on $[0, 3^{-n}]$.

§4. Continuous mappings

4.1. Prove that the boundedness of the sequence $\{z_n\}$ is equivalent to the boundedness of the sequence $\{P(z_n)\}$. This is not true for arbitrary polynomials in two variables (for example, for $Q(x, y) = x$). Verify that $Q(F) = \mathbb{R}\setminus\{0\}$, where $F = \{(x, y) | xy \geq 1\}$ is a closed set.

4.3. To construct one of the possible examples, consider two-sided numerical sequences $\{x_k\}_{k\in\mathbb{Z}}$ and $\{y_k\}_{k\in\mathbb{Z}}$ satisfying the conditions

$$\cdots < x_k < y_k < x_{k+1} < \cdots, \ \inf x_k = 0, \ \sup x_k = 1,$$

and form the sets $A = \{0\} \cup \bigcup_{k\in\mathbb{Z}} [x_k, y_k)$ and $B = \{1\} \cup \bigcup_{k\in\mathbb{Z}} [y_k, x_{k+1})$. The desired homeomorphism can be obtained by mapping each interval $[x_k, y_k) \subset A$ onto the interval $[y_{-k}, x_{-k+1}) \subset B$ and 0 into 1.

4.4. a) The unambiguity of the definition can be verified by a direct computation of φ and ψ at ternary-rational points. The continuity of φ and ψ follows from the coincidence of any previously specified number of first digits obtained in the representation of sufficiently close points $t, t' \in [0, 1]$ as ternary fractions (without the digit 2 in a period).

b) Using the expansion of numbers in $[0, 1]$ as ternary fractions $0.\alpha_1\alpha_2\alpha_3\cdots = \sum \alpha_k 3^{-k}$, where $\alpha_k = 0, 1$, or 2, show that if $\xi, \eta \in [0, 1]$, $\xi = 0.\beta_1\beta_2\beta_3\cdots$, and $\eta = 0.\gamma_1\gamma_2\gamma_3\cdots$, then $\xi = \varphi(t)$ and $\eta = \psi(t)$, where $t = 0.\alpha_1\alpha_2\alpha_3\ldots$, and the digits $\alpha_1, \alpha_2, \ldots$ are determined successively as follows:

$$\alpha_1 = \beta_1,$$
$$\alpha_2 = \begin{cases} \gamma_1 & \text{if } \alpha_1 \text{ is even}, \\ 2-\gamma_1 & \text{if } \alpha_1 \text{ is odd}; \end{cases}$$
$$\cdots\cdots\cdots\cdots\cdots$$
$$\alpha_{2n-1} = \begin{cases} \beta_n & \text{if } \alpha_2+\alpha_4+\cdots+\alpha_{2n-2} \text{ is even}, \\ 2-\beta_n & \text{if } \alpha_2+\alpha_4+\cdots+\alpha_{2n-2} \text{ is odd}; \end{cases}$$
$$\alpha_{2n} = \begin{cases} \gamma_n & \text{if } \alpha_1+\alpha_3+\cdots+\alpha_{2n-1} \text{ is even}, \\ 2-\gamma_n & \text{if } \alpha_1+\alpha_3+\cdots+\alpha_{2n-1} \text{ is odd}. \end{cases}$$

c) Different solutions of the system $\varphi(t)=\xi$, $\psi(t)=\eta$ can be obtained only due to nonuniqueness of the representation of the numbers $\xi, \eta \in [0, 1]$ as ternary fractions. Verify that for $\xi = 2/9 = 0.02000\ldots = 0.01222\ldots$ and $\eta = 1/3 = 0.1000\ldots = 0.0222\ldots$ the system has four solutions: $5/108 = 0.001020202\ldots$, $11/108 = 0.00220202\ldots$, $13/108 = 0.010020202\ldots$, and $19/108 = 0.011202020\ldots$.

d) Let $t = 0.\alpha_1\alpha_2\ldots \in \varphi^{-1}(\{c\})$. If infinitely many of the digits α_{2k} are even, then, replacing α_{2k} by $2-\alpha_{2k}$ for sufficiently large k, we get a point in $\varphi^{-1}(\{c\})$ arbitrarily close to t. If $\alpha_{2k}=1$ for $k \geq k_0$, then consider the point $(k \geq k_0)$ $0.\alpha_1\alpha_2\cdots\alpha_{2k-1}0(2-\alpha_{2k+1})0\alpha_{2k+3}1\alpha_{2k+5}\cdots$ obtained from t by replacing the digits $\alpha_{2k}=1$ and $\alpha_{2k+2}=1$ by 0's, and α_{2k+1} by the number $2-\alpha_{2k+1}$. The arguments are similar for the consideration of the set $\psi^{-1}(\{c\})$.

e) Prove that $\varphi \in \mathrm{Lip}_{1/2}([0, 1])$. For $2 < m \in \mathbb{N}$ let $p = [m/2]$, and take $t, t' \in [0, 1]$, $t = 0.\alpha_1\alpha_2\alpha_3\ldots$, $t' = 0.\alpha_1'\alpha_2'\alpha_3'\ldots$, such that $3^{-m} \leq t-t' < 3^{-m+1}$. If $\alpha_1 = \alpha_1', \ldots, \alpha_{m-1} = \alpha_{m-1}'$, then

$$\varphi(t) = 0.\beta_1\cdots\beta_p\beta_{p+1}\cdots, \qquad \varphi(t') = 0.\beta_1\beta_2\cdots\beta_p\beta_{p+1}'\cdots,$$

and hence $|\varphi(t)-\varphi(t')| \leq 3^{-p} \leq \sqrt{3}|t-t'|^{1/2}$. Assume now that there is an index $k \leq m-1$ such that $\alpha_k \neq \alpha_k'$. Let k is the smallest index with this property. Then in view of the inequalities $3^{-m} \leq t-t' < 3^{-m+1}$ the expansions for t and t' must have the form $t = 0.\alpha_1\cdots\alpha_{k-1}\alpha_k0\cdots 0\alpha_m\alpha_{m+1}\cdots$ and $t' = 0.\alpha_1\cdots\alpha_{k-1}(\alpha_k-1)22\cdots2\alpha_m'\alpha_{m+1}'\cdots$ (if at least one of the digits $\alpha_{k+1}, \ldots, \alpha_{m-1}$ is not 0 or at least one of the digits $\alpha_{k+1}', \ldots, \alpha_{m-1}'$ is less than 2, then $t-t' \geq 3^{-m+1}$). The following four cases are possible:

1) $k=2l$ is even, and $\alpha_2+\alpha_4+\cdots+\alpha_k$ is even;
2) $k=2l$ is even, and $\alpha_2+\alpha_4+\cdots+\alpha_k$ is odd;
3) $k=2l-1$ is odd, and $\alpha_1+\alpha_3+\cdots+\alpha_k$ is even;
4) $k=2l-1$ is odd, and $\alpha_1+\alpha_3+\cdots+\alpha_k$ is odd.

In the first case we see that $\varphi(t) = 0.\beta_1\beta_2\cdots\beta_l0\cdots0\beta_p\beta_{p+1}\cdots$ and $\varphi(t') =$

$0.\beta_1\beta_2\cdots\beta_l\cdots 0\beta'_p\beta'_{p+1}\cdots$. Consequently, $|\varphi(t) - \varphi(t')| \leq 3.3^{-p} \leq 3\sqrt{3}|t'-t|^{1/2}$. The remaining cases are handled similarly.

4.5. Let $\varepsilon > 0$ and $N = 1 + [1/\varepsilon]$. The points $t_k = k/N$ $(k = 1, \dots, N)$ form an ε-net* for $[0, 1]$. Verify that the points $(u(t_k), v(t_k))$ form a δ-net for the square $[0, 1] \times [0, 1]$, where $\delta \leq C\varepsilon^\alpha$. Prove that the number of points in a δ-net for the square is not less than $C_0\delta^{-2}$.

4.6. Let Ξ be the set of all binary sequences, and let $\{\Delta_{\varepsilon_1\cdots\varepsilon_n} | n \in \mathbb{N}, \varepsilon_1, \dots, \varepsilon_n \in \{0, 1\}\}$ and $\{\widetilde{\Delta}_{\varepsilon_1\cdots\varepsilon_n} | n \in \mathbb{N}, \varepsilon_1, \dots, \varepsilon_n \in \{0, 1\}\}$ be the families used to define the generalized Cantor sets K and $\widetilde{K}$ (see problem I.1.31). Define the mappings $\varphi: \Xi \to K$ and $\tilde{\varphi}: \Xi \to \widetilde{K}$ by the equalities $\varphi(\varepsilon) = t$ and $\tilde{\varphi}(\varepsilon) = \tilde{t}$, where $t \in \bigcap_{n=1}^{\infty} \Delta_{\varepsilon_1\cdots\varepsilon_n}$ and $\tilde{t} \in \bigcap_{n=1}^{\infty} \widetilde{\Delta}_{\varepsilon_1\cdots\varepsilon_n}$ $(\varepsilon = \{\varepsilon_k\} \in \Xi)$. Prove that φ and $\tilde{\varphi}$ are bijections, and $f = \tilde{\varphi} \circ \varphi^{-1}$ is a homeomorphism of K onto $\widetilde{K}$.

§5. Functional equations

5.2. c) It is evident that f is not linear, and hence $f(x_0) \neq x_0 f(1)$ for some $x_0 \in \mathbb{R}$. This inequality is equivalent to the linear independence of the vectors $e' = (1, f(1))$ and $e'' = (x_0, f(x_0))$. Therefore, the set $\{se' + te'' | s, t \in \mathbb{R}\}$ coincides with $\mathbb{R}^2$, and the set $\{se' + te'' | s, t \in \mathbb{Q}\}$ is dense in $\mathbb{R}^2$. It remains to see that the points $se' + te''$ belong to the graph of f for any rational s and t (this follows from a)).

5.3. Apply the result of the preceding problem to the function $g(x) = f(e^x)$.

5.4. Show that $f(1) = f(-1) = 0$, and prove that f is odd. Consider the function $g(x) = \frac{1}{x}f(x)$ $(x > 0)$ and use the arguments in the solution of problem 5.3.

5.5. Use the result of problem 5.2.

5.6. Prove successively that $f(0) = 0$, $f(x^n) = (f(x))^n$, and $f(x+y) = f(x) + f(y)$, for $x, y \in \mathbb{R}$. Using the equality $f((rx+y)^n) = (rf(x)+y)^n$, where $r \in \mathbb{Q}$, show that $f(x^2) = \pm(f(x))^2$ for any $x \in \mathbb{R}$; use the result of problem 5.2.

5.7. Prove by induction that

$$\left| \sum_{1\leq k\leq p} f(x_k) - f\left(\sum_{1\leq k\leq p} x_k \right) \right| \leq \sigma(p-1) \qquad (p \in \mathbb{N}). \qquad (*)$$

Derive from this that the ratio $(1/n)f(\pm n)$ $(n \in \mathbb{N})$ is bounded. Fix strictly increasing sequences $(n_k) \subset \mathbb{N}$ and $\{m_k\} \subset \mathbb{N}$ such that the limits

$$A = \lim \frac{f(n_k)}{n_k} \quad \text{and} \quad B = \lim \frac{f(-m_k)}{-m_k}.$$

*The definition of an ε-net is given in VIII.5

exist. We prove that

$$\lim_{x\to+\infty}\frac{f(x)}{x}=A \quad\text{and}\quad \lim_{x\to-\infty}\frac{f(x)}{x}=B.$$

For any $\varepsilon>0$ fix an index n_k such that $\sigma/n_k<\varepsilon$ and $|f(n_k)/n_k-A|<\varepsilon$. Let $x>0$ and $p=[x/n_k]$. Then $x=pn_k+s$, $0\le s<n_k$. It follows from $(*)$ that $|pf(n_k)+f(s)-f(x)|\le\sigma p$, and hence

$$\left|\frac{p}{pn_k+s}f(n_k)+\frac{f(s)}{x}-\frac{f(x)}{x}\right|\le\frac{\sigma p}{x}<\varepsilon.$$

For sufficiently large x the last inequality implies that $|f(x)/x-A|<3\varepsilon$. It is proved similarly that $f(x)/x\to B$ as $x\to-\infty$. Passing to the limit as $x\to+\infty$ in the inequality

$$\left|\frac{f(2x)}{x}+\frac{f(-x)}{x}-\frac{f(x)}{x}\right|\le\frac{\sigma}{x},$$

we see that $A=B$. Consider now the function $\tilde f(x)=f(x)-Ax$ and verify that $|\tilde f(x)|\le\sigma$ for any $x\in\mathbb{R}$. Indeed, it is clear that $\tilde f(x)/x\to 0$ as $x\to\infty$, and $\tilde f$, like f, satisfies the inequality $(*)$. Therefore, $|\tilde f(x)-\tilde f(nx)/n|\le\sigma$ for any $x\in\mathbb{R}$ and $n\in\mathbb{N}$. Assuming that $x\ne 0$ and passing to the limit in the last inequality, we get what is required.

We remark that the estimate for $\tilde f$ cannot be improved, as we can see from the example of the function $f(x)=\max(1-|x|,0)$.

5.8. Show that $f(0)=0$ and that f is even. Prove that if $f\not\equiv 0$, then $f(x)\ne 0$ for $x\ne 0$. To do this verify that if $f(a)\ne 0$ $(a>0)$, then $f(x)\ne 0$ on $(0,a)$. Indeed, suppose that this is not so, and let c be the largest zero of f on $(0,a)$. Then $f(c/2^n)=0$ for any $n\in\mathbb{N}$, and hence $f(c\sqrt{1+2^{-2n}})=f(c)+f(c/2^n)=0$. Since $c\sqrt{1+2^{-2n}}\in(c,a)$ for sufficiently large n, we arrive at a contradiction to the choice of the number c. Note further that a can be chosen arbitrarily large, because $f(2^na)=4^nf(a)\ne 0$. Accordingly, $f(x)\ne 0$ for $x\ne 0$. Assume that $f(1)=1$ and consider the set $A=\{x>0 | f(x)=x^2\}$. Verify that its closure coincides with $[0,+\infty)$ (see problem I.1.21).

5.9. Prove that $f(0)=1$ and that f is even. Show that if $f(x_0)=0$, then $f(t)=0$ for $|t|\ge|x_0|$. Derive from this that $f(x)>0$ for any $x\in\mathbb{R}$. Verify that the function $g(x)=\ln f(x)$ $(x\in\mathbb{R})$ satisfies the condition of the preceding problem.

5.10. Prove that the function $g(x)=f(e^x)$ is the sum of a linear function and a periodic function.

5.11. a) Consider the functions $\psi(x)=\varphi(x)/\ln x$ $(x\in(0,1))$ and $H(u)=\psi(e^{-e^u})$ $(u\in\mathbb{R})$. .

b) Verify that $0 \leq f \leq 1$ and that if $f(x_0) = 0$ $(f(x_0) = 1)$ at some point $x_0 \in (0, 1)$, then $f \equiv 0$ $(f \equiv 1)$. Assuming that $0 < f < 1$ in $(0, 1)$, prove that $\lim_{x\to+0} f(x) = 0$ and $\lim_{x\to1-0} f(x) = 1$ and that f is strictly increasing. Consider the function $g(t) = \ln|\ln f(e^{-t})|$ and use the result of problem 5.10.

5.12. Assuming that the desired function h is odd, show that the function $\varphi(t) = \ln(h(e^t))$ $(t \in \mathbb{R})$ is strictly increasing and satisfies the equation $\varphi(2t) = 4\varphi(t)$. Find a solution of this equation.

5.13. Verify that $f(0) = 0$ and $f(1) = 1$. For $z \neq 0$ consider the function $g(z) = f(z)/f(|z|)$ and by using the fact that g is one-to-one on each circle $|z| = r$ prove successively that (for $r \in (0, 1]$) $g(r) = 1$, $g(-r) = -1$, $g(ir) = \pm i$, $g(re^{i\pi/4}) = e^{\pm i\pi/4}$, etc. Deduce from this that $g(z)$ coincides either with $z/|z|$ or with $\overline{z}/|z|$. Prove that the function $h(t) = |f(t)|$ is increasing on $(0, 1)$. Represent $f(t)$ in the form $h(t)e^{i\varphi(t)}$ and use the result of problem 5.11.

5.14. Prove successively that $f(0) = 0$, $f(1) = 1$ (and hence f is increasing), $f(1/2) = 1/2$, and $f(x) = 1/2$ for $x \in [1/3, 2/3]$. Use induction to see that f coincides with the Cantor function (see problem 3.17) on the open intervals complementary to the Cantor set.

5.15. Prove successively that $\partial^2 f/\partial x\partial y \equiv \text{const}$, $f(x, y) = Axy + g(x) + g(y)$, and $g'' \equiv \text{const}$.

5.16. Verify that if $f(a) = f(b)$ for $|a| \neq |b|$ and the function g exists, then $f(t) = f(ct)$ for some c with $|c| < 1$ and for any $t \in \mathbb{R}$. Deduce from this that $f \equiv \text{const}$.

5.17. a) Verify that for each $\xi \in S^1$ the mapping $\varphi_\xi \colon \mathbb{Z} \to S^1$ defined by $\varphi_\xi(k) = \xi^k$ is a character. Prove that each character φ has the form φ_ξ, where $\xi = \varphi(1)$.

5.18. a) Verify that for each $t \in \mathbb{R}$ the mapping $\varphi_t \colon \mathbb{R} \to S^1$ defined by $\varphi_t(x) = e^{itx}$ $(x \in \mathbb{R})$ is a character. If φ is an arbitrary continuous character, then $\operatorname{Re}\varphi(2^{-n}) > 0$ for $n \geq N$. Let $\varphi(2^{-N}) = e^{i\theta_0}$, $|\theta_0| < \pi$, $\theta = 2^N\theta_0$. Verify that $\varphi(2^{-n}) = e^{i\theta 2^{-n}}$ for any $n \in \mathbb{N}$, and prove that $\varphi = \varphi_\theta$.

b)–d). Use a) and the analogy with problem 5.17b).

e) Verify that for each $n \in \mathbb{Z}$ the mapping $\varphi_n \colon S^1 \to S^1$ defined by $\varphi_n(\xi) = \xi^n$ $(\xi \in S^1)$ is a character. To prove that each character φ has this form, study the set $\varphi^{-1}(\{1\})$. Prove that it is finite if $\varphi \not\equiv 1$, and that for some $m \in \mathbb{N}$ the equality $\xi^m = 1$ holds for any $\xi \in \varphi^{-1}(\{1\})$. Prove that if $n \in \mathbb{N}$ is the smallest such m, then $\varphi = \varphi_n$ or $\varphi = \varphi_{-n}$.

f) Use the fact that the groups $\mathbb{R}_+$ and $\mathbb{R}$ are isomorphic.

g) Use the fact that the groups $\dot{\mathbb{C}}$ and $\mathbb{R}_+ \times S^1$ are isomorphic.

5.19. a) See problem 5.2.

b) Consider $\ln\varphi(x)$.

c) Consider $\varphi(e^x)$.

d) Consider $\ln \varphi(e^x)$.

e) See problem 5.18a).

f) Use the fact that the groups $\dot{\mathbb{C}}$ and $\mathbb{R}_+ \times S^1$ are isomorphic. Prove that if $\varphi\colon S^1 \to \mathbb{R}_+$ is a continuous homeomorphism, then $\varphi \equiv 1$. Use part d) and the result of problem 5.18g).

CHAPTER IV

Series

§1. Convergence

1.1. Find the number of indices $n \in [10^N, 10^{N+1})$ whose decimal expansions do not contain the digit 9, and get an upper estimate for the sum $\sum_{10^N \le n < 10^{N+1}} \varepsilon_n/n$.

1.2. a), b) Consider the numerators of adjacent terms and show that they cannot be simultaneously close to zero.

c) It suffices to prove that the series

$$\sum \tfrac{1}{n}(|\sin(n-1)^2| + |\sin n^2| + |\sin(n+1)^2|)$$

diverges. To do this, show that the numerator is bounded below by a positive number. Indeed, if not, then for arbitrarily large indices n we have that $n^2 = \pi k + \varepsilon$, $(n-1)^2 = \pi k' + \varepsilon'$, and $(n+1)^2 = \pi k'' + \varepsilon''$, where $k, k', k'' \in \mathbb{Z}$ and $|\varepsilon|, |\varepsilon'|, |\varepsilon''| < \frac{1}{4}$. Therefore, $2 = (n-1)^2 - 2n^2 + (n+1)^2 = \pi(k' + k'' - 2k) + \varepsilon' + \varepsilon'' - 2\varepsilon$, which is impossible.

1.3. For $a \le 1$ consider the sums $\sigma_m = \sum_{n \in \mathrm{N}_m} n^{-a}\cos(b \ln n)$, where $N_m = \{n \in \mathbb{N} \mid 2\pi m \le b \ln n \le 2\pi m + \pi/4\}$, and show that $\sigma_m \not\to 0$.

1.4. a) Consider separately the cases $p > 1$ and $0 < p < 1$.

b) To study the sums $\sum_{m^2 \le n < (m+1)^2} (-1)^m n^{-p}$ use the result of problem VI.3.13 for $1/2 < p \le 1$.

c) Use the relation $\arccos x \sim \sqrt{2(1-x)}$ for $x \to 1 - 0$.

1.5. a) Obviously, $\tau(n) \le 2\sqrt{n}$.

b) Use the result of problem I.2.5b).

c) Use the equality $\varphi(p_n) = p_n - 1$.

d) Prove the following more general assertion: if $a_n > 0$, $a_n \uparrow$, and $\sum 1/a_n = +\infty$, then $\sum 1/(A_n - A_{n-1}) = +\infty$, where $A_n = na_n$ $(A_0 = 0)$. Let

$$\sigma_m = 2^{-m} \sum_{2^m < n \le 2^{m+1}} \frac{1}{A_n - A_{n-1}}.$$

Since the harmonic mean does not exceed the arithmetic mean, it follows that

$$\frac{1}{\sigma_m} \le \frac{1}{2^m} \sum_{2^m < n \le 2^{m+1}} (A_n - A_{n-1}) = \frac{1}{2^m}(A_{2^{m+1}} - A_{2^m}) \le 2^{-m} A_{2^{m+1}}.$$

Consequently,

$$\sum_{2^m<n\le 2^{m+1}} \frac{1}{A_n - A_{n-1}} \ge \frac{2^{2m}}{A_{2^{m+1}}} = \frac{1}{4}\frac{2^{m+1}}{a_{2^{m+1}}}.$$

Therefore,

$$\sum \frac{1}{A_n - A_{n-1}} = \sum_{m\ge 0} \sum_{2^m<n\le 2^{m+1}} \frac{1}{A_n - A_{n-1}} \ge \frac{1}{4}\sum_{m\ge 0} \frac{2^{m+1}}{a_{2^{m+1}}} = +\infty$$

(the last series diverges according to the Cauchy theorem; see problem 2.2a)).

1.6. Let $\alpha_n = \min_{m\in\mathbb{N}} |\sqrt{2} - m/n|$. Since $\alpha_n < 1/2n$, it must be shown that $\sum 1/(\alpha_n n^{1+a}) < +\infty$. Let $N_m = \{n \in \mathbb{N} \mid c_{m-1}n^{-2} \le \alpha_n < c_m n^{-2}\}$, where $c_0 = 1/4$ and $c_m \uparrow +\infty$ (the choice of the sequence $\{c_m\}$ will be made more precise below). Since $\alpha_n > c_0 n^{-2}$ (see problem I.3.11a)), it follows that $\mathbb{N} = \bigcup N_m$. Let $N_m = \{n_1, n_2, \dots\}$. Since $c_{m-1}/n_1^2 \le \alpha_{n_1} < 1/2n_1$, it follows that $n_1 > 2c_{m-1}$. Using the result of problem I.3.11b), we get that

$$n_{j+1} > n_1 \left(1 + \frac{1}{8c_m}\right)^j > 2c_{m-1}\left(1 + \frac{1}{8c_m}\right)^j.$$

Consequently,

$$\begin{aligned} s_m = \sum_{n\in\mathbb{N}_m} \frac{1}{\alpha_n n^{1+a}} &\le \frac{1}{c_{m-1}} \sum_{n\in N_m} \frac{1}{n^{a-1}} \\ &\le \frac{2}{(2c_{m-1})^a} \sum_{j\ge 0} \left(1 + \frac{1}{8c_m}\right)^{j(1-a)} \le \operatorname{const} c_m (c_{m-1})^{-a}. \end{aligned}$$

Taking $C_m = 2^{m-2}$, we get that

$$\sum \frac{1}{\alpha_n n^{1+a}} = \sum s_m \le \operatorname{const} \sum 2^{m(1-a)} < +\infty.$$

1.7. Instead of double series it is convenient to consider sums of the form $\sum_{m\ge1}(\sum_{1\le n\le m} \cdots)$. In examples e) and g) consider such a sum, extended only over the prime values of m, and use the result of problem 1.5.b).

f) Use the inequality

$$\sum_{\mathrm{GCD}(m,n)=l} \frac{\mathrm{GCD}(m,n)}{n^3 + m^3} \le \sum_{k,j\ge1} \frac{l}{(lk)^3 + (lj)^3}.$$

h) Use the equality $\mathrm{LCM}(n,m)\,\mathrm{GCD}(n,m) = nm$ and the method used in the solution of problem f).

1.8. Each fraction $1/2^n$ $(n = 0, 1, \dots)$ appears in the series 2^n times with the sign $+$ and 2^n times with the sign $-$. The desired rearrangement is obtained by multiplying the series by -1.

§2. Properties of numerical series connected with monotonicity

2.1. If $a_n/a_{n+1} \not\to 1$, then the series diverges. Suppose that $a_n/a_{n+1} \to 1$. Since $\arccos x \sim \sqrt{2(1-x)}$ as $x \to 1 - 0$, the given series converges simultaneously with the series $\sum(1 - a_n/a_{n+1})$, and the latter converges simultaneously with the series $\sum \ln(a_n/a_{n+1}) = \lim_{N\to\infty} \ln(a_1/a_N)$.

2.2. a) Use the inequalities $2^m a_{2^{m+1}} \le \sum_{2^m<n\le 2^{m+1}} a_n \le 2^m a_{2^m}$.

b), c) Use a).

2.3. To prove the relation $a_n = o(1/n)$ use the result of problem 2.2a).

2.4. To prove the relation $a_n = o(1/\ln n)$ use the results of problems 2.2a) and 2.3.

2.5. Using the result of problem 2.3, show that $a_n^p = o(n^{-1/p} a_n^{p-1})$. The assertion is false for nonmonotone sequences; a counterexample is given by the sequence $a_n = 1$ for $n = 2^k$, $a_n = 0$ for $n \ne 2^k$.

2.6. a) Let $\sigma_m = \sum_{1\le n\le m} x_n a_n$ ($\sigma_0 = 0$). By changing x_1 it can be assumed that $\sigma_m \to 0$. Let $\varepsilon > 0$ and let $|\sigma_m| < \varepsilon$ for $m > M$ ($M \in \mathbb{N}$). Then setting $\sigma = \max_{m\in\mathbb{N}} |\sigma_m|$, we get that

$$\begin{aligned} a_m\left|\sum_{1\le n\le m} x_n\right| &= a_m\left|\sum_{1\le n\le m} \frac{\sigma_n - \sigma_{n-1}}{a_n}\right| \\ &= a_m\left|\frac{\sigma_m}{a_m} + \sum_{1\le n<m} \sigma_n\left(\frac{1}{a_n} - \frac{1}{a_{n+1}}\right)\right| \\ &\le \varepsilon + a_m\sigma \sum_{1\le n<M}\left(\frac{1}{a_{n+1}} - \frac{1}{a_n}\right) \\ &\quad + \varepsilon a_m \sum_{M\le n<m}\left(\frac{1}{a_{n+1}} - \frac{1}{a_n}\right) \\ &\le \varepsilon + \sigma\frac{a_m}{a_M} + \varepsilon. \end{aligned}$$

Since $a_m \to 0$, it follows that $\overline{\lim}\, a_m|\sum_{1\le n\le m} x_n| \le 2\varepsilon$. The required assertion follows from this, since $\varepsilon > 0$ is arbitrary.

For nonmonotone sequences $\{a_n\}$ the assertion is false; a corresponding counterexample is easy to construct by taking $x_n = 1$ for $n \in \mathbb{N}$.

b) Take a subsequence $\{n_k\}$ of indices such that $a_{n_k}/a_{n_{k-1}} \le 1/2$, and let $x_n = 1/ka_{n_k}$ for $n = n_k$ and $x_n = 0$ for $n \ne n_k$. It is clear that $\sum a_n x_n = \sum 1/k = +\infty$. Let $n_k \le m < n_{k+1}$. Then

$$\begin{aligned} a_m \sum_{1\le n\le m} x_n \le a_{n_k} \sum_{1\le n\le m} x_n &= a_{n_k} \sum_{1\le j\le k} \frac{1}{ja_{n_j}} \\ &\le \sum_{1\le j\le k} \frac{1}{j}\left(\frac{1}{2}\right)^{k-j} \to 0. \end{aligned}$$

To prove b) for a nonmonotone sequence $\{a_n\}$, take a subsequence $\{a_{p_n}\}$ such that $a_{p_n} = \max_{j\ge p_n} a_j$, and repeat the argument above for it.

2.7. Apply the Abel transformation to the series $\sum_{k\geq n} a_k x_k = \sum_{k\geq n} a_k(r_k - r_{k+1})$, where $r_k = \sum_{j\geq k} x_j$, and use the fact that $r_k \underset{k\to\infty}{\to} 0$.

2.8. a) $\Leftrightarrow$ b). Since

$$\sum_{n>1} a_n n^{-p} = \sum_{k\geq 1}\left(\sum_{k^q<n\leq(k+1)^q} a_n n^{-p}\right),$$

it suffices to prove that the series $\sum_{\mathrm{I}} = \sum_{k\geq 1} k^{-pq}(A_{(k+1)^q} - A_{k^q})$ and $\sum_{\mathrm{II}} = \sum_{k\geq 1} k^{-(1+pq)} A_{k^q}$ converge or diverge simultaneously. Applying the Abel transformation to the series $\sum_{\mathrm{I}}$, we get that this assertion is equivalent to the relation $A_{n^q} = o(n^{pq})$. If the series $\sum_{\mathrm{I}}$ converges, then this relation follows from the result in problem 2.6a). And if the series $\sum_{\mathrm{II}}$ converges, then $\sum_{n/2\leq m\leq n} m^{-(1+pq)} A_{m^q} \to 0$, and hence $n^{-pq} A_{n^q} \to 0$.

The assertion a) $\Leftrightarrow$ c) can be proved similarly.

2.9. Use the Abel transformation. The series $\sum |x_n|/n^{1+\varepsilon}$ diverges, for example, for $x_n = (-1)^n\sqrt{n}$ and $0 < \varepsilon \leq 1/2$. The series $\sum x_n/n$ diverges if $x_n = 1/\ln(1+n)$.

2.10. It is not hard to construct positive but nonmonotone sequences $\{\alpha_n\}$ and $\{\beta_n\}$ such that $\sum \alpha_n = \sum \beta_n = +\infty$ and $\sum \min(\alpha_n, \beta_n) < +\infty$. The required monotone series are obtained by considering the series $\alpha_1 + \cdots + \alpha_k/n_k + \cdots + \alpha_k/n_k + \cdots$ and $\beta_1 + \cdots + \beta_k/n_k + \cdots + \beta_k/n_k + \cdots$ (the terms of the form α_k/n_k and β_k/n_k are repeated n_k times) for a suitable choice of the sequence $\{n_k\}$.

2.11. a) Since $\sum a_n/S_n = \sum(1 - S_{n-1}/S_n)$ $(S_0 = 0)$, the series diverges if $S_{n-1}/S_n \not\to 1$. But if $S_{n-1}/S_n \to 1$, then divergence follows from the divergence of the series $\sum_{n\geq 2} \ln(S_{n-1}/S_n) = \lim \ln(S_1/S_N) = -\infty$.

b) Compare the given series with the series $\sum(S_{n-1}^{-\varepsilon} - S_n^{-\varepsilon})$.

2.12. c) Use the inequality $a_n e^{-\alpha_n} \leq \int_{\alpha_{n+1}}^{\alpha_n} e^{-t}\,dt$.

d) Consider the counterexample $a_n = 2^{-n!}$.

2.13. a) Use the inequality

$$\sum a_n \leq 2\sum_{n\geq 1}\left(\frac{a_n}{n^2}\sum_{1\leq k\leq n} k\right) = 2\sum_{k\geq 1}\left(k\sum_{n\geq k}\frac{a_n}{n^2}\right).$$

Estimate the inside sum with the help of the Hölder inequality.

b) Apply the assertion a) to the remainder of the series $\sum_{k\geq 1} a_k^q$.

2.14. Consider first the case $a > 0$. Let $\alpha_n = a_n - a$. It is clear that the original series converges simultaneously with the series

$$\sum(\alpha_{n-1} - \alpha_n)/\ln(a + \alpha_n).$$

If $a \neq 1$, then this series converges, but if $a = 1$, then it diverges, since the series $\sum(\alpha_{n-1} - \alpha_n)/\alpha_n$ diverges (see problem 2.12a)). Suppose now that

$a_n \downarrow 0$. It can be assumed that $a_n < 1$ for all $n \in \mathbb{N}$. It follows from the divergence of the series $\sum(a_{n-1}/a_n - 1)$ that the series

$$\sum \frac{a_{n-1}/a_n - 1}{\sum_{1\le k\le n} a_{k-1}/a_k - n}.$$

diverges too (cf. problem 2.11a)). Using the inequality between the arithmetic mean and the geometric mean, we get that the series

$$\sum \frac{1}{n} \frac{a_{n-1}/a_n - 1}{1 - \sqrt[n]{a_n}}$$

diverges.

If $a_n \to +\infty$, then the series under consideration can converge (for example, $a_n = n^n$) or it can diverge (for example, $a_n = 2^n$).

2.15. Note that it follows from convergence of each of the series that $x/f(x) \to 0$ as $x \to +\infty$ (for a proof, get a lower estimate of the sum $\sum_{n\le m\le 2n}$). The series

$$\sum \frac{1}{n^2} f^{-1}(n) = \sum_{k\ge 1} \left(\sum_{k<f^{-1}(n)\le k+1} \frac{1}{n^2} f^{-1}(n) \right)$$

converges simultaneously with the series

$$\sum_{k\ge 1} k \left(\sum_{f(k)<n\le f(k+1)} \frac{1}{n(n+1)} \right) = \sum_{k\ge 1} k \left(\frac{1}{1+[f(k)]} - \frac{1}{1+[f(k+1)]} \right).$$

Using the Abel transformation and the relation $x/f(x) \to 0$ as $x \to +\infty$, we get that this series converges simultaneously with the series $\sum 1/(1+[f(k)])$, that is, simultaneously with the series $\sum 1/f(k)$.

2.16. Let $S_n = a_1 + \cdots + a_n$, $S_0 = 0$. If $\sum f(n) < +\infty$, then $\int_0^\infty f(t)\,dt < +\infty$. Consequently,

$$\sum a_n f(S_n) \le \sum \int_{S_{n-1}}^{S_n} f(t)\,dt = \int_0^{+\infty} f(t)\,dt < +\infty.$$

But if $\sum f(n) = +\infty$, then $\int_0^{+\infty} f(t)\,dt = \infty$. Breaking up this integral into the sum of the integrals over the intervals $[S_{n-1}, S_n]$, we get that the series $\sum a_n f(S_{n-1})$ diverges. Therefore, the series $\sum a_n f(S_n)$ also diverges—for a proof it suffices to observe that the series $\sum a_n(f(S_{n-1}) - f(S_n))$ converges (this follows from the boundedness of the sequence $\{a_n\}$).

2.17. a) $\Rightarrow$ b). If $a_{n_0} > \sum_{n>n_0} a_n$ for some index n_0, then $\alpha = \sum_{n\ne n_0} a_n < \beta = \sum_{1\le n\le n_0} a_n$. It is easy to see that the sums of all the partial series do not fall in the interval (α, β).

b) $\Rightarrow$ a). Fixing a number $s \in (0, A]$, use induction to construct a sequence $\nu = \{n_1, n_2, \dots\}$ of indices such that

$$n_k = \min\{n \in \mathbb{N} \mid n > n_{k-1},\ a_{n_1} + \cdots + a_{n_{k-1}} + a_n \le s\}.$$

Verify that $A(\nu) = s$.

2.18. Consider the series $\sum 2/3^n$.

2.19. Take the n_k as small as possible, that is, $n_1 = \min\{n \in \mathbb{N} \mid a_n < 1\}$, and $n_k = \min\{n \in \mathbb{N} \mid n > n_{k-1},\ a_n < 1/k\}$ for $k \geq 2$. Assume that $\sum_{k\geq 1} a_{n_k} < +\infty$. Then $a_{n_k} = o(1/k)$. Consequently, there exists an index N such that $a_{n_k} < 1/(k+1)$ for $k \geq N$. Then in view of the definition of the sequence $\{n_k\}$, we have that $n_{k+1} = 1 + n_k$ for all $k \geq N$. Therefore, the sufficiently remote remainders of the series $\sum_{n\geq 1} a_n$ and $\sum_{k\geq 1} a_{n_k}$ coincide, which is incompatible with the condition of the problem.

The assertion is false for nonmonotone sequences. The following example was suggested by A. A. Shul′man.

We consider the sequence $\{x_k\}$, where $x_k = 1/(n^2(2^n - k))$ if $2^{n-1} \leq k < 2^n$. It is clear that $x_k \to 0$. Moreover, the series $\sum x_k$ diverges:

$$\sum x_k = \sum_{n\geq 1}\left(\frac{1}{n^2}\sum_{2^{n-1}\leq k<2^n}\frac{1}{2^n-k}\right) = \sum_{n\geq 1}\left(\frac{1}{n^2}\sum_{1\leq j\leq 2^{n-1}}\frac{1}{j}\right)$$

$$\geq \sum_{n\geq 1}\left(\frac{1}{n^2}\sum_{1\leq j\leq 2^{n-1}}\ln\left(1+\frac{1}{j}\right)\right) \geq \sum_{n\geq 1}\frac{n-1}{n^2}\ln 2 = +\infty.$$

We now consider a sequence $\{k_j\}$ of indices such that $x_{k_j} < 1/j$ for $j \in \mathbb{N}$. For arbitrary $n \in \mathbb{N}$ let $P_n = \{j \in \mathbb{N} \mid 2^{n-1} \leq k_j < 2^n\}$. Let $m = \operatorname{card} P_n$, $M = \max P_n$, and $\mu = 2^n - k_M = \min_{j\in P_n}(2^n - k_j)$. Then

$$\sum_{j\in P_n} x_{k_j} = \frac{1}{n^2}\sum_{j\in P_n}\frac{1}{2^n-k_j} \leq \frac{1}{n^2}\sum_{\mu\leq j<\mu+m}\frac{1}{j}$$

$$\leq \frac{2}{n^2}\sum_{\mu\leq j<\mu+m}\ln\left(1+\frac{1}{j}\right) = \frac{2}{n^2}\ln\left(1+\frac{m}{\mu}\right).$$

On the other hand,

$$\frac{1}{m} > \frac{1}{M} > x_{k_M} = \frac{1}{n^2(2^n-k_M)} = \frac{1}{n^2\mu},$$

therefore, $n^2 > m/\mu$, and hence $\sum_{j\in P_n} x_{k_j} \leq \frac{2}{n^2}\ln(1+n^2)$. Thus,

$$\sum_{j\geq 1} x_{k_j} = \sum_{n\geq 1}\left(\sum_{j\in P_n} x_{k_j}\right) \leq \sum \frac{2}{n^2}\ln(1+n^2) < +\infty.$$

2.20. a) If $a_n \geq 0$, then it is possible to use the result of problem 2.12b). In the general case we take two sequences $\varepsilon_k \downarrow 0$ and $\mu_k \uparrow +\infty$ such that $\sum \varepsilon_k\mu_k < +\infty$. From the sequence $\{\varepsilon_k\}$ we construct a sequence of indices $N_k \uparrow +\infty$ such that $|\sum_{m<j\leq n} a_j| < \varepsilon_k$ for $n > m > N_k$. Let $\lambda_j = \mu_k$ if

$N_k < j \le N_{k+1}$. Then for $N_k < m \le N_{k+1}$ we have:

$$\left|\sum_{m<j\le n} \lambda_j a_j\right| \le \left|\sum_{m<j\le N_{k+1}} \lambda_j a_j\right| + \left|\sum_{N_{k+1}<j\le N_{k+2}} \lambda_j a_j\right| + \cdots$$
$$\le \mu_k \varepsilon_k + \mu_{k+1}\varepsilon_{k+1} + \cdots \to 0.$$

b) Consider the counterexample $\lambda_n = (1+(-1)^n(\ln n)/\sqrt{n})^{-1}\ln n$, $a_n\lambda_n = (-1)^n/\sqrt{n}$. But if $\lambda_n \uparrow +\infty$, then the series $\sum \lambda_n a_n$ diverges, since otherwise the series $\sum a_n$ would converge by the Dirichlet test.

2.21. a) Using the inequality between the arithmetic mean and the geometric mean, we get that

$$\sum_{N\le n<M} \frac{1}{n}\frac{1+t_{n+1}}{t_n} > \frac{1}{M}\sum_{N\le n<M}\frac{1+t_{n+1}}{1+t_n}$$
$$\ge \frac{M-N}{M}\left(\frac{1+t_M}{1+t+N}\right)^{1/(M-N)}$$
$$\ge \frac{M-N}{M}(1+t_N)^{1/(N-M)}.$$

Therefore,

$$\sum_{N\le n<M} \frac{1}{n}\frac{1+t_{n+1}}{t_n} \ge \frac{1}{2}$$

for sufficiently large $M \in \mathbb{N}$.

b) Get a lower estimate of the partial sums of the series

$$\sum a_n(1+t_{n+1})/(1+t_n)$$

with the help of the result of problem I.2.8c).

§3. Various assertions about series

3.1. For example, $a_n = (-1)^n/n$. But if $a_n > 0$, then

$$\sum\left(a_n + \frac{1}{n^2 a_n}\right) \ge \sum \frac{2}{n} = +\infty.$$

3.2. For example, $\sum(\cos(\frac{2}{3}\pi n))/\ln(1+n)$. Moreover, there exists such convergent series $\sum a_n$ that for any $m = 2, 3, \ldots$ the series $\sum_{n\ge1} a_n^m$ is divergent. For example, $a_{3k} = 2/\ln k$, $a_{3k-1} = a_{3k+1} = -1/\ln k$ $(k = 2, 3, \ldots)$.

3.3. a) Fixing a parameter $\lambda > 1$ (its choice will be made more precise below), break up $\sum(a_n)^{1-1/n}$ into two sums: in the first the summation is over those indices n such that $a_n \ge \lambda^{-n}$, and in the second over the remaining indices. It is easy to see that the first sum does not exceed $\lambda\sum a_n$, and the second does not exceed $\sum\lambda^{1-n}$. Minimizing the quantity $\lambda/(\lambda-1) + \lambda\sum a_n$ over all $\lambda > 1$, we get that $\sum(a_n)^{1-1/n} \le (1+\sqrt{\sum a_n})^2$.

b) Get an upper estimate of the sum of those terms for which $a_n \le n^{-3}$.

3.4. Let $b_n = a_n/\ln(1/a_n)$. Then $\ln b_n \sim \ln a_n$, and hence $a_n \sim b_n \ln(1/b_n)$. By assumption, $\sum b_n < +\infty$, which by the result of problem 3.3b) gives us that

$$\sum_{n>1} \frac{b_n}{\ln n} \ln \frac{1}{b_n} < +\infty,$$

and thus $\sum_{n>1} a_n/\ln n < +\infty$.

The converse assertion is false—consider a sequence $\{a_n\}$ such that $a_n = 1/2$ for $n = 2^{k^2}$ and $\sum a_n/\ln(1+n) < +\infty$. But if $a_n \downarrow 0$, then it follows from the convergence of the series $\sum a_n/\ln(1+n)$ that $a_n/\ln(1+n) = o(1/n)$, which implies that $a_n/\ln(1/a_n) = O(a_n/\ln n)$.

3.5. It follows from the result of problem 2.6a) that $\frac{1}{n}(a_1 + \cdots + a_n) \to 0$. It remains to use the proof of problem II.2.4b).

3.6. b) $\Rightarrow$ a). It can be assumed that $a_n \ge 0$. Supposing that $\sum a_n = +\infty$, we construct a sequence $\{n_k\}$ with density 0 such that the series $\sum a_{n_k}$ diverges. Since at least one of the series $\sum a_{2m}$ and $\sum a_{2m+1}$ diverges, there exists a set $N_1 \subset \mathbb{N}$ with density $1/2$ such that $\sum_{n\in N_1} a_n = +\infty$. Therefore, $\sum_{n\in N_1\cap[1,P_1]} a_n \ge 1$ for a sufficiently large $P_1 \in \mathbb{N}$. Similarly, we construct a set $N_2 \subset N_1$ with density $1/4$ and an index $P_2 > P_1$ such that $\sum_{n\in N_2} a_n = +\infty$ and $\sum_{n\in N_2\cap(P_1,P_2]} a_n \ge 1$. By induction we construct a sequence $\{N_j\}$ of nested sets and a strictly increasing sequence $\{P_j\}$ of indices such that the density of N_j is equal to 2^{-j} and $\sum_{n\in N_j\cap(P_{j-1},P_j]} a_n \ge 1$. The desired sequence $\{n_1, n_2, \dots\}$ is obtained by numbering the elements of the set $A = \bigcup_{j\ge1}(N_j \cap (P_{j-1}, P_j])$ in increasing order. The density $\theta(A)$ of A is equal to zero, since $\theta(A) \le \theta(N_j) = 2^{-j}$ for any $j \in \mathbb{N}$.

3.7. For $m, n \in \mathbb{N}$ let $S_n(m) = \sum_k a_{mk}$, where the summation is only over the indices $k \ge 1$ relatively prime to all numbers not exceeding n. Using the equality $S_{n+1}(m) = S_n(m)$ if $n+1$ is not a prime number, and the equality $S_{n+1}(m) = S_n(m) - S_n((n+1)m)$ if $n+1$ is a prime number, show by induction on n that $S_n(m) = 0$ for any n. Derive from this that $|a_m| \le \sum_{k>mn} |a_k|$ for any $m, n \in \mathbb{N}$.

3.8. Assume that $\sum |a_n| = +\infty$. Let $\sum_{n_k<n\le n_{k+1}} |a_n| \ge k$ $(k \in \mathbb{N})$. Let $t_n = (1/k)\operatorname{sgn} a_n$ if $n_k < n \le n_{k+1}$. Then $t_n \to 0$, and it is obvious that $\sum t_n a_n = +\infty$.

3.9. a) $\Rightarrow$ b). Use the Abel transformation to prove that the series $\sum \lambda_n a_n$ converges.

b) $\Rightarrow$ a). Assume that $\sum |\lambda_n - \lambda_{n+1}| = +\infty$. Let

$$S_n = \sum_{1\le k\le n} |\lambda_k - \lambda_{k+1}|.$$

Then $\sum \frac{1}{S_n}|\lambda_n - \lambda_{n+1}| = +\infty$ (see problem 2.11a)). Consequently,

$$\sum \alpha_n(\lambda_n - \lambda_{n+1}) = +\infty,$$

where $\alpha_n = \frac{1}{S_n}\operatorname{sgn}(\lambda_n - \lambda_{n+1})$. Using the Abel transformation, we get that $\sum(\alpha_n - \alpha_{n-1})\lambda_n = +\infty$, although the series $\sum(\alpha_n - \alpha_{n-1})$ converges.

3.10. It suffices to consider the case $\varphi(+0) = 0$. To estimate the sum $\sum \varphi(\varepsilon n)a_n$, break it up into two sums: $\sum_{1\le n\le N}$ and $\sum_{n>N}$. Using the Abel transformation, show that the second sum is small for all $\varepsilon > 0$ if N is sufficiently large.

3.11. In the sequence $\{1/n\}$ rearrange the terms with $2^{k-1} < n \le 2^k$ in increasing order.

3.12. In the sequence $\{1/n\}$ rearrange the terms with $(k-1)! < n \le k!$ in increasing order.

3.13. a) $\Leftrightarrow$ c). The inequality $A_n \le Ca_n = C(A_n - A_{n-1})$ is equivalent to the inequality $A_{n-1} \le (1 - 1/C)A_n$. It can be proved similarly that b) $\Leftrightarrow$ d). a) and c) $\Rightarrow$ b):

$$a_n\alpha_n \le A_n \sum_{k\ge n} \frac{1}{a_k} \le C \sum_{k\ge n} \frac{A_n}{A_k} \le C \sum_{k\ge n} q_1^{k-n} = \frac{C}{1-q_1}.$$

b) and d) $\Rightarrow$ a):

$$\frac{1}{a_n}A_n \le \alpha_n \sum_{1\le k\le n} a_k \le C \sum_{1\le k\le n} \frac{\alpha_n}{\alpha_k} \le C \sum_{1\le k\le n} q_2^{n-k} \le \frac{C}{1-q_2}.$$

Thus, the first four assertions are equivalent. Let us now prove that a) $\Rightarrow$ g): since $a_n/a_{n+L} \le CA_n/A_{n+L} \le Cq_1^L$, it follows that $a_n/a_{n+L} \le 1/Q$ for a sufficiently large index L. The assertion g) $\Rightarrow$ a) is obvious if $a_n \uparrow$. The example of the sequence $2, 3, 2^2, 3^2, 2^3, 3^3, \dots$ shows that it is not true for nonmonotone sequences.

It remains to prove that a) $\Leftrightarrow$ e) and b) $\Leftrightarrow$ f). We prove the first assertion (the second has an analogous proof). It suffices to prove that a) $\Rightarrow$ e) (the converse assertion is obtained by applying this assertion to the sequence $\{\tilde{a}_n\} = \{a_n^p\}$). Since a $\Rightarrow$ g), it follows that

$$\begin{aligned} a_1^p + \dots + a_n^p &= O(a_{n-L+1}^p + \dots + a_n^p) = O((a_{n-L+1} + \dots + a_n)^p) \\ &= O(A_n^p) = O(a_n^p). \end{aligned}$$

3.14. The solution is analogous to that of the preceding problem.

3.15. a) Let $\alpha_n = \frac{p}{p-1}a_nA_n^{p-1} - A_n^p$ and prove that

$$\alpha_n \ge \frac{1}{p-1}(nA_n^p - (n-1)A_{n-1}^p). \tag{1}$$

Indeed, since $a_n = nA_n - (n-1)A_{n-1}$ (take $A_0 = 0$), it follows that

$$\alpha_n = \left(\frac{np}{p-1} - 1\right)A_n^p - \frac{p(n-1)}{p-1}A_{n-1}A_n^{p-1}. \tag{2}$$

Finding the minimum of the function

$$\varphi(x) = \left(1 - \frac{1}{p}\right) A_n^p + \frac{x^p}{p} - A_n^{p-1} x$$

for $x \geq 0$, we see that

$$A_{n-1} A_n^{p-1} \leq \left(1 - \frac{1}{p}\right) A_n^p + \frac{1}{p} A_{n-1}^p.$$

Together with (2) this yields (1). Thus,

$$\sum_{1 \leq n \leq N} \left(\frac{p}{p-1} a_n A_n^{p-1} - A_n^p\right)$$
$$= \sum_{1 \leq n \leq N} \alpha_n \geq \sum_{1 \leq n \leq N} \frac{1}{p-1} (n A_n^p - (n-1) A_{n-1}^p) = \frac{N A_N^p}{p-1} \geq 0,$$

that is,

$$\sum_{1 \leq n \leq N} A_n^p \leq \frac{p}{p-1} \sum_{1 \leq n \leq N} a_n A_n^{p-1}$$

for all $N \in \mathbb{N}$.

b) Using the Hölder inequality (problem VII.1.29), we get from the inequality a) that

$$\sum A_n^p \leq \frac{p}{p-1} \left(\sum a_n^p\right)^{1/p} \left(\sum A_n^p\right)^{1-1/p},$$

which implies that $\sum A_n^p \leq (\frac{p}{p-1})^p \sum a_n^p$—the Hardy-Landau inequality.

3.16. The divergence of the series $\sum \frac{1}{n}(a_1 + \cdots + a_n)$ follows from the inequality $(a_1 + \cdots + a_n)/n \geq a_1/n$ and the divergence of the harmonic series. To prove convergence of the series $\sum \sqrt[n]{a_1 \cdots a_n}$, observe that

$$\sqrt[n]{a_1 \cdots a_n} \leq \left(\frac{a_n^{1/p} + \cdots + a_n^{1/p}}{n}\right)^p$$

for $p > 1$ (see problem VII.1.30). It follows from the Hardy-Landau inequality (see problem 3.15b)) that

$$\sum \sqrt[n]{a_1 \cdots a_n} \leq \left(\frac{p}{p-1}\right)^p \sum a_n.$$

Letting p go to infinity, we get the Carleman inequality:

$$\sum \sqrt[n]{a_1 \cdots a_n} \leq e \sum a_n.$$

3.17. The assertions a) $\Rightarrow$ b) and b) $\Rightarrow$ c) are obvious. We prove that c) $\Rightarrow$ a). Assume the opposite: there exists a sequence $a_n \to 0$ such that $|f(a_n)/a_n| \to +\infty$. It can be assumed without loss of generality that $0 < a_n < 1/2n^2$ and $\lambda_n = f(a_n)/a_n > n$. Consider the series obtained from

$\sum a_n$ by repeating each term P_n times, where $P_n = [1/(n^2 a_n)]$. This series converges:

$$\sum_{n\geq 1}\left(\sum_{1\leq k\leq P_n} a_n\right) = \sum a_n P_n \leq \sum \frac{1}{n^2}.$$

However, the series consisting of the values of f diverges:

$$\sum_{n\geq 1}\left(\sum_{1\leq k\leq P_n} f(a_n)\right) = \sum P_n f(a_n) = \sum P_n \lambda_n a_n$$

$$\geq \frac{1}{2}\sum \frac{\lambda_n}{n^2} \geq \frac{1}{2}\sum \frac{1}{n} = +\infty.$$

3.18. Prove first that f is odd in some neighborhood of zero. Assume not. Then there exists a sequence $a_n \to 0$ such that $f(a_n)+f(-a_n) = \Delta_n \neq 0$. Take a sequence $\{P_n\} \subset \mathbb{N}$ such that $P_n\Delta_n \not\to 0$. Then the series obtained from the series $a_1 - a_1 + a_2 - a_2 + \cdots$ by repeating the nth pair of terms P_n times is convergent. However, the series consisting of the values of f diverges, since $P_n\Delta_n \not\to 0$.

It is clear that f is continuous at zero (see problem 3.17). Prove that f is continuous in some neighborhood of zero. Assume the opposite: there exists a sequence $\{a_n\}$ of points of discontinuity such that $a_n \to 0$. For any $n \in \mathbb{N}$ there exists a number $\varepsilon_n > 0$ such that $\sup_{|t-a_n|<\delta} |f(a_n) - f(t)| > \varepsilon_n$ for any $\delta > 0$. Take sequences $\{P_n\} \subset \mathbb{N}$ and $\{t_n\} \subset \mathbb{R}$ such that $P_n\varepsilon_n \not\to 0$, $|f(a_n) - f(t_n)| > \varepsilon_n$, and $\sum P_n|a_n - t_n| < +\infty$. The series obtained from the series $a_1 - t_1 + a_2 - t_2 + \cdots$ by repeating P_n times the pair $a_n - t_n$ converges, and the series of corresponding values of f diverges, because $\sum_{n\geq 1}\left(\sum_{1\leq k\leq P_n}(f(a_n) + f(-t_n))\right) = \sum P_n(f(a_n) - f(t_n))$ and $P_n|f(a_n) - f(t_n)| \geq P_n\varepsilon_n \not\to 0$. Accordingly, f is continuous and odd in some neighborhood of zero.

Prove now that $f(x+h) + f(x-h) = 2f(x)$ in some neighborhood of zero. If this is not so, then there are sequences $x_n \to 0$ and $h_n \to 0$ such that $f(x_n + h_n) + f(x_n - h_n) - 2f(x_n) = \Delta_n \neq 0$. Take a sequence of integers $P_n \uparrow +\infty$ such that $P_n\Delta_n \not\to 0$. The series obtained from the series $(x_1 + h_1) + (x_1 - h_1) - x_1 - x_1 + \cdots + (x_n + h_n) + (x_n - h_n) - x_n - x_n + \cdots$ by repeating each group $(x_n + h_n) + (x_n - h_n) - x_n - x_n$ of terms P_n times obviously converges. However, the series consisting of the corresponding values of f diverges, because

$$P_n(f(x_n + h_n) + f(x_n - h_n) - 2f(x_n)) = P_n\Delta_n \not\to 0.$$

Thus, in some interval $[-\delta, \delta]$ the function f is odd and continuous, and satisfies the equality $f(a) + f(b) = 2f((a+b)/2)$ for $|a|, |b| \leq \delta$. Consequently (see problem III.5.2), f is linear on $[-\delta, \delta]$.

3.19. Obviously, if $\{q_n\}$ is the required sequence, then the numbers $r_m = x - \sum_{1\leq k\leq m} q_k 2^{-k}$ ($r_0 = x$) satisfy the inequality $0 < r_m < 2^{-m}$. Therefore, in selecting the number q_{m+1} (after the numbers $q_1, \dots, q_m$ have already been found) it is necessary to ensure the inequalities $0 < r_m - 2^{-m-1}q_{m+1} < 2^{-m-1}$, that is, the inequalities

$$2^{m+1}r_m - 1 < q_{m+1} < 2^{m+1}r_m. \qquad (*)$$

It is clear that the satisfaction of this inequality for all $m \in \mathbb{N}$ suffices for the relation $r_m \to 0$, that is, for the equality $x = \sum q_k 2^{-k}$. It is not hard to verify that if $(*)$ holds for some index, then it holds also for the preceding index, and hence for all the preceding indices.

The sequence $\{q_n\}$ will be constructed by induction. It can be assumed here that Q has already been numbered in some way. Therefore, we can refer to the first element in any nonempty subset of it.

As q_1 take the first element of Q satisfying the inequality $(*)$ for $m = 0$, that is, $2x - 1 < q_1 < 2x$. It follows from the conditions $\inf Q = 0$ and $\sup Q = 1$ that such elements exist. Assume that $q_1, \dots, q_n$ have already been constructed and that $(*)$ holds for $m = 0, \dots, n-1$. Let q be the first element in the set $Q \setminus \{q_1, \dots, q_n\}$. If $q \in (2^{n+1}r_n - 1, 2^{n+1}r_n)$, then we let $q_{n+1} = q$. The inequality $(*)$ will be satisfied for $m = n$. If $q \notin (2^{n+1}r_n - 1, 2^{n+1}r_n)$, then we consider two cases: a) $q \geq 2^{n+1}r_n$; b) $q \leq 2^{n+1}r_n - 1$.

In case a) let $M = [\log_2(q/r_n)]$. Then $r_n 2^M \leq q < r_n 2^{M+1}$. In particular, $M > n$. Select numbers $q_{n+1}, \dots, q_M$ from $Q \setminus \{q_1, \dots, q_n\}$ small enough that $r_M = r_n - (q_{n+1}2^{-n-1} + \dots + q_M 2^{-M}) > 2^{-M-1}$. Taking $q_{M+1} = q$, we get that $r_M > 2^{-M-1}q_{M+1}$ and $2^{M+1}r_M - 1 < 2^{M+1}r_n - 1 < q_{M+1}$, that is, the number q_{M+1} satisfies the inequality $(*)$ for $m = M$, and hence $(*)$ holds for $m = 0, \dots, M$.

In case b) consider the first index $M > n$ such that $r_n - (2^{-n-1} + \dots + 2^{-M}) < (q+1)2^{-M-1}$. Such an index exists, because as $M \to +\infty$ the left-hand side tends to $r_n - 2^{-n} = r_{n-1} - (q_n + 1)2^{-n} < 0$ (see the inequality $(*)$ with $m = n-1$). Now select numbers $q_{n+1}, \dots, q_M$ from $Q \setminus \{q_1, \dots, q_n\}$ close enough to 1 that $r_M = r_n - (q_{n+1}2^{-n-1} + \dots + q_M 2^{-M}) < (q+1)/2^{1+M}$. Taking $q_{M+1} = q$, we get that $2^{M+1}r_M - 1 < q_{M+1}$. By the choice of M,

$$(1 + q_{M+1})2^{-M} \leq r_n - (2^{-n-1} + \dots + 2^{-M+1}) < r_{M-1},$$

that is, $q_{M+1} < -1 + 2^M r_{M-1} = -1 + q_M + 2^M r_M < 2^{M+1} r_{M+1}$. Thus, $(*)$ holds for $m = M$, and hence also for $m = 0, \dots, M$.

§4. Computation of sums of series

4.1. Use the equality $3\sin\varphi - \sin 3\varphi = 4\sin^3\varphi$.

4.2. Use the equality $3\cos\varphi + \cos 3\varphi = 4\cos^3\varphi$.

4.3. Use the equality

$$\binom{n+1}{r+1}^{-1} = \frac{r+1}{r}\left(\binom{n+r-1}{r}^{-1} - \binom{n+r}{r}^{-1}\right).$$

4.4. Use the equality $\int_0^{\pi/2} \cos^{2n+1} t\,dt = (2n)!!/(2n+1)!!$.

4.5. Consider the partial sums S_{nm}.

4.6. Use the result of problem VII.1.22c).

4.7. Change the order of summation.

4.8. Show that

$$\sum_{1\le n\le 2N} (-1)^n \frac{\ln n}{n} = \ln 2 \sum_{1\le n\le N} \frac{1}{n} - \sum_{N<n\le 2N} \frac{1}{n}\ln n$$

and replace the sums by the corresponding asymptotic expressions (see problems II.2.9 and II.2.13a)).

4.9. Show that

$$\sum_{1\le n<2^N} \frac{(-1)^n}{n}[\log_2 n] = -\sum_{1\le k<N} \sigma_k + (N-1)\sigma_N,$$

where $\sigma_k = \sum_{1\le j<2^k} (-1)^j/j$, and, transforming σ_k to the form $\sigma_k = -\sum_{2^{k-1}\le j<2^k} 1/j$, use the asymptotic behavior of the partial sums of the harmonic series.

4.10. Reduce the integration in each integral to the interval $[2\pi, +\infty)$. To justify changing the order of summation and integration, represent the integral $\int_{2\pi}^{+\infty} (\sin nx)/x\,dx$ in the form

$$\int_{2\pi}^{2\pi N} \frac{\sin nx}{x}\,dx + O\left(\frac{1}{nN}\right),$$

where $N \in \mathbb{N}$. Then use the results of problems 6.16 and 6.8a) and Stirling's formula.

4.11. The solution is analogous to that of the preceding problem.

4.12. Denote the right-hand side of the equality to be proved by I_N. It suffices to prove that $I_{N-1} - I_N = 1/N^2$ and $I_N \to 0$. To prove the first assertion use the known equality

$$\int_0^{\pi/2} \cos^{2N} t\,dt = \frac{\pi}{2}\frac{(2N-1)!!}{(2N)!!}$$

and integrate twice by parts. To prove the second assertion note that

$$I_N = \frac{4}{\pi}\frac{(2N)!!}{(2N-1)!!}\left(\int_0^{\varepsilon} + \int_{\varepsilon}^{\pi/2}\right) \le 2\varepsilon^2 + O(N\cos^N \varepsilon).$$

Consequently, $\overline{\lim}\, I_N \le 2\varepsilon^2$. Since $\varepsilon > 0$ is arbitrary, $I_N \to 0$.

4.13. a) Using the equality $(1-x)^{-1} = 1 + x + x^2 + \cdots$ for $x = p_1^{-s}, p_2^{-s}, \ldots$, prove that

$$\sum_{1\le n\le m} n^{-s} \le (1-p_1^{-s})^{-1}(1-p_2^{-s})^{-1}\cdots(1-p_m^{-s})^{-1} \le \sum n^{-s}.$$

b) The solution is analogous to that of problem a).

4.14. Use the results of problems 4.12 and 4.13a).

4.15. Representing the fraction $t^n/(1-t^n)$ as the sum of a geometric progression, prove the identity $\sum a_n t^n/(1-t^n) = \sum b_n t^n$, where $b_n = \sum_{k\,|\,n} a_k$ ($k\,|\,n$ means that n is divisible by k).

§5. Function series

5.2. If $|\lambda_n| \le 1$, then, taking $x = m + \theta$, $m \in \mathbb{Z}$, and $\theta \in [0, 1)$, we get that

$$\left|\sum \lambda_n e^{-|x-n|}\right| \le \sum e^{-|x-n|} \le e^{-\theta}\sum_{n\le 0} e^n + e^{\theta}\sum_{n\ge 1} e^{-n} \le \frac{e+1}{e-1}.$$

5.3. Since

$$\frac{x^n}{(1+x)\cdots(1+x^n)} = \frac{1}{(1+x)\cdots(1+x^{n-1})} - \frac{1}{(1+x)\cdots(1+x^n)},$$

the partial sums of the series are equal to $S_n(x) = 1 - \prod_{1\le k\le n}(1+x^k)^{-1}$. Obviously, $S_n(x) \to 1$ as $n\to\infty$ if $x \ge 1$, and $\sup_{x\ge 1}|S_n(x) - 1| = 2^{-n}$. If $0 \le x < 1$, then $S_n(x) \underset{n\to\infty}{\to} S(x) = 1 - \prod_{k\ge 1}(1+x^k)^{-1}$. Thus, $\sup_{x\ge 0} S(x) = 1$. The uniform convergence on $[0, 1]$ follows from the inequalities

$$\frac{x^{2m+1}}{(1+x)\cdots(1+x^{2m+1})} \le \frac{x^{2m}}{(1+x)\cdots(1+x^{2m})} \le \frac{x^{2m}}{(1+x)\cdots(1+x^m)}$$
$$\le \frac{x^{2m}}{(1+x^m)^m} \le \frac{2}{m(m-1)}.$$

5.4. The convergence of the series for $x > 0$ is obvious. The convergence is not uniform, because the general term of the series $u_n(x)$ does not tend uniformly to zero: $u_n(1/n!) = 1$. To estimate the sum of the series we break it up into two sums: $\sum_{1\le n\le N} + \sum_{n>N}$, where $N \in \mathbb{N}$ is such that $N! \le 1/x < (N+1)!$ (if $x > 1$, then the first sum is equal to zero, and the second does not exceed $\sum 1/n! = e - 1$). Then

$$\sum_{1\le n\le N} u_n(x) = x\sum_{1\le n\le N} n! \le 2xN! \le 2;$$
$$\sum_{n>N} u_n(x) = \frac{1}{x}\sum_{n>N}\frac{1}{n!} < \frac{1}{x(N+1)!}\sum_{k\ge 0}\frac{1}{(N+2)^k} < \frac{N+2}{N+1} \le \frac{3}{2}.$$

Consequently, $\sum \min(n!x, 1/n!x) \le 7/2$.

5.5. Prove that $\lim_{t\to 1-0} B(t) = +\infty$ and that as $t \to 1-0$ the limits superior (inferior) of the ratios $A(t)/B(t)$ and $\sum_{n\geq N} a_n t^n / B(t)$ are the same for any $N \in \mathbb{N}$.

5.6. By changing a_0 it is possible to assume that $S_n = a_0 + \cdots + a_n \to 0$. Then from the equality $\sum_{n\geq 0} a_n t^n = (1-t)\sum_{n\geq 0} S_n t^n$ we get that

$$\left|\sum_{n\geq 0} a_n t^n\right| \leq (1-t)\sum_{0\leq n\leq N} |S_n| t^n + (1-t)\sum_{n>N} |S_n| t^n$$
$$\leq (1-t)\sum_{0\leq n\leq N} |S_n| + \max_{n>N} |S_n|.$$

By selecting N to be large we make the second term small, and then by selecting t to be close to 1 we make the first term small.

Another solution can be obtained with the help of the equality

$$\sum_{n\geq 0} a_n t^n = (1-t)\sum_{n\geq 0} S_n t^n = \sum_{n\geq 0} S_n t^n \Big/ \sum_{n\geq 0} t^n$$

and the result in the preceding problem.

5.7. Convergence of the series $\sum a_n t^n$ follows from the relation $a_n = S_n - S_{n-1} = O(n)$. Since

$$\sum_{n\geq 0} a_n t^n = (1-t)\sum_{n\geq 0} S_n t^n = (1-t)^2 \sum_{n\geq 0} (S_0 + \cdots + S_n) t^n$$
$$= \sum_{n\geq 0} (S_0 + \cdots + S_n) t^n \Big/ \sum_{n\geq 0} (n+1) t^n,$$

the use of the result in problem 5.5 gives what is required.

5.8. Use the equality

$$A(t)/B(t) = \sum_{n\geq 0} (a_0 + \cdots + a_n) t^n \Big/ \sum_{n\geq 0} (b_0 + \cdots + b_n) t^n$$

and the result in problem 5.5.

5.9. Convergence of the product $\prod_{n\geq 1}(1 + a_n t^n)$ is equivalent to convergence of the series $\sum \ln(1 + a_n t^n)$. Since $|\ln(1 + a_n t^n)| = o(t^n)$, this series (and with it also the product) converges for all $t \in (-1, 1)$. The relation $A(t) \to A$ as $t \to 1-0$ is equivalent to the relation $\sum \ln((1+a_n t^n)/(1+a_n)) \to 0$ as $t \to 1-0$. By Taylor's formula, it reduces to the equality

$$\lim_{t\to 1-0} \sum (a_n(t^n - 1) + O(a_n^2)(1 - t^n)) = 0,$$

which follows from the Abel theorem (see problem 5.6).

The example $a_n = (-1)^n\sqrt{2/n} + 1/n$ shows that the condition $\sum a_n^2 < +\infty$ is essential. Indeed, in this case

$$\ln \frac{1 + a_n t^n}{1 + a_n} = (-1)^n \sqrt{\frac{2}{n}}(t^n - 1) + \frac{1}{n}(t^n - t^{2n}) + O\left(\frac{1 - t^n}{n^{3/2}}\right)$$

and hence

$$\frac{A(t)}{A} = \exp\left(\sum \frac{t^n - t^{2n}}{n} + o(1)\right) \to 2$$

as $t \to 1-0$.

5.10. Show that instead of the given series it is possible to consider the alternating series $\sum S_m(x)$, where

$$\begin{aligned} S_m(x) &= \sum_{m \le x\sqrt{n} < m+1} n^{-\alpha} \sin(\pi x \sqrt{n}) \\ &= \left(\frac{m}{x}\right)^{-2\alpha} \sum_{m \le x\sqrt{n} < m+1} \sin(\pi x\sqrt{n}) + O(m^{-2\alpha}). \end{aligned}$$

To investigate the sum $\sum_{m \le x\sqrt{n} < m+1} \sin(\pi x\sqrt{n})$ use the result of problem VI.3.14.

5.11. For $x > 4$ let $N = [\log_4 x]$. Then

$$|f(x)| \le \sum_{1 \le n \le N} \frac{1}{n} + \sum_{n > N} \frac{x}{n4^n} < 1 + \ln N + \frac{x}{N4^N} = O(\ln\ln x).$$

On the other hand, taking $x_m = \pi(4^m + 4^{m-1} + \cdots + 4 + 1)$ $(m \in \mathbb{N})$, we get that

$$\begin{aligned} f(x_m) &= \sum_{1 \le n \le m} \frac{1}{n} \sin \pi \left(\frac{1}{4} + \frac{1}{16} + \cdots\right) + O(1) \\ &\ge \frac{\sqrt{2}}{2} \sum_{1 \le n \le m} \frac{1}{n} + O(1) \ge \text{const} \ln\ln x_m. \end{aligned}$$

The equality $\underline{\lim}_{x \to +\infty} f(x) = 0$ follows from the easily verified relation $f(\pi 4^m) = O(1/m)$.

5.12. If $f \in \mathrm{Lip}_1$, then $f(\pi/2N) = O(1/N)$, and hence

$$\sum_{1 \le n \le N} a_n \sin\frac{\pi n}{2N} = O(1/N).$$

From this, $\sum_{1 \le n \le N} na_n = O(1)$.

5.13. Use the fact that the coefficients of the polynomial P_n satisfy a linear system of order $m+1$, and prove that they can be expressed in terms of the values of P_n at the fixed points $t_k \in [\alpha, \beta]$ $(k = 0, 1, \ldots, m)$. Derive from this that the coefficients of P_n converge.

5.14. Prove that the series $\sum f_n$ converges uniformly on any interval $[\alpha, \beta] \subset (a, b)$ (Dini's theorem).

§6. Trigonometric series

6.1. Construct a sequence $\{n_k\}$ of indices and a sequence of nested intervals on which $\cos(A_{n_k} x + \varphi_k) \ge 1 - 1/k$.

6.2. Represent $a_n \cos nx + b_n \sin nx$ in the form $\rho_n \cos(nx + \varphi_n)$, and, assuming that $\rho_n \not\to 0$, use the result in the preceding problem.

6.3. Compare with problem II.1.10a).

6.4. Integrate the partial sums of the series over the interval $[\alpha, \beta]$ and show that for some $\gamma > 0$

$$\int_\alpha^\beta |a_n \cos nx + b_n \sin nx|\, dx \geq \gamma\sqrt{a_n^2 + b_n^2}.$$

6.6. Use the boundedness of the function $(1/(\sin x) - 1/x)$ on the interval $(0, \pi/2]$ and the result of problem VI.1.10a) and c).

6.7. Compare the sums

$$\sum_{k^2 \leq n < (k+1)^2} n^{-1/2} \exp(in\pi(x - \ln[\sqrt{n}]))$$

and

$$\sigma_k(x) = \frac{1}{k} \sum_{k^2 \leq n < (k+1)^2} \exp(i\pi n(x - \ln k)).$$

Show that $\sigma_{k_j}(x) \not\to 0$ if the sequence $\{k_j\}$ is such that $\ln k_j < x + 2m_j < \ln(1 + k_j)$ $(m_j \in \mathbb{Z})$ for all $j \in \mathbb{N}$.

6.8. Find the sum of the series $\sum(1/n)z^n$ for $z \in \mathbb{C}$, $|z| < 1$, and use the Abel theorem (see problem 5.6).

6.9. The problem can be reduced to the preceding problem with the help of the equality

$$\frac{1}{4n^2 - 1} = \frac{1}{2}\left(\frac{1}{2n-1} - \frac{1}{2n+1}\right).$$

6.10–6.12. Using the Abel transformation, employ the result of problem 6.5.

6.13. Use the result of the preceding problem and the equality

$$\int_0^\pi \sin nx \sin nmx\, dx = \begin{cases} 0 & \text{for } m \neq n, \\ \frac{\pi}{2} & \text{for } m = n. \end{cases}$$

6.14. To prove the relation $\lambda_n = o(1/n)$ consider the sum $\sum_{n/2 \leq k \leq n} \lambda_k \sin kx$ for $x = \pi/2n$. If $\lambda_n = o(1/n)$, then for a uniform estimate of the remainder break this sum up into two sums, the first containing the terms with "small" indices (not exceeding $\pi/2x$). In estimating it, use the inequality $|\sin t| \leq |t|$. To estimate the second sum use the Abel transformation and the result in problem 6.5.

6.15. The solution is analogous to that of the preceding problem.

6.16. It can be assumed that $[a, b] \subset [0, \pi]$ (otherwise, break up $[a, b]$ into several intervals). If $a > 0$, then it is possible to use the result of problem 6.10. If $a = 0$, then, using the results of problems 6.10 and 6.5, estimate the integrals of the remainder of the series over $[0, \varepsilon]$ and $[\varepsilon, b]$ for small $\varepsilon > 0$.

6.17. With the help of the Abel transformation and the results in problems 6.6 and 2.4, show that there is a $C > 0$ such that

$$\int_0^\pi \left| \sum_{1\le n\le N} \lambda_n \cos nx \right| dx \le C$$

for all $N \in \mathbb{N}$. Using the result of problem 6.10, show that the integrals $\int_\varepsilon^\pi |f(x)|\, dx$ are bounded uniformly with respect to $\varepsilon \in (0, \pi)$.

6.18. The proof of the absolute integrability of g is analogous to the solution of the preceding problem. If $\int_0^\pi |g(x)|\, dx < +\infty$, then integrate the series (see problem 6.16) $\sum g(x)\frac{1}{n} \sin nx$ termwise over $[0, \pi]$ and use the result of problem 6.13.

6.19. Repeating the solution of problem 6.17, prove that for $M \ge N$

$$\int_0^\pi \left| \varphi(x) \sum_{N\le n\le M} \lambda_n \cos nx \right| dx \le \rho_N,$$

where $\rho_N \to 0$. Since the series $\sum_{k\ge N} \lambda_n \cos nx$ converges uniformly on $[\varepsilon, \pi]$ (see problem 6.10), this implies that $\int_\varepsilon^\pi \left|\varphi(x) \sum_{n\ge N} \lambda_n \cos nx\right| dx \le \rho_N$, and hence $\int_0^\pi |\varphi(x) \sum_{n\ge N} \lambda_n \cos nx|\, dx \le \rho_N$.

6.20. a) Arguing by contradiction, consider a point $x_0 \in (0, \pi)$ at which the sum $S_n(x) = \sum_{1\le k\le n} (\sin kx)/k$ has a nonpositive minimum. Using the necessary condition for an extremum $(S_n'(x_0) = 0)$, show that $\sin nx_0 \ge 0$, and hence the sum S_{n-1} also takes nonpositive values. Continuation of this argument leads to a contradiction of the fact that $S_1(x) = \sin x > 0$ on $(0, \pi)$.

b) Prove that the local minima of the function $\sum_{1\le k\le n} \frac{1}{k} \cos kx$ on $[0, \pi]$ form an increasing sequence.

6.21. a) Use the Abel transformation and the result of problem 6.20a).

b) Let

$$\mathscr{D}_n(x) = \frac{1}{2} + \cos x + \cdots + \cos nx = \frac{\sin(n+1/2)x}{2\sin(x/2)} \qquad \left(\mathscr{D}_0 = \frac{1}{2}\right).$$

Then

$$\begin{aligned} f(x) &= \frac{\lambda_0}{2} + \sum_{n\ge 1} \lambda_n (\mathscr{D}_n(x) - \mathscr{D}_{n-1}(x)) \\ &= \sum_{n\ge 0} (\lambda_n - \lambda_{n+1}) \mathscr{D}_k(x). \end{aligned}$$

Since

$$\mathscr{D}_0(x) + \mathscr{D}_1(x) + \cdots + \mathscr{D}_{n-1}(x) = \left(\frac{\sin(nx/2)}{2\sin(x/2)}\right)^2,$$

the Abel transformation gives us that

$$f(x) = \sum_{n\ge 1} (\lambda_{n-1} - 2\lambda_n + \lambda_{n+1}) \left(\frac{\sin(nx/2)}{2\sin(x/2)}\right)^2 \ge 0.$$

Note that

$$\int_0^\pi \left(\frac{\sin(nx/2)}{2\sin(x/2)}\right)^2 dx = \int_0^\pi (\mathscr{D}_0(x) + \cdots + \mathscr{D}_{n-1}(x))\,dx = \frac{\pi}{2}n.$$

Therefore (see problems 6.10 and 5.14),

$$\int_0^\pi f(x)\,dx = \frac{\pi}{2}\sum n(\lambda_{n-1} - 2\lambda_n + \lambda_{n+1}) = \frac{\pi}{2}\sum(\lambda_{n-1} - \lambda_n) = \frac{\pi}{2}\lambda_0.$$

6.22. Prove that the local maxima of the function S_n on $[0, \pi]$ form a decreasing sequence.

6.23. Use the fact that

$$\begin{aligned} c_k = \frac{1}{2\pi}\int_{-\pi}^\pi f(x)e^{-ikx}\,dx &= -\frac{1}{2\pi}\int_{-\pi}^\pi f\left(x + \frac{\pi}{k}\right)e^{-ikx}\,dx \\ &= \frac{1}{4\pi}\int_{-\pi}^\pi \left(f(x) - f\left(x + \frac{\pi}{k}\right)\right)e^{-ikx}\,dx. \end{aligned}$$

6.24. In studying the quantity $|f(x)-f(y)|$ represent the difference $x-y$ (assuming that $x > y$) in the form $x - y = 2\pi\sum_{k\geq p}\varepsilon_k 4^{-k}$, where $\varepsilon_k = 0, 1, 2, 3$ and $\varepsilon_p \neq 0$. Break up the series

$$f(x) - f(y) = \sum 2^{1-n}\sin\left(4^n\frac{x-y}{2}\right)\sin\left(4^n\frac{x+y}{2}\right)$$

into two sums: $\sum_{1\leq n\leq p}$ and $\sum_{n>p}$, and estimate each of them separately. To prove that $f \notin \mathrm{Lip}_\alpha$ for $\alpha > 1/2$, get a lower estimate of the difference $f(0) - f(x)$ for $x = 2\pi/4^m$ $(m \in \mathbb{N})$ or use the result of the preceding problem.

6.25. The solution is analogous to that of the preceding problem.

6.26. Use the same device as in the solution of problem 6.24. The function f is not in the class Lip_1, since $f(x) > (m/2)x$ for $x = \pi 3^{-m-1}$ $(m \in \mathbb{N})$.

6.27. a) Get a lower estimate of $f(x)$ for $x = 1/m!$ $(m \in \mathbb{N})$.

b) For an arbitrary point $x \in \mathbb{R}$ consider the intervals $[\alpha_n, \beta_n]$ $(n \geq n_0)$, containing it, where $\alpha_n = p_n/n!$, $\beta_n = (p_n + s_n)/n!$, $p_n = [xn!]$, $s_n \in \mathbb{N}$, and $s_n \leq n/2$. To get a lower estimate of the quantity $|f(\beta_n) - f(\alpha_n)|$ use the inequality

$$\begin{aligned} |f(\beta_n) - f(\alpha_n)| &\geq \left|\sum_{1\leq k<n} 2^{1-k}\sin\left(\pi k!\frac{s_n}{n!}\right)\cos(\pi k!(\alpha_n + \beta_n))\right| \\ &\geq \frac{4}{2^n}\left|\sin\frac{\pi s_n}{n}\right|\left|\cos\frac{\pi}{n}(2p_n + s_n)\right| \\ &\quad - \sum_{1\leq k\leq n-2} 2^{1-k}\sin(\pi k! s_n/n!). \end{aligned}$$

Show that s_n can be chosen so that $|\cos\frac{\pi}{n}(2p_n + s_n)| \geq \frac{1}{2}$. Prove that in this case $|f(\beta_n) - f(\alpha_n)| \geq 1/(n2^n) \geq \sqrt{\beta_n - \alpha_n}$.

6.28. b) $\Rightarrow$ c). Use the fact that for $N > N_0$ the difference $S_n(x) - S_{N_0}(x)$ has period $\pi/2^{N_0}$, and hence $\max_{a\le x\le b}|S_N(x) - S_{N_0}(x)| = \max_{x\in\mathbb{R}}|S_N(x) - S_{N_0}(x)|$, if N_0 is sufficiently large.

c) $\Rightarrow$ a). Write the sum S_{2N} in the form

$$S_{2N}(x) = \sum_{0\le n\le 2N} c_n \cos(2^n x + \varphi_n),$$

where $c_n = \sqrt{a_n^2 + b_n^2}$ and $\varphi_n \in [-\pi, \pi]$ for $n = 0, \ldots, 2N$, and consider the nonnegative trigonometric polynomial

$$R_N(x) = \prod_{0\le n\le N} (1 + \cos(2^{2n} x + \varphi_{2n}))$$

(*the F. Riesz product*). It is easy to see that R_N has the form

$$R_N(x) = 1 + \sum_{0\le n\le N} \cos(2^{2n} x + \varphi_{2n}) + \sum \alpha_k \cos(kx + \psi_k),$$

where $\alpha_1 = \alpha_2 = \alpha_4 = \cdots = \alpha_{2^n} = \cdots = 0$. Using the equalities

$$\int_{-\pi}^{\pi} \cos(mx + \varphi)\cos(kx + \psi)\, dx = 0, \quad m, k \in \mathbb{Z}, \quad |m| \ne |k|$$

(orthogonality of the trigonometric system), we get that

$$\begin{aligned}\int_{-\pi}^{\pi} S_{2N}(x) R_N(x)\, dx &= \sum_{0\le n\le N} c_{2n} \int_{-\pi}^{\pi} \cos^2(2^{2n} x + \varphi_{2n})\, dx \\ &= \pi \sum_{0\le n\le N} \sqrt{a_{2n}^2 + b_{2n}^2}.\end{aligned}$$

Let $S = \sup_{x\in\mathbb{R}, M\in\mathbb{N}} S_M(x)$. By assumption, $S < +\infty$. Therefore,

$$\begin{aligned}\sum_{0\le n\le N} (|a_{2n}| + |b_{2n}|) &\le \sqrt{2} \sum_{0\le n\le N} \sqrt{a_{2n}^2 + b_{2n}^2} \\ &= \frac{\sqrt{2}}{\pi} \int_{-\pi}^{\pi} S_{2N}(x) R_N(x)\, dx \\ &\le \frac{\sqrt{2}}{\pi} S \int_{-\pi}^{\pi} R_N(x)\, dx = 2\sqrt{2} S.\end{aligned}$$

Thus, $\sum_{n\ge 0}(|a_{2n}| + |b_{2n}|) \le 2\sqrt{2}S$. The sum $\sum_{n\ge 0}(|a_{2n+1}| + |b_{2n+1}|)$ can be estimated similarly, except that the product

$$\widetilde{R}_N(x) = \prod_{0\le n\le N} (1 + \cos(2^{2n+1} x + \varphi_{2n+1}))$$

is considered instead of the polynomial $R_N(x)$.

In conclusion note that the proof of the assertion c) $\Rightarrow$ a) used only the boundedness of the sums $S_N(x)$ from above. Therefore, the series

$\sum(|a_n| + |b_n|)$ converges if the sums $S_N(x)$ are uniformly bounded above on some interval $[a, b]$, $a < b$.

6.29. Verify the proof by induction.

6.30. The assertion is obvious for $N = 0, 1$. Assume that the inequalities are true for $N < 2^m$ for some $m \in N$, and show that they are true also for $2^m \le N < 2^{m+1}$. Note first of all that $R_N(P_n) = P_n$ and $R_N(Q_n) = Q_n$ for $n \le m$, and in this case the validity of the inequalities to be proved follows from the result of problem 6.29c). Verify that for $n \ge m+2$

$$R_N(P_n) = R_N(Q_n) = R_N(P_{m+1}).$$

Therefore, it suffices to estimate $R_N(P_{m+1})$ and $R_N(Q_{m+1})$. It is clear that $R_N(P_{m+1})(z) = P_m(z) + z^{2^m} R_M(Q_m)(z)$, where $M = N - 2^m$. If $M < 2^{m-1} \le N/2$, then we get from the induction hypothesis that $|R_M(Q_m)(z)| \le 10\sqrt{M} \le 10\sqrt{N/2}$, and hence $|R_N(P_{m+1})(z)| \le |P_m(z)| + 10\sqrt{N/2} \le \sqrt{2N} + 10\sqrt{N/2} \le 10\sqrt{N}$. But if $M \ge 2^{m-1}$, then $R_M(Q_m)(z) = P_{m-1}(z) + z^{2^{m-1}} R_L(Q_{m-1})(z)$, where $L = M - 2^{m-1} \le 2^{m-1}$. Therefore,

$$\begin{aligned}|R_N(P_{m+1})(z)| &\le |P_m(z)| + |R_M(Q_m)(z)| \\ &\le |P_m(z)| + |P_{m-1}(z)| + |R_L(Q_{m-1})(z)| \\ &\le 2^{(m+1)/2} + 2^{m/2} + 10\sqrt{L} \\ &\le \sqrt{2N} + \sqrt{N} + 10\sqrt{N/2} \le 10\sqrt{N}.\end{aligned}$$

The polynomial $R_N(Q_{m+1})$ can be estimated similarly.

6.31. Consider a sequence $\{\varepsilon_k\}_{k\ge 0}$ whose terms for $k < 2^n$ coincide with the coefficients of the polynomial P_n in problem 6.29. To prove the inequality a) use the result of problem 6.30. To prove the inequality b) use the relations

$$2\pi(m+1) = \int_0^{2\pi} |S_m(\theta)|^2\, d\theta \le \max_{0\le\theta\le 2\pi} |S_m(\theta)| \int_0^{2\pi} |S_m(\theta)|\, d\theta.$$

6.32. Consider the series

$$\sum_{k\ge 2} \frac{\varepsilon_k}{\sqrt{k}\ln^2 k} e^{ik\theta},$$

where the sequence $\{\varepsilon_k\}$ is defined in the hint for problem 6.31. To prove uniform convergence of the series use the Abel transformation.

6.33. a), b). Use the Abel transformation.

c) Let $f(\theta) = S_N(\theta) + R_N(\theta)$, where $S_N(\theta)$ is the Nth partial sum of the series defining f. Then

$$|f(\theta) - f(\theta')| \le |S_N(\theta) - S_N(\theta')| + |R_N(\theta)| + |R_N(\theta')|.$$

Taking $N = [1/|\theta - \theta'|]$, estimate the remainders R_N with the help of the assertion b). To estimate the difference $S_N(\theta) - S_N(\theta')$ use the inequality $|S_N(\theta) - S_N(\theta')| \le |\theta - \theta'| \max_{x\in\mathbb{R}} |S'_N(x)|$ and the assertion a).

6.34. Use the results of problems 6.31 and 6.33c) for $\alpha = 1/2$.

CHAPTER V

Integrals

§1. Improper integrals of functions of a single variable

1.1. Consider the integrals over the intervals $[k\pi, (k+1)\pi]$.

1.2. To estimate the integrals over the intervals $[k\pi, (k+1)\pi]$ use the inequalities $\frac{2}{\pi}x < \sin x < x$ for $x \in (0, \pi/2)$.

1.3–1.7. The solutions of these problems are analogous to the solution of problem 1.2.

1.8. Use the result of problem VI.2.9.

1.9. Integrating by parts,

$$\int_1^A \cos(x^3 - x)\,dx = \int_1^A \frac{d\sin(x^3-x)}{3x^2-1}$$
$$= o(1) + 6\int_1^A \frac{x}{(3x^2-1)^2}\sin(x^3-x)\,dx,$$

reduce the original integral to an absolutely convergent integral.

1.10–1.11. The solutions of these problems are analogous to the solution of problem 1.9.

1.12. a) Consider the function with $f(x) = n$ for $x \in [n, n+n^{-3}]$ $(n \in \mathbb{N})$, and $f(x) = 0$ otherwise.

b) Consider the function $f(x) = x^2\sin(x^p)$, where $p > 3$.

1.13. In investigating convergence of the integral at infinity with the help of the result of problem 1.16 and integrating by parts, show that $f(x) \sim (\ln(x^2-1))/x^{1+\varepsilon}$ as $x \to +\infty$.

1.14. a) Find the limit of the integral

$$\int_\varepsilon^{+\infty} \frac{f(ax) - f(bx)}{x}\,dx = \int_{\varepsilon a}^{\varepsilon b} \frac{f(t)}{t}\,dt \quad \text{as } \varepsilon \to +0.$$

b) If $\int_0^T f(t)\,dt = 0$, then see problem a). Otherwise, consider the function $\tilde{f}(x) = f(x) - \frac{1}{T}\int_0^T f(t)\,dt$.

c) Find the limit of the integral

$$\int_\varepsilon^{A} \frac{f(ax) - f(bx)}{x}\,dx = \int_{\varepsilon a}^{\varepsilon b} \frac{f(t)}{t}\,dt - \int_{Aa}^{Ab} \frac{f(t)}{t}\,dt \quad \text{as } A \to +\infty,\ \varepsilon \to 0.$$

1.15. a) It follows from the second mean value theorem that for any numbers A and B with $0 < A < B$ there is a number $C \in [A, B]$ such that $\int_A^B e^{-\varepsilon x} f(x)\,dx = e^{-A\varepsilon} \int_A^C f(x)\,dx$. Since the integral $\int_0^{+\infty} f(x)\,dx$ converges, all the integrals of the form $\int_A^C f(x)\,dx$ are small if A is sufficiently large. Therefore, the integrals $\int_A^B e^{-\varepsilon x} f(x)\,dx$ are small, and this is equivalent to the convergence of the integral $\int_0^{+\infty} f(x)e^{-\varepsilon x}\,dx$.

We remark that if f is continuous on $[0, +\infty)$, then the assertion to be proved is easy to get by integrating by parts: $\int_0^A e^{-\varepsilon x} f(x)\,dx = e^{-\varepsilon A}F(A) + \varepsilon \int_0^A e^{-\varepsilon x} F(x)\,dx$ (here $F(x) = \int_0^x f(t)\,dt$).

b) We represent the difference between the integrals in the form

$$\int_0^{+\infty} f(x)\,dx - \int_0^{+\infty} e^{-\varepsilon x} f(x)\,dx$$
$$= \int_0^A (1 - e^{-\varepsilon x}) f(x)\,dx + \int_A^{+\infty} f(x)\,dx - \int_A^{+\infty} e^{-\varepsilon x} f(x)\,dx.$$

For large A the last two integrals are small for all $\varepsilon > 0$ (the estimate of the third integral follows from the second mean value theorem). The first integral can be made small by letting ε go to zero for a fixed A.

1.16. With the help of integration by parts, compute the sum of the integrals over the intervals $[0, a]$ and $[1/a, +\infty)$ for $a \in (0, 1)$.

1.17. Since $I = \int_0^{\pi/2} \ln(\sin x)\,dx = \int_0^{\pi/2} \ln(\cos x)\,dx$, it follows that

$$2I = \int_0^{\pi/2} \ln\left(\frac{\sin 2x}{2}\right) dx = \frac{1}{2}\int_0^{\pi} \ln\left(\frac{\sin t}{2}\right) dt$$
$$= \int_0^{\pi/2} \ln\left(\frac{\sin t}{2}\right) dt = I - \frac{\pi}{2}\ln 2.$$

1.18. After expanding $(1-x)^{-1}$ in a series, integrate it termwise and use the result of problem IV.4.12. To justify termwise integration of the sum of the series, show that

$$\lim_{N\to+\infty} \int_0^1 \left(\sum_{n\geq N} x^n \ln x\right) dx = \lim_{N\to+\infty} \int_0^1 \frac{x^N \ln x}{1-x}\,dx = 0.$$

To prove the last equality use the estimate

$$\int_0^1 \frac{x^N \ln x}{1-x}\,dx = \int_0^{1/2} + \int_{1/2}^1 = O\left(\frac{1}{2^N}\right) + O\left(\int_{1/2}^1 x^N\,dx\right) = o(1).$$

1.19. In the preceding problem make the change of variable $x \mapsto x^2$.

1.20. Integrating twice by parts, we get

$$\int_0^{+\infty} \frac{x^2\,dx}{\cosh^2 x} = \int_0^{\infty} x^2\,d(\tanh x - 1) = 4\int_0^{+\infty} \frac{xe^{-2x}\,dx}{1+e^{-2x}}$$
$$= 2\int_0^{+\infty} \ln(1+e^{-2x})\,dx = \int_0^1 \frac{\ln t}{t-1}\,dt.$$

It remains to use the result of problem 1.18.

1.21. Differentiate the integral with respect to the parameter a.

1.22. Use the results of problems 1.21 and 1.15b).

1.23. By integrating by parts, reduce the problem to the preceding one.

1.24. Taking the factor $e^{-\varepsilon x}$ under the integral sign and computing the integral obtained by means of differentiation with respect to the parameter ε, use the result of problem 1.15b).

1.25. Integrating by parts, we get that

$$\begin{aligned}
&\int_{-\infty}^{\infty} \frac{(e^{iax}-1)(e^{ibx}-1)}{x^2}\,dx \\
&\quad = \int_{-\infty}^{+\infty} \frac{1}{x}(e^{i(a+b)x}i(a+b) - iae^{iax} - ibe^{ibx})\,dx \\
&\quad = \int_{-\infty}^{+\infty} (a\sin ax + b\sin bx - (a+b)\sin(a+b)x)\frac{dx}{x} \\
&\quad = (|a|+|b|-|a+b|)\int_{-\infty}^{+\infty} \frac{\sin x}{x}\,dx.
\end{aligned}$$

It remains to use the result of problem 1.22.

1.26, 1.27. Replace e^{-x} by $(1-x/n)^n$ and use the inequality in problem I.2.16a).

1.28. Reduce the original integral to the difference

$$\int_0^1 \frac{1-e^{-x}}{x}\,dx - \int_1^{+\infty} \frac{e^{-x}}{x}\,dx.$$

with the help of the result in problem 1.24. Use integration by parts and the integral in problem 1.26.

1.29. Reduce the problem to the preceding one.

1.30. Reduce the given integral to the difference

$$\int_0^1 \frac{1-e^{-x}}{x}\,dx - \int_1^{+\infty} \frac{e^{-x}}{x}\,dx,$$

and use the result of problem 1.26.

1.31. Using integration by parts and a change of variable, reduce the given integral to the Frullani integral (see problem 1.14).

1.32. Consider the integral over the interval $[\Delta, +\infty)$ $(\Delta > 0)$ and prove that it is equal to $\int_\Delta^{+\infty} \frac{1}{x}\ln(1+e^{-x})\,dx$. Using the equality $\ln(1+e^{-x}) = \ln(1-e^{-2x}) - \ln(1-e^{-x})$, reduce the last integral to the integral $\int_\Delta^{2\Delta} \frac{1}{x}\ln(1-e^{-x})\,dx$ and show that it is equal to the integral $\int_\Delta^{2\Delta} \frac{1}{x}\ln x\,dx$ up to a quantity infinitesimally small as $\Delta \to 0$.

1.34. For $k = 1$, 3, and 5 reduce the given integral to the Euler-Poisson integral. For $k = 4$ use the result of problem 1.14.

1.35. Integrating by parts in the indefinite integral, we get that

$$\int \frac{e^{-x^2}\,dx}{(x^2+1/2)^2} = -\int \frac{e^{-x^2}}{x}\,d\frac{1}{2x^2+1} = -\frac{e^{-x^2}}{x(2x^2+1)} - \int \frac{e^{-x^2}}{x^2}\,dx$$
$$= \frac{xe^{-x^2}}{x^2+1/2} + 2\int e^{-x^2}\,dx.$$

Consequently,

$$\int_0^{+\infty} \frac{e^{-x^2}\,dx}{(x^2+1/2)^2} = 2\int_0^{+\infty} e^{-x^2}\,dx = \sqrt{\pi}.$$

1.36. Denote the desired integral by $I(a)$. Differentiating with respect to a for $a > 0$ and making the change of variable $x \mapsto a/x$, we get that $I'(a) = -2I(a)$, and hence $I(a) = Ce^{-2a}$. It remains to use the Euler-Poisson integral: $C = I(0) = \sqrt{\pi}/2$.

1.37. a) By the result in problem 1.15b), $\int_0^{+\infty} \frac{1}{\sqrt{x}} \sin x\,dx = \lim_{\varepsilon\to+0} I(\varepsilon)$, where $I(\varepsilon) = \int_0^{+\infty} \frac{1}{\sqrt{x}}(\sin x)e^{-\varepsilon x}\,dx$. To compute the integral $I(\varepsilon)$ for $\varepsilon > 0$, use the equality $\frac{1}{\sqrt{x}} = \frac{2}{\sqrt{\pi}}\int_0^{+\infty} e^{-xy^2}\,dy$ (see problem 1.33):

$$I(\varepsilon) = \frac{2}{\sqrt{\pi}}\int_0^{+\infty} e^{-\varepsilon x}\sin x\left(\int_0^{+\infty} e^{-xy^2}\,dy\right)dx$$
$$= \frac{2}{\sqrt{\pi}}\int_0^{+\infty}\left(\int_0^{+\infty}\frac{\sin x\,dx}{e^{(\varepsilon+y^2)x}}\right)dy = \frac{2}{\sqrt{\pi}}\int_0^{+\infty}\frac{dy}{1+(\varepsilon+y^2)^2}.$$

Therefore,

$$\int_0^{+\infty}\frac{\sin x}{\sqrt{x}}\,dx = \lim_{\varepsilon\to+0} I(\varepsilon) = \frac{2}{\sqrt{\pi}}\int_0^{+\infty}\frac{dy}{1+y^4} = \sqrt{\frac{\pi}{2}}.$$

b) The solution is analogous to that of problem a).

1.38. Computing the integrals over the intervals $[p_k, p_{k+1}]$ (where $p_1 = 2$, $p_2 = 3$, ... are the prime numbers, numbered in increasing order), show that

$$\int_2^{+\infty}\frac{\pi(x)\,dx}{x^3-x} = -\frac{1}{2}\lim_{k\to+\infty}\ln\left(\left(1-\frac{1}{2^2}\right)\left(1-\frac{1}{3^2}\right)\cdots\left(1-\frac{1}{p_k^2}\right)\right)$$

and use the result of problem IV.4.14a).

§2. Computation of multiple integrals

2.1. Represent the given integral in the form

$$2\int_0^{+\infty}\left(\int_0^x\left(\ln\frac{x}{y}\right)e^{-(x+y)}\,dy\right)dx.$$

In the inside integral make the change of variable $y = tx$ $(t \in (0, 1))$ and change the order of integration.

2.2. Using the symmetry of the domains of integration and of the integrands with respect to arbitrary permutations of the variables, show that:

a) $\int_{[0,1]^n} \max(x_1, \dots, x_n)\, dx_1 \cdots dx_n = n! \int_0^1 x_1\, dx_1 \int_0^{x_1} dx_2 \cdots \int_0^{x_{n-1}} dx_n$;

b) $\int_{[0,1]^n} \min(x_1, \dots, x_n)\, dx_1 \cdots dx_n = n! \int_0^1 dx_1 \int_0^{x_1} dx_2 \cdots \int_0^{x_{n-1}} x_n\, dx_n$;

c)

$$\int_{[1,+\infty)^n} \frac{dx_1 \cdots dx_n}{x_1 \cdots x_n (\max(x_1, \dots, x_n))^{\varepsilon}} = n! \int_1^{+\infty} \frac{dx_1}{x_1} \int_{x_1}^{+\infty} \frac{dx_2}{x_2} \cdots \int_{x_{n-1}}^{+\infty} \frac{dx_n}{x_n^{1+\varepsilon}}.$$

2.3. Use the Euler-Poisson integral (see problem 1.33).

2.4. a) Assuming that $a^2 + b^2 > 0$, let $u = (ax + by)/\sqrt{a^2 + b^2}$ and $v = cx + dy$, where c and d are chosen so that $(x, y) \to (u, v)$ is an orthogonal transformation. Then $x^2 + y^2 = u^2 + v^2$, and hence

$$\begin{aligned}
&\iint_{\mathbb{R}^2} |ax + by| e^{-(x^2+y^2)/2}\, dx\, dy \\
&= \iint_{\mathbb{R}^2} |u| e^{-(u^2+v^2)/2} \sqrt{a^2 + b^2}\, du\, dv \\
&= 2\sqrt{a^2 + b^2} \int_{-\infty}^{+\infty} e^{-v^2/2}\, dv \int_0^{+\infty} u e^{-u^2/2}\, du \\
&= 2\sqrt{a^2 + b^2}\sqrt{2\pi}.
\end{aligned}$$

b) Pass to a new coordinate system with one axis directed along the vector a.

2.5. Let $\|x\| = a$. It can be assumed without loss of generality that $x = (0, 0, a)$. Passing to spherical coordinates, we get that

$$\begin{aligned}
\iiint_{\|y\| \le 1} \frac{dy}{\|x - y\|} &= \int_0^1 r^2\, dr \int_0^{2\pi} d\varphi \int_0^{\pi} \frac{\sin\theta\, d\theta}{\sqrt{r^2 - 2ar\cos\theta + a^2}} \\
&= \frac{2\pi}{a} \int_0^1 r \sqrt{r^2 - 2ar\cos\theta + a^2}\Big|_{\theta=0}^{\theta=\pi} \\
&= \frac{2\pi}{a} \int_0^1 r(r + a - |r - a|)\, dr.
\end{aligned}$$

We leave it to the reader to finish the computations.

2.6. It is clear that

$$\begin{aligned}
&\iiiint_{x^2+y^2+u^2+v^2 \le 1} e^{x^2+y^2-u^2-v^2}\, dx\, dy\, du\, dv \\
&= \iint_{x^2+y^2 \le 1} e^{x^2+y^2} \left(\iint_{u^2+v^2 \le 1-x^2-y^2} e^{-(u^2+v^2)}\, du\, dv \right) dx\, dy.
\end{aligned}$$

To compute the double integrals use polar coordinates.

2.7. Choosing an orthogonal basis in $\mathbb{R}^4$ in which the matrix A has diagonal form, we get

$$\int_{(Ax,x) \le 1} e^{(Ax,x)}\, dx = \iiiint_{\sum_{1 \le k \le \nu} \lambda_k t_k^2 \le 1} e^{\sum_{1 \le k \le 4} \lambda_k t_k^2}\, dt_1\, dt_2\, dt_3\, dt_4$$

with the help of the corresponding orthogonal transformation, where λ_k $(k = 1, 2, 3, 4)$ are the eigenvalues of A. Making the change of variables $u_k = \sqrt{\lambda_k} t_k$, we see that

$$\int_{(Ax, x)\le 1} e^{(Ax, x)}\, dx = \frac{1}{\sqrt{\lambda_1\lambda_2\lambda_3\lambda_4}} \iiiint_{\sum_{1\le k\le 4} u_k^2 \le 1} e^{\sum_{1\le k\le 4} u_k^2}\, du_1\, du_2\, du_3\, du_4.$$

It remains to see that $\lambda_1\lambda_2\lambda_3\lambda_4 = \det A$, and the integral can be computed with the help of polar coordinates (see problem 2.6).

2.8. Let $t = (t_1, t_2, t_3, t_4)$. It is obvious that

$$K(t) = \int_0^{+\infty} e^{-t_1x_1} \left(\iiint_{\sqrt{x_2^2+x_3^2+x_4^2}\le x_1} e^{-t_2x_2-t_3x_3-t_4x_4}\, dx_2\, dx_3\, dx_4 \right) dx_1. \qquad (*)$$

Let $s = (t_2, t_3, t_4)$ and $y = (x_2, x_3, x_4)$. Then the inside integral in $(*)$ can be written in the form $\int_{\|y\|\le x_1} e^{-(y,s)}\, dy = \mathscr{J}(s)$. With the help of an orthogonal transformation of variables the problem can be reduced to the case when the vector s has the form $(a, 0, 0)$, $a = \|s\|$, and thus

$$\mathscr{J}(s) = \iiint_{\|y\|\le x_1} e^{-ax_2}\, dx_2\, dx_3\, dx_4 = \int_{-x_1}^{x_1} e^{-ax_2} \pi(x_1^2 - x_2^2)\, dx_2.$$

Computing the last integral, we see that it is equal to

$$\frac{4\pi x_1}{a^2}\cosh(ax_1) - \frac{4\pi}{a^3}\sinh(ax_1).$$

After substitution of this result in $(*)$, $K(t)$ takes the form

$$K(t) = \frac{4\pi}{a^2}\int_0^{\infty} \left(x_1\cosh(ax_1) - \frac{1}{a}\sinh(ax_1) \right) e^{-t_1x_1}\, dx_1.$$

This integral is finite if and only if $a < t_1$, that is (since $a = \|s\| = \sqrt{t_2^2 + t_3^2 + t_4^2}$), for $t \in \operatorname{Int} A$. We leave it to the reader to compute the last integral.

2.9. Show that

$$M = \frac{1}{(b-a)^2}\int_a^b \int_a^b |x-y|\, dx\, dy \quad \text{and} \quad P = \frac{S(E)}{(b-a)^2},$$

where $E = \{(x, y) | x, y \in [a, b],\ |x-y| > (b-a)/3\}$, and $S(E)$ is the area of E.

2.10. Suppose for definiteness that $x \le y \le z$. These numbers are the sides of some triangle if and only if $z \le x + y$. Consequently, to find the desired probability it is necessary to compute the volume of the pyramid $\{(x, y, z) \,|\, 0 \le x \le y \le z \le a,\ z \le x + y\}$.

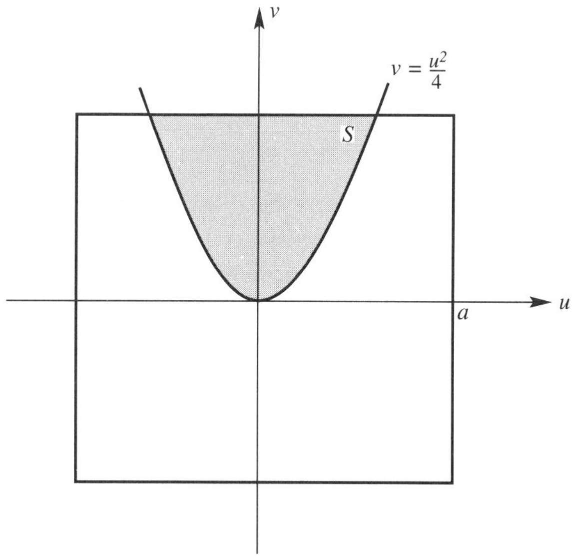

FIGURE 12

2.11. a) The realness of the roots is determined by the sign of the discriminant, which is equal to $u^2 - 4v$. For $a > 4$ consider the square $[-a, a]\times[-a, a]$ and the parabola $v = u^2/4$ in the (u, v)-plane (see Figure 12). Let S be the set of points in the square that lie above the parabola. The equation $z^2+uz+v = 0$ has real roots if and only if $(u, v) \notin S$. Therefore, $P_1(a)$ is always $> 1/2$. Since the area of S is equal to $(4/3)a^{3/2}$, it follows that $P_1(a) = 1 - 1/(3\sqrt{a}) \to 1$ as $a \to +\infty$.

b) A biquadratic equation has both real and complex roots if and only if $v < 0$. Therefore, the corresponding probability is equal to $1/2$.

2.12. Show that

$$S = \frac{1}{\pi^2}\iint_{x^2+y^2\leq 1}\left(\iint_{u^2+v^2\leq 1}\frac{\pi}{4}((x-u)^2+(y-v)^2)\,du\,dv\right)dx\,dy.$$

2.13. Show that:

a)

$$L = \frac{R}{(2\pi)^2}\int_0^{2\pi}\int_0^{2\pi}|e^{i\varphi} - e^{i\psi}|\,d\varphi\,d\psi = \frac{R}{2\pi}\int_0^{2\pi}|e^{i\theta}-1|\,d\theta;$$

b)

$$\alpha = \frac{2}{(2\pi)^2}\left(\iint_E|\varphi-\psi|\,d\varphi\,d\psi + \int_\pi^{2\pi}\left(\int_0^{\pi}|\varphi-\psi-\pi|\,d\psi\right)d\varphi\right),$$

where $E = \{(\varphi, \psi)\,|\,0 \leq \psi \leq \varphi \leq 2\pi,\ \varphi - \psi \leq \pi\}$.

2.14. It is easy to see that $\Phi(x, y) = f(a)$ and $\Psi(x, y) = g(a)$, where $a = \sqrt{x^2+y^2}$, $f(a) = \iint_{u^2+v^2\leq 1}((u-a)^2+v^2)^{-p/2}\,du\,dv$, and $g(a) =$

$\iint_{u^2+v^2\le 1}\ln((u-a)^2+v^2)\,du\,dv$. Prove that $f\in C^1([0,+\infty))$ and $f'(0)=0$, which clearly implies that the function Φ is smooth. Show that

$$f'(a)=\iint_{u^2+v^2\le 1}(((u-a)^2+v^2)^{p/2})'_a\,du\,dv. \qquad (**)$$

For $a>1$ this equality is easily established by considering the relation $\frac{1}{h}(f(a+h)-f(a))$. To justify passing to the limit as $h\to 0$ under the integral sign it suffices to use the mean value theorem and the inequality $(u-a)^2+v^2\ge(a-1)^2>0$.

Suppose now that $a\in[0,1)$. Note that the function under the double integral sign in $(**)$ is absolutely integrable on the disk $\{(u,v)\mid u^2+v^2\le 1\}$:

$$\begin{aligned}\iint_{u^2+v^2\le 1}|(\cdots)'_a|\,du\,dv&=|p|\iint_{u^2+v^2\le 1}\frac{|u-a|\,du\,dv}{((u-a)^2+v^2)^{1+p/2}}\\&=|p|\iint_{(u+a)^2+v^2\le 1}\frac{|u|\,du\,dv}{(u^2+v^2)^{1+p/2}}\\&\le|p|\iint_{u^2+v^2\le 4}\frac{du\,dv}{(u^2+v^2)^{(1+p)/2}}\\&=2\pi|p|\int_0^4\frac{dr}{r^p}<+\infty.\end{aligned}$$

We consider the set $E=\{(u,v)\mid u^2+v^2\ge 1,\ |u|\le 1,\ |v|\le 1\}$ and represent f in the form

$$\begin{aligned}f(a)&=\int_{-1}^{+1}\int_{-1}^{1}((u-a)^2+v^2)^{-p/2}\,du\,dv-\iint_E((u-a)^2+v^2)^{-p/2}\,du\,dv\\&=f_1(a)-f_2(a).\end{aligned}$$

For $a\in[0,1)$ the inequality $(u-a)^2+v^2\ge(a-1)^2>0$ holds on the set E, therefore, $f_2'(a)=\iint_E(((u-a)^2+v^2)^{-p/2})'_a\,du\,dv$. Since $f_1(a)=\int_{-1-a}^{1-a}\int_{-1}^{1}(u^2+v^2)^{-p/2}\,du\,dv$, it follows that $f_1\in C^1([0,+\infty))$, and

$$\begin{aligned}f_1'(a)&=-\int_{-1}^{1}((1-a)^2+v^2)^{-p/2}\,dv+\int_{-1}^{1}((1+a)^2+v^2)^{-p/2}\,dv\\&=-\int_{-1}^{1}\int_{-1}^{1}(((u-a)^2+v^2)^{-p/2})'_u\,du\,dv\\&=\int_{-1}^{1}\int_{-1}^{1}(((u-a)^2+v^2)^{-p/2})'_a\,du\,dv.\end{aligned}$$

The equality $(**)$ follows from the representations obtained for $f_1'(a)$ and $f_2'(a)$.

Thus, for all $a \geq 0$ with $a \neq 1$

$$\begin{aligned} f'(a) &= \iint_{u^2+v^2\leq 1} ((u-a)^2+v^2)^{-p/2})'_a \, du\, dv \\ &= -\iint_{u^2+v^2\leq 1} (((u-a)^2+v^2)^{-p/2})'_u \, du\, dv \\ &= -\int_{-1}^{1} ((u-a)^2+v^2)^{-p/2}\Big|_{u=-\sqrt{1-v^2}}^{u=\sqrt{1-v^2}} dv \\ &= -\int_{-\pi}^{\pi} ((\cos\varphi - a)^2 + \sin^2\varphi)^{-p/2} \cos\varphi \, d\varphi. \end{aligned}$$

This implies that f' is continuous on $[0, 1) \cup (1, +\infty)$ and has a finite limit as $a \to 1$. Since f is clearly continuous on $[0, +\infty)$, it follows that $f \in C^1([0, +\infty))$. It is clear that $f'(0) = 0$.

It is proved similarly that g is smooth and

$$g'(a) = -\int_{-\pi}^{\pi} \cos\varphi \ln(1 - 2a\cos\varphi + a^2)\, d\varphi.$$

Therefore,

$$\begin{aligned} g'(a) &= 4a\int_0^{\pi} \frac{\sin^2\varphi\, d\varphi}{1-2a\cos\varphi + a^2} = 4a\left(\int_0^{\pi/2} + \int_{\pi/2}^{\pi}\right) \\ &= 4a(1+a^2)\int_0^{\pi/2} \frac{\sin^2\varphi\, d\varphi}{(1+a^2)^2 - 4a^2\cos^2\varphi} \\ &= \frac{\pi}{a}(1+a^2 - |1-a^2|) = 2\pi\min(a, a^{-1}). \end{aligned}$$

Since

$$g(0) = \iint_{u^2+v^2\leq 1} \ln(u^2+v^2)\, du\, dv = 2\pi\int_0^1 r\ln r^2\, dr = -\pi,$$

it follows that $g(a) = \pi(a^2 - 1)$ for $a \in [0, 1]$, and $g(a) = 2\pi \ln a$ for $a \geq 1$.

2.15. Use induction.

2.16. Let $J_n(a) = \int_{[0,1]^n} (a + x_1 + \cdots + x_n)^{-1}\, dx$. It is clear that

$$J_n^{(k)}(a) = (-1)^k k! \int_{[0,1]^n} \frac{dx}{(a+x_1+\cdots+x_n)^{k+1}}.$$

On the other hand,

$$\begin{aligned} J_n(a) &= \int_0^1 \left(\int_{[0,1]^{n-1}} \frac{dx_2\cdots dx_n}{(a+x_1)+x_2+\cdots+x_n}\right) dx_1 \\ &= \int_0^1 J_{n-1}(a+t)\, dt. \end{aligned}$$

Therefore, $J_n'(a) = \int_0^1 J_{n-1}'(a+t)\,dt = \Delta J_{n-1}(a)$. By induction, $J_n^{(k)}(a) = \Delta_k J_{n-k}(a)$ $(k = 1, 2, \dots, n-1)$. In particular,

$$J_n^{(n-1)}(a) = \Delta_{n-1} J_1(a) = \Delta_{n-1}(\ln(a+1) - \ln a) = \Delta_n(\ln a),$$

that is,

$$\int_{[0,1]^n} \frac{dx}{(a + x_1 + \dots + x_n)^n} = \frac{(-1)^{n-1}}{(n-1)!}\Delta_n(\ln a). \tag{1}$$

Thus, the required equality is proved for $m = n$. Let us prove it for $1 \le m < n$ by induction from $m+1$ to m. The equality (1) is the base for the induction. Assume that $m < n$ and

$$\int_{[0,1]^n} (a + x_1 + \dots + x_n)^{-m-1}\,dx = \frac{(-1)^m}{m!(n-m-1)!}\Delta_n(a^{n-m-1}\ln a).$$

Integrating this equality over the integral $[\alpha, \beta]$ $(0 < \alpha < \beta)$, we get that

$$\int_{[0,1]^n} \left(\int_\alpha^\beta (a + x_1 + \dots + x_n)^{-m-1}\,da \right) dx$$
$$= \frac{(-1)^m}{m!(n-m-1)!}\Delta_n \left(\int_\alpha^\beta a^{n-m-1} \ln a\,da \right). \tag{2}$$

We remark that

$$\int_\alpha^\beta t^{n-m-1} \ln t\,dt = \frac{1}{n-m}(\beta^{n-m}\ln\beta - \alpha^{n-m}\ln\alpha) - \frac{\beta^{n-m} - \alpha^{n-m}}{(n-m)^2},$$

and, moreover, $\Delta_n(\beta^{n-m} - \alpha^{n-m}) = 0$, because $\Delta_k(p(x)) \equiv 0$, if the degree of the polynomial p is less than n. Therefore, it follows from (2) that

$$\int_{[0,1]^n} ((\alpha + x_1 + \dots + x_n)^{-m} - (\beta + x_1 + \dots + x_n)^{-m})\,dx$$
$$= \frac{(-1)^m}{(m-1)!(n-m)!}\Delta_n(\beta^{n-m}\ln\beta - \alpha^{n-m}\ln\alpha). \tag{3}$$

Since

$$\Delta_n(\beta^k \ln\beta) \to 0 \quad \text{as } \beta \to +\infty \text{ if } k < n, \tag{4}$$

we get the required result by passing to the limit in (3) as $\beta \to +\infty$. The relation (4) follows from the equalities $\Delta_n f(x) = f^{(n)}(x + \theta_n n)$, where $0 < \theta_n < 1$, and

$$(x^k \ln x)^{(n)} = \frac{k!(n-k-1)!}{x^{n-k}}(-1)^{n-k-1} \quad \text{for } n > k.$$

2.17. a) In the integral $\int_{Q_n} f(a_0 + (a_1 - a_0)x_1 + \dots + (a_n - a_0)x_n)\,dx_1 \cdots dx_n$ make the change of variables $y_1 = 1 - x_1$, $y_2 = x_2/y_1$, $\dots$, $y_n = x_n/y_1$. Use induction.

b) Both integrals (see [31], p. 77, about *Feynman integrals*) can be computed by induction.

2.18. a) Assuming that $ab \neq 0$, make the change of variables $x = u$, $ax + by = cv$, where $c = |a| + |b|$. This leads to the equality

$$\frac{1}{\pi} \iint_{\mathbb{R}^2} f(ax+by) \frac{dx\,dy}{(1+x^2)(1+y^2)} = \frac{1}{\pi} \int_{-\infty}^{\infty} f(cv)\,dv \int_{-\infty}^{\infty} \frac{c\,du}{|b|(1+u^2)(1+\frac{a^2}{b^2}(u-\frac{c}{a}v)^2)}.$$

Show that the inside integral is equal to $\pi/(1+v^2)$.

b) Using a) for the induction step, prove by induction that

$$\frac{1}{\pi^n} \int_{\mathbb{R}^n} \left| c + \sum_{1 \le k \le n} a_k x_k \right|^p \frac{dx_1 \cdots dx_n}{(1+x_1^2)\cdots(1+x_n^2)} = \frac{1}{\pi} \int_{-\infty}^{\infty} \left| c + t \sum_{1 \le k \le n} |a_k| \right|^p \frac{dt}{1+t^2}.$$

For $c = 0$ this leads to the equality

$$\frac{1}{\pi^n} \int_{\mathbb{R}^n} \left| \sum_{1 \le k \le n} a_k x_k \right|^p \frac{dx_1 \cdots dx_n}{(1+x_1^2)\cdots(1+x_n^2)} = \frac{1}{\pi} \left(\sum_{1 \le k \le n} |a_k| \right)^p \int_{-\infty}^{\infty} \frac{|t|^p\,dt}{1+t^2}.$$

The last integral can be expressed in terms of the Euler Γ-function:

$$\begin{aligned}\int_{-\infty}^{\infty} \frac{|t|^p\,dt}{1+t^2} &= 2\int_0^{\pi/2} \cos^{-p}\varphi \sin^p \varphi\,d\varphi \\ &= \int_0^{\pi/2} \cos^{-p-1}\varphi \sin^{p-1}\varphi\,d\sin^2\varphi \\ &= \int_0^1 s^{(p-1)/2}(1-s)^{-(p+1)/2}\,ds = B\left(\frac{p+1}{2}, \frac{1-p}{2}\right) \\ &= \Gamma\left(\frac{p+1}{2}\right)\Gamma\left(\frac{1-p}{2}\right) = \frac{\pi}{\sin((p+1)\pi/2)} = \frac{\pi}{\cos(\pi p/2)}.\end{aligned}$$

2.19. Represent the integral as the sum of the integrals over the squares

$$\left[\frac{k-1}{n}, \frac{k}{n}\right] \times \left[\frac{j-1}{n}, \frac{j}{n}\right] \qquad (1 \le k, j \le n).$$

To compute these integrals with an accuracy of $o(n^{-4})$, replace the function f by its second-degree Taylor polynomial constructed for the points $(\frac{k-1/2}{n}, \frac{j-1/2}{n})$.

CHAPTER VI

Asymptotics

§1. Asymptotics of integrals

1.2. a) Use the relation $1-x^a \sim a(1-x)$ as $x \to 1$.

b) Compare the integral $\mathscr{K}(t)$ with the integral

$$\int_0^1 (2(1-x)(1-tx)(1+t))^{-1/2}\,dx.$$

c) Compare the integral $\mathscr{L}(t)$ with the integral obtained from it by replacing $\cos x$ by $1-x^2/2$.

1.3. a) Use the relation

$$\frac{1}{\ln x} = \frac{1}{\ln A + O(1)} = \frac{1}{\ln A} + O\left(\frac{1}{\ln^2 A}\right)$$

for $x \in [A, 2A]$.

b), c) Use the result of problem 1.1a).

d), e) Integrate by parts and use the result of problem 1.1d).

1.4. a) Make the change of variables $y = x^p/A$.

b) Integrate by parts after the change of variables $y = Ax^2$.

c) Complete the square in the exponent.

d) Represent the integral in the form

$$A\ln^p A \int_{2/A}^{+\infty} \left(1+\frac{\ln t}{\ln A}\right)^p \frac{dt}{e^t}$$

and prove that

$$\int_{2/A}^{+\infty} \left(1+\frac{\ln t}{\ln A}\right)^p \frac{dt}{e^t} = \int_{2/A}^{+\infty} \frac{dt}{e^t} + o(1).$$

To do this for $p>0$ use the relation $|(1+z)^p - 1| = O(|z|+|z|^p)$ $(z>-1)$. For $p<0$ verify that

$$\int_{2/A}^{1/\sqrt{A}} \left(1+\frac{\ln t}{\ln A}\right)^p \frac{dt}{e^t} = O\left(\frac{1}{\sqrt{A}\ln^p A}\right)$$

and estimate the integral

$$\int_{1/\sqrt{A}}^{+\infty} \left|\left(1+\frac{\ln t}{\ln A}\right)^p - 1\right| \frac{dt}{e^t},$$

using the fact that $|(z+1)^p - 1| = O(z)$ $(z > -1/2)$.

1.6. a, b). Integrating by parts, prove the boundedness of the integral $\int_1^{+\infty} (\cos 2x)/x^{1+\varepsilon}\, dx$ for $\varepsilon \geq 0$.

c) To estimate the integral over $[\pi/2, +\infty)$ integrate twice by parts.

d) Representing the integral in the form $\varepsilon(1+\varepsilon)\int_0^{+\infty}(1-\cos x)/x^{2+\varepsilon}\, dx$, prove that it is possible to pass to the limit under the integral sign. Use the equality (see problem V.1.22) $\int_0^{+\infty}(1-\cos x)/x^2\, dx = \int_0^{+\infty}(\sin x)/x\, dx = \pi/2$.

1.7. a) After the change of variable $y = x^p$ the solution is analogous to that of problem 1.6d).

b) With the help of the change of variable $y = x^p$ and integration by parts prove that the integral over $[1, +\infty)$ is infinitesimally small.

1.8. Using the boundedness of the integrals $\int_1^t (\varphi(u) - c_\varphi)\, du$ show that the integrals $\int_1^{+\infty}(\varphi(t) - c_\varphi)/t^{1+\varepsilon}\, dt$ are bounded for $\varepsilon > 0$.

1.9. a) Use the boundedness of the function $x^{-2} - \sin^{-2} x$ on $(0, \pi/2)$.

b) Make the change of variable $y = Ax$ and use the result of problem V.1.37b). Integrate by parts to estimate the integral $\int_1^{+\infty}(\cos y)/\sqrt{y}\, dy$.

c) Let

$$I(A) = \int_0^{\pi/2} \cos^2 x\,/(1+\cos^2 Ax)\, dx$$

and

$$J(A) = \int_0^{\pi/2} \sin^2 x\,/(1+\cos^2 Ax)\, dx.$$

It is clear that

$$\begin{aligned} I(A) + J(A) &= \int_0^{\pi/2} \frac{dx}{1+\cos^2 Ax} \\ &= \frac{1}{A}\int_0^{A\pi/2} \frac{dx}{1+\cos^2 x} \underset{A\to+\infty}{\longrightarrow} \int_0^{\pi/2} \frac{dx}{1+\cos^2 x} = \frac{\pi}{2\sqrt{2}}. \end{aligned}$$

Prove that $I(A) - J(A) \to 0$ as $A \to +\infty$. See also problem 1.14.

d) Use the relation $x^{-4} - \sin^{-4} x = O(x^{-2})$ and the result of problem a).

1.10. Represent the integral as the sum of the integrals over the intervals $[\pi n, \pi(n+1)]$ $(n \in \mathbb{N})$ and estimate each of them from above and from below. See also problem 1.13.

1.11. Use the relation

$$\begin{aligned} \left|\int_\alpha^{\alpha+T} f(t)\varphi(t)\, dt\right| &= \left|\int_\alpha^{\alpha+T} (f(t) - f(\alpha))\varphi(t)\, dt\right| \\ &\leq (f(\alpha) - f(\alpha+T))\int_0^T |\varphi(t)|\, dt. \end{aligned}$$

1.12. Use the result of the preceding problem.

1.13. The solution is analogous to that of problem 1.10.

1.14. It can be assumed that $c_\varphi = 0$ (otherwise consider the function $\tilde{\varphi} = \varphi - c_\varphi$). Representing the given integral as a sum of integrals over intervals of length T/A, use the relation

$$\left|\int_t^{t+T/A} f(x)\varphi(Ax)\,dx\right| = \left|\int_t^{t+T/A} (f(x)-f(t))\varphi(Ax)\,dx\right| \leq \omega_f\left(\frac{T}{A}\right)\frac{1}{A}\int_0^T |\varphi(x)|\,dx,$$

where $\omega_f(\delta) = \sup_{|t'-t''|\leq\delta} |f(t') - f(t'')|$ is the modulus of continuity of f.

1.15. Let Δ_k be the intervals of the form $[2\pi k/A, 2\pi(k+1)/A]$, $k = p, \dots, q$, contained in $[a, b]$; ω_f is the modulus of continuity of f. Then

$$\begin{aligned} I(A) &= O\left(\frac{1}{A}\right) + \sum_{p\leq k\leq q}\int_{\Delta_k} f(x, \sin Ax)\,dx \\ &= O\left(\frac{1}{A}\right) + O\left(\omega_f\left(\frac{2\pi}{A}\right)\right) + \sum_{p\leq k\leq q}\int_{\Delta_k} f\left(\frac{2\pi k}{A}, \sin Ax\right)\,dx. \end{aligned}$$

Consequently,

$$\begin{aligned} I(A) &= o(1) + \frac{1}{2\pi}\sum_{p\leq k\leq q}\frac{2\pi}{A}\int_0^{2\pi} f\left(\frac{2\pi k}{A}, \sin t\right)\,dt \\ &\underset{A\to+\infty}{\longrightarrow} \frac{1}{2\pi}\int_a^b\left(\int_0^{2\pi} f(x, \sin t)\,dt\right)\,dx. \end{aligned}$$

In conclusion note that under the conditions of the problem any continuous periodic function on $\mathbb{R}$ can be taken instead of $\sin Ax$.

1.16. a) Make the change of variable $x = t/\Delta$, where $\Delta = \Delta_\varepsilon$ is a small positive parameter. We have that

$$\int_0^{+\infty} e^{-x(1-x^{-\varepsilon})}\,dx = \frac{1}{\Delta}\int_0^{+\infty}\exp\left(\left(\frac{t}{\Delta}\right)^{1-\varepsilon} - \frac{t}{\Delta}\right)\,dt.$$

Let us now select the parameter Δ_ε so that as $\varepsilon\to+0$ the integrand tends to a positive integrable function. It is not hard to see that by taking $\Delta_\varepsilon = \varepsilon\ln(1/\varepsilon)$, we get that

$$\exp\left(\left(\frac{t}{\Delta_\varepsilon}\right)^{1-\varepsilon} - \frac{t}{\Delta_\varepsilon}\right) \underset{\varepsilon\to+0}{\longrightarrow} e^{-t}.$$

Therefore, to prove the relation

$$\int_0^{+\infty} e^{-x(1-x^{-\varepsilon})}\,dx \underset{\varepsilon\to+0}{\sim} \frac{1}{\varepsilon|\ln\varepsilon|}$$

it suffices to justify passage to the limit under the integral sign:

$$\int_0^{+\infty}\exp\left(\left(\frac{t}{\Delta_\varepsilon}\right)^{1-\varepsilon} - \frac{t}{\Delta_\varepsilon}\right)\,dt \underset{\varepsilon\to+0}{\longrightarrow} \int_0^{+\infty} e^{-t}\,dt = 1.$$

To study the integral on the left-hand side, we break it up into the sum of the integrals over the intervals $[0, 1/|\ln\varepsilon|]$, $[\ln^2\varepsilon, +\infty)$, and $[1/|\ln\varepsilon|, \ln^2\varepsilon]$. Let us estimate the first two integrals from above. It follows from the inequalities $u^{1-\varepsilon} - u \le \varepsilon(1-u)$ for $u > 0$ and $0 < \varepsilon < 1$ that

$$\exp\left(\left(\frac{t}{\Delta}\right)^{1-\varepsilon} - \frac{t}{\Delta}\right) \le \exp\left(\varepsilon - \frac{\varepsilon}{\Delta}t\right).$$

Therefore,

$$\int_{\ln^2\varepsilon}^{+\infty} \exp\left(\left(\frac{t}{\Delta}\right)^{1-\varepsilon} - \frac{t}{\Delta}\right) dt \le \int_{\ln^2\varepsilon}^{+\infty} e^{\varepsilon - \varepsilon t/\Delta}\, dt = e^{\varepsilon}\varepsilon \ln\frac{1}{\varepsilon} \underset{\varepsilon\to+0}{\to} 0.$$

Since the integrand is uniformly bounded above for $t > 0$ and $0 < \varepsilon < 1$, the integral over $[0, 1/|\ln\varepsilon|]$ is infinitesimally small. To compute the integral over $[1/|\ln\varepsilon|, \ln^2\varepsilon]$ we represent the difference $(t/\Delta)^{1-\varepsilon} - t/\Delta$ in the form

$$\frac{t}{\Delta}\left(e^{-\varepsilon\ln(t/\Delta)} - 1\right) = \frac{t}{\Delta}\left(-\varepsilon\ln\frac{t}{\Delta} + O\left(\varepsilon^2\ln^2\frac{t}{\Delta}\right)\right) = -t(1+o(1)).$$

Consequently,

$$\int_{1/|\ln\varepsilon|}^{\ln^2\varepsilon} \exp\left(\left(\frac{t}{\Delta}\right)^{1-\varepsilon} - \frac{t}{\Delta}\right) dt = \int_{1/|\ln\varepsilon|}^{\ln^2\varepsilon} \frac{dt}{e^{t(1+o(1))}} \sim \int_{1/|\ln\varepsilon|}^{\ln^2\varepsilon} \frac{dt}{e^t} \underset{\varepsilon\to+0}{\to} 1.$$

b) The solution of this example is analogous to that of example a).

1.17. a) Use the relation

$$x^{\varepsilon x} = e^{\varepsilon x\ln x} = 1 + \varepsilon x\ln x + \frac{\varepsilon^2}{2}x^2\ln^2 x + O(\varepsilon^3), \qquad 0 < x \le 1.$$

b) Making the change of variable $t = \varepsilon/\sqrt{x}$, we get that

$$\begin{aligned}\int_0^1 e^{-\varepsilon/\sqrt{x}}\, dx &= 2\varepsilon^2\int_{\varepsilon}^{+\infty} \frac{e^{-t}}{t^3}\, dt \\ &= 2\varepsilon^2\int_{\varepsilon}^{1} \frac{e^{-t}}{t^3}\, dt + O(\varepsilon^2).\end{aligned}$$

It remains to use the Taylor expansion of the exponential.

c) Break up the interval $[0, 1]$ into the intervals $[0, \theta]$ and $[\theta, 1]$, where $\theta = \theta_\varepsilon = o(\varepsilon/\ln\varepsilon)$. The integral over $[0, \theta]$ does not exceed θ. On $[\theta, 1]$ replace the exponential by its Taylor expansion, verifying first with the help of integration by parts that

$$\int_\theta^1 \frac{dx}{x^2\ln^4 x/e} = O\left(\frac{1}{\theta\ln^4\theta}\right).$$

Show that for $\theta_\varepsilon = \varepsilon/\ln^2\varepsilon$ the remainder term in the formula to be proved has the estimate $O(\varepsilon\ln|\ln\varepsilon|/\ln^2\varepsilon)$.

1.18. Use the inequality

$$|e^{\varepsilon u} - (1+\varepsilon u)| \le \begin{cases} \varepsilon^2 e^u & \text{for } u > 0,\ 0 < \varepsilon < 1; \\ C_p \varepsilon^p |u|^p & \text{for } u < 0. \end{cases}$$

1.19. a) Make the change of variables $t = nx$ and estimate the integrals over the intervals $[(k-1)\pi, k\pi]$ from above and from below.

b)–c) We prove the following more general assertion:

$$\int_0^\pi \max_{1\le k\le n} a_k |\sin kx| \frac{dx}{x} = \sum_{1\le k\le n} \frac{a_k}{k} + O(1)$$

if the nonincreasing sequence $\{a_k\}$ of positive numbers is such that the sequence $\{ka_k\}$ is nondecreasing.

Let $M(x) = \max_{1\le k\le n} a_k |\sin kx|$. Since $|\sin t| \le |t|$, it follows that $M(x) \le xna_n$, and hence the integral over $[0, 1/n]$ gives a bounded contribution. To get an upper estimate of the integral use the fact that

$$M(x) \le \max(a_1 x, 2a_2 x, \dots, ka_k x, a_{k+1}, \dots, a_n) = ka_k x$$

if $x \ge 1/k$. Consequently,

$$\begin{aligned} \int_0^\pi \frac{M(x)}{x}\,dx &\le O(1) + \sum_{2\le k\le n} \int_{1/k}^{1/(k-1)} ka_k x \frac{dx}{x} \\ &= O(1) + \sum_{1\le k\le n} \frac{a_k}{k}. \end{aligned}$$

To get a lower estimate of the integral note that $M(x) \ge a_k \sin kx \ge \frac{2}{\pi} kxa_k$ on the intervals $[\pi/2(k+1), \pi/2k]$ $(k = 2, \dots, n)$, and thus

$$\begin{aligned} \int_0^\pi \frac{M(x)}{x}\,dx &\ge O(1) + \sum_{2\le k\le n} \int_{\pi/2(k+1)}^{\pi/2k} \frac{2}{\pi} kxa_k \frac{dx}{x} \\ &= O(1) + \sum_{1\le k\le n} \frac{a_k}{k}. \end{aligned}$$

§2. The Laplace method

2.1. Since $(n+1)\int_0^{\pi/2} \sin x \cos^n x\,dx = 1$, it suffices to prove that

$$\int_0^{\pi/2} (xf(x) - f(0)\sin x)\cos^n x\,dx = o(1/n).$$

Fixing an arbitrary number $\varepsilon > 0$, we get the estimate $O(\cos^n \varepsilon) = o(1/n)$ for the integral over $[\varepsilon, \pi/2]$. To estimate the integral over $[0, \varepsilon]$ note that $|xf(x) - f(0)\sin x| \le \delta_\varepsilon \sin x$ on this interval, where

$$\delta_\varepsilon = \sup_{0<x\le\varepsilon} |f(0) - xf(x)/\sin x| \underset{\varepsilon\to+0}{\to} 0.$$

2.2. Make the change of variable $y = x\sqrt{A}$.

2.4. Verify that the integral over each interval $[\sigma, \theta]$, $0 < \sigma < \theta$, decreases exponentially as $A \to +\infty$.

2.5. a) It is clear that

$$A^{1/p} \int_0^1 (1 - x^p)^A \, dx = \int_0^{A^{1/p}} \left(1 - \frac{t^p}{A}\right)^A dt.$$

Therefore,

$$\varlimsup_{A\to+\infty} A^{1/p} \int_0^1 (1 - x^p)^A \, dx \le \int_0^{+\infty} e^{-t^p} \, dt = \frac{1}{p} \int_0^{+\infty} x^{(1/p)-1} e^{-x} \, dx$$
$$= \frac{1}{p} \Gamma\left(\frac{1}{p}\right) = \Gamma\left(1 + \frac{1}{p}\right).$$

On the other hand, for any positive number a

$$\varliminf_{A\to+\infty} A^{1/p} \int_0^1 (1 - x^p)^A \, dx \ge \varliminf_{A\to+\infty} \int_0^a \left(1 - \frac{t^p}{A}\right)^A dt = \int_0^a e^{-t^p} \, dt.$$

Consequently,

$$\varliminf_{A\to+\infty} A^{1/p} \int_0^1 (1 - x^p)^A \, dx \ge \int_0^{+\infty} e^{-t^p} \, dt = \Gamma(1 + 1/p).$$

It follows from the solution given that the integral over any interval $[0, \theta]$ with $0 < \theta < 1$ has the same asymptotics.

b), c). The solution is analogous to the solution of problem a). In problem c) make a preliminary change of variable $y = x^{1+p}$.

2.6. It can be assumed without loss of generality that $a = 0$ and $C = 1$ (otherwise make the change of variable $y = C^p(x - a)$). Fix the numbers l and L, $0 < l < 1 < L$, and choose a positive number θ such that $lx^{1/p} \le 1 - \varphi(x) \le Lx^{1/p} < 1$ for $x \in (0, \theta)$. Break up the integral $\Phi(A)$ into the sum of the integrals over $[0, \theta]$ and $[\theta, b)$. For the first of them we the two-sided estimate

$$\int_0^\theta (1 - Lx^{1/p})^A \, dx \le \int_0^\theta \varphi^A(x) \, dx \le \int_0^\theta (1 - lx^{1/p})^A \, dx,$$

while the second is easily estimated from above:

$$\int_\theta^b \varphi^A(x) \, dx \le \varphi^{A-1}(\theta) \int_\theta^b \varphi(x) \, dx = o(A^{-p}).$$

Therefore,

$$\varliminf_{A\to+\infty} A^p \int_0^\theta (1 - Lx^{1/p})^A \, dx \le \varliminf_{A\to+\infty} A^p \Phi(A) \le \varlimsup_{A\to+\infty} A^p \Phi(A)$$
$$\le \varlimsup_{A\to+\infty} A^p \int_0^\theta (1 - lx^{1/p})^A \, dx.$$

Consequently (see the solution of problem 2.5a)),

$$L^{-p}\Gamma(1+p) \le \varliminf_{A\to+\infty} A^p\Phi(A) \le \varlimsup_{A\to+\infty} A^p\Phi(A) \le l^{-p}\Gamma(1+p).$$

Since l and L are arbitrary numbers, $0 < l < 1 < L$, this implies the equality $\lim_{A\to+\infty} A^p\Phi(A) = \Gamma(1+p)$.

2.7. The solution is analogous to that of the preceding problem.

2.8. Let $h = g - f$ and prove that

$$\int_a^b h(x)\varphi^A(x)\,dx = o\left(\int_a^b f(x) \times \varphi^A(x)\,dx\right).$$

Fix an arbitrary number $\varepsilon > 0$ and let β be a number in (a, b) such that $|h(x)| \le \varepsilon f(x)$ for $x \in (a, \beta)$. Then $|\int_a^\beta h(x)\varphi^A(x)\,dx| \le \varepsilon\int_a^\beta f(x)\varphi^A(x)\,dx$ and $|\int_\beta^b h(x)\varphi^A(x)\,dx| \le \varphi^A(\beta)\int_a^b |h(x)|\,dx$. Consequently,

$$\left|\int_a^b h(x)\varphi^A(x)\,dx\right| \le \varepsilon\int_a^b f(x)\varphi^A(x)\,dx + O(\varphi^A(\beta)).$$

Show now that $\varphi^A(\beta) = o(\int_a^b f(x)\varphi^A(x)\,dx)$. For this fix a number α in (a, β) and note that

$$\int_a^b f(x)\varphi^A(x)\,dx \ge \int_a^\alpha f(x)\varphi^A(x)\,dx \ge \varphi^A(\alpha)\int_a^\alpha f(x)\,dx.$$

Since $\varphi(\alpha) > \varphi(\beta)$ and $\int_a^\alpha f(x)\,dx > 0$, it follows that

$$\varphi^A(\beta) = o(\varphi^A(\alpha)) = o\left(\int_a^b f(x)\varphi^A(x)\,dx\right).$$

Thus, $|\int_a^b h(x)\varphi^A(x)\,dx| \le 2\varepsilon\int_a^b f(x)\varphi^A(x)\,dx$ if A is sufficiently large.

2.9. Use the asymptotic formula of Laplace (problem 2.6).

b) Using the result of problem 2.8, replace $\cos(ax)$ by 1.

c) Make the change of variable $x = Ay$.

d) The main contribution is given by the integral over $[0, 1]$. Using the result of problem 2.8, replace $\sin x$ by x in it and then make the change of variable $y = x^{2-p}$.

f) Verify that the main contribution is given by the integral over $[0, \pi/4]$. In this integral replace $\cos x$ by $\cos(2x)$.

h) By means of the result of problem 2.7 show that the main contribution is given by the integral over $[1/2, 1]$.

2.10. c) Make the change of variable $x = y^2$.

2.11. a) It is easy to see that $x_A \to 1$ as $A \to +\infty$. Making the change of variable $x = 1 + t$, we get that

$$e^{\sqrt{A}}\Phi(A) = \int_{-1}^{+\infty} \left(\frac{2+2t}{2+2t+t^2}\right)^A \frac{dt}{e^{\sqrt{A}t}}$$
$$= \int_{-\varepsilon}^{\varepsilon} + \int_{\varepsilon}^{+\infty} + \int_{-1}^{-\varepsilon} = I_1 + I_2 + I_3.$$

Since

$$I_1 = \int_{-\varepsilon}^{\varepsilon} \exp\left(-\frac{A}{2}t^2 - \sqrt{A}t + O(At^3)\right) dt$$

in the case when $A\varepsilon^3 \to 0$ we have that

$$I_1 = (1+o(1))\sqrt{\frac{e}{A}} \int_{1-\varepsilon\sqrt{A}}^{1+\varepsilon\sqrt{A}} e^{-u^2/2}\, du.$$

Therefore, $I_1 \sim \sqrt{2\pi e/A}$ if $\varepsilon\sqrt{A} \to +\infty$ (for example, $\varepsilon = A^{-2/5}$). Integral I_2 is easy to estimate from above:

$$I_2 \leq \int_{\varepsilon}^{+\infty} e^{-\sqrt{A}t}\, dt = o\left(\frac{1}{\sqrt{A}}\right).$$

For an estimate of I_3 verify that the integrand is nondecreasing on $[-1, -\varepsilon]$ if $\varepsilon = A^{-2/5}$. Therefore,

$$I_3 \leq \left(\frac{2-2\varepsilon}{2-2\varepsilon+\varepsilon^2}\right)^A e^{\varepsilon\sqrt{A}} = \exp(\varepsilon\sqrt{A} - A\varepsilon^2/2 + O(A\varepsilon^3)) = o\left(\frac{1}{\sqrt{A}}\right).$$

b) Using problem 2.8, replace the function under the integral by A^{px}.

c) Make the change of variable $x = Ay$. Using Taylor's formula, study the integral over $[0, A^{-2/3}]$, and get an upper estimate of the integral over $[A^{-2/3}, 1]$ by using the fact that the integrand is decreasing on $[A^{-2/3}, 1]$ for large A.

2.12. Get an upper estimate of the integrals over the intervals $[0, 1/(A\ln A)]$ and $[(2\ln A)/A, 1]$. To compute the integral over the remaining interval use the relation $|\ln x| = (1+o(1))\ln A$ for $x \in [1/(A\ln A), (2\ln A)/A]$.

2.13. a) The solution is similar to the solution of problem 2.11b).

b) It is clear that the contribution of the integral over the interval [1/3, 1/2] is small. To study the integral over the interval $[0, 1/3]$, make the change of variables $t = \varphi(x) = -x\ln x$. Then

$$\int_0^{1/3} x^{Ax}\, dx = -\int_0^c \frac{e^{-At}\, dt}{1+\ln\psi(t)},$$

where $c = (\ln 3)/3$ and ψ is the function inverse to φ. Since $\ln x \sim \ln\varphi(x)$ as $x \to +0$, it follows that $\ln t \sim \ln\psi(t)$ as $t \to +0$. Using the result of problem 2.8, we get

$$\int_0^{1/3} x^{Ax}\, dx \sim -\int_0^c \frac{e^{-At}}{\ln t}\, dt = \frac{1}{A}\int_0^{cA} \frac{e^{-u}\, du}{\ln A - \ln u}$$
$$\sim \frac{1}{A}\int_0^{\sqrt{A}} \frac{e^{-u}\, du}{\ln A - \ln u} \sim \frac{1}{A}\int_0^{\sqrt{A}} \frac{e^{-u}\, du}{\ln A} \sim \frac{1}{A\ln A}.$$

2.14. Writing $\Gamma(1+x)$ in the form

$$x\left(\frac{x}{e}\right)^x \int_{-1}^{+\infty}\left(\frac{1+t}{e^t}\right)^x dt,$$

use the asymptotic formula of Laplace (problem 2.6).

2.15. Use Stirling's formula.

2.16. The solution is analogous to that of problem 2.14.

2.17. a) Let $\varphi(x) = x + \theta\sqrt{x}$ $(\theta = \theta(x))$. By the result of the preceding problem the equality

$$\lim_{x\to+\infty}\frac{1}{\Gamma(1+x)}\int_0^{\varphi(x)} t^x e^{-t}\,dt = \frac{1}{2}$$

is equivalent to the equality

$$\lim_{x\to+\infty}\frac{1}{\Gamma(1+x)}\int_x^{x+\theta\sqrt{x}} t^x e^{-t}\,dt = 0.$$

Making the change of variable $t = x + \sqrt{x}u$ and using Stirling's formula, we get that it is equivalent to the relation

$$\int_0^{\theta(x)}\left(\frac{1+u/\sqrt{x}}{e^{u/\sqrt{x}}}\right)^x du = o(1),$$

which is valid only in the case when $\theta(x)\to 0$ as $x\to+\infty$.

b), c) The proofs of these assertions are analogous to the proof of assertion a).

§3. Asymptotics of sums

3.2. Obtain a lower estimate of the sum $\sum_{n/2\le k\le n} k^\alpha a_k$.

3.3. Express the sums S_n and σ_n in terms of the numbers $t_m = \sum_{1\le k\le m} k^\alpha x_k$ and use the results of problems II.2.7 b) and c).

3.4. Express the sums S_n in terms of the numbers $t_m = \sum_{k\ge m} k^\alpha x_k$ and use the results of problems II.2.7 b) and c).

3.5. a) Fix an arbitrary number $M > 1$. It is clear that

$$S_{[Mn]} - S_n = n^{1-\gamma}\left(\frac{M^{1-\gamma}-1}{1-\gamma} + o(1)\right).$$

Since

$$S_{[Mn]} - S_n = a_{[Mn]} + \cdots + a_{n+1} \le a_n([Mn]-n) \le (M-1)na_n,$$

it follows that for any $M > 1$

$$\varliminf_{n\to\infty} n^\gamma a_n \ge \varliminf_{n\to\infty}\frac{n^{\gamma-1}}{M-1}(S_{[Mn]} - S_n) = \frac{M^{1-\gamma}-1}{(M-1)(1-\gamma)}.$$

Therefore,

$$\varliminf_{n\to\infty} n^\gamma a_n \ge \lim_{M\to1+0}\frac{M^{1-\gamma}-1}{(M-1)(1-\gamma)} = 1.$$

To prove the inequalities $\overline{\lim}_{n\to\infty} n^\gamma a_n \le 1$, show that for $m \in (0, 1)$

$$\overline{\lim_{n\to\infty}}\, n^\gamma a_n \le \overline{\lim_{n\to\infty}} \frac{n^{\gamma-1}}{1-m}(S_n - S_{[mn]}) \le \frac{1-m^{1-\gamma}}{(1-m)(1-\gamma)}.$$

b) Consider the sequence $a_n = (\ln 2)/2^k$ for $2^k \le n < 2^{k+1}$.

3.6. The solution is analogous to the solution of problem 3.5a).

3.7. For $\alpha \le 1$ use the relation $(1-1/n)^k = 1 + O(k/n)$, and for $\alpha > 1$ the equality

$$\begin{aligned}\ln n &= \sum_{k\ge 1} \frac{1}{k}\left(1-\frac{1}{n}\right)^k = x_n + \sum_{k>n^\alpha} \frac{1}{k}\left(1-\frac{1}{n}\right)^k \\ &= x_n + O(n^{1-\alpha}).\end{aligned}$$

3.8. See [5]. It is easy to see that $\tau(k)$ is equal to the number of points $(u, v) \in \mathbb{N}^2$ lying on the hyperbola $uv = k$. Therefore, the sum $T(n)$ is equal to the number of points $(u, v) \in \mathbb{N}^2$ not lying above the hyperbola $uv = n$. Since the set $\{(u, v) \in \mathbb{N}^2 | uv \le n\}$ is symmetric with respect to the line $u = v$, to compute $T(n)$ it suffices to calculate the number $\widetilde{T}(n)$ of points in the set $E = \{(u, v) \in \mathbb{N}^2 | uv \le n,\ u \le \sqrt{n}\}$ and take into account that the points in the square $[0, \sqrt{n}] \times [0, \sqrt{n}]$ are also in the set symmetric to E (make a sketch). This gives us that

$$\begin{aligned}T(n) &= 2\widetilde{T}(n) - [\sqrt{n}]^2 = 2 \sum_{1\le k\le\sqrt{n}} \left[\frac{n}{k}\right] - [\sqrt{n}]^2 \\ &= 2n \sum_{1\le k\le\sqrt{n}} \frac{1}{k} - n + O(\sqrt{n}).\end{aligned}$$

To compute this sum we can use the following refinement of the result of problem II.2.9:

$$\sum_{1\le k\le N} \frac{1}{k} - \ln N - \gamma \sim \frac{1}{2N},$$

which is easy to verify with the help of the theorem of Stoltz (problem II.2.6).

3.9. a) Let $a = \sup n\sqrt{\alpha_n}$. Since $\alpha_n \le a^2/n^2$, for any index N

$$\begin{aligned}f(t) &\le t\left(\sum_{1\le k\le N} \frac{1}{t^2} + \sum_{k>N} \alpha_k\right) \\ &\le \frac{N}{t} + ta^2 \sum_{k>N} \frac{1}{n^2} < \frac{N}{t} + \frac{ta^2}{N}.\end{aligned}$$

Taking $N = 1 + [at]$, we get for $t \ge 1/a$ that

$$f(t) \le \frac{1+at}{t} + \frac{ta^2}{at} \le 3a.$$

But if $t \in [0, 1/a)$, then $f(t) \leq \frac{1}{a}\sum \alpha_k \leq a\sum \frac{1}{n^2} \leq 2a$. The inequality b) can be proved analogously.

c) To get a lower estimate of the quantity $\underline{\lim}_{t\to\infty} f(t)$ fix a large index n. Then for all t in the interval $[1/\sqrt{\alpha_{n-1}}, 1/\sqrt{\alpha_n}]$ we have that

$$\begin{aligned} f(t) &\geq t \sum_{1\leq k<n} \frac{\alpha_k}{1+t^2\alpha_k} \geq t \sum_{1\leq k<n} \frac{\alpha_k}{2t^2\alpha_k} \\ &= \frac{n-1}{2t} \geq \frac{n-1}{2}\sqrt{\alpha_n}. \end{aligned}$$

Consequently, $\underline{\lim}_{t\to+\infty} f(t) \geq \frac{1}{2}\underline{\lim}\, n\sqrt{\alpha_n}$.

3.10. Use the inequality $\int_h^\infty f(t)\,dt \leq \varphi(h) \leq \int_0^\infty f(t)\,dt$. The equality is not true for nonmonotone functions. For example, if $f(n) = 1$ for any $n \in \mathbb{N}$, then $\varphi(1/n) = +\infty$. Here it is not hard to ensure that the integral $\int_0^{+\infty} f(t)\,dt$ converges.

3.11. Use the result of the preceding problem.

3.12. It is clear that

$$\begin{aligned} f(p) &= \sum_{k\geq 0}\left(\frac{1}{(2k+1)^p} - \frac{1}{(2k+2)^p}\right) \\ &= \sum_{k\geq 0}\frac{1}{(2k+1)^p}\left(1-\left(1+\frac{1}{2k+1}\right)^{-p}\right). \end{aligned}$$

Using Taylor's formula, we get that $f(p) = O(p) + p\sum_{k\geq 0} 1/(2k+1)^{1+p}$. Since

$$\frac{1}{(2k+1)^p} = \int_k^{k+1}\frac{dt}{(2t+1)^{1+p}} + O\left(\frac{1}{k^{2+p}}\right),$$

it follows that

$$f(p) = p\int_0^\infty \frac{dt}{(2t+1)^{1+p}} + O(p) = \frac{1}{2} + O(p).$$

3.13. Use the inequalities $f(k+1) \leq \int_k^{k+1} f(t)\,dt \leq f(k)$.

3.14. See the hint for the preceding problem.

3.15. It suffices to estimate the quantity $\Delta_k = f(k) - \int_{k-1/2}^{k+1/2} f(t)\,dt$ $(k = M, \ldots, N)$. In part a) use the equality

$$\Delta_k = \int_{k-1/2}^{k}(f(k)-f(t))\,dt + \int_k^{k+1/2}(f(k)-f(t))\,dt.$$

In part b) use integration by parts to prove that

$$\Delta_k = -\frac{1}{2}\int_{k-1/2}^{k}\left(t-k+\frac{1}{2}\right)^2 f''(t)\,dt - \frac{1}{2}\int_k^{k+1/2}\left(t-k-\frac{1}{2}\right)^2 f''(t)\,dt.$$

We remark that by using integration by parts it is possible to get from the equality in b) estimates of the difference Δ that contain higher-order derivatives.

The equation b) is actually a special case of the classical Euler-Maclaurin formula. On this topic see, for example, [29].

3.16. Use the result of problem 3.13a) and Taylor's formula.

3.17. By using the recursion formula

$$\binom{n-\alpha-1}{n} = \frac{\alpha}{\alpha-n}\binom{n-\alpha}{n}$$

it is possible to reduce the problem to the case $\alpha < 0$. Then $\ln\binom{n-\alpha}{n} = \sum_{1\le k\le n}\ln(1-\alpha/k)$. It remains to use the result of problem 3.13a). We remark that the equality $c_\alpha = 1/(\Gamma(1-\alpha))$ follows from the properties of the Γ-function (see problem 2.15).

3.18. After taking logarithms in parts a) and b), use the result of problem 3.14, and after taking logarithms in part c), use the result in problem 3.15b).

3.19. a)–c). Using the result of problem 3.14, replace the summation by integration. The integrals thus obtained were considered in problems 2.11b) and 2.13a). In part c), first use Stirling's formula.

d) Find the asymptotics of the sum of the last $[\sqrt[3]{n}]$ terms and show that the sum of the remaining terms is sufficiently small.

3.20. It suffices to prove that $\sum_{k>n} n^k/k! \sim e^n/2$. To do this, use the integral representation of the remainder in Taylor's formula for the exponential along with the result of problem 2.9c).

3.21. Find the asymptotics of the sums $S_n^{(p)} = \sum_{1\le k\le n}\binom{n}{k}^p$, where p is a fixed positive number. Since the numbers $\binom{n}{k}$ are rapidly increasing for $1 \le k \le n/2$, the main contribution to the sum under consideration comes from the terms with indices close to $n/2$ (as is clear from the following, it suffices to take into account about $n^{2/3}$ central terms).

Assuming that $n = 2m$, we have that

$$S_n^{(p)} = \binom{2m}{m}^p\left(1 + 2\sum_{1\le k<m}\left(\binom{2m}{m-k}\Big/\binom{2m}{m}\right)^p\right).$$

To find the principal part of the sum obtained, note that for $k > 1$

$$\frac{\binom{2m}{m-k}}{\binom{2m}{m}} = \frac{m}{m+k}\prod_{1\le j<k}\frac{1-j/m}{1+j/m} = \frac{m}{m+k}\exp\left(\sum_{1\le j<k}\ln\frac{1-j/m}{1+j/m}\right).$$

It follows from the relation $0 < -2t - \ln((1-t)/(1+t)) = O(t^3)$ $(0 < t < 1/2)$ that

$$\frac{\binom{2m}{m-k}}{\binom{2m}{m}} = \begin{cases} (1+O(m^{-1/3}))e^{-k^2/m} & \text{for } k \le m^{2/3}, \\ O(e^{-m^{1/3}}) & \text{for } k > m^{2/3}. \end{cases}$$

Thus (see problem 3.14),

$$\sum_{1\le k<m}\left(\frac{\binom{2m}{m-k}}{\binom{2m}{m}}\right)^p = (1+O(m^{-1/3}))\sum_{1\le k\le m^{2/3}} e^{-pk^2/m} + o(1)$$

$$= (1+O(m^{-1/3}))\int_1^{m^{2/3}} e^{-pu^2/m}\,du + O(1) \sim \frac{1}{2}\sqrt{\frac{\pi}{p}m}.$$

To obtain the final result it remains to verify that $\binom{2m}{m}\sim 4m/\sqrt{\pi m}$, which implies that

$$S_n^{(p)} \sim \frac{1}{\sqrt{p}}\left(\frac{\pi}{2}n\right)^{(1-p)/2} 2^{np}.$$

3.22. Denote the sum of the series by $\varphi_N(z)$. It is clear that

$$\varphi_N(z) = \sum_{n\ge N}\int_0^1 e^{-z\sqrt{n+t}}\frac{dt}{n+t} + \sum_{n\ge N}\int_0^1\left(\frac{e^{-z\sqrt{n}}}{n} - \frac{e^{-z\sqrt{n+t}}}{n+t}\right)dt.$$

Therefore,

$$\varphi_N(z) = \int_N^\infty \frac{e^{-z\sqrt{t}}}{t}\,dt + \sum_{n\ge N} e^{-z\sqrt{n}}O\left(\frac{1}{n^2}\right)$$

$$+\sum_{n\ge N}\int_0^1 \frac{e^{-z\sqrt{n}} - e^{-z\sqrt{n+t}}}{n+t}\,dt$$

$$= \frac{2}{z}e^{-z\sqrt{N}} + O\left(\frac{e^{-z\sqrt{N}}}{N}\right) + \sum_{n\ge N} e^{-z\sqrt{n}}\int_0^1 O\left(\frac{zt}{n^{3/2}}\right)dt$$

$$= e^{-z\sqrt{N}}\left(\frac{2}{z} + O\left(\frac{1}{N} + \frac{|z|}{\sqrt{N}}\right)\right).$$

3.23. Using the result of problem 3.15b), we get that

$$\sum_{n\ge N} e^{-z\sqrt{n}} = \int_{N-1/2}^\infty e^{-z\sqrt{t}}\,dt - \frac{1}{2}\int_{N-1/2}^\infty\left(t-[t]-\frac{1}{2}\right)^2 (e^{-z\sqrt{t}})''_t\,dt = I - J.$$

The integral I is easy to compute. To estimate J we write it in the form

$$J = \frac{1}{2}\int_{N-1/2}^\infty\left(t-[t]-\frac{1}{2}\right)^2\left(\frac{z}{4t^{3/2}} + \frac{z^2}{4t}\right)e^{-z\sqrt{t}}\,dt$$

$$= O\left(\frac{z}{\sqrt{N}}e^{-z\sqrt{N-1/2}}\right) + \frac{z^2}{8}\int_{N-1/2}^\infty \frac{(t-[t]-1/2)^2}{t}e^{-z\sqrt{t}}\,dt.$$

The integral obtained can be estimated by breaking it up into the sum of the integrals over the intervals $[k-1/2, k+1/2]$ ($k\in\mathbb{N}$, $k\ge N$); to estimate this sum use the result in the preceding problem.

3.24. Using the identity

$$\cos nx = \frac{x}{2\sin x/2}\int_{n-1/2}^{n+1/2}\cos(xt)\,dt,$$

we get that

$$\sum_{n\geq 1}\frac{\cos nx}{n^\alpha}=\frac{x}{2\sin(x/2)}\left(\int_{1/2}^{+\infty}\frac{\cos(xt)}{t^\alpha}\,dt+O(1)\right)$$
$$=x^{\alpha-1}\int_0^\infty\frac{\cos v}{v^\alpha}\,dv+O(1).$$

It can be proved similarly that

$$\sum_{n\geq 1}\frac{\sin nx}{n^\alpha}=x^{\alpha-1}\int_0^{+\infty}\frac{\sin v}{v^\alpha}\,dv+O(1).$$

It follows from the result in problem VII.1.26 that the integrals $A=\int_0^{+\infty}(\cos v)/v^\alpha\,dv$ and $B=\int_0^{+\infty}(\sin v)/v^\alpha\,dv$ are positive.

3.25. With the help of the device used in the solution of the preceding problem, we get that for $x\in(0,\pi)$

$$\sum_{1\leq n\leq N}\frac{\cos nx}{n^\alpha}=\int_0^N\frac{\cos(xt)}{t^\alpha}\,dt+O(1)$$
$$=x^{\alpha-1}\int_0^{Nx}\frac{\cos v}{v^\alpha}\,dv+O(1).$$

From this, uniform (with respect to $n\in\mathbb{N}$ and $x\in[0,\pi]$) boundedness below of the sums $\sum_{1\leq n\leq N}n^{-\alpha}\cos nx$ is equivalent to nonnegativity of the integral $I(u)=\int_0^u(\cos v)/v^\alpha\,dv$ for $u\geq 0$. It is clear that $\min_{u\geq 0}I(u)=I(3\pi/2)=\int_0^{3\pi/2}(\cos v)/v^\alpha\,dv$. It remains to show that the function $J(\alpha)=\int_0^{3\pi/2}(\cos v)/v^\alpha\,dv$ changes sign once on the interval $(0,1)$. Since $J(0)=-1$ and $J(1-0)=+\infty$, it suffices to see that the function J is monotone. It is easy to see that

$$J'(\alpha)=\int_0^{3\pi/2}\frac{\cos v}{v^\alpha}\ln\frac{1}{v}\,dv>\int_0^{\pi/2}\frac{\cos v}{v^\alpha}\ln\frac{1}{v}\,dv.$$

Using the decrease of the function $v^{-\alpha}\cos v$ on $(0,\pi/2)$, we get that

$$J'(\alpha)>\int_0^{\pi/2}\frac{\cos v}{v^\alpha}\ln\frac{1}{v}\,dv\geq\cos 1\int_0^{\pi/2}\ln\frac{1}{v}\,dv>0.$$

Similarly, the study of the sums $\sum(\sin nx)/n^\alpha$ reduces to the study of the integral $\widetilde{J}(\alpha)=\int_0^{2\pi}(\sin v)/v^\alpha\,dv$, which is positive for all $\alpha>0$.

3.26. Let $A=\psi^{-1}(1/\varepsilon)$ and $C=\sup_{t\geq a}\psi(t)/\psi'(t)$. Then

$$\int_a^{+\infty}e^{-\varepsilon\psi(t)}\,dt-A=-a+\int_a^A(e^{-\varepsilon\psi(t)}-1)\,dt+\int_A^{+\infty}e^{-\varepsilon\psi(t)}\,dt$$
$$=-a+I_1+I_2.$$

Let us estimate the integrals I_1 and I_2:

$$\begin{aligned}|I_1| &\le \int_a^A (1-e^{-\varepsilon\psi(t)})\,dt \le \varepsilon\int_a^A \psi(t)\,dt\\ &\le \varepsilon C\int_a^A \psi'(t)\,dt < \varepsilon C\psi(A) = C;\\ 0 \le I_2 &\le \int_A^{+\infty} \frac{C}{\psi(t)}\psi'(t)e^{-\varepsilon\psi(t)}\,dt\\ &\le \frac{C}{\psi(A)}\int_A^{+\infty} \psi'(t)e^{-\varepsilon\psi(t)}\,dt\\ &= \frac{C}{\varepsilon\psi(A)}e^{-\varepsilon\psi(A)} = \frac{C}{e}.\end{aligned}$$

3.27. a)–e). Using the monotone decrease of the general term, show that the summation can be replaced by integration (see problem 3.14). To investigate the integrals obtained use the results of problems 2.9g) and 3.26.

f) With the help of the inequality $(n/3)^n < n! < (n/2)^n$ reduce the problem to problem e).

3.28. Since for fixed $t \in (0, 1)$ the general term of the series does not tend monotonically to zero, the inequality of problem 3.14 is not applicable to these series. One should consider separately the partial sum in which the general term increases, and the remainder of the series, in which the general term decreases.

In b) use the result of problem 1.4d).

3.29. a)–f), h) The solutions are analogous to the solutions of problems 3.27 and 3.28. In b) and f) use the results of problems V.1.18 and V.1.19.

g), i)–l) Expand the general term of the series in a power series and change the order of summation.

3.30. See the hints for problems 3.27–3.29.

3.31. Use the results of problems IV.4.15 and 3.29e), f). Another solution of problem a) is obtained by using the results of problems 3.8, IV.5.8, and 3.28b) for $p = 1$.

§4. Asymptotics of implicit functions and recursive sequences

4.1. With the help of integration by parts, prove that $z(t) = (1 + o(1)) \times e^{-t^2/2}/t$ as $t \to +\infty$. Therefore,

$$\ln z = -\frac{t^2}{2} - \ln t + o(1) \sim -\frac{t^2}{2}.$$

Consequently, $t(z) \sim \sqrt{2\ln 1/z}$ as $z \to +0$.

4.2. Show that $y_n = x_n - \pi n - \pi/2$ is an infinitesimally small quantity. To obtain the asymptotics of the sequence $\{y_n\}$, replace x_n in the equality $x_n = \tan x_n$ by $y_n + \pi n + \pi/2$ and pass to the equivalent quantities.

4.7. It is not hard to see that for any number $a > 1$

$$\varphi(x) = \frac{x}{1+ax} \le x(1-x) < \frac{x}{1+x} = \psi(x) \quad \text{for } x \in \left(0\,,\, 1 - \frac{1}{a}\right].$$

Iterating the functions φ and ψ, we get the two-sided estimate

$$\frac{x_0}{1+nax_0} \le x_n < \frac{x_0}{1+nx_0} \quad \text{if } x_0 \in \left(0\,,\, 1 - \frac{1}{a}\right].$$

For $n = 1000$ and $a = 1/0.999$, we have that

$$10^{-3}\frac{0.999}{1.999} \le x_{1000} < \frac{1}{2}10^{-3}.$$

Consequently,

$$0 < \frac{1}{2}10^{-3} - x_{1000} < \frac{1}{3}10^{-6}.$$

4.8. It is clear that $x_n \downarrow 0$. The asymptotic behavior of the sequence $\{x_n\}$ can be found with the help of the device used in the solution of the preceding problem. For this, get upper and lower estimates of the function f in a neighborhood of zero by the functions considered in problem 4.6, and then compare their iterates.

Another solution is based on the following considerations, which can be used also in more complicated situations (see problems 4.12–4.14). The recursive sequence $x_n = f(x_{n-1})$, $x_0 > 0$, clearly converges to zero if the function f is such that $0 < f(t) < t$ for all $t > 0$. With the help of a figure it is easy to see that the sequence $\{x_n\}$ converges rapidly to zero if the difference $t - f(t)$ is large, and converges slowly otherwise. Therefore, to find the asymptotics of the sequence $\{x_n\}$ it is expedient to consider the function $\varphi(t) = t - f(t)$. Then the recursive formula takes the form

$$x_n - x_{n-1} = -\varphi(x_{n-1}).$$

Assuming that x_n is the value at n of some smooth function θ defined on $[1\,,+\infty)$, we get that

$$\theta(n) - \theta(n-1) = -\varphi(\theta(n-1)).$$

If as an approximation the difference $\theta(n) - \theta(n-1)$ is replaced by the derivative $\theta'(n-1)$, then it turns out that the function θ satisfies the "differential equation" $\theta' \approx -\varphi(\theta)$. Solving it, we find that $\int_{\theta(n)}^{\theta(1)} 1/\varphi(t)\,dt \approx n$, that is, $x_n = \theta(n) \approx \Phi^{-1}(n)$, where $\Phi(y) = \int_y^{\theta(1)} 1/\varphi(t)\,dt$.

To justify these heuristic considerations we must study the difference $\Phi(x_n) - \Phi(x_{n-1})$ and, using the given recursion relation, show that they tend to 1. Then, adding the equalities $\Phi(x_k) - \Phi(x_{k-1}) = 1 + o(1)$ for $k = 1\,,\dots,\,n$, we get that $\Phi(x_n) = n + o(n)$, that is, $x_n = \Phi^{-1}(n + o(n))$. To simplify the computations it is sometimes more convenient to consider instead of Φ a function $\widetilde{\Phi}$ equivalent to it. For example, in the case

$f(x) - x \underset{x\to+0}{\sim} -Cx^{1+p}$, where C and p are fixed positive constants, we have that

$$\Phi(y) = \int_y^{x_1} \frac{dt}{t - f(t)} \underset{y\to+0}{\sim} \int_y^{x_1} \frac{dt}{Ct^{1+p}} \sim \frac{1}{Cpy^p} = \widetilde{\Phi}(y).$$

Considering the difference $\widetilde{\Phi}(x_n) - \widetilde{\Phi}(x_{n-1})$, we get that

$$\widetilde{\Phi}(x_n) - \widetilde{\Phi}(x_{n-1}) = \frac{1}{Cp}(x_n^{-p} - x_{n-1}^{-p}) \sim \frac{x_{n-1} - x_n}{Cx_{n-1}^{1+p}} \to 1.$$

From this, $1/Cpx_n^p = \widetilde{\Phi}(x_n) \sim n$, that is, $x_n \sim 1/(Cpn)^{1/p}$.

4.9. a) Use the idea described in the solution of problem 4.7.

b) Refine the result in a). Since

$$x_n^p - x_{n-1}^p = x_{n-1}^p \left(\left(1 + \frac{x_n - x_{n-1}}{x_{n-1}} \right)^p - 1 \right) = p + O\left(\frac{1}{n}\right),$$

it follows that $x_n^p = np + O(\ln n) = np(1 + O((\ln n)/n))$. Thus, $x_n = (np)^{1/p}(1 + y_n)$, where $y_n = O((\ln n)/n)$. Substituting the expressions obtained for x_n and x_{n-1} into the recursion relation, we get by means of the Taylor formula that

$$y_n = y_{n-1}\left(1 - \frac{1}{n}\right) + \frac{1}{n^2}\frac{p-1}{2p^2} + O\left(\frac{\ln^2 n}{n^3}\right),$$

that is,

$$ny_n - (n-1)y_{n-1} = \frac{1}{n}\frac{p-1}{2p^2} + O\left(\frac{\ln^2 n}{n^2}\right).$$

This implies that

$$ny_n = \frac{p-1}{2p^2}\ln n + O(1).$$

4.10. It suffices to solve the following problem: construct on $(0, +\infty)$ a function f such that the sequence $x_n = f(x_{n-1})$ coincides with a previously given sequence $\{\alpha_n\}$ strictly decreasing to zero. It can be assumed here that $\alpha_n = \varphi(n)$, where φ is a strictly decreasing function on $[0, +\infty)$ with $\lim_{t\to+\infty} \varphi(t) = 0$. Then the equality $f(x_{n-1}) = x_n = \varphi(n)$ is equivalent to the equality $f(\varphi(n-1)) = \varphi(n)$, which is satisfied, for example, by the function $f(t) = \varphi(1 + \varphi^{-1}(t))$.

4.11. Consider the sequence $\{x_{2n}\}$ and $\{x_{2n-1}\}$.

4.12. Prove first that $a_n \uparrow +\infty$. Using for $\varepsilon \in (0, 1)$ the inequality

$$0 < a_n - a_{[\varepsilon n]} = \sum_{\varepsilon n \le k < n} \frac{1}{a_0 + \cdots + a_k} \le \frac{n}{[\varepsilon n]},$$

prove that $a_n \sim a_{[\varepsilon n]}$. Deduce from this that $\underline{\lim}(a_1 + \cdots + a_n)/na_n \ge 1$, and hence $a_1 + a_2 + \cdots + a_n \sim na_n$. Therefore,

$$a_{n+1}^2 - a_n^2 \sim 2a_n(a_{n+1} - a_n) = \frac{2a_n}{a_0 + \cdots + a_n} \sim \frac{2}{n},$$

which implies that $a_n^2 \sim 2\ln n$.

4.13. Prove first that $a_n \uparrow +\infty$, $a_{n+1} \sim a_n$, and $a_n = o(A_n)$, where $A_n = a_0 + \cdots + a_n$. For $0 \le p < 1$ the last two relations are obvious. For $-1 < p < 0$ they follow from the two-sided estimate

$$0 < mn^{(1-p)/(1+p)} \le a_n \le Mn^{(1-p)/(1+p)} \qquad (n \in \mathbb{N}).$$

The upper estimate follows from the obvious inequality

$$a_{n+1} \le 1 + \sum_{1 \le k \le n} (ka_k)^{-p} \le 1 + a_{n+1}^{-p} n^{1-p}.$$

The lower estimate can be proved by induction if the number $m \in (0, 1)$ is chosen so that

$$m^{1+p} \le \left(\frac{2}{1+p}\right)^p n\left(\left(1+\frac{1}{n}\right)^{(1-p)/(1+p)} - 1\right) \quad \text{for all } n \in \mathbb{N}.$$

From these relations obtain successively that

$$a_{n+1}^2 - a_n^2 \sim 2\frac{A_n^{1-p} - A_{n-1}^{1-p}}{1-p}, \qquad A_n \sim \left(\frac{1-p}{2}a_n^2\right)^{1/(1-p)},$$

$$a_{n+1} - a_n \sim \left(\frac{1-p}{2}\right)^{p/(p-1)} a_n^{2p/(p-1)},$$

and, finally, $a_{n+1}^{(1+p)/(1-p)} - a_n^{(1+p)/(1-p)} = K_p^{(1+p)/(1-p)} + o(1)$, which implies that $a_{n+1}^{(1+p)/(1-p)} \sim nK_p^{(1+p)/(1-p)}$.

4.14. Prove successively that $a_n \uparrow +\infty$, $a_{n+1} \sim a_n$, and $a_{n+1}^p = o(S_n)$, where $S_n = a_0^p + \cdots + a_n^p$. The last relation follows from the inequality

$$0 \le \overline{\lim}\, \frac{a_{n+1}^p}{S_n} = \overline{\lim}\, \frac{a_{n-k}^p}{S_n} \le \overline{\lim}\, \frac{a_{n-k}^p}{ka_{n-k}^p} = \frac{1}{k},$$

which is valid for any fixed index k. For $p \ne -1$ deduce from this that

$$a_{n+1}^{p+1} - a_n^{p+1} \sim (p+1)a_n^p/S_n \sim (p+1)(\ln S_{n+1} - \ln S_n).$$

Therefore, for $p < -1$ the limit $\lim S_n = \sigma_p$ exists and is finite, that is, $a_{n+1} - a_n = (1/\sigma_p) + o(1)$, from which $a_n \sim n/\sigma_p = C_p \cdot n$. For $p > -1$ the relation $a_n^{p+1} \sim (p+1)\ln S_n \to +\infty$ is valid. Using the mean value theorem, prove that this implies the equality

$$S_{n+1}(\ln S_{n+1})^{-p/(p+1)} - S_n(\ln S_n)^{-p/(p+1)} = (p+1)^{p/(p+1)} + o(1),$$

that is, $S_n(\ln S_n)^{-p/(p+1)} \sim n(p+1)^{p/(p+1)}$. Therefore, $\ln S_n \sim \ln n$, and hence $a_n^{p+1} \sim (p+1)\ln S_n \sim (p+1)\ln n$.

For $p = -1$, prove by induction that $a_n = (2n)!!/(2n-1)!!$.

CHAPTER VII

Functions (Continuation)

§1. Convexity

1.1. Use induction (verify first that a convex combination $\sum_{1\le j\le n}\lambda_j x_j$ falls in $\langle a, b\rangle$).

1.2. Let

$$\lambda_1 = \frac{x_3 - x_2}{x_3 - x_1} \quad \text{and} \quad \lambda_2 = \frac{x_1 - x_1}{x_3 - x_1}$$

and write Jensen's inequality for $x_2 = \lambda_1 x_1 + \lambda_2 x_3$.

1.3. a) Since some of the numbers $x_j \in \Delta$ can be the same, it suffices to prove Jensen's inequality for the arithmetic mean of n numbers. Prove this first for $n = 2^k$. Verify that if the inequality is true for the arithmetic mean of n arbitrary numbers, then it remains true also when n is replaced by $n-1$.

b) Use a).

1.4. It can be assumed that $p, q \in (a, b)$. Let $K = \sup_{p<x<q} f(x)$. First verify that f is bounded above on any interval $[c, d]$, $(p, q) \subset [c, d] \subset (a, b)$. To do this, consider the set $E = \{x \in [c, d] | f(x) \le C\}$, where $C = \max(f(c), f(d), K)$, and, using the result of problem I.1.19, show that $E = [c, d]$. If f is bounded above on $(p, q)\backslash\mathbb{Q}$, then show that f is bounded also on (p, q), because each point of this interval is the midpoint of some interval with endpoints in $(p, q)\backslash\mathbb{Q}$.

The continuity of f at a point $x \in (c, d)$ follows from the inequalities

$$\frac{1}{m}(f(x) - C) \le f(x) - f(x-\delta) \le f(x+\delta) - f(x) \le \frac{1}{n}(C - f(x)),$$

where δ is an arbitrarily small positive number, and m and n are the largest integers for which $x - m\delta$, $x + n\delta \in [c, d]$. The middle inequality is obvious, the right-hand one can be proved with the help of Jensen's inequality (see 1.3a)) for the points x, $x+\delta$, and $x + n\delta$, and the left-hand one can be proved similarly.

1.5. a) $\Leftrightarrow$ b) is obvious. To prove the implication b) $\Rightarrow$ c) consider the set $K = \bigcup_{\lambda>0}\lambda(\Gamma^+ - c)$, where $c \in \mathbb{R}^2$ is a point of the graph. Show that K is a convex cone $\ne \mathbb{R}^2$, and take a straight line containing one of its boundary rays. To prove that c) $\Rightarrow$ b) verify that Γ^+ is an intersection of half-planes.

1.6. Let $0 \le x < y \le 1/2$. Represent y and $1-y$ as convex combinations of the points x and $1-x$, and use the convexity.

1.7. Use Borel's covering lemma.

1.8. Use the three chords lemma (problem 1.2).

1.9. Consider first the case when $L_f(x) > 0$ for all $x \in (a, b)$. Suppose that $x_1, x_2 \in (a, b)$, $x_1 < x_2$, and $y = g(x)$ is the equation of the chord between the graph points with abscissae x_1 and x_2. Let $f_1 = f - g$. It is clear that $L_{f_1}(x) = L_f(x) > 0$, and for f to be convex it suffices that $f_1(x) \le 0$ for all $x \in [x_1, x_2]$. Assuming the contrary, show that $L_{f_1}(\overline{x}) \le 0$ at a point $\overline{x}$ where f_1 attains its largest value on $[x_1, x_2]$. In the general case ($L_f(x) \ge 0$) consider the functions $f_\varepsilon(x) = f(x) + \varepsilon x^2$, $\varepsilon > 0$, and pass to the limit as $\varepsilon \to 0$.

1.10. The necessity is obvious. To prove sufficiency rewrite the inequality in the form $\frac{1}{2h}\int_0^h (f(x+t) + f(x-t) - 2f(x))\,dt \ge 0$ and use the preceding problem.

1.11. a) Use the uniform continuity and "rectify" the graph by chords.

b) Using a), show that it suffices to consider only functions of the form $|x-a|$.

1.12, 1.13. Use the three chords lemma (problem 1.2).

1.14. Verify that Jensen's inequality for one function is carried by the transformation $x \mapsto 1/x$ into Jensen's inequality (with different coefficients) for the other.

1.15. Use the existence and monotonicity of one-sided derivatives (problem 1.8).

1.16. Reduce the problem to the case $a = 0$, $f(0) = 0$, and verify that the function $f(x)/x$ is monotone.

1.18. Use the three chords lemma (problem 1.2) for the points 0, $x-h$, and x $(0 < h < x)$.

1.19. It suffices to verify that $\varphi(x_1) \le \varphi(x_2)$ for all $x_1 < x_2$, $x_2 - x_1 < b$. This follows from the three chords lemma (problem 1.2), applied to the points $x_1, x_2, x_1 + b$ and $x_2, x_1 + b, x_2 + b$:

$$\begin{aligned}\varphi(x_1) &= b\frac{f(x_1+b) - f(x_1)}{b} \le b\frac{f(x_1+b) - f(x_2)}{x_1 + b - x_2} \\ &\le b\frac{f(x_2+b) - f(x_2)}{b} = \varphi(x_2).\end{aligned}$$

1.20. It will be assumed that $n = 2m+1$ is an odd number, since zero can always be added as the last point. For $a \le b$ the function $x \mapsto f(b+x) - f(a+x)$, $x \ge 0$, is increasing (see problem 1.19). Consequently,

$$f(a_{2i}) - f(a_{2i+1}) \geq f\left(\sum_{i\leq k\leq m}(a_{2k} - a_{2k+1})\right) - f\left(\sum_{i<k\leq m}(a_{2k} - a_{2k+1})\right),$$
$$i = 0, \dots, m-1,$$

$$f(a_{2m}) - f(a_{2m+1}) \geq f(a_{2m} - a_{2m+1}).$$

Adding these inequalities, we get what is required.

1.21. Using the inequality $a_{n+1} - a_n \leq a_{n+2} - a_{n+1}$, prove that

$$\frac{a_{n+p+q} - a_n}{p+q} \leq \frac{a_{n+p+q} - a_{n+p}}{q}.$$

Interpret this inequality geometrically.

1.22. See problem IV.2.3.

1.23. Verify that only piecewise linear functions in the indicated classes may be considered (see problem 1.11a)). We explain the subsequent arguments for two inequalities.

a$'$) Partition the subgraph of the function into triangles with the common vertex $(0, 0)$ by joining this point to the node points of the graph (see Figure 13a). Suppose that the kth triangle is bounded by the graphs of the functions φ_k and φ_{k+1}, $0 \equiv \varphi_1 \leq \varphi_k \leq \varphi_{k+1} \leq \varphi_N \equiv f$, and $\varphi_{k+1}(x) = \varphi_k(x)$ outside the projection of the kth triangle on the x-axis. Then $f = \sum_{1\leq k<N}(\varphi_{k+1} - \varphi_k)$. Prove the inequality a$'$) for each function $\varphi_{k+1} - \varphi_k$, and then add these inequalities. Show that the convexity requirement is essential.

d$'$) Continuing the rectilinear segments of the graph until they intersect the x-axis, we get a partition of the subgraph into triangles (Figure 13b). We introduce the functions φ_k $(1 \leq k \leq N)$ just as in the preceding case. Then $f^2 = \sum_{1\leq k<N}(\varphi_{k+1}^2 - \varphi_k^2)$. Verify that

$$\int_0^1 (\varphi_{k+1}^2(x) - \varphi_k^2(x))\,dx = \frac{2}{3}M_k \int_0^1 (\varphi_{k+1}(x) - \varphi_k(x))\,dx,$$

where M_k is the height of the kth triangle.

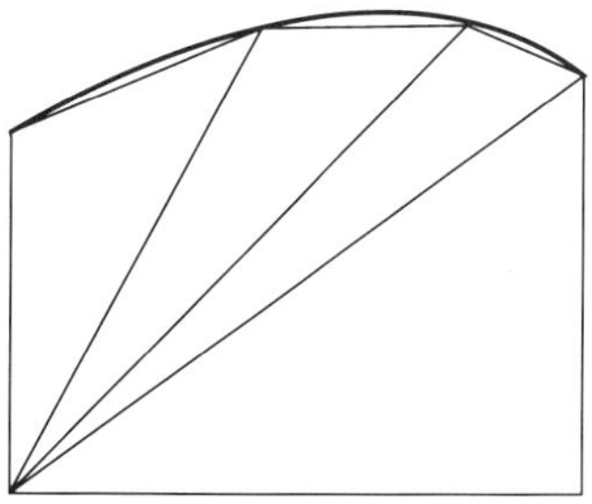

FIGURE 13a

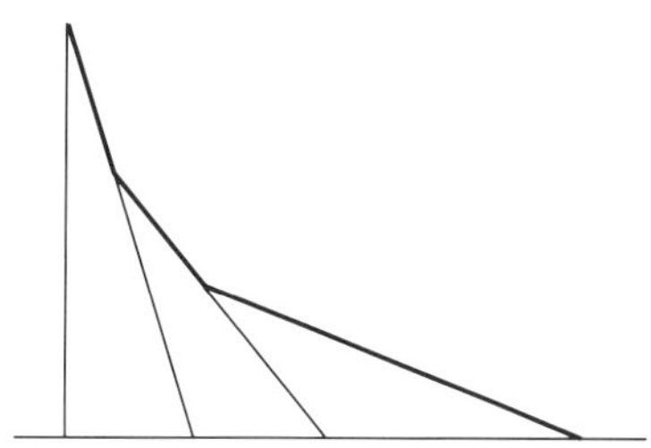

FIGURE 13b

1.24. Replace the integration over $[0, 1]$ by integration over $[0, 1/2]$. Use the results of problems 1.6 and I.2.22.

1.25. To verify necessity, consider the functions $\varphi(x) = C$, $\varphi(x) = Cx$ $(C \in \mathbb{R})$, and $\varphi(x) = \max(0, a - x)$ $(a \in (0, 1))$. To prove sufficiency use the fact that a piecewise linear convex function is a sum, with positive coefficients, of functions of the form $\varphi(x) = \max(0, a - x)$ and a linear function.

1.26. a), b) Represent the given integrals as the sums of the corresponding integrals over intervals of length $2\pi/a$. In b) use the results of problem 1.6.

1.27. a) Using the Cauchy-Schwarz-Bunyakovskiĭ inequality, verify that $f''f - f'^2 \geq 0$.

b) With the help of problem 1.11b) we can confine ourselves to the case of a twice differentiable function. The logarithmic convexity of f means that $f''f - f'^2 \geq 0$. Verify that the sum of two such functions f and g satisfies the inequalities

$$(f'' + g'')(f + g) \geq (\sqrt{f''f} + \sqrt{g''g})^2 \geq (f' + g')^2.$$

1.28. (See [18].) We confine ourselves to a proof that 3) and 4) are equivalent.

Since f is convex (see problem 1.3b)), the one-sided derivatives f'_- and f'_+ exist at each point x, $f'_-(x) \leq f'_+(x)$, and $f'_-(x) = f'_+(x)$ everywhere except on a countable set of points (see problem 1.8). We prove that $f'(1)$ exists. Let $h > 0$. Then

$$\begin{aligned} 1 = f(1) &\geq f(1+h)f(1/(1+h)) \\ &= [f(1) + f'_+(1)h + o(h)][f(1) - f'_-(1)h + o(h)] \\ &= 1 + (f'_+(1) - f'_-(1))h + o(h), \end{aligned}$$

which implies that $f'_+(1) \leq f'_-(1)$, and (considering the opposite inequality) $f'_+(1) = f'_-(1)$. Let $p = f'(1)$. Fix an x such that $f'(x)$ exists. Consider the function $\phi(t) = f(t)f(x/t)$, $t > 0$. It is differentiable at 1 and satisfies the relation $\phi(t) \leq f(x) = \phi(1)$ for all t. Therefore, $\phi'(1) = 0$, that is

$$\begin{gathered} f'(1)f(x) - f(1)\frac{x}{1^2}f'(x) = 0, \\ pf(x) = xf'(x). \end{gathered} \tag{$*$}$$

Thus, the derivative $f'(x)$, which exists on a dense subset E of $(0, +\infty)$, coincides on it with the function $pf(x)/x$, which is continuous on $(0, +\infty)$. From this, since f'_+ and f'_- are monotone (see problem 1.8), it follows that $f \in C^1(0, +\infty)$. It now follows easily from $(*)$ that $f(x) = Ax^p$, where $A = 1$ since $f(1) = 1$, and the restriction on p is due to the convexity of f.

In connection with the condition 1′), consider the function $f(x) = \min(\sqrt{x}, 1)$ $(x \in (0, +\infty))$.

1.29. Write Jensen's inequality for the function $f(x) = x^p$ in the form

$$\left(\frac{1}{c}\sum_{1\le j\le n} c_j x_j\right)^p \le \frac{1}{c}\sum_{1\le j\le n} c_j x_j^p,$$

where $c = \sum_{1\le j\le n} c_j$, $x_j = a_j/t_j$, $c_j = b_j t_j$, and choose the numbers $t_j > 0$ suitably.

1.30. a) If $0 < r < s$, then use the Hölder inequality with $p = s/r$ (see the preceding problem). If $r < s < 0$, then use the relation $(\kappa(-t))^{-1} = (\frac{1}{n}\sum_{1\le j\le n} x_j^{-t})^{1/t}$. If $r < 0 < s$, then use b).

e) Jensen's inequality for the function $\ln\varphi$ at the points $r, s, \lambda r + \mu s$ ($\lambda, \mu > 0$, $\lambda + \mu = 1$) becomes the Hölder inequality if we set $a_j = x_j^{\lambda r}$, $b_j = x_j^{\mu s}$, $p = 1/\lambda$, and $q = 1/\mu$.

For a continuous positive function f we should define $\kappa(p)$ to be $(\int_0^1 f^p(x)\,dx)^{1/p}$, and consider

$$\exp\left(\int_0^1 \ln f(x)\,dx\right)$$

instead of $(x_1, \dots, x_n)^{1/n}$. Assertions a)–e) remain in force.

1.31. The inequalities a) and b) are continuous analogues of Jensen's inequality and can be obtained from it by passing to the limit. Another way of proving the inequality b) (which is more general than a)) is as follows. By 1.11, we can assume that φ'' exists. Setting $I = \int_a^b f(x)p(x)\,dx$, we get from Taylor's formula that

$$\begin{aligned}\varphi(f(x)) &= \varphi(I) + \varphi'(I)(f(x) - I) + \tfrac{1}{2}\varphi''(c)(f(x) - I)^2\\ &\ge \varphi(I) + \varphi'(I)(f(x) - I).\end{aligned}$$

The required result is obtained after multiplication of this inequality by p and integration over $[a, b]$.

c) Use b) with $\varphi(t) = \ln(1/t)$.

1.32. Both quantities are as large as possible in the case when the points $A_1, \dots, A_{n-1}$ divide the arc joining A_0 and A_n into equal parts. For a proof use the concavity of the sine function on $[0, \pi]$.

1.33. Consider an arbitrary ray emanating from the point $S = (0, b)$, $0 < b < f(0)$. Let $A_i = (x_i, y_i)$, $i = 1, 2, \dots$, be the successive points of reflection of the ray from the graph, and let B_i be the points of reflection from the x-axis. Then to each ray there corresponds a (finite or possibly infinite) sequence (trajectory) $SA_1B_1A_2B_2\cdots$ or $SB_1A_2B_2A_3\cdots$ (the point A_1 is absent if the first reflection is from the x-axis). We prove the existence of a number M such that for all rays $\sup_i x_i \le M$ (this means that only a bounded part of the subgraph is illuminated).

1) Note first that if $x_{n+1} < x_n$ for some n, then $x_{n+k} < x_n$ for all $k > 0$ (the ray emanates from the domain).

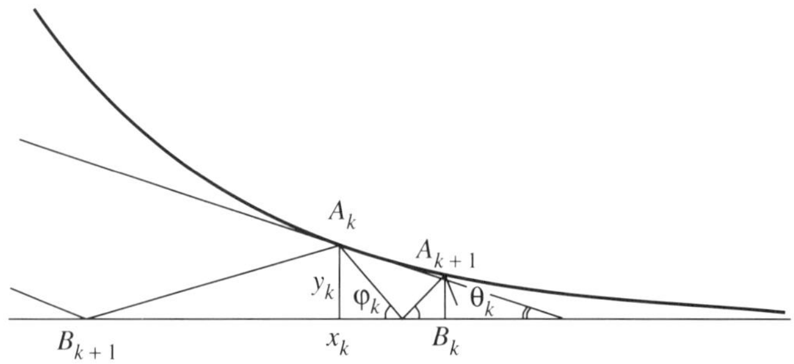

FIGURE 14

2) Let $SA_1^0B_1^0A_2^0\cdots$ be the unique trajectory for which the three points S, A_1^0, and B_1^0 lie on a single line, and let $A_i^0 = (x_i^0, y_i^0)$. Then for any other trajectory containing A_1 the inequality $x_2 < x_2^0$ holds (it is very easy to see this).

3) Suppose that $n \geq 2$, $x_{n+1} > x_n$, $\varphi_k \in (0, \pi/2)$ is the angle between the components of the polygonal line that are incident to B_k and the x-axis, and $\theta_k = |\arctan f'(x_k)|$ (see Figure 14). Then, since f is convex,

$$y_n - y_{n+1} < (\tan\theta_n)(x_{n+1} - x_n).$$

Taking into account also that

$$2y_n > y_n + y_{n+1} = (\tan\varphi_n)(x_{n+1} - x_n) \quad \text{and} \quad \varphi_n - \varphi_{n-1} = 2\theta_n,$$

we get the inequality

$$\frac{y_n - y_{n+1}}{y_n} < 2\frac{\tan\theta_n}{\tan\varphi_n} < \frac{\tan 2\theta_n}{\tan\varphi_n} = \frac{\tan(\varphi_n - \varphi_{n-1})}{\tan\varphi_n},$$

from which

$$\begin{aligned}\frac{y_{n+1}}{y_n} &> 1 - \frac{\tan(\varphi_n - \varphi_{n-1})}{\tan\varphi_n} \\ &= \frac{\sin\varphi_{n-1}}{\sin\varphi_n\cos(\varphi_n - \varphi_{n-1})} > \frac{\sin\varphi_{n-1}}{\sin\varphi_n}.\end{aligned}$$

Consequently,

$$\frac{y_{n+1}}{y_2} > \frac{\sin\varphi_1}{\sin\varphi_n} > \sin\varphi_1, \qquad y_{n+1} > y_2\sin\varphi_1.$$

Note that $\varphi_1 \geq \theta_1^0 = |\arctan f'(x_1^0)|$, $y_2 \geq y_2^0$. Consequently, the number $f^{-1}(y_2^0\sin\theta_1^0)$ can be taken as M.

1.34. a) Observe that f^* is the pointwise supremum of the family of linear functions $h_{t,a}(x) = xt - a$ over all pairs (t, a) such that $a \geq f(t)$.

b) Use Jensen's inequality and problem 1.5.

c) Use b) and the fact that $g_{x,b}(t) = xt - b \leq f(t)$ for all $t \in \mathbb{R}$ if and only if $b \geq \sup_{t\in\mathbb{R}}(xt - f(t)) = f^*(t)$, along with the hint for a).

e) Obviously, for a convex function φ differentiable at the point t the equality $\varphi'(t) = 0$ is a sufficient condition for a minimum. Therefore, if $f'(t) = x$ exists at the point t, then the supremum in the definition of $f^*(x)$ is attained at t. Let $a < b$, suppose that $\alpha = f'(a)$ and $\beta = f'(b)$ exist and are finite, and let $x \in (\alpha, \beta)$. Then the values of f^* at the points α and β are finite, and hence f^* is finite on (α, β) (see b)).

g) The function f' cannot have jumps and is strictly monotone, hence continuous, and its range really is an interval. The failure of f^* to be differentiable at the point x would mean that $t_- = (f^*)'_-(x) < t_+ = (f^*)'_+(x)$, and then the function $f^{**} = f$ would be linear on $[t_-, t_+]$ (see f)), which contradicts the strict convexity of f. Similarly, differentiability of f implies strict convexity of f^*. For a strictly convex differentiable function f and for any $x \in \Delta$ the equality $f^*(x) = xt - f(t)$ is attained only for those t such that $f'(t) = x$. Analogously, for $t \in (a, b)$ the equality $f(t) = f^{**}(t) = xt - f^*(x)$ is attained only for those x such that $(f^*)'(x) = t$. This means that f' and $(f^*)'$ are mutual inverses.

1.35. Use problem 1.34.

1.38. See problem 1.36a) and f). To prove uniqueness use problem 1.34d) and h).

§2. Smooth functions

2.2. Integrate the given identity with respect to y over the interval $[0, 1]$.

2.3. Approximate $f^{(n)}$ by a polynomial uniformly on $[a, b]$ with an arbitrary degree of accuracy. (The possibility of such an approximation follows, for example, from 3.8b)).

2.4. Study the Taylor series of f at an endpoint of an interval on which $f(x) \equiv 0$.

2.5. Use induction on n. Employ the formula

$$f = e^{Q_n}(P_n + \sum_{1\leq k<n} P_k e^{Q_k - Q_n})$$

and, arguing by contradiction, study the derivatives of the function $P_n + \sum_{1\leq k<n} P_k e^{Q_k - Q_n}$.

2.6. Prove that either $f'(x) \to 0$ as $x \to +\infty$ or f' has infinitely many extrema.

2.7. Let $x, \delta \in (0, 1)$. Then

$$f(\delta x) = f(x) + (\delta - 1)xf'(x) + \frac{(\delta - 1)^2}{2}x^2 f''(\overline{x})$$

where $\overline{x} \in (\delta x, x)$. Therefore,

$$\begin{aligned} xf'(x) &= \frac{f(x) - f(\delta x)}{1-\delta} + \frac{1-\delta}{2} x^2 f''(\overline{x}) \\ &= \frac{f(x) - f(\delta x)}{1-\delta} + O\left((1-\delta)\frac{x^2}{\overline{x}^2}\right) \\ &= \frac{f(x) - f(\delta x)}{1-\delta} + O\left(\frac{1-\delta}{\delta^2}\right). \end{aligned}$$

By choosing δ close to 1, we make the second term small, and then by suitably choosing x we make the first term small.

2.8. Verify that

$$\begin{aligned} -h'(t) &= -\int_{-T}^{T} f'(x) g(t-x)\,dx \\ &= \int_0^T f'(x)(g(x+t) - g(x-t))\,dx \\ &= \int_0^T \int_{x-t}^{x+t} f'(x) g'(y)\,dx\,dy \\ &= \left(\iint_P + \int_0^t \int_{x-t}^{x+t} + \int_{T-t}^{T} \int_{2T-x-t}^{x+t}\right) f'(x) g'(y)\,dx\,dy, \end{aligned}$$

where $t \in (0, T)$ and P is the rectangle with vertices $(0, t)$, $(t, 0)$, $(T, T-t)$, and $(T-t, T)$. Since $P \subset [0, T]^2$, the first integral is nonnegative, and the remaining two integrals are equal to zero because g' is odd.

2.9. Write $f(1/2)$ according to Taylor's formula $f(1/2) = f(x_0) + \cdots$, setting x_0 equal to 0 and then 1.

2.10. a) Starting from the identity

$$f'(x) = \frac{1}{2h} \int_{-h}^{h} (f'(x) - f'(x+t))\,dt + \frac{f(x+h) - f(x-h)}{2h}$$

and applying the mean value theorem to the integrand, we arrive at the inequality

$$|f'(x)| \le \frac{1}{2h}\left(\int_{-h}^{h} |t| M_2\,dt + 2M_0\right) = \frac{h}{2} M_2 + \frac{M_0}{h}$$

for all $h > 0$. The expression on the right-hand side has a minimum equal to $2M_0M_2$. The inequality becomes an equality for the piecewise smooth function $f(x) = \int_0^x (\int_0^t \varphi(u)\,du + 2)\,dt$, where $\varphi(u) = -2\operatorname{sgn} u$ for $|u| \le 1$ and $\varphi(u) = 0$ for $|u| > 1$. Replacing φ by a continuous function φ_ε as shown in Figure 15a), we get a C^2-smooth function f_ε, and for it $M_1/\sqrt{M_0M_2} \to \sqrt{2}$ as $\varepsilon \to 0$.

The inequality proved admits the following mechanical interpretation: if the motion of a point takes place on the interval $[-M_0, M_0]$, and the acceleration never exceeds M_2 in absolute value, then the velocity of the point

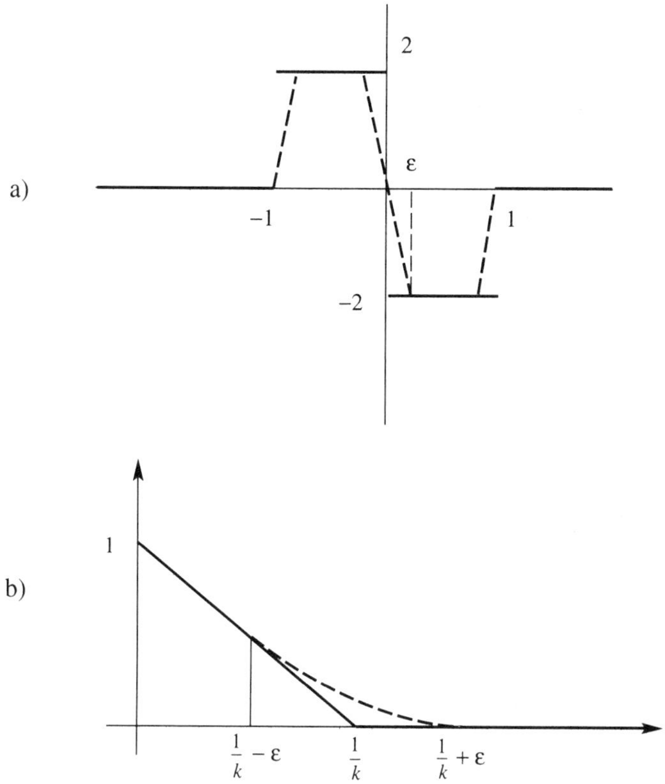

FIGURE 15

cannot be too large at any time. More precisely, it does not exceed $\sqrt{2M_0M_2}$. All are familiar with an analogue of this fact from a basic physics course: in a free fall (with zero initial velocity) the instantaneous velocity of a body is equal to $\sqrt{2gS}$, where S is the length of the path passed by the body and g is the acceleration of gravity.

b) Replace $[x-h, x+h]$ by $[x, x+h]$ in the preceding argument. As a result the inequality $M_1 \leq 2\sqrt{M_0M_2}$ is obtained.

d) Use an analogous device, choosing h in $(0, (b-a)/2]$ and integrating over $[x, x+h]$ or over $[x-h, h]$, depending on the sign of $x-(a+b)/2$.

c) In the case it is convenient to use another argument. Adding the equalities

$$f(x) - f(0) = xf'(x) - \tfrac{1}{2}x^2 f''(c_1),$$
$$f(2) - f(x) = (2-x)f'(x) + \tfrac{1}{2}(2-x)^2 f''(c_2)$$

termwise, we get that

$$2f'(x) = f(2) - f(0) + \tfrac{1}{2}(x^2 f''(c_1) - (2-x)^2 f''(c_2)),$$

which at once gives the desired estimate. That it is sharp is demonstrated by the example $f(x) = (x^2/2) - 1$.

2.11. The convergence of the integral $\int_0^\infty |f''(x)|\,dx$ (if at least one integral diverges, then there is nothing to prove) implies that the limit $c = \lim_{x\to\infty} f'(x)$ exists, and the convergence of $\int_0^\infty |f(x)|\,dx$ implies that $c = 0$. Consider the function g determined by the conditions $g'(x) = -\int_x^\infty |f''(x)|\,dx$ and $g(0) = f(0)$ (for definiteness assume that this number is positive). It is decreasing and convex, and it satisfies the inequality $g(x) \le f(x)$ (a consequence of the inequality $g'(x) \le f'(x) = -\int_x^\infty f''(t)\,dt$). Draw the tangent to the graph of g at the point 0, and let α be the point where it intersects the x-axis, $\alpha = f(0)/g'(0)$. Then we get from a comparison of areas that

$$\frac{1}{2}f(0)\alpha \le \int_0^\infty \max(g(x), 0)\,dx \le \int_0^\infty |f(x)|\,dx,$$

which implies that

$$\frac{1}{2}(f(0))^2 \le \int_0^\infty |f(x)|\,dx \cdot |g'(0)|.$$

First replacing $|g'(0)|$ by $\int_0^\infty |f''(x)|\,dx$ and then the geometric mean by the arithmetic mean, we get a) and b).

To prove that the constants 2 and $\sqrt{2}$ cannot be improved, we approximate the function $h(x) = \max(1 - kx, 0)$ $(k > 0)$ by a convex function $f_\varepsilon \in C^2([0, +\infty))$ as shown in Figure 15b).

Then

$$\int_{1/k-\varepsilon}^{1/k+\varepsilon} f''(x)\,dx = \left|f'\left(\frac{1}{k} - \varepsilon\right)\right| = k,$$

$$\int_0^\infty f(x)\,dx \cong \frac{1}{2k}, \qquad f(0) = 1,$$

and hence

a) $1 = (f(0))^2 = 2 \cdot \frac{1}{2k} \cdot k \cong 2\int_0^\infty |f(x)|\,dx \int_0^\infty |f''(x)|\,dx$;

b) $\int_0^\infty (|f(x)| + |f''(x)|)\,dx \cong \frac{1}{2k} + k = \sqrt{2} = \sqrt{2}f(0)$ for $k = 1/\sqrt{2}$.

2.12. By Taylor's formula, $f(1) = f'(0) + \int_0^1 f''(t)(1-t)\,dt$, which implies that $|\int_0^1 f''(t)(1-t)\,dt| = |a|$. Applying the Cauchy-Schwarz-Bunyakovskiĭ inequality to the integral, we get a lower estimate:

$$|a| \le \left(\int_0^1 (f''(t))^2\,dt\right)^{1/2} \cdot \frac{1}{\sqrt{3}}.$$

Equality is attained in the case when the functions $f''(t)$ and $1 - t$ are proportional, which implies that $f(t) = at(t-1)(t-2)/2$.

2.13. Let $F_n = \{x | f^{(n)}(x) = 0\}$, $G_n = \mathbb{R}\setminus F_n$. Prove the auxiliary assertion (A): If f is not a polynomial on some open interval Δ, then for any n there is a constituent interval Δ_n of the nonempty open set $G_n \cap \Delta$ on which f is not a polynomial.

The assertion of the problem is easily derived from (A): assuming that f is not a polynomial on $\mathbb{R}$, we construct a sequence of nested open intervals $\Delta_n = (a_n, b_n) \subset G_n$, $[a_{n+1}, b_{n+1}] \subset \Delta_n$, with nonempty intersection, and this contradicts the assumption.

Assume that (A) does not hold, i.e., that for some n and Δ the function f is not a polynomial on Δ, but its restriction to any constituent interval of the set $G_n \cap \Delta$ is a polynomial. Let $\sigma(x)$ $(x \in \Delta)$ be a maximal interval with the properties that $x \in \sigma(x) \subset \Delta$ and $f|_{\sigma(x)}$ is a polynomial. Verify that if the restrictions of f to two adjacent intervals are polynomials, then f is a polynomial on their union. Show that if a is an endpoint of $\sigma(x)$, then: 1) there exists a sequence $x_k \in F_n$, $x_k \neq a$, $x_k \to a$; 2) $f^{(m)}(a) = 0$ for all $m \geq n$; 3) $f^{(n)}$ vanishes on $\sigma(x)$. Then, in particular, $f^{(n)}(x) = 0$ for all $x \in G_n \cap \Delta$, which is impossible (the solution of D. Yu. Burago).

2.14. By formal termwise differentiation obtain a recursion relation that can be used to successively define α_n and β_n. Note that by suitably choosing β_n it is possible to ensure the possibility of termwise differentiation of the series.

2.15. It can be assumed that $a = 0$. Then

$$f(x) = \int_0^1 \frac{d}{dt} f(tx)\,dt = \int_0^1 \sum_j \frac{\partial f}{\partial x_j}(tx) x_j\,dt$$
$$= \sum_j x_j \int_0^1 \frac{\partial f}{\partial x_j}(tx)\,dt,$$

and $\int_0^1 (\partial f/\partial x_j)(tx)\,dt$ can be taken as $g_j(x)$.

2.16. Note that if f is a constant, then b) and c) imply that $F(f) = 0$. Two-fold use of the result of the preceding problem gives the representation

$$f(x) = f(a) + \sum_j (x_j - a_j) g_j(a)$$
$$+ \sum_{j,k} (x_j - a_j)(x_k - a_k) h_{jk}(x),$$

where $g_j(a) = (\partial f/\partial x_j)(a)$. It remains to use a), b), and c) once more.

2.17. Rejecting the trivial case when $f \equiv 0$, we assume that the number $a = \inf\{t | f(t) > 0\}$ is equal to zero (otherwise, we make the substitution $u = t - a$). Then it follows from the assumption that for $t > 0$

$$\frac{f(t)g(t)}{C + \int_0^t f(x)g(x)\,dx} \leq g(t).$$

Passing to primitives, we get the inequality

$$\ln\left(C + \int_0^t f(x)g(x)\,dx\right) \leq \ln C + \int_0^t g(x)\,dx.$$

Then

$$C+\int_0^t f(x)g(x)\,dx \le C\exp\left(\int_0^t g(x)\,dx\right),$$

which immediately implies what is required.

To prove the corollary observe that, since $|h(t)| = |\int_0^t h'(x)\,dx| \le \int_0^t |h'(x)|\,dx$, the condition of the first part of the problem is satisfied by the functions $f(t)=|h'(t)|$ and $g(t)\equiv M$ and the number $C=0$.

2.18. a) Use the fact that $f(t_j-t_k)=f(t_j)\overline{f(t_k)}$.

b) Use the fact that $\cos(a, t)$ is half the sum of two positive-definite functions.

2.20. Represent f and g in terms of their Fourier transforms.

2.21. b) Use the identity

$$-\|x\|^\alpha = c_\alpha \int_{\mathbb{R}^n} \frac{\cos(x, t)-1}{\|t\|^{n+\alpha}}\,dt,$$

where $c_\alpha>0$ (cf. problem 1.26b)). Then (see problem 2.18b))

$$\begin{aligned} &-\sum_{\substack{1\le j\le m\\1\le k\le m}} \|x_j-x_k\|^\alpha z_j\overline{z}_k \\ &\quad = c_\alpha \int_{\mathbb{R}^n} \frac{1}{\|t\|^{n+\alpha}} \sum_{\substack{1\le j\le m\\1\le k\le m}} (\cos(t, x_j-x_k)-1) z_j\overline{z}_k\,dt \ge 0. \end{aligned}$$

2.22. b) Use the fact that, as follows from the inequality (1), the matrix

$$\begin{pmatrix} f(0) & f(t)\\ f(-t) & f(0)\end{pmatrix}$$

is positive-definite for any $t\in\mathbb{R}$.

c) Use the positive definiteness of the matrix

$$\begin{pmatrix} f(0) & f(t_1) & f(t_2)\\ \overline{f(t_1)} & f(0) & f(t_2-t_1)\\ \overline{f(t_2)} & \overline{f(t_2-t_1)} & f(0)\end{pmatrix}$$

for any $t_1, t_2\in\mathbb{R}$.

2.23. Suppose that $z_1, \dots, z_m\in\mathbb{C}$ and $z_1+\dots+z_m=0$. Then for $s>0$

$$\begin{aligned} 0 &\le \frac{1}{s}\sum_{\substack{1\le j\le m\\1\le k\le m}} e^{s\varphi(t_j-t_k)} z_j\overline{z}_k \\ &= \sum_{\substack{1\le j\le m\\1\le k\le m}} \frac{e^{s\varphi(t_j-t_k)}-1}{s} z_j\overline{z}_k. \end{aligned}$$

Passing to the limit in this inequality as $s\to 0$, we get the inequality

$$\sum_{\substack{1\le j\le m\\1\le k\le m}} \varphi(t_j-t_k) z_j\overline{z}_k \ge 0.$$

2.24. The assertions c) and d) follow from b), b) follows from a) and the fact that $Sh(x) = 0 \ (x \neq -\delta/\gamma)$, a) can be verified directly, and e) follows from the equality $g'' = -gSf/2$.

To prove f) use the fact that, by the conditions $f''(x_0) = 0$ and $Sf(x_0) < 0$, $f'(x)f'''(x) < 0$ in some neighborhood of the point x_0, and consider the possible combinations of signs of $f'(x)$, $f''(x)$, and $f'''(x)$ on both sides of x_0.

2.25. b) Let $a_1, \dots, a_{n-1}$ be the roots of the polynomial $f'(x)$. Then

$$Sf(x) = 2\sum_{i<j} \frac{1}{(x-a_i)(x-a_j)} - \frac{3}{2}\left(\sum_i \frac{1}{x-a_i}\right)^2 < 0.$$

2.26. Assume that $a = f'(c) > 0$ is the minimal value of f' in a neighborhood of the point c. Then the graph of f looks as pictured in Figure 16. Choosing x_1, x_2, x_3, and x_4 as shown in the figure (with x_2 and x_3 sufficiently close to c), we have that

$$\frac{f(x_4)-f(x_1)}{x_4-x_1}\cdot\frac{f(x_3)-f(x_2)}{x_3-x_2} - \frac{f(x_4)-f(x_3)}{x_4-x_3}\cdot\frac{f(x_2)-f(x_1)}{x_2-x_1}$$
$$= b(a+\varepsilon_1) - (b+\varepsilon_2)(b+\varepsilon_3),$$

where ε_1, ε_2, and ε_3 are small numbers, and b denotes the expression $(f(x_4)-f(x_1))/(x_4-x_1)$. Passing in this equality to the limit as $x_2, x_3 \to c$, we get that $b(a-b) < 0$, which contradicts the assumption. It is useful to observe that if a function $f \in C^3(\Delta)$ is monotone, then the validity of the

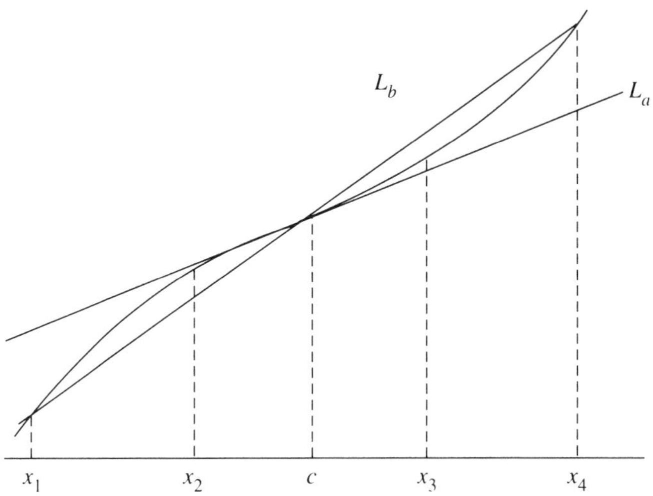

FIGURE 16

inequality in the condition of the problem for an arbitrary quadruple of points $x_1 < x_2 < x_3 < x_4$ is equivalent to the condition $Sf < 0$.

§3. Bernstein polynomials

3.1. b) To compute the sums $S_{n,0}$ and $S_{n,1}$ use the binomial formula and the equality

$$\sum_{0\le k\le n} k\binom{n}{k}x^k(1-x)^{n-k} = \sum_{0\le k<n} nx\binom{n-1}{k}x^k(1-x)^{n-1-k} = nx.$$

Compute the sums $S_{n,2}$, $S_{n,3}$, and $S_{n,4}$ with the help of the recursion formula a).

3.2. a) Comparing the sums $\sigma_{n,\delta}$ and $S_{n,2}$ (see problem 3.1), we get that

$$\begin{aligned}\sigma_{n,\delta}(x) &\le \frac{1}{\delta^2}\sum_{0\le k\le n}\binom{n}{k}x^k(1-x)^{n-k}\left(\frac{k}{n}-x\right)^2\\ &= \frac{1}{\delta^2 n^2}S_{n,2}(x) = \frac{x(1-x)}{\delta^2 n} \le \frac{1}{4\delta^2 n}.\end{aligned}$$

b) Compare the sums $\sigma_{n,\delta}$ and $S_{n,4}$ and use the inequality $S_{n,4}(x) < n^2/4$.

3.3. a) Use the result in problem 3.1.

3.4. b) Use the inequality

$$B_n((f-A)^2, x) = B_n(f^2, x) - 2AB_n(f, x) + A^2 \ge 0 \quad \text{for } A = B_n(f, x).$$

3.5. Show that

a)

$$B_n'(f, x) = n\sum_{0\le k\le n-1}\binom{n-1}{k}x^k(1-x)^{n-k-1}\left(f\left(\frac{k+1}{n}\right) - f\left(\frac{k}{n}\right)\right);$$

b)

$$\begin{aligned}B_n''(f, x) = n(n-1)&\sum_{0\le k\le n-2}\binom{n-2}{k}x^k(1-x)^{n-k-2}\\ &\times\left(f\left(\frac{k+2}{n}\right) - 2f\left(\frac{k-1}{n}\right) + f\left(\frac{k}{n}\right)\right).\end{aligned}$$

3.6. a) Use the equality $\sum_{0\le k\le n}\binom{n}{k}x^k(1-x)^{n-k} = 1$.

b) It suffices to consider the case when $\sup_\Delta f = 0$ (otherwise take the function $\tilde{f} = f - \sup_\Delta f)$. Then $f(k/n) \le 0$ if $k/n \in \Delta$, and hence

$$B_n(f, x) \le \sum_{\substack{0\le k\le n\\ k/n\notin\Delta}} \binom{n}{k} f\left(\frac{k}{n}\right) x^k (1-x)^{n-k}$$
$$\le (\sup_{[0,1]} f) \sum_{\substack{0\le k\le n\\ k/n\notin\Delta}} \binom{n}{k} x^k(1-x)^{n-k}.$$

It remains to use the inequality a) in problem 3.2.

c) Let $\sup_\Delta f = 0$. It can be assumed that $\sup_{\Delta'\setminus\Delta} f \le -1$ (this can be ensured by multiplying f by a positive constant). Let $\Delta = (a, b)$, $\Delta' = (\alpha, \beta)$, $0 \le \alpha < a < b < \beta \le 1$, and $M = \sup_{[0,1]} f$. Then

$$B_n(f, x) \le \sum_{0\le k\le \alpha n} M\binom{n}{k} x^k(1-x)^{n-k} - \sum_{\alpha n<k\le an} \binom{n}{k} x^k(1-x)^{n-k}$$
$$- \sum_{bn\le k\le \beta n} \binom{n}{k} x^k(1-x)^{n+k} + \sum_{\beta n\le k\le n} M\binom{n}{k} x^k(1-x)^{n-k}$$
$$= \sum\nolimits_1 - \sum\nolimits_2 - \sum\nolimits_3 + \sum\nolimits_4.$$

Prove that $\sum_1 < \sum_2$ for sufficiently large $n \in \mathbb{N}$ (the inequality $\sum_4 < \sum_3$ is proved similarly). Since the terms in the sum $\sum_1$ are nondecreasing, it suffices to prove that

$$Mn\binom{n}{[\alpha n]} x^{[\alpha n]}(1-x)^{n-[\alpha n]} < \binom{n}{[an]} x^{[an]}(1-x)^{n-[an]},$$

that is,

$$Mn\binom{n}{[\alpha n]}\left(\frac{x}{1-x}\right)^{[\alpha n]-[an]} < \binom{n}{[an]}.$$

The left-hand side of this inequality decreases as x increases. Therefore, we need only verify it for $x = a$. Since $\sqrt[n]{\binom{n}{[pn]}} \underset{n\to+\infty}{\to} p^{-p}(1-p)^{p-1}$ (see problem II.1.8), it suffices to show that $a^\alpha(1-a)^{1-\alpha} < \alpha^\alpha(1-\alpha)^{1-\alpha}$ for $\alpha < a$, but this is obvious, because the left-hand side of the last inequality decreases as a increases, and coincides with the right-hand side for $a = \alpha$.

3.7. Applying the inequality b) of problem 3.6 to the functions $f - g$ and $g - f$, show that $B_n(f, x_0) - B_n(g, x_0) \to 0$.

3.8. a) Apply the inequality b) of problem 3.6 to the functions $f(x) - f(x_0)$ and $f(x_0) - f(x)$.

b) Fix an $\varepsilon > 0$ and choose a number $\delta = \delta_\varepsilon > 0$ such that $|f(t) - f(x)| < \varepsilon$ if $x \in \Delta$, $t \in [0, 1]$, and $|t - x| < \delta$. Let $M = \sup_{[0,1]} |f|$. With the help of the inequality a) in problem 3.2 we get that

$$\begin{aligned}|B_n(f, x) - f(x)| &= \left|\sum_{0\le k\le n}\left(f\left(\frac{k}{n}\right) - f(x)\right)\binom{n}{k}x^k(1-x)^{n-k}\right| \\ &\le \sum_{\substack{0\le k\le n\\ |k/n-x|<\delta}} \varepsilon\binom{n}{k}x^k(1-x)^{n-k} \\ &\quad + \sum_{\substack{0\le k\le n\\ |k/n-x|\ge\delta}} 2M\binom{n}{k}x^k(1-x)^{n-k} \\ &\le \varepsilon + \frac{M}{2\delta^2 n}.\end{aligned}$$

Thus, $|B_n(f, x) - f(x)| < 2\varepsilon$ if $n > M/(2\varepsilon\delta^2)$ and $x \in \Delta$.

c) It suffices to consider the case when $f(x) = 1$ for $x \le x_0$ and $f(x) = 0$ for $x > x_0$. Then $B_n(f, x_0) = \sum_{0\le k\le nx_0}\binom{n}{k}x_0^k(1-x_0)^{n-k}$. To compute this sum, break it up into two parts S_1 and S_2 by singling out the contribution of the terms with "small" indices:

$$S_1 = \sum_{0\le k\le n(x_0-\delta)}\binom{n}{k}x_0^k(1-x_0)^{n-k},$$

$$S_2 = \sum_{n(x_0-\delta)<k\le nx_0}\binom{n}{k}x_0^k(1-x_0)^{n-k}$$

(here $\delta = \delta_n$ is a small positive parameter whose choice will be made more precise below). Estimating the sum S_1 with the help of the inequality in 3.2a), we see that $S_1 = O(1/n\delta^2)$. With the help of Stirling's formula and Taylor's formula we get that for the terms in S_2

$$\binom{n}{k}x_0^k(1-x_0)^{n-k} = \frac{1+O(1/n+\delta)}{\sqrt{2\pi n x_0(1-x_0)}}\exp\left(-\frac{n(x_0-k/n)^2}{2x_0(1-x_0)} + O(n\delta^3)\right).$$

We now assume that $\delta = \delta_n$ has been chosen in such a way that $n\delta_n^2 \to +\infty$ and $n\delta_n^3 \to 0$ (for example, $\delta_n = n^{-2/5}$). Then $S_1 = o(1)$, and

$$\begin{aligned}B_n(f, x_0) &= o(1) + \frac{1+o(1)}{\sqrt{2\pi n x_0(1-x_0)}}\sum_{n(x_0-\delta)<k\le nx_0}\exp\left(-\frac{n(x_0-k/n)^2}{2x_0(1-x_0)}\right) \\ &= o(1) + (1+o(1))\sqrt{\frac{n}{2\pi x_0(1-x_0)}} \\ &\quad\times\left(\int_{x_0-\delta}^{x_0}\exp\left(-\frac{n(x_0-t)^2}{2x_0(1-x_0)}\right)dt + O\left(\frac{1}{n}\right)\right) \\ &= o(1) + \frac{1}{\sqrt{\pi}}\int_0^{\delta\sqrt{n/2x_0(1-x_0)}} e^{-t^2}\,dt = \frac{1}{2} + o(1).\end{aligned}$$

3.9. a) From the assumption it follows that $\int_0^1 f(x)B_n(f, x)\,dx = 0$ for $n \in \mathbb{N}$. Since $B_n(f) \rightrightarrows f$ on $[0, 1]$ (see problem 3.8b) for $\Delta = [0, 1]$), it follows that $\int_0^1 f^2(x)\,dx = \lim_{n\to\infty} \int_0^1 f(x)B_n(f, x)\,dx = 0$, and hence $f \equiv 0$.

b) It is clear that $\int_0^1 (ax + b)(x^{n+1}(1 - x)^2)''\,dx = 0$ for all $a, b \in \mathbb{R}$ and $n \in \mathbb{N}$. Let $\tilde{f}(x) = f(x) - (ax + b)$, where a and b are chosen so that $\int_0^1 \tilde{f}(x)\,dx = 0$ and $\int_0^1 \tilde{f}(x)x\,dx = 0$. Since $\int_0^1 \tilde{f}(x)(x^{n+1}(1-x)^2)''\,dx = 0$, the numbers $M_n = \int_0^1 \tilde{f}(x)x^n\,dx$ satisfy the recursion relation

$$(n + 3)(n + 2)M_{n+1} = 2(n + 2)(n + 1)M_n - (n + 1)nM_{n-1}.$$

Using the equalities

$$M_0 = \int_0^1 \tilde{f}(x)\,dx = 0, \qquad M_1 = \int_0^1 \tilde{f}(x)x\,dx = 0,$$

we get from this that $M_n = 0$ for all $n = 0, 1, \ldots$. It remains to use the result of problem 3.9a).

3.10. Show that it is possible to weaken the conditions on g, requiring instead of membership in $C^2([0, 1])$ that its first two derivatives vanish at the endpoints of $[0, 1]$. Indeed, for any such function g and for any number $\Delta \in (0, 1/2)$ the function g_Δ equal to $g((x - \Delta)/(1 - 2\Delta))$ for $x \in [\Delta, 1 - \Delta]$ and equal to zero otherwise satisfies the conditions of the problem. Therefore,

$$0 = \int_0^1 f(x)g_\Delta''(x)\,dx = \int_\Delta^{1-\Delta} f(x)\left(g\left(\frac{x - \Delta}{1 - 2\Delta}\right)\right)''dx$$
$$= \frac{1}{1 - 2\Delta}\int_0^1 f(\Delta + x(1 - 2\Delta))g''(x)\,dx.$$

This implies that $\int_0^1 f(x)g''(x)\,dx = 0$. In particular, this equality is valid for functions g of the form $g(x) = x^{p+n}(1 - x)^p$, where $p > 2$ and $n \geq 0$. Passing to the limit as $p \to 2$, we get that $\int_0^1 f(x)(x^{2+n}(1-x)^2)''\,dx = 0$ for any $n \geq 0$, and this is equivalent (see problem 3.9b)) to the linearity of f.

3.11. Apply the result of problem 3.8b) for $\Delta = [0, 1]$ to the function $\tilde{f}(x) = f(\arccos x)$.

3.12. a) Formulate and prove the two-dimensional analogue of the inequality a) in problem 3.2, and then modify the solution of problem 3.8b).

3.13. Prove by induction the identity

$$B_n^{(r)}(f, x) = n(n-1)\cdots(n-r+1)\sum_{0\leq k\leq n-r}\binom{n-r}{k}x^k(1-x)^{n-k-r}\Delta_r^{(n)}f\left(\frac{k}{n}\right),$$

where $\Delta_1^{(n)}f(y) = f(y + \frac{1}{n}) - f(y)$ and $\Delta_j^{(n)}f(y) = \Delta_1^{(n)}(\Delta_{j-1}^{(n)}f(y))$. Use the equality

$$\Delta_r^{(n)}f(y) = \frac{1}{n^r}f^{(r)}\left(y + \frac{r}{n}\theta\right), \quad \text{where } \theta \in (0, 1).$$

3.14. a) Suppose that the number $M > 0$ is such that $|f(x) - f(y)| \leq M|x-y|^\alpha$ for all $x, y \in [0, 1]$. Then

$$|B_n(f, x) - f(x)| = \left| \sum_{0\leq k\leq n} \left(f\left(\frac{k}{n}\right) - f(x) \right) \binom{n}{k} x^k (1-x)^{n-k} \right|$$

$$\leq M \sum_{0\leq k\leq n} \left| \frac{k}{n} - x \right|^\alpha \binom{n}{k} x^k (1-x)^{n-k}.$$

Let $p_k = \binom{n}{k} x^k (1-x)^{n-k}$. It follows from the result of problem 3.1b) that

$$\sum_{0\leq k\leq n} p_k = 1 \quad \text{and} \quad \sum_{0\leq k\leq n} p_k \left(\frac{k}{n} - x \right)^2 = \frac{x(1-x)}{n}.$$

Using the Hölder inequality (see problem VII.1.29), we get that

$$|B_n(f, x) - f(x)| \leq M \left(\sum_{0\leq k\leq n} p_k \right)^{(2-\alpha)/2} \left(\sum_{0\leq k\leq n} p_k \left(\frac{k}{n} - x \right)^2 \right)^{\alpha/2}$$

$$= M \left(\frac{x(1-x)}{n} \right)^{\alpha/2}.$$

b) To get a lower estimate of $B_n(f_\alpha, 1/2)$ consider the sums

$$s_a(x) = \sum_{0\leq k\leq n} \binom{n}{k} x^k (1-x)^{n-k} |k - nx|^a.$$

It follows from the Hölder inequality that $s_{a+b}(x) \leq (s_{pa}(x))^{1/p} (s_{qb}(x))^{1/q}$, if $p, q > 1$ and $1/p + 1/q = 1$. Choose parameters a, b, and p such that $a + b = 2$, $ap = \alpha$, $bq = 4$ ($a = 2\alpha/(4-\alpha)$, $b = 4(2-\alpha)/(4-\alpha)$, $p = (4-\alpha)/2$, $q = (4-\alpha)/(2-\alpha)$). In this case the last inequality takes the form

$$s_2(x) \leq s_\alpha^{2/(4-\alpha)}(x) s_4^{(2-\alpha)/(4-\alpha)}(x).$$

Since (see problem 3.1b)) $s_2(x) = S_{n,2}(x) = nx(1-x)$ and $s_4(x) = S_{n,4}(x) \leq n^2 x(1-x)$, it follows that $n^{\alpha/2} x(1-x) \leq s_\alpha(x)$. Consequently, $n^{-\alpha} s_\alpha(x) \geq x(1-x)/n^{\alpha/2}$. For $x = 1/2$ we get that

$$B_n(f_\alpha, \tfrac{1}{2}) - f_\alpha(\tfrac{1}{2}) = B_n(f_\alpha, \tfrac{1}{2}) = n^{-\alpha} s_\alpha(\tfrac{1}{2}) \geq \frac{1}{4n^{\alpha/2}}.$$

3.15. a) We write $B_n(f, x)$ in the form $B_n(f, x) = \sum_{0\leq k\leq n} p_k f(k/n)$, where $p_k = \binom{n}{k} x^k (1-x)^{n-k}$. Since $\sum_{0\leq k\leq n} p_k = 1$ and $\sum_{0\leq k\leq n} kp_k = nx$ (see problem 3.1b)), we get with the help of Jensen's inequality (problem VII.1.1) that $B_n(f, x) \geq f(\sum_{0\leq k\leq n} p_k k/n) = f(x)$.

b) Assume the opposite: there exist numbers a, b, and x such that $0 \leq a < x < b \leq 1$ and

$$f(x) > \frac{x-a}{b-a} f(b) + \frac{b-x}{b-a} f(a).$$

It will be assumed that $f(a) = f(b) = 0$ (this can be ensured by subtracting a linear function from f). Moreover, it can be assumed that $f(x) = \max_{[a,b]} f$. It follows from the result of problem 3.6c) that $B_n(f, x) < f(x)$ for sufficiently large n, but this contradicts the assumption.

3.16. To estimate the quantity

$$\Delta_n(x) = B_n(f, x) - f(x) - \frac{x(1-x)}{2n} f''(x)$$

write the difference $B_n(f, x) - f(x)$ in the form

$$B_n(f, x) - f(x) = \sum_{0 \le k \le n} \binom{n}{k} x^k (1-x)^{n-k} \left(f\left(\frac{k}{n}\right) - f(x) \right). \qquad (*)$$

Use the following variant of Taylor's formula:

$$f(t) = f(x) + f'(x)(t-x) + \tfrac{1}{2} f''(x)(t-x)^2 + \varphi(t, x)(t-x)^2,$$

where the function φ is continuous on the square $[0, 1] \times [0, 1]$ and equal to zero for $t = x$. Setting $t = k/n$, we get that

$$f\left(\frac{k}{n}\right) = f(x) + f'(x)\left(\frac{k}{n} - x\right) + \frac{f''(x)}{2}\left(\frac{k}{n} - x\right)^2 + \varphi\left(\frac{k}{n}, x\right)\left(\frac{k}{n} - x\right)^2.$$

Substituting this expression in $(*)$ and using the result of problem 3.1b), we arrive at the equality

$$\Delta_n(x) = \sum_{0 \le k \le n} \binom{n}{k} x^k (1-x)^{n-k} \varphi\left(\frac{k}{n}, x\right)\left(\frac{k}{n} - x\right)^2.$$

Now fix an arbitrary number $\varepsilon > 0$, and let $\delta = \delta_\varepsilon > 0$ be such that $|\varphi(t, x)| < \varepsilon$ if $|t - x| < \delta$. Then

$$\begin{aligned} |\Delta_n(x)| &\le \sum_{0 \le k \le n} \binom{n}{k} x^k (1-x)^{n-k} \left|\varphi\left(\frac{k}{n}, x\right)\right| \left(\frac{k}{n} - x\right)^2 \\ &\le \sum_{|k/n - x| \le \delta} + \sum_{|k/n - x| > \delta} \\ &\le \frac{\varepsilon}{n^2} S_{n,2}(x) + \frac{M}{n^4 \delta^2} S_{n,4}(x) \\ &\le \varepsilon \frac{x(1-x)}{n} + \frac{M}{\delta^2} \frac{x(1-x)}{n^2}, \end{aligned}$$

where $M = \sup|\varphi|$. Thus, $|\Delta_n(x)| \le 2\varepsilon x(1-x)/n$, if $n > M/\varepsilon\delta^2$.

3.17. See [40] about problems 3.17 and 3.18. Let

$$S_n(f) = \sum_{0 \le k \le n} \left(B_n\left(f, \frac{k}{n}\right) - f\left(\frac{k}{n}\right) \right) g\left(\frac{k}{n}\right),$$

$$I(f) = \frac{1}{2} \int_0^1 f(x)(x(1-x)g(x))''\, dx.$$

It follows from problem 3.16 that $S_n(\varphi) \to I(\varphi)$ for any function $\varphi \in C^2([0, 1])$. To prove convergence in the general case, write the sum $S_n(f)$ in the form $S_n(f) = \sum_{0\le j\le n} f(j/n)a_{n,j}$, where

$$a_{n,j} = \binom{n}{j}\sum_{0\le k\le n} g\left(\frac{k}{n}\right)\left(\frac{k}{n}\right)^j\left(1-\frac{k}{n}\right)^{n-j} - g\left(\frac{j}{n}\right).$$

It suffices to see that the sums $\sum_{0\le j\le n}|a_{n,j}|$ are bounded. Indeed, suppose that $\sum_{0\le j\le n}|a_{n,j}| \le C$, and let $M = \max_{0\le x\le 1}|(x(1-x)g(x))''|$. Fix an arbitrary number $\varepsilon > 0$ and choose a function $\varphi \in C^2([0, 1])$ such that $\max_{0\le x\le 1}|f(x)-\varphi(x)| < \varepsilon$ (a Bernstein polynomial $B_N(f)$ with sufficiently large index N can be taken as φ—see problem 3.8b) with $\Delta = [0, 1]$). Then

$$\begin{aligned}|S_n(f) - I(f)| &\le |S_n(f-\varphi)| + |I(f-\varphi)| + |S_n(\varphi) - I(\varphi)|\\ &\le \sum_{0\le j\le n}\varepsilon|a_{n,j}| + M\varepsilon + |S_n(\varphi) - I(\varphi)|\\ &\le (M+C)\varepsilon + |S_n(\varphi) - I(\varphi)|.\end{aligned}$$

Consequently, $S_n(f) \to I(f)$ for any function $f \in C([0, 1])$ if $\sum_{0\le j\le n}|a_{n,j}| = O(1)$. To estimate this sum use the smoothness of the function g:

$$g\left(\frac{k}{n}\right) = g\left(\frac{j}{n}\right) + g'\left(\frac{j}{n}\right)\frac{k-j}{n} + O\left(\left(\frac{k-j}{n}\right)^2\right).$$

This gives us that

$$\begin{aligned}|a_{n,j}| \le \left|g\left(\frac{j}{n}\right)\right|&\left|\binom{n}{j}\sum_{0\le k\le n}\left(\frac{k}{n}\right)^j\left(1-\frac{k}{n}\right)^{n-j} - 1\right|\\ &+\left|g'\left(\frac{j}{n}\right)\right|\binom{n}{j}\left|\sum_{0\le k\le n}\left(\frac{k}{n}\right)^j\left(1-\frac{k}{n}\right)^{n-j}\frac{k-j}{n}\right|\\ &+O(1)\binom{n}{j}\sum_{0\le k\le n}\left(\frac{k}{n}\right)^j\left(1-\frac{k}{n}\right)^{n-j}\left(\frac{k-j}{n}\right)^2.\end{aligned}$$

Since g is equal to zero outside the interval (a, b), this implies that

$$\sum_{0\le j\le n}|a_{n,j}| \le A_n \max|g| + B_n \max|g'| + O(1)C_n,$$

where

$$A_n = \sum_{an\le j\le bn} \left| \binom{n}{j} \sum_{0\le k\le n} \left(\frac{k}{n}\right)^j \left(1-\frac{k}{n}\right)^{n-j} - 1 \right|,$$

$$B_n = \sum_{an\le j\le bn} \binom{n}{j} \left| \sum_{0\le k\le n} \left(\frac{k}{n}\right)^j \left(1-\frac{k}{n}\right)^{n-j} \frac{k-j}{n} \right|,$$

$$C_n = \sum_{0\le j\le n} \sum_{0\le k\le n} \binom{n}{j} \left(\frac{k}{n}\right)^j \left(1-\frac{k}{n}\right)^{n-j} \left(\frac{k-j}{n}\right)^2.$$

Changing the order of summation and using the result of problem 3.1b), we get that

$$C_n = \frac{1}{n^2} \sum_{0\le k\le n} S_{n,2}\left(\frac{k}{n}\right) = \frac{1}{n^2} \sum_{0\le k\le n} k\left(1-\frac{k}{n}\right) = \frac{n^2-1}{6n^2}.$$

It remains to prove that the sums A_n and B_n are bounded. To estimate A_n, write it in the form

$$A_n = \sum_{an\le j\le bn} \left| n^{-n} \binom{n}{j} \sum_{0\le k\le n} \varphi_j(k) - 1 \right|,$$

where $\varphi_j(t) = t^j(n-t)^{n-j}$ for $t \in [0, n]$. Using the result of problem VI.3.15b), we get that

$$\sum_{0\le k\le n} \varphi_j(k) = \int_0^n \varphi_j(t)\,dt + O\left(\int_0^n |\varphi_j''(t)|\,dt\right).$$

Since the function φ_j'' changes sign on $[0, n]$ only twice,

$$\int_0^n |\varphi_j''(t)|\,dt \le 4 \max_{0\le t\le n} |\varphi_j'(t)|.$$

Uncomplicated computations show that

$$\max_{0\le t\le n} \varphi_j(t) = \varphi_j(j) \quad \text{and} \quad \max_{0\le t\le n} |\varphi_j'(t)| = |\varphi_j'(j + O(\sqrt{n}))|.$$

It follows from the equality $\varphi_j'(t) = n\varphi_j(t)(j-t)/(t(n-t))$ that for $j \in [an, bn]$

$$\max_{0\le t\le n} |\varphi_j'(t)| \le \frac{\text{const}}{\sqrt{n}} \max_{0\le t\le n} \varphi_j(t) = O\left(\frac{1}{\sqrt{n}} j^j (n-j)^{n-j}\right).$$

By the equality

$$\int_0^n \varphi_j(t)\,dt = n^{n+1} \int_0^1 t^j(1-t)^{n-j}\,dt = n^{n+1} \frac{1}{(n+1)\binom{n}{j}},$$

this gives us that

$$A_n = \sum_{an \le j \le bn} \left| n^{-n} \binom{n}{j} \left(\frac{n^{n+1}}{(n+1)\binom{n}{j}} + O\left(\frac{1}{\sqrt{n}} j^j (n-j)^{n-j} \right) \right) - 1 \right| = O(1).$$

The sum B_n is estimated similarly.

3.18. Use the results of problems 3.17 and 3.10a).

3.19. The necessity of the condition $f(0)$, $f(1) \in \mathbb{Z}$ is obvious. But if it is satisfied, then the polynomial $\sum_{0 \le k \le n} [\binom{n}{k} f(k/n)] x^k (1-x)^{n-k}$ differs little from the polynomial $B_n(f, x)$:

$$\left| \sum_{0 \le k \le n} \left[\binom{n}{k} f\left(\frac{k}{n}\right) \right] x^k (1-x)^{n-k} - \sum_{0 \le k \le n} \binom{n}{k} f\left(\frac{k}{n}\right) x^k (1-x)^{n-k} \right|$$
$$\le \sum_{1 \le k < n} x^k (1-x)^{n-k} \le \frac{1}{n} \sum_{1 \le k < n} \binom{n}{k} x^k (1-x)^{n-k} < \frac{1}{n}.$$

It remains to use the result of problem 3.8b) with $\Delta = [0, 1]$.

3.20. To compute $B_n^{(k)}(f, 0)$ use the identity given in the solution of problem 3.13.

3.21. a) Show that $\Delta_k f \equiv 0$ for $k > m$ and use the formula for $B_n(f, x)$ in the preceding problem. The uniform convergence follows from the result of problem IV.5.13.

b) Write $B_n(f, x)$ in the form

$$B_n(f, x) = n^{-m} \sum_{0 \le k \le n} k^m \binom{n}{k} x^k (1-x)^{n-k} = n^{-m} \frac{d^m}{dt^m} F_x(0),$$

where $F_x(t) = \sum_{0 \le k \le n} \binom{n}{k} e^{kt} x^k (1-x)^{n-k} = (xe^t + 1 - x)^n$. It is easy to see that

$$F_x(t) = \left(1 + x \sum_{j \ge 1} \frac{t^j}{j!} \right)^n = 1 + \sum_{j \ge 1} \frac{t^j}{j!} \left(\sum_{1 \le i \le j} C_{i,j}^{(n)} x^i \right),$$

and all the coefficients $C_{i,j}^{(n)}$ are nonnegative. Therefore,

$$\left| \frac{d^m}{dt^m} F_x(0) \right| = \left| \sum_{1 \le i \le m} C_{i,m}^{(n)} x^i \right| \le \sum_{1 \le i \le m} C_{i,m}^{(n)} |x|^i$$
$$\le \max(1, |x|^m) \sum_{1 \le i \le m} C_{i,m}^{(n)} = \max(1, |x|^m) \frac{d^m F_1}{dt^m}(0).$$

It remains to observe that $F_1(t) = e^{nt}$, and hence $d^m F_1(0)/dt^m = n^m$. Regarding this problem see also Bernstein's paper [2].

3.22. a) Obviously, we can assume that $r > 1$. Fixing an arbitrary number $\varepsilon > 0$, we choose an index $N = N_\varepsilon$ large enough that $\sum_{m > N} |c_m| r^m < \varepsilon$. Let $P(x) = \sum_{0 \le m \le N} c_m x^m$ and let $\rho(x) = f(x) - P(x)$. Then $B_n(f) - f =$

$B_n(P) - P + B_n(\rho) - \rho$. The difference $B_n(P) - P$ is uniformly small on $[-r, r]$ if n is sufficiently large (see problem 3.21a)). Since

$$|\rho(x)| \le \sum_{m>N} |c_m|\,|x|^m \le \sum_{m>N} |c_m| r^m < \varepsilon,$$

it remains to estimate $B_n(\rho, x)$. To do this, use the inequality in 3.11b):

$$\begin{aligned} |B_n(\rho, x)| &\le \sum_{m>N} |c_m| \max(1, |x|^m) \\ &\le \sum_{m>N} |c_m| r^m < \varepsilon. \end{aligned}$$

A sharper result was obtained by Kantorovich in [9].

b) Use the inequalities 3.15a) and 3.21b).

3.23. a) Make the change of variable $y = x/(1-x)$.

b) Use the fact that multiplication by $x + (1-x)$ does not change the polynomial.

c) The necessity of the condition is obvious. To prove sufficiency compute the coefficients a_n^k. Let $P(x) = \sum_{0\le j\le m} b_j x^j$. Make the change of variable $y = x/(1-x)$. Then it follows from the equality $\sum_{0\le j\le m} b_j x^j = \sum_{0\le k\le n} a_n^k x^k (1-x)^{n-k}$ that $\sum_{0\le j\le m} b_j y^j (1+y)^{n-j} = \sum_{0\le k\le n} a_n^k y^k$. Equating the coefficients of equal powers of y, we get that

$$\begin{aligned} a_n^k &= \sum_{0\le j\le \min(m,k)} b_j \binom{n-j}{k-j} \\ &= \binom{n}{k} \left(\sum_{0\le j\le \min(m,k)} b_j \left(\frac{k}{n}\right)^j + O\left(\frac{1}{n}\right) \right) \\ &= \binom{n}{k} \left(P\left(\frac{k}{n}\right) + O\left(\frac{1}{n}\right) \right). \end{aligned}$$

Therefore, $a_n^k > 0$ for $k = 0, \dots, n$ for all sufficiently large indices n if the polynomial P is positive on the whole interval $[0, 1]$. In the general case we represent P in the form $P(x) = x^a \widetilde{P}(x)(1-x)^b$, where $\widetilde{P} > 0$ on $[0, 1]$. Since the coefficients of the expansions of the polynomial $\widetilde{P}$ in the basis $\{x^k (1-x)^{n-a-b-k}\}_{0\le k\le n-a-b}$ are positive, this representation implies that the coefficients $\{a_n^k\}_{0\le k\le n}$ of the expansion of P in the basis $\{x^k(1-x)^{n-k}\}_{0\le k\le n}$ are positive for $k = a, a+1, \dots, n-b$ and are equal to zero for the remaining k.

§4. Almost periodic functions and sequences

4.1. Let $\delta(x) = x - [x]$ and let $p_n = \delta(n\lambda)$. Since λ is irrational, $p_n \ne p_m$ for $n \ne m$. Note that $p_{n+m} = (p_n + p_m) \bmod 1$ for any $n, m \in \mathbb{Z}$. For any $N > 1/\varepsilon$ there are two points among $p_1, \dots, p_N$ separated by a

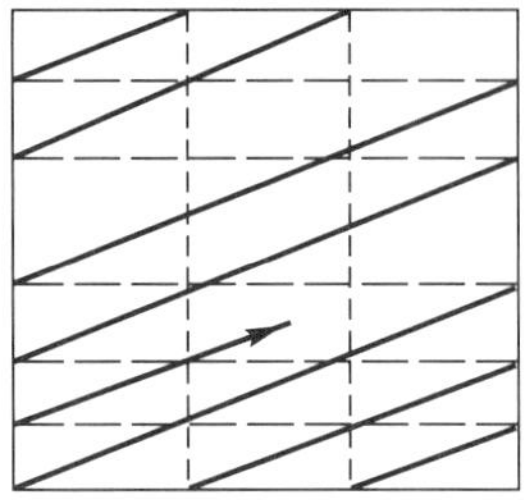

FIGURE 17

distance less than ε. Let k_1, k_2 $(k_1, k_2 \leq N)$ be indices such that $0 < p_{k_1} - p_{k_2} < \varepsilon$, and let $m = k_1 - k_2$. Then $0 < p_m < \varepsilon$. Let $r \in \mathbb{N}$ be such that $1 - \varepsilon < p_{rm} < 1$. In this case if $p_n \in (a - \varepsilon, a + \varepsilon)$, then either p_{n+m} or p_{n+rm} falls in the same interval. The same can be said of one of the points p_{n-m} and p_{n-rm}. Thus, the distance between adjacent solutions of the equation $n\lambda \stackrel{\varepsilon}{=} a \pmod 1$ does not exceed $L = r|m|$, which proves the second assertion.

4.2. a) Consider the set $\Gamma = \{(x(t) = \delta(\lambda_1 t),\ y(t) = \delta(\lambda_2 t)) | t \in \mathbb{R}\}$. Verify that for irrational λ_1/λ_2 the sets $A_x = \{x(t) | y(t) = 0\}$ and $A_y = \{y(t) | x(t) = 0\}$ are dense in $[0, 1)$ (use 4.1). The "curve" Γ consists of segments (see Figure 17) obtained by intersecting the square $[0, 1) \times [0, 1)$ with the family of parallel lines having slope λ_2/λ_1 and passing through all possible points of the sets $A_x \times \{0\}$ and $\{0\} \times A_y$. It is called an irrational winding of the torus $T^2 = [0, 1) \times [0, 1) \cong \mathbb{R}^2/\mathbb{Z}^2$. We suggest that the reader follow the motion of a point $(x(t), y(t))$ with, for example, $x(t) = t \bmod 1$, $y(t) = t\sqrt{2} \bmod 1$ on a sufficiently large interval of the parameter t. It follows from the structure of Γ described above that this set is dense in $[0, 1) \times [0, 1)$, which is equivalent to solvability of the system (1).

Consider the sequences $t_k = (k + a_1)/\lambda_1$ and $y_k = y(t_k)$. As established in problem 4.1, there exists a relatively dense set of $k \in \mathbb{Z}$ such that $y_k \stackrel{\varepsilon}{=} a_2$. At the same time, $x_k = x(t_k) = a_1$.

b) We pass to integer solutions of the system (1). The irrationality of λ_1 and λ_2 leads to the fact that the points $p_k = (x_k, y_k)$ are pairwise distinct, where this time $x_k = x(k)$ and $y_k = y(k)$. Then the p_k have an accumulation point. Consequently, for any $\varepsilon > 0$ there exist $k_1, k_2 \in \mathbb{Z}$ such that the coordinates ξ and η of the point $p_{k_1} - p_{k_2} \in \mathbb{R}^2$ satisfy the condition $|\xi| < \varepsilon$ and $|\eta| < \varepsilon$. It follows form the irrationality of λ_2/λ_1 that η/ξ is irrational, and hence the winding $\widetilde{\Gamma} = \{(\delta(\xi t), \delta(\eta t)) | t \in \mathbb{R}\}$ of the torus is dense in the square T^2. Let $m = k_1 - k_2$. Then the points p_{rm}, $r \in \mathbb{Z}$, form an $\varepsilon\sqrt{2}$-net on each segment that is a part of $\widetilde{\Gamma}$, and hence also in the square. Let $p_n = (x_n, y_n)$ be a point such that $|x_n - a_1| < \varepsilon$ and $|y_n - a_2| < \varepsilon$. For the four open squares $Q_1, \dots, Q_4$ in T^2 with sides of length ε and

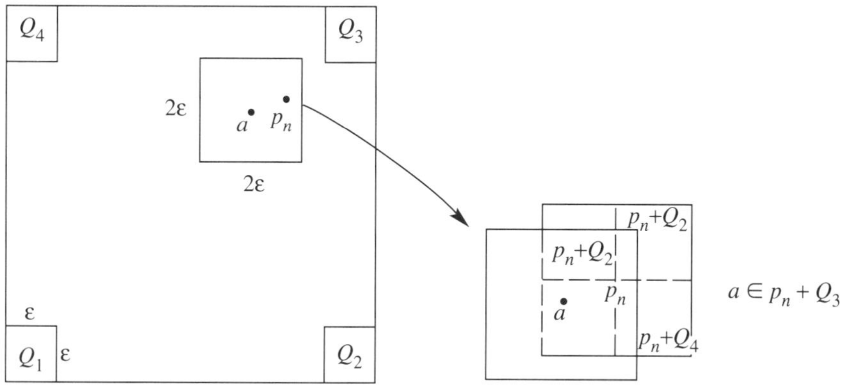

FIGURE 18

abutting on the angles in T^2 choose positive integers $r_1, \ldots, r_4$ such that the point $p_{r_i m}$ falls in Q_i (one of the numbers r_i can be taken equal to 1). Then for any n that is a solution of (1), one of the numbers $n + r_i m$ is also a solution of (1) (as is one of the numbers $n - r_i m$) (see Figure 18). This implies that in each interval of length $L = \max_{1 \le i \le 4} r_i |m|$ on the line there is at least one integer solution of our system.

4.3. Prove that the system is solvable for any $\lambda_1, \ldots, \lambda_n$. Let $t_n = n/\lambda_1$, and $x_i(t_n) = \delta(\lambda_i t_n)$, $i = 1, \ldots, k$, $n \in \mathbb{Z}$. Then $x_1(t_n) = 0$, and the points $p_n = (x_2(t_n), \ldots, x_k(t_n)) \in T^{k-1} = \underbrace{[0, 1) \times \cdots \times [0, 1)}_{k-1 \text{ times}}$ have an accumulation point. Arguing as in the solution of the preceding problem, find an $n_0 \ge 1$ such that $x_i(t_{n_0}) < \varepsilon$.

4.4. Prove that for the real solvability of the system for any $a_1, \ldots, a_k$ and ε the numbers $\lambda_1, \ldots, \lambda_k$ must be rationally independent (incommensurable), that is, the relation $n_1\lambda_1 + \cdots + n_k\lambda_k = 0$ ($n_i \in \mathbb{Z}$) must hold only for $n_1 = \cdots = n_k = 0$. Let $n_1, \ldots, n_k \in \mathbb{Z}$, and suppose that the system has a solution t for any values of $a_1, \ldots, a_k$ and ε. Then $|\lambda_i t - a_i - m_i| < \varepsilon$, $i = 1, \ldots, k$, for some $m_1, \ldots, m_k \in \mathbb{Z}$. Consequently, $|\sum n_i\lambda_i t - \sum n_i a_i - \sum n_i m_i| < \varepsilon \sum |n_i|$. If $\sum n_i \lambda_i = 0$, then, since ε is arbitrary, this implies that $\sum n_i a_i \in \mathbb{Z}$. Since also $a_1, \ldots, a_k$ are arbitrary, $n_1 = \cdots = n_k = 0$. Using problem 4.2a), prove by induction that the rational incommensurability of the numbers $\lambda_1, \ldots, \lambda_k$ ensures the existence of a relatively dense set of solutions.

4.5. Show that a necessary and sufficient condition for a positive answer to the questions a) and b) is the rational independence of the numbers $\lambda_1, \ldots, \lambda_k$ mod 1. Assuming the existence of an integer solution m and fixing arbitrary numbers $n_1, \ldots, n_k \in \mathbb{Z}$, we arrive as in the preceding problem at the inequality

$$\left|\sum n_i\lambda_i m - \sum n_i a_i - \sum n_i m_i\right| < \varepsilon \sum |n_i|.$$

If $(\sum n_i\lambda_i) \bmod 1 = 0$, then $\sum n_i\lambda_i m \in \mathbb{Z}$, and hence $\sum n_i a_i \in \mathbb{Z}$ in view of the arbitrariness of ε; as earlier, this implies the equalities $n_1 = \dots = n_k = 0$. The proof of sufficiency is obtained by generalizing the arguments in the solution of problem 4.2b).

4.6. If a and b are commensurable (i.e., the number a/b is rational), then the function f is periodic, and the mean value of f on a period is equal to zero. If a/b is irrational, then there are arbitrarily close maximum (minimum) points of the two terms (see problem 4.2a)).

4.7. To verify the first assertion consider the system of equations $\lambda_k t/2\pi \overset{\varepsilon}{=} 0 \pmod 1$, $k = 1, \dots, n$, and use problem 4.3 (see also problem 4.15).

The assumption of periodicity for f implies the identity $\sum a_k(e^{i\lambda_k\tau} - 1)e^{i\lambda_k x} \equiv 0$ (τ is a period). Setting $b_k = a_k(e^{i\lambda_k\tau} - 1)$ and differentiating this identity at zero, we get the equalities $\sum \lambda_k^m b_k = 0$, $m = 0, 1, 2, \dots$. Since the λ_k are pairwise distinct, the Vandermonde determinant $\det(\lambda_k^m)_{k=1}^{n}{}_{m=0}^{n-1}$ is nonzero, and hence all the b_k are equal to zero. But this is possible only if $\lambda_k\tau/2\pi \in \mathbb{Z}$, that is, the λ_k are pairwise commensurable.

4.8. Use problem 4.4.

4.9. Suppose that there is an $L > 0$ such that for any $\varepsilon \in (0, \varepsilon_0)$ the interval $[1, 1+L]$ contains a τ_ε for which

$$|f(x) - f(x+\tau_\varepsilon)| < \varepsilon, \qquad x \in \mathbb{R}. \tag{1}$$

Since the closed interval is compact, there exist a sequence $\varepsilon_n \downarrow 0$ and a τ such that $\tau_{\varepsilon_n} \to \tau$. Passing to the limit in (1), we see that τ is a period of f.

4.10. Uniform continuity and boundedness are proved in almost the same way as in the periodic case. For example, to estimate $|f(x_1) - f(x_2)|$ use an almost period τ taken in the interval $(-x_1, -x_1 + L)$. The answers to the remaining questions are as in the periodic case: they exist; it does not follow.

4.12. Consider the functions $\varphi_h(x) = h^{-1}(f(x+h) - f(x))$ $(h > 0)$. They are uniformly almost periodic and tend uniformly on $\mathbb{R}$ to f' as $h \to 0$.

4.13. For simplicity suppose that F is real, let $M = \sup_{\mathbb{R}} F(x)$, $m = \inf_{\mathbb{R}} F(x)$, and $\varepsilon > 0$, and let x_1 and x_2 be such that $F(x_1) < m + \varepsilon/6$ and $F(x_2) > M - \varepsilon/6$. Define $d = |x_1 - x_2|$ and let $L = L(\varepsilon/3d)$ be a number chosen according to the definition of almost periodicity for f. If τ is an

$(\varepsilon/3d)$-almost period of f, then for $z_i = x_i + \tau$, $i = 1, 2$,

$$\begin{aligned}F(z_2) - F(z_1) &= F(x_2) + \int_{x_2}^{z_2} f(x)\,dx - F(x_1) - \int_{x_1}^{z_1} f(x)\,dx \\ &= F(x_2) - F(x_1) + \int_{x_1}^{x_2} (f(u+\tau) - f(u))\,du \\ &\geq M - m - \frac{2\varepsilon}{3} - d\frac{\varepsilon}{3d} = M - m - \varepsilon.\end{aligned}$$

This implies that each interval of length $L + d$ contains points z_1 and z_2 such that $F(z_1) < m + \varepsilon$ and $F(z_2) > M - \varepsilon$. Then for any x and any $z_1 \in (x, x + L + d)$ with $F(z_1) < m + \varepsilon$ we have that

$$\begin{aligned}&F(x+\tau) - F(x) \\ &\quad = F(z_1 + \tau) - F(z_1) + \int_x^{z_1} f(x)\,dx - \int_{x+\tau}^{z_1+\tau} f(x)\,dx \\ &\quad = F(z_1 + \tau) - F(z_1) - \int_x^{z_1} (f(u+\tau) - f(u))\,du \\ &\quad > m - (m + \varepsilon) - (L + d)\varepsilon = -\varepsilon(1 + L + d).\end{aligned}$$

It can be proved similarly that $F(x+\tau) - F(x) < \varepsilon(1 + L + d)$ for all x, and this yields what is required.

4.14. *Necessity.* Consider the sequences $\{f(x + h_n)\}$, $x \in \mathbb{Q}$, and, using the diagonal process, choose a subsequence $k_i = h_{n_i}$ such that the sequence $\{f(x + k_i)\}$ converges for all $x \in Q$. To prove the uniform convergence of the sequence $\{f_{k_i}\}$ on $\mathbb{R}$ use the uniform continuity of f.

Sufficiency. Assume the contrary: there exist an $\varepsilon_0 > 0$ and a sequence $\{\Delta_n\}$ of open intervals with lengths d_n tending to infinity that do not contain ε_0-almost periods of f. Choose a sequence $\{h_k\}$ and a subsequence $\{\Delta_{n_k}\}$ of intervals in such a way that $d_{n_k} > \max_{1\leq i,j\leq k} |h_i - h_j|$, $h_{k+1} - h_i \in \Delta_{n_k}$ for $i = 1, \ldots, k$. Considering $\sup_{x\in\mathbb{R}} |f_{h_i}(x) - f_{h_j}(x)|$, $i < j$, prove that the sequence $\{f_{h_k}\}$ cannot contain a uniformly convergent subsequence.

4.15. For the case of a sum use the criterion of almost periodicity from the preceding problem. In the case of a product use the equality $fg = ((f+g)^2 - (f-g)^2)/4$.

4.16. Let $L = L(\varepsilon)$ be a number such that any interval of length L contains an ε-almost period of f, and let $C = \sup_{t\in\mathbb{R}} |f(t)|$. Using an almost period τ taken from the interval $(a, a+L)$, $a \in \mathbb{R}$, prove the inequality

$$\left| \frac{1}{2T}\int_{-T}^{T} f(x)\,dx - \frac{1}{2T}\int_{a-T}^{a+T} f(x)\,dx \right| < \varepsilon + \frac{CL}{T}.$$

Deduce from this that the estimate $|\varphi(nT) - \varphi(T)| < \varepsilon + (CL/T)$ $(n \in \mathbb{N})$ holds for $\varphi(T) = \frac{1}{2T}\int_{-T}^{T} f(x)\,dx$. Then prove that $|\varphi(T_1) - \varphi(T_2)| < 2\varepsilon + CL(\frac{1}{T_1} + \frac{1}{T_2})$ for T_1 and T_2 with $T_1/T_2 \in \mathbb{Q}$.

4.17. a) We explain only why the equality $\langle f, f\rangle = 0$ implies that $f \equiv 0$. Suppose that $|f(x)|^2 > \varepsilon > 0$ in some interval (a, b). Then in each interval of length $L = L(\varepsilon/2)$ there is a τ such that $|f(x)|^2 > \varepsilon/2$ for $x \in (a+\tau, b+\tau)$. Consequently,

$$\frac{1}{2T}\int_{-T}^{T}|f(x)|^2\,dx \geq \frac{1}{2T}\left[\frac{2T}{L}\right](b-a)\frac{\varepsilon}{2} \underset{T\to\infty}{\longrightarrow} \frac{b-a}{2L}\varepsilon > 0.$$

c) Let $g = \sum_{1\leq k\leq n} d_k\varphi_{\lambda_k}$, $d_k \in \mathbb{C}$, $k = 1, \dots, n$. Then

$$\begin{aligned} 0 &\leq \langle f-g, f-g\rangle \\ &= \langle f, f\rangle + \langle f, -g\rangle + \overline{\langle f, -g\rangle} + \langle g, g\rangle \\ &= \langle f, f\rangle - \sum c_{\lambda_k}\overline{d}_k - \sum \overline{c_{\lambda_k}} d_k + \sum |d_k|^2 \\ &= \langle f, f\rangle + \sum (d_k - c_{\lambda_k})\overline{d}_k - \sum \overline{c_{\lambda_k}}(d_k - c_{\lambda_k}) - \sum c_{\lambda_k}\overline{c_{\lambda_k}} \\ &= \langle f, f\rangle + \sum |d_k - c_{\lambda_k}|^2 - \sum |c_{\lambda_k}|^2. \end{aligned}$$

Setting $d_k = c_{\lambda_k}$, we prove the inequality.

The countability of the set of nonzero "Fourier coefficients" c_λ follows from the fact that, since $M(|f|^2)$ is finite, the set $\{\lambda \,|\, |c_\lambda| \geq 1/n\}$ is finite for each $n \in \mathbb{N}$. On the other hand, for each countable collection $\lambda_1, \lambda_2, \dots$ and for the arbitrary absolutely convergent series $\sum a_k$ the series $\sum a_k e^{i\lambda_k x}$ is uniformly convergent on $\mathbb{R}$, and its sum f is a UAP function (see problems 4.7 and 4.11). In view of uniform convergence,

$$M(f\overline{\varphi_\lambda}) = \sum a_k M(\varphi_{\lambda_k}\overline{\varphi_\lambda}) = \begin{cases} a_k & \text{for } \lambda = \lambda_k, \\ 0 & \text{for } \lambda \neq \lambda_k. \end{cases}$$

The reader can become more familiar with the details of these questions, for example, in [17].

4.18. Use the equality $a_n = b_n + c_n$.

4.19. The uniform almost periodicity of f implies the periodicity of $\{\varepsilon_k\}_{k\in\mathbb{Z}}$.

4.20. False. It is convenient to describe a corresponding example in the notation of problem 4.23. Let $A_0 = 0$,

$$A_p = A_{p-1}\underbrace{A^1_{p-1}\cdots A^1_{p-1}}_{2^{p+1}-1} \qquad (p \in \mathbb{N}).$$

4.21. Let $p = (m, n) \in \Sigma$. Consider the points $(m \pm 1, n)$, $(m, n \pm 1)$ adjacent to p. Note that the projections of the segments joining $(m+1, n)$ to $(m, n+1)$ and $(m-1, n)$ to $(m, n-1)$ on the line perpendicular to the sides of the strip are equal to the width of the strip, which implies that one and only one of the endpoints of each of these segments belongs to $\sum$. Joining each point in $\sum$ to the two neighbors by segments, we get an infinite

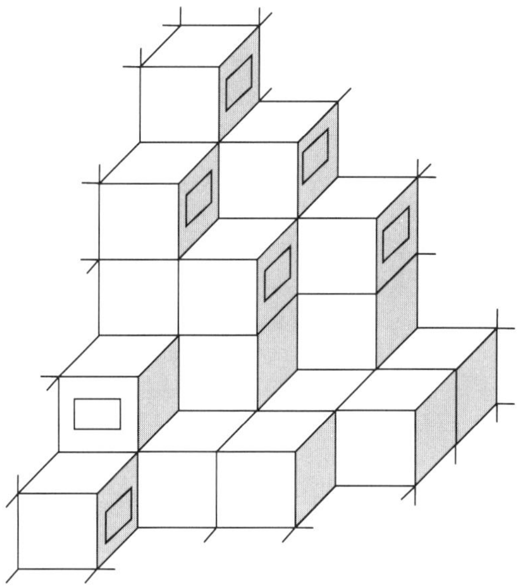

FIGURE 19

polygonal curve lying in S with components parallel to the coordinate axes, which easily implies the first assertion.

Let us prove the second assertion. Considering the distances from the successive vertices p_i of the polygonal curve to the line $y = kx$, divided by $\sqrt{1+k^2}$, we get a sequence $\{x_i\}_{i\in\mathbb{Z}}$, $x_i \in [0, 1)$. Then $x_{i+1} - x_i = 1/(k+1) = \alpha$ if the edge $p_i p_{i+1}$ is vertical (it will be assumed that in this case $\varepsilon_i = 1$), and $x_{i+1} - x_i = -k/(k+1) = \alpha - 1$, if $p_i p_{i+1}$ is horizontal ($\varepsilon_i = 0$). Thus, if it is assumed that $p_0 = (0, 0)$, then $x_i = i\alpha \bmod 1$. Let $\varepsilon_{s+1} \cdots \varepsilon_{s+l}$ be an arbitrary word, and let $x_{s+j} = \max(x_{s+1}, \dots, x_{s+l})$, $\delta = 1 - x_{s+j}$. It is clear that each sufficiently long segment of $\mathbb{Z}$ contains a number t such that $x_{t+j} \in (1-\delta, 1)$ (see problem 4.1). It is easy to see that the differences $x_{t+i} - x_{s+i}$, $i = 1, \dots, l$, are then the same, and hence the "words" $\varepsilon_{s+1} \cdots \varepsilon_{s+l}$ and $\varepsilon_{t+1} \cdots \varepsilon_{t+l}$ coincide.

The analogous problem in space leads to the construction of an interesting partition of the plane. Suppose now that S_0 is the open layer in $\mathbb{R}^3$ located between two parallel planes P and Q with $P \cap \mathbb{Z}^3 = \{(0, 0, 0)\}$ and $Q \cap \mathbb{Z}^3 = \{(1, 1, 1)\}$, and let $S = S_0 \cup P$. Projecting the segments joining pairs of adjacent points in $S \cap \mathbb{Z}^3$ orthogonally on P, we get segments bounding certain parallelograms. These parallelograms cover P, and their interiors are pairwise disjoint (see Figure 19). The covering obtained is remarkable in that it consists of parallelograms of all three sorts, is not periodic (that is, does not pass into itself under some translation), but is quasiperiodic in the sense that any finite part of it occurs infinitely many times. The proofs of these facts are left to the reader.

4.22. First show that if $A = A_n = \varepsilon_1 \cdots \varepsilon_{2^n}$, then ε has the form $A^0 A^1 A^1 A^0 A^1 A^0 A^0 A^1 \cdots$, where $A^0 = A$, $A^1 = (1 - \varepsilon_1, \dots, 1 - \varepsilon_{2^n})$, that is, the upper indices again form the sequence ε. Let B be any word in ε. If n is large enough that the word A_n contains B, then, as is easy to see, B is contained in each word of length 2^{n+2}.

4.23. It is useful to single out the case when there exists an n such that $A_{p+1} = A_p A_p \cdots A_p$ for all $p \geq n$. In this case ε has the form $A_n A_n A_n \cdots$, and hence turns out to be periodic.

In the opposite case, for a word B in ε choose some word of the form A_n that contains B. Then find a $k > 0$ such that the words A_n and A_n^1 necessarily appear in A_{n+k}. Since ε is obtained by concatenating A_{n+k} and A_{n+k}^1, every word whose length L is double the length of A_{n+k} contains A_n, and hence B.

It is interesting to note that besides the case mentioned above, ε is periodic only if $A_{p+1} = A_p A_p^1 A_p A_p^1 \cdots A_p^1 A_p$ for all sufficiently large p (verify this).

4.24. Obviously, s has the form $As_{2k}\overline{A}s_{4k}As_{6k}\overline{A}s_{8k}\cdots$, where $k = 2^n$ $(n \in \mathbb{N})$, $A = A_n = s_1 \cdots s_{2k-1}$, $\overline{A} = \overline{s}_{2k-1} \cdots \overline{s}_1$. This at once yields almost periodicity.

If $\sigma(n)$ denotes $(-1)^{s_n}$ $(n \in \mathbb{N})$, then the sequence $\{\sigma(n)\}$ is characterized by the following properties:

$$1)\ \sigma(2n+1) = (-1)^n \sigma(1); \qquad 2)\ s_{2n} = \tfrac{1}{2}(\sigma(2n) + 1)$$

is again a sequence of folds. Assume that the sequence $\{s_n\}_{n \geq n_0}$ has period $t = 2^a(2b+1)$. Then $\sigma(n + 2^a(2b+1)) = \sigma(n)$ for all $n \geq n_0$. Setting n equal to $2^{a+1}m$, we arrive at the equalities $\sigma(2^a(2m+2b+1)) = \sigma(2^{a+1}m)$, $m \geq m_0$. Denoting $\sigma(2^a m)$ by $\sigma_a(m)$, $m \geq 1$, we get a sequence σ_a satisfying 1) and 2). Consequently,

$$\sigma_a(2m + 2b + 1) = (-1)^{m+b}\sigma_a(1) = \sigma_{a+1}(m), \qquad m \geq m_0.$$

But then $\sigma_{a+2}(m) = (-1)^{2m+b}\sigma_a(1) = \text{const}$, which is impossible.

As a consequence of aperiodicity we remark that if with a number $x \in [0, 1)$ whose infinite binary expansion has the form $x = 0.\varepsilon_1\varepsilon_2\ldots$ (for such a representation to be unique we assume that among the ε_n there are always infinitely many zeros) we associate $y = 0.s_1 s_2 \ldots$, where $\{s_n\}$ is the sequence of folds given by the equalities $s_{2^n} = \varepsilon_{n+1}$, $n = 0, 1, \dots$, then we get a function on $[0, 1)$ that is continuous at irrational points and right-continuous at dyadic rational points, and takes only irrational values.

4.25. Take numbers of the form $m2^l$, $m \in \mathbb{N}$, for τ_m (these numbers satisfy the condition a)) and write s in the form $As_k\overline{A}s_{2k}As_{3k}\overline{A}s_{4k}\cdots$ with $k = 2^{l-1}$, where A and $\overline{A}$ are words of length $k-1$ (see the solution of problem 4.24). Since the length of the word $As_k\overline{A}s_{2k}$ is equal to 2^l, at least $(1 - 1/k)n$ of the n numbers $a_j = |s_{j+\tau_m} - s_j|$, $j = 1, \dots, n$, will be zero (corresponding to those s_j falling in some word of the form A or $\overline{A}$). Consequently,

$$\frac{1}{n}\sum_{1\le j\le n} a_j < \frac{1}{k} < \varepsilon$$

for sufficiently large l, which implies the properties b) and c).

4.27. Each finite word w in r is contained in some word A_n. Verify that r has the form $B_1B_2\cdots$ where each word B_i has length 2^{n+3} and contains A_n. Therefore, any word in r of length 2^{n+4} contains A_n, and hence w.

We mention some curious properties of sequences s of folds (see problem 4.24) and the Rudin-Shapiro sequence r.

If s is defined by the equalities $s_{2^k} = k \bmod 2$, $k = 0, 1, 2, \dots$, then r_{n+1} coincides with $(\sum_{1\le j\le 2n} s_j - n) \bmod 2$, $n \in \mathbb{N}$ $(r_1 = 0)$. An interesting polygonal curve on the plane is also connected with these sequences (see Figure 20, the solid curve). It is composed of alternating "horizontal" and "vertical" components of equal length, and the turns (by 90°) of the polygonal curve to the right and to the left correspond, as in problem 4.24, to the 0's and 1's of the sequence s. This polygonal curve is connected with the sequence r as follows: the numbers $r_1, r_2, \dots$ correspond to the horizontal components of the polygonal curve, taken in order, where $r_n = 0$ if the nth horizontal component goes from left to right, and $r_n = 1$ if from right to left (the sequence of "up-down" passages of vertical components is also reminiscent of the Rudin-Shapiro sequence).

With the polygonal curve corresponding to the sequence s it is possible to associate a new polygonal curve, pictured in Figure 20 by the dotted line. It

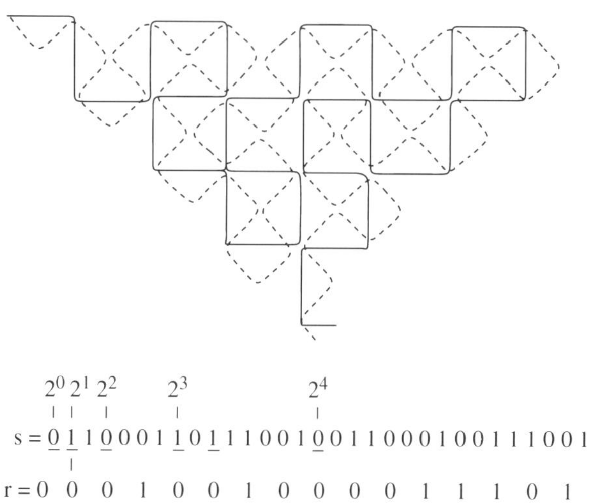

FIGURE 20

is obtained from the solid line as follows: we replace each component by two new components serving as the legs of an isosceles triangle with the original component as hypotenuse. Further, the vertex of this triangle deviates from the solid polygonal curve alternately to the right and to the left. Note that the new polygonal curve was obtained from the old one by rotation and contraction by a factor $\sqrt{2}$. From the second polygonal curve it is possible to construct a third one in an analogous way, then a fourth, and so on. In the limit such a sequence of polygonal curves fills a sector in the plane bounded by two sides of a 45° angle, that is, they determine a curve of Peano type. Try to carry out an independent investigation of curves arising in a similar way from sequences of folds with a different alternation of 0's and 1's in the sequence $\{s_{2^n}\}$ (for example, $s_{2^n} \equiv 1$), and also of the corresponding limit sets on the plane. More details on the questions touched upon in problems 4.24–4.26 can be found in the papers [35] and [42].

CHAPTER VIII

Lebesgue Measure and the Lebesgue Integral

§1. Lebesgue measure

1.1. Use the equality $\bigcap_{1\le k\le m} E_k = (0, 1)\setminus\bigcup_{1\le k\le m} E'_k$, where $E'_k = (0, 1)\setminus E_k$. Prove that $\lambda(\bigcup_{1\le k\le m} E'_k) < 1$.

1.2. Consider the sets $E_k = z_k^{-1}E$ $(k = 1, \dots, m)$ and, using problem 1.1, show that $\bigcap_{1\le k\le m} E_k \ne \varnothing$. Take a point z_0 in the set $\bigcap_{1\le k\le m} E_k$.

1.3. a) Partition the space $\mathbb{R}^m$ into cubes with vertices at the points of the lattice $\mathbb{Z}^m$: $\mathbb{R}^m = \bigcup_{l\in\mathbb{Z}^m} Q_l$, where $Q_l = \{l + y | y \in [0, 1)^m\}$. Then "translate" all the sets $E \cap Q_l$ into the cube[1] $[0, 1)^m$, that is, consider the sets $E_l = \{x - l | x \in E \cap Q_l\}$. Since $\sum_l \lambda_m(E_l) = \sum_l \lambda_m(E\cap Q_l) = \lambda_m(E) > 1$, the sets E_l cannot be pairwise disjoint. Consequently, there are distinct vectors $l', l'' \in \mathbb{Z}$ such that $E_{l'} \cap E_{l''} \ne \varnothing$. Let $a \in E_{l'} \cap E_{l''}$. Then $a = x' - l' = x'' = l''$, where $x', x'' \in E$. Thus, $0 \ne x' - x'' = l' - l'' \in \mathbb{Z}^m$.

b) Consider the set $E = \{x/2 | x \in V\}$ and use a).

c) Consider the set $E = \{x/2 | x \in V\}$, and, arguing by analogy with the solution of a), show that there is a point belonging to at least $N + 1$ of the sets E_l. Deduce from this that E contains distinct points $x_1, \dots, x_{N+1}$ such that $x_k - x_j \in \mathbb{Z}^m$.

Consider an index s such that x_s does not belong to the convex hull of the remaining points. Verify that the points $\pm(x_k - x_s)$ with $k \ne s$ are all distinct, and hence are what is required.

1.5. Prove that the measure of the set of points in $(0, 1)$ with 0 not among the first n digits of their decimal expansions is equal to $(9/10)^n$.

1.6. a) Consider sets E_{n_k} such that $\sum(1 - \lambda(E_{n_k})) < 1$. Use the same idea as in the solution of problem 1.1.

b) Consider the sets $E_n \subset (0, 1)$ consisting of the points with nth decimal digits nonzero, and verify that $\lambda(\bigcap_{k=1}^m E_{n_k}) = (9/10)^m$ (cf. the solution of problem 1.5).

1.7. Consider the sets $H_q = \bigcup_{p=0}^q ((p - \alpha_q)/q, (p + \alpha_q)/q)$ $(q = 1, 2, \dots)$ and verify that $A \subset \bigcup_{q\ge m} H_q$ for any $m \in \mathbb{N}$.

1.8. Since Lebesgue measure is regular, there is an open set $G \supset E$ such

[1] In what follows we shall call *cubic cells* the translator of cubes of the form $[0, a)^m$.

that

$$\theta\lambda(G) < \lambda(E). \tag{$*$}$$

Let $G = \bigcup \Delta_n$, where Δ_n are disjoint cubic cells. It follows from $(*)$ that $\theta \sum \lambda(\Delta_n) < \sum \lambda(E \cap \Delta_n)$. Therefore, for at least one index $n = n_0$ the inequality $\theta\lambda(\Delta_{n_0}) < \lambda(E \cap \Delta_{n_0})$ must hold. Setting $\Delta = \Delta_{n_0}$, we get the required result.

1.9. Let $H = E - E$, and let Δ be an interval such that $3\lambda(\Delta)/4 < \lambda(E \cap \Delta)$ (see problem 1.8). Prove that $H \supset (-\lambda(\Delta)/2, \lambda(\Delta)/2)$.

Let $|x| < \lambda(\Delta)/2$. Verify that $(E\cap\Delta)\cap(x+E\cap\Delta) \neq \varnothing$. Indeed, otherwise,

$$\begin{aligned} \lambda((E \cap \Delta) \cup (x + E \cap \Delta)) &= \lambda(E \cap \Delta) + \lambda(x + E \cap \Delta) \\ &= 2\lambda(E \cap \Delta) > 3\lambda(\Delta)/2. \end{aligned} \tag{$*$}$$

On the other hand, since $|x| < \lambda(\Delta)/2$, we have that $\lambda(\Delta\cup(x+\Delta)) < 3\lambda(\Delta)/2$ and hence $\lambda((E\cap\Delta)\cup(x+E\cap\Delta)) \leq \lambda(\Delta\cup(x+\Delta)) \leq 3\lambda(\Delta)/2$, which contradicts the inequality $(*)$.

Let $y \in (E\cap\Delta)\cap(x+E\cap\Delta)$. Since $y \in x+E\cap\Delta$, there is a point $x' \in E\cap\Delta$ such that $y = x + x'$. Thus, $x = y - x'$, where $y, x' \in E \cap \Delta \subset E$, that is, $x \in H$. This proves the inclusion $(-\lambda(\Delta)/2, \lambda(\Delta)/2) \subset H$.

b) This assertion follows immediately from a).

1.10. Show with the help of problem 1.9c) that $\operatorname{Int} E \neq \varnothing$, and use the result of problem I.1.19.

1.11. a) If $\lambda(E_\nu) = 0$, then $\operatorname{Int} E_\nu = \varnothing$, and hence (see problem I.1.34) $n_{k+1} > n_k + 1$ infinitely many times. But if $n_{k_j+1} > n_{k_j} + 1$ for $j \in \mathbb{N}$, then in the binary expansions of the points in E_ν the digits at the places with indices $n_{k_j} + 1$ are equal to 0, and this implies that $\lambda(E_\nu) = 0$ (cf. the solution of problem 1.5).

b) Verify that if $n_{k_j+1} - n_{k_j} \geq 2 + \log_2 m$, then in the binary expansions of the points in $E^{(m)}$ the digits having indices $n_{k_j} + 1$ are equal to 0.

c) Use the representation

$$H = \bigcup_{m,n=1}^{\infty} \frac{1}{n}(E^{(2m)} - ma), \quad \text{where } a = \sum_{k\geq 1} 2^{-n_k}.$$

1.12. Verify that for any positive integer n the set E can be covered by 2^{n+1} intervals of length $1/n!$.

1.14. It follows from the definition of the set E of Cantor type on $[0, 1]$ with defining sequence $\{l_n\}_{n\geq 0}$ that $\lambda(E) + \sum_{n\geq 0} 2^n(l_n - 2l_{n+1}) = 1$, where $l_0 = 1$. Therefore, $\lambda(E) > 0$ if and only if $\sum_{n\geq 0} 2^n(l_n - 2l_{n+1}) < 1$. Since the last sum can be arbitrarily small, the measure of a set of Cantor type constructed on $[0, 1]$ can be arbitrarily close to 1.

1.15. Using the hint for problem 1.14, prove the following.

LEMMA. *Let Δ be an arbitrary finite (nonempty) open interval, and let $0 < \theta < 1$. There exists a compact set $K \subset \Delta$ such that:*

a) $\operatorname{Int} K = \varnothing$; b) $\lambda(K) = \theta\lambda(\Delta)$; c) $\Delta \backslash K$ *consists of disjoint open intervals with lengths not exceeding half the length of* Δ.

To get the set A required in the problem, carry out the following construction. Let $\Delta_0 = (0, 1)$, $\theta_j > 0$, and $\sum_{j\geq 0} \theta_j < 1$. Use the lemma (with $\theta = \theta_0$) to construct a compact subset K_0 of Δ_0. Let $\Delta_n^{(1)}$ $(n \in \mathbb{N})$ be the open intervals making up the set $\Delta_0 \backslash K_0$, and let $K_n^{(1)}$ be compact subsets of these intervals constructed according to the lemma (with $\theta = \theta_1$). Next, let $\Delta_n^{(2)}$ $(n \in \mathbb{N})$ be the open intervals making up the set $\Delta_0 \backslash (K_0 \cup \bigcup_{n\geq 1} K_n^{(1)})$, and let $K_n^{(2)}$ be compact subsets of these intervals constructed according to the lemma (with $\theta = \theta_2$). Continuing this process, use induction to get a family of sets $\{K_n^{(j)}\}_{n,j} \geq 1$.

Let $A = K_0 \cup \bigcup_{j=1}^{\infty} \bigcup_{n=1}^{\infty} K_n^{(j)}$. Verify that

$$\lambda(\Delta_0 \backslash A) = \prod_{j=0}^{\infty} (1 - \theta_j). \qquad (*)$$

Show that any (nonempty) interval $\Delta \subset \Delta_0$ contains some interval $\Delta_m^{(l)}$, and hence

$$\Delta \cap A \supset \Delta_m^{(l)} \cap A \supset K_m^{(l)} \quad \text{and} \quad \lambda(\Delta \cap A) \geq \lambda(K_m^{(l)}) > 0.$$

Verify, moreover, that

$$\Delta \backslash A \supset \Delta_m^{(l)} \backslash \left(K_m^{(l)} \cup \bigcup_{j=l+1}^{\infty} \bigcup_{n=1}^{\infty} K_n^{(j)} \right),$$

and, replacing Δ by $\Delta_m^{(l)}$ and A by $A' = K_m^l \cup \bigcup_{j=l+1}^{\infty} \bigcup_{n=1}^{\infty} K_n^{(j)}$, prove that the equality $\lambda(\Delta_m^{(l)} \backslash A') = \lambda(\Delta_m^{(l)}) \cdot \prod_{j=l}^{\infty} (1 - \theta_j)$ holds together with the equality $(*)$, which implies that $\lambda(\Delta \backslash A) > 0$.

1.16. Assume first that A is a convex polygon with vertices $M_1, \ldots, M_n$ and let L be the length of the boundary polygonal curve. Together with each segment $M_j M_{j=1}$ $(j = 1, 2, \ldots, n,\ M_{n+1} = M_1)$ of the boundary polygonal curve consider the open rectangle P_j lying outside A such that the length of the side perpendicular to $M_j M_{j+1}$ is equal to ε. It is clear that

$$A_\varepsilon = A \cup \bigcup_{j=1}^{n} P_j \cup \bigcup_{j=1}^{n} S_j,$$

where S_j is the sector of the disk $B(M_j, \varepsilon)$ included between the rectangles P_{j-1} and P_j for $j > 1$, and between P_1 and P_n for $j = 1$. It is easy to see that the sum of the central angles corresponding to these sectors is equal to 2π. Therefore,

$$\begin{aligned} \lambda_2(A_\varepsilon) &= \lambda_2(A) + \sum_{1\leq j\leq n} \lambda_2(P_j) + \sum_{1\leq j\leq n} \lambda_2(S_j) \\ &= \lambda_2(A) + L\varepsilon + \pi\varepsilon^2. \end{aligned}$$

The case of an arbitrary compact convex set is exhausted by approximating it by convex polygons.

1.17. Obviously, it suffices to consider the case when $\lambda_n(G) < \infty$. Let the set G be represented as a union of pairwise disjoint cubic cells: $G = \bigcup_{j=1}^{\infty} Q_j$. In each cell we inscribe a ball $B_j^{(0)}$. It is clear that $\lambda_n(B_j^{(0)})/\lambda_n(Q_j) = \alpha_n/2^n$, where $\alpha_n = \lambda_n(B(0, 1))$. Therefore,

$$\lambda_n\left(G\setminus\bigcup_{j=1}^{\infty} B_j^{(0)}\right) = \sum \lambda_n(Q_j\setminus B_j^{(0)})$$
$$= (1 - \alpha_n/2^n)\lambda_n(G) = q\lambda_n(G),$$

where $q = 1 - \alpha_n 2^{-n} < 1$. Let $G_1 = G\setminus\bigcup_{j=1}^{\infty} \overline{B_j^{(0)}}$. The set G_1 is open, and by repeating the procedure described above, we get balls $B_j^{(1)}$ such that

$$\lambda_n\left(G_1\setminus\bigcup_{j=1}^{\infty} B_j^{(1)}\right) = q\lambda_n(G_1) = q^2\lambda_n(G).$$

Induction gives a family $\{B_j^{(k)}\}_{j,k=1}^{\infty}$ of disjoint balls such that

$$\lambda\left(G\setminus\bigcup_{k=0}^{\infty}\bigcup_{j=1}^{\infty} B_j^{(k)}\right) = 0.$$

1.18. Denote by σ Lebesgue measure on S^1. If measurability of at least one of the sets E_n is assumed, then in view of their congruence and the invariance of the measure σ with respect to rotations all the sets will be measurable, and the equality $\sigma(E_n) = \sigma(E_m)$ is valid for any $n, m \in \mathbb{N}$. Using the countable additivity of the measure, we get that

$$\sigma(S^1) = \sigma\left(\bigcup_{n=1}^{\infty} E_n\right) = \sum_{n\geq 1} \sigma(E_n) = \sum_{n\geq 1} \sigma(E_1).$$

Thus, $0 < \sum_{n\geq 1} \sigma(E_1) < \infty$, which is impossible both in the case when $\sigma(E_1) > 0$ and in the case when $\sigma(E_1) = 0$.

1.19. Partition the numbers in $[0, 1)$ into equivalence classes by regarding two numbers as equivalent if their difference is rational. Let C be a set having precisely one point in each equivalence class, and let C_θ be the image of C under translation by θ modulo 1. We number all the rational numbers in $[0, 1)$ in the sequence $\{\theta_n\}$ and let $E_n = C_{\theta_n}$. Since translation modulo 1 preserves the measurability and the measure of a set, the sets E_n are measurable or not measurable simultaneously, and they have the same measures if they are measurable. Assuming that the E_n are measurable and arguing as in the solution of problem 1.18, we again arrive at the impossible inequality $0 < \sum_{n\geq 1} \lambda(E_1) < \infty$.

1.20. It will be assumed without loss of generality that $E \subset (0, 1)$. Let the E_n be the sets constructed in the solution of problem 1.19, and let $H_n =$

$E \cap E_n$. It is clear that $E = \bigcup_{n=1}^{\infty} H_n$. Assume that all the sets H_n are measurable. Then at least one of them has positive measure, and hence contains (distinct) points separated by a rational distance (see problem 1.9b)), which contradicts the construction of the sets E_n.

1.21. Show that if $x \in (0, 1)\backslash\mathbb{Q}$, then $x \in E$ if and only if $1 - x \notin E$. Prove the following more general result: if $x \in \Delta\backslash\mathbb{Q}$, where Δ is an arbitrary interval of the form $(k/2^n, (k+1)/2^n)$, $n = 0, 1, 2, \dots, k = 0, 1, \dots, 2^n - 1$, and x' is the point symmetric to x with respect to the midpoint of Δ, then $x \in E$ if and only if $x' \notin E$.

Assuming that E is measurable, deduce from this property that $\lambda(E \cap \sigma) = \lambda(\sigma)/2$ for any open interval $\sigma \subset (0, 1)$. The latter contradicts the result of problem 1.8.

1.22. Let $Q = [0, 1)^m$, $Q_l = l + Q$, and $E_l = -l + E \cap Q_l$ $(l \in \mathbb{Z}^m)$. Verify that $\bigcup_{l \in \mathbb{Z}^m} E_l = Q$ and $E_l \cap E_{l'} = \varnothing$ for $l \neq l'$.

1.23. Let $c > 0$ and $P = [-c, c)^m$. It suffices to verify that $\lambda_m(P \backslash \bigcup_{a \in A}(a + E)) = 0$. Let ε be an arbitrary positive number, and let Q be a cube such that (see (problem 1.8)

$$\lambda_m(Q\backslash E) < \varepsilon\lambda_m(Q). \tag{1}$$

Show that there exist points $a_1, a_2, \dots, a_N \in A$ such that

$$P \subset \bigcup_{1 \le n \le N} (a_n + Q) \quad \text{and} \quad \sum_{1 \le n \le N} \lambda_m(a_n + Q) \le 2\lambda_m(P). \tag{2}$$

Then

$$P \backslash \bigcup_{a \in A} (a + E) \subset \bigcup_{1 \le n \le N} (a_n + Q)\backslash(a_n + E), \tag{3}$$

and in view of (1) and (2)

$$\begin{aligned} \sum_{1 \le n \le N} \lambda_m((a_n + Q)\backslash(a_n + E)) &< \sum_{1 \le n \le N} \varepsilon\lambda_m(a_n + Q) \\ &\le 2\varepsilon\lambda_m(P). \end{aligned} \tag{4}$$

Since ε is arbitrary, (3) and (4) give us the equality

$$\lambda_m\left(P \backslash \bigcup_{a \in A}(a + E)\right) = 0.$$

§2. Measurable functions

2.1. Prove that for some $\varepsilon > 0$ the set $E_\varepsilon = \{x \in E \mid |f(x)| \le 1 - \varepsilon\}$ has positive measure, and consider the functions $f_1 = f + \varepsilon\chi_{E_\varepsilon}$ and $f_2 = f - \varepsilon\chi E_\varepsilon$.

2.2. Consider the function $f(x) = \sum 2^{-n}\varphi_0(x - r_n)$, where $\varphi_0 = \chi_{[0, +\infty)}$, and $\{r_n\}$ is the set of rational numbers, enumerated in an arbitrary way.

2.3. Consider the characteristic function of the set in problem 1.15.

2.4. b) Find the lengths of the images of the intervals of nth rank used in the construction of the Cantor set (see problem I.1.27).

c) Use the result of problem 1.20, which ensures the existence of a non-measurable subset of the set $g(K)$.

2.13. We prove the first of the required equalities. Suppose that $c_1 < c_2 < \cdots < c_n$, and let $c_{n+1} = +\infty$. Since the function f is strictly decreasing from $+\infty$ to $-\infty$ on each of the open intervals (c_k, c_{k+1}) $(k = 1, 2, \ldots, n-1)$ and from $+\infty$ to 0 on $(c_n, +\infty)$, the set $E_t = \{x \in \mathbb{R} | f(x) > t\}$ is the union of the intervals (c_k, x_k), where x_k is a point such that $f(x_k) = t$, $c_k < x_k < c_{k+1}$. Thus,

$$\lambda(E_t) = \sum_{1 \le k \le n} (x_k - c_k) = \sum_{1 \le k \le n} x_k - \sum_{1 \le k \le n} c_k.$$

To find the sum $\sum_{1 \le k \le n} x_k$ note that the points x_k are roots of the equation $f(x) = t$ or of the equivalent equation

$$\sum_{1 \le k \le n} a_k \frac{P(x)}{x - c_k} = tP(x), \quad \text{where } P(x) = \prod_{1 \le k \le n} (x - c_k).$$

The last equation can obviously be written in the form

$$tx^n - \left(t \sum_{1 \le k \le n} c_k + A \right) x^{n-1} + Q(x) = 0,$$

where the degree of the polynomial $Q(x)$ is less than $n - 1$. By Vieta's theorem, $\sum_{1 \le k \le n} x_k = \sum_{1 \le k \le n} c_k + A/t$, which leads to the required result. The second equality can be proved analogously.

§3. Integrable functions

3.1. It is clear that

$$\begin{aligned}
\mathscr{I} &= \int_0^1 \sum_{1 \le l \le N} \chi^3_{E_l}(x)\,dx \\
&\quad + 3 \int_0^1 \sum_{1 \le l < j \le N} (\chi^2_{E_l}(x)\chi_{E_j}(x) + \chi_{E_l}(x)\chi^2_{E_j}(x))\,dx \\
&\quad + 6 \int_0^1 \sum_{1 \le l < j < k \le N} \chi_{E_l}(x)\chi_{E_j}(x)\chi_{E_k}(x)\,dx \\
&= S_1 + 6S_2 + 6S_3,
\end{aligned}$$

where $S_3 = \sum_{1 \le l < j < k \le N} \lambda(E_l \cap E_j \cap E_k)$. To compute S_3 note that since $E_l \cup E_j \cup E_k = [0, 1]$ for $1 \le l < j < k \le N$, it follows that

$$\chi_{[0,1]} = \chi_{E_l} + \chi_{E_j} + \chi_{E_k} - \chi_{E_l}\chi_{E_j} - \chi_{E_l}\chi_{E_k} - \chi_{E_j}\chi_{E_k} + \chi_{E_l}\chi_{E_j}\chi_{E_k}.$$

Consequently,

$$\begin{aligned}1 = \lambda(E_l) + \lambda(E_j) + \lambda(E_k) - \lambda(E_l \cap E_j) \\ - \lambda(E_l \cap E_k) - \lambda(E_j \cap E_k) + \lambda(E_l \cap E_j \cap E_k).\end{aligned}$$

Summing these equalities over $1 \le l < j < k \le N$, we find S_3.

3.2. It follows from the condition of the problem that $\sum_{1\le j\le N} \chi_{E_j}(x) \ge k$ for any $x \in [0, 1]$.

3.5. a) Prove the convergence of the series $\sum \int_\Delta 2^{-n}|x-r_n|^{-1/2}\,dx$, where Δ is an arbitrary finite interval.

b) Let $\varphi(x) = \sum 2^{-n}|x-r_n|^{-1/2}$, where $\{r_n\}$ is the sequence of all rational numbers, enumerated arbitrarily. Consider the function φ^2.

3.6. Consider the function

$$f(x) = \begin{cases} 0 & \text{for } x = 0, \\ x^2 \sin(1/x^2) & \text{for } x \ne 0. \end{cases}$$

3.10. a) Let $E_x = \{x - y | y \in E\}$. It is clear that $\int_E \|x-y\|^{-p}\,dy = \int_{E_x} \|z\|^{-p}\,dz$. Let $A = E_x \cap B(0, r)$ and $C = E_x \setminus A$. Then $\lambda_m(C) = \lambda_m(B(0, r)\setminus A)$ and

$$\sup\{\|z\|^{-p} | z \in C\} \le r^{-p} = \inf\{\|z\|^{-p} | z \in B(0, r)\}.$$

Therefore, $\int_C \|z\|^{-p}\,dz \le r^{-p}\lambda_m(C) \le \int_{B(0,r)\setminus A} \|z\|^{-p}\,dz$ and

$$\begin{aligned}\int_E \|x-y\|^{-p}\,dy &= \int_A \|z\|^{-p}\,dz + \int_C \|z\|^{-p}\,dz \\ &\le \int_A \|z\|^{-p}\,dz + \int_{B(0,r)\setminus A} \|z\|^{-p}\,dz \\ &= \int_{B(0,r)} \|z\|^{-p}\,dz.\end{aligned}$$

b) Show that $|\int_E e^{ix}\,dx| = \int_E \cos(x-\alpha)\,dx$ for some $\alpha \in [0, 2\pi]$ and use the same idea as in a).

c) Show that

$$\left|\iint_E \frac{dxdy}{x+iy}\right| = \iint_{E'} \frac{xdxdy}{x^2+y^2},$$

where E' is a set obtained from E by a rotation. To estimate the integral $\iint_{E'} x\,dxdy/(x^2+y^2)$ consider a set $E_t = \{(x, y) \in \mathbb{R}^2 | x/(x^2+y^2) \ge t\}$, on which the function $x/(x^2+y^2)$ is large and choose t such that the measure of E_t is equal to that of E'. Prove as in a) that

$$\iint_{E'} \frac{xdxdy}{x^2+y^2} \le \iint_{E_t} \frac{xdxdy}{x^2+y^2}$$

for such a choice of t, and compute the last integral.

3.15. It is clear that the integral $\int_E g(x)\ln f(x)\,dx$ exists. In the case when it is finite use the inequality $|(e^{pt}-1)/p| \leq |t| + e^{|t|}$ $(t \in \mathbb{R},\ 0 < p < 1)$ to prove that

$$\int_E \frac{f^p(x)-1}{p} g(x)\,dx \underset{p\to+0}{\longrightarrow} \int_E g(x)\ln f(x)\,dx.$$

If $\int_E g(x)\ln f(x)\,dx = -\infty$, then consider the functions $f_\delta = \max(\delta, f)$, where $0 < \delta < 1$, and pass in the inequality

$$\left(\int_E g(x) f^p(x)\,dx\right)^{1/p} \leq \left(\int_E g(x) f_\delta^p(x)\,dx\right)^{1/p}$$

first to the limit superior as $p \to +0$, and then to the limit as $\delta \to +0$.

3.16. b) Use the pairwise orthogonality of the functions $\varepsilon_k - 1/2$ $(k = 1, 2, \ldots, n)$.

3.20. Represent the set $(a, b)\setminus K$ as a union $\bigcup_k (\alpha_k, \beta_k)$ of disjoint open intervals. Then

$$\mathcal{I}_s(x) = \sum_k \int_{\alpha_k}^{\beta_k} \frac{\rho^s(y)}{|x-y|^{s+1}}\,dy.$$

Therefore,

$$\begin{aligned}\int_K \mathcal{I}_s(x)\,dx &= \sum_k \int_K \left(\int_{\alpha_k}^{\beta_k} \frac{\rho^s(y)}{|x-y|^{s+1}}\,dy\right) dx \\ &= \sum_k \int_{\alpha_k}^{\beta_k} \rho^s(y) \left(\int_K \frac{dx}{|x-y|^{s+1}}\right) dy.\end{aligned} \tag{1}$$

Let us estimate the inside integral on the right-hand side of (1). Assuming that $y \in (\alpha_k, \beta_k)$, we get that

$$\begin{aligned}\int_K \frac{dx}{|x-y|^{s+1}} &\leq \int_a^{\alpha_k} \frac{dx}{|x-y|^{s+1}} + \int_{\beta_k}^b \frac{dx}{|x-y|^{s+1}} \\ &\leq \frac{1}{s}\left(\frac{1}{(y-\alpha_k)^s} + \frac{1}{(\beta_k - y)^s}\right) \leq \frac{2}{s}\rho^{-s}(y).\end{aligned}$$

On the basis of this inequality it follows from (1) that

$$\int_K \mathcal{I}_s(x)\,dx \leq \sum_k \int_{\alpha_k}^{\beta_k} \frac{2}{s}\,dy \leq \frac{2}{s}(b-a) < \infty.$$

3.21. It is clear that a) follows from b). To prove b) partition the cube $[-a, a]^m$ into the $2m$ pyramids with the faces of the cube as bases and with vertices at the origin of coordinates. To compute the integral over each of these pyramids use Fubini's theorem.

3.22. To avoid the difficulties associated with verifying that the function is measurable on subsets of dimension less than n it will be assumed that the function is continuous. (The general case can be exhausted with the help of approximation.) By using polar and spherical coordinates it is easy to establish the induction base. Suppose that the formulas a) and b) are valid for S^{n-2} and $\mathbb{R}^{n-1}$, respectively. Replacing $f(x)$ by $f(x)\chi_{[0,1]}(\|x\|)$, where $\chi_{[0,1]}$ is the characteristic function of $[0, 1]$, we see that

$$\int_{B^{n-1}} f(x)\,dx = \int_0^1 t^{n-2}\left\{\int_{S^{n-2}} f(t\omega)\,d\mu_{n-2}(\omega)\right\} dt. \tag{1}$$

We proceed to the proof of the equality in a). Let

$$S_+^{n-1} = \{(x_1, \dots, x_n) \in S^{n-1} | x_n \geq 0\},$$
$$S_-^{n-1} = \{(x_1, \dots, x_n) \in S^{n-1} | x_n \leq 0\}.$$

We prove that

$$\int_{S_+^{n-1}} f(x_1, \dots, x_{n-1}, x_n)\,d\mu_{n-1}(x)$$
$$= \int_0^{\pi/2} \sin^{n-2}\theta \left\{\int_{S^{n-2}} f(u_1\sin\theta, \dots, u_{n-1}\sin\theta, \cos\theta)\,d\mu_{n-2}(u)\right\} d\theta. \tag{2}$$

Indeed,

$$\int_{S_+^{n-1}} f(x_1, \dots, x_{n-1}, x_n)\,d\mu_{n-1}(x)$$
$$= \int \cdots \int_{B^{n-1}} f(y_1, \dots, y_{n-1}, \sqrt{1-\|y\|^2})\frac{dy_1\cdots dy_{n-1}}{\sqrt{1-\|y\|^2}}.$$

Using (1), we get that

$$\int_{S_+^{n-1}} f(x_1, \dots, x_{n-1}, x_n)\,d\mu_{n-1}(x)$$
$$= \int_0^1 \frac{t^{n-2}}{\sqrt{1-t^2}}\left\{\int_{S^{n-2}} f(t\omega_1, \dots, t\omega_{n-1}, \sqrt{1-t^2})\,d\mu_{n-2}(\omega)\right\} dt.$$

After the change of variable $t = \sin\theta$ we arrive at (2). It can be proved similarly that

$$\int_{S_-^{n-1}} f(x_1, \dots, x_{n-1}, x_n)\,d\mu_{n-1}(x)$$
$$= \int_{\pi/2}^{\pi} \sin^{n-2}\theta \left\{\int_{S^{n-2}} f(u_1\sin\theta, \dots, u_{n-1}\sin\theta, \cos\theta)\,d\mu_{n-2}(u)\right\} d\theta.$$

Let us proceed to the proof of b). By Fubini's theorem,

$$\int_{\mathbb{R}^n} f(x)\,dx = \int_{\mathbb{R}^1}\left\{\int \cdots \int_{\mathbb{R}^{n-1}} f(x_1, \dots, x_{n-1}, x_n)\,dx_1\cdots dx_{n-1}\right\} dx_n.$$

We transform the inside integral by using the induction hypothesis, after which we get that

$$\int_{\mathbb{R}^n} f(x)\,dx = \int_{\mathbb{R}^1}\left\{\int_0^\infty r^{n-2}\left\{\int_{S^{n-2}} f(ru_1, \dots, ru_{n-1}, x_n)\,d\mu_{n-2}(u)\right\} dr\right\} dx_n.$$

We now pass to the polar coordinates $x_n = t\cos\theta$, $r = t\sin\theta$ in the half-plane (x_n, r), $r \geq 0$. This gives us that

$$\int_{\mathbb{R}^n} f(x)\,dx = \int_0^\infty t^{n-1}\left\{\int_0^\pi \sin^{n-2}\theta\left\{\int_{S^{n-2}} f\left(tu_1\sin\theta, \dots, tu_{n-1}\sin\theta, t\cos\theta\right) d\mu_{n-2}(u)\right\} d\theta \int_0^\pi\right\} dt.$$

Using the equality a) already proved, we see that

$$\int_0^\pi \sin^{n-2}\theta\left\{\int_{S^{n-2}} f(tu_1\sin\theta, \dots, tu_{n-1}\sin\theta, t\cos\theta)\,d\mu_{n-2}(u)\right\} d\theta = \int_{S^{n-1}} f(t\omega)\,d\mu_{n-1}(\omega),$$

and thus

$$\int_{\mathbb{R}^n} f(x)\,dx = \int_0^\infty t^{n-1}\left\{\int_{S^{n-1}} f(t\omega)\,d\mu_{n-1}(\omega)\right\} dt.$$

3.23. a) Apply the formula b) in problem 3.22 to the characteristic function of the set T, changing the order of integration on the right-hand side.

b) Use a).

3.24. Use the formula given before problem 3.24, assuming that f is the characteristic function of the set E.

3.25. Using the formula before problem 3.24, we get that

$$I_n(r) = (2\pi)^{-n/2}\int_0^r n\alpha_n e^{-t^2/2}t^{n-1}\,dt = \frac{1}{\Gamma(n/2)}\int_0^r \left(\frac{t^2}{2}\right)^{n/2-1} e^{-t^2/2}\,d\left(\frac{t^2}{2}\right).$$

The function $I_n(r)$ is called the *χ-square distribution* with n degrees of freedom, and is frequently used in mathematical statistics.

3.26. By Fubini's theorem,

$$\gamma_4(K_a) = \frac{1}{(2\pi)^2}\int_{-\infty}^{+\infty} e^{-x^2/2}\left\{\iiint_{(K_a)_x} e^{-(y^2+z^2+t^2)/2}\,dy\,dz\,dt\right\} dx,$$

where $(K_a)_x = \{(y, z, t) \in \mathbb{R}^3 | \sqrt{y^2+z^2+t^2} \le a|x|\}$. Therefore,

$$\gamma_4(K_a) = \frac{2}{(2\pi)^2}\int_0^\infty e^{-x^2/2}\left(\int_0^{ax} 4\pi r^2 e^{-r^2/2} dr\right) dx$$
$$= \frac{2}{\pi}\int_0^{\arctan a}\left\{\int_0^\infty \rho^3 \sin^2\varphi e^{-\rho^2/2} d\rho\right\} d\varphi.$$

We leave the rest of the computations to the reader.

3.27. In the section of the set T by the hyperplane $x_1 = x$ we get the set $T_x = \{(x_2, x_3, x_4) \in \mathbb{R}^3 | \sqrt{x_2^2+x_3^2+x_4^2} \in B_x\}$, where

$$B_x = \{y \in \mathbb{R}^1 | (x, y) \in B\} = [a - \sqrt{R^2-x^2}, a + \sqrt{R^2-x^2}].$$

Therefore,

$$\lambda_3(T_x) = \int_{a-\sqrt{R^2-x^2}}^{a+\sqrt{R^2-x^2}} 4\pi y^2\, dy,$$

and hence

$$\lambda_4(T) = \int_{-R}^{R} \lambda_3(T_x)\, dx = \int_{-R}^{R}\left\{\int_{a-\sqrt{R^2-x^2}}^{a+\sqrt{R^2-x^2}} 4\pi y^2\, dy\right\} dx.$$

We leave the final computations to the reader.

3.28. Let A' be the projection of A on the x_1-axis, and let $A_x = \{y | (x, y) \in A\}$. It is clear that $\lambda_4(T) = \int_{A'} \lambda_3(T_x)\, dx$, where $T_x = \{(x_2, x_3, x_4) | \sqrt{x_2^2+x_3^2+x_4^2} \in A_x\}$. Since $\lambda_3(T_x) = \int_{A_x} 4\pi y^2\, dy$, it follows that

$$\lambda_4(T) = \int_{A'}\left\{\int_{A_x} 4\pi y^2\, dy\right\} dx = 4\pi \iint_A y^2\, dx\, dy,$$

which is what was required.

3.29. Use the equalities

$$\lambda_n(T) = \int_{A'} \lambda_{n-1}(T_x)\, dx,$$
$$\gamma_n(T) = \frac{1}{\sqrt{2\sigma}}\int_{A'} e^{-x^2/2}\gamma_{n-1}(T_x)\, dx,$$

where A' is the projection of A on the x_1-axis, $A_x = \{y | (x, y) \in A\}$, and

$$T_x = \{(x_2, \dots, x_n\} \in \mathbb{R}^{n-1} | \sqrt{x_2^2 + \cdots + x_n^2} \in A_x\},$$

and show by using the spherical invariance of the set T_x that

$$\lambda_{n-1}(T_x) = \int_{A_x} \sigma_{n-2} y^{n-2}\, dy,$$
$$\gamma_{n-1}(T_x) = (2\pi)^{-(n-1)/2}\int_{A_x} \sigma_{n-2} y^{n-2} e^{-y^2/2}\, dy.$$

3.30. Let $V_n(t) = \lambda_n\{(x_1, \dots, x_n) \in \mathbb{R}^n \mid \sum_{1\le k\le n} |x_k|^p \le t^p\}$, $t > 0$. Verify that $V_n(t) = V_n(1)t^n$. Using the equality and Fubini's theorem, prove that $V_n(1) = 2V_{n-1}(1)\int_0^1 (1-t^p)^{(n-1/p} dt$. Expressing the integral in terms of the function Γ, prove the desired formula by induction.

3.31. a) Since the measure μ_{n-1} is rotation-invariant, it can be assumed that $a = \|a\|e_1$, where $e_1 = (1, 0, \dots, 0)$. Therefore

$$\begin{aligned}
\mathcal{I} &= \frac{\|a\|^p}{\sigma_{n-1}} \int_{S^{n-1}} |x_1|^p \, d\mu_{n-1}(x) \\
&= 2\|a\|^p \sigma_{n-1}^{-1} \int_{x_1^2+\cdots+x_{n-1}^2\le 1} |x_1|^p \frac{dx_1 \cdots dx_{n-1}}{\sqrt{1-x_1^2-\cdots-x_{n-1}^2}} \\
&= \|a\|^p \sigma_{n-1}^{-1} \int_{-1}^{+1} |x_1|^p \left\{ \int\cdots\int_{x_2^2+\cdots+x_{n-1}^2\le 1-x_1^2} \frac{2dx_2\cdots dx_{n-1}}{\sqrt{(1-x_1^2)-x_2^2-\cdots-x_{n-1}^2}} \right\} dx_1.
\end{aligned}$$

The inside integral on the right-hand side is the area of the sphere $S^{n-2}(\sqrt{1-x_1^2})$, and hence is equal to $\sigma_{n-2}(1-x_1^2)^{(n-2)/2}$. Consequently,

$$\begin{aligned}
\mathcal{I} &= \frac{\sigma_{n-2}}{\sigma_{n-1}} \|a\|^p \cdot 2\int_0^1 t^p (1-t^2)^{(n-2)/2} \, dt \\
&= \frac{\sigma_{n-2}}{\sigma_{n-1}} \|a\|^p \int_0^1 u^{(p+1)/2-1} (1-u)^{n/2-1} \, du \\
&= \frac{\sigma_{n-2}}{\sigma_{n-1}} \cdot \frac{\Gamma((p+1)/2)\cdot\Gamma(n/2)}{\Gamma((n+p+1)/2)} \cdot \|a\|^p.
\end{aligned}$$

Substituting the values of σ_{n-2} and σ_{n-1} in this equality, we get the desired result.

3.32. (See [36].) Since the measure is spherically invariant, we can assume that $a = e_1$ and $b = \cos\theta e_1 + \sin\theta e_2$, where $e_1 = (1, 0, 0, \dots, 0)$ and $e_2 = (0, 1, 0, \dots, 0)$. Thus, we should compute the integral $\mathcal{I} = \int_{S^{n-1}} f(x_1, x_2)\, d\mu_{n-1}(x)$, where $f(u, v) = (\operatorname{sgn} u)\cdot \operatorname{sgn}(u\cos\theta + v\sin\theta)$. It is clear that

$$\begin{aligned}
\mathcal{I} &= 2\int\cdots\int_{x_1^2+\cdots+x_{n-1}^2\le 1} f(x_1, x_2) \cdot \frac{dx_1\cdots dx_{n-1}}{\sqrt{1-x_1^2-\cdots-x_{n-1}^2}} \\
&= \iint_{x_1^2+x_2^2\le 1} \frac{f(x_1, x_2)}{\sqrt{1-x_1^2-x_2^2}} \\
&\quad \times \left\{ \int\cdots\int_{x_3^2+\cdots+x_{n-1}^2\le 1-x_1^2-x_2^2} \frac{2\sqrt{1-x_1^2-x_2^2}\, dx_3\cdots dx_{n-1}}{\sqrt{(1-x_1^2-x_2^2)-x_3^2-\cdots-x_{n-1}^2}} \right\} dx_1\, dx_2.
\end{aligned}$$

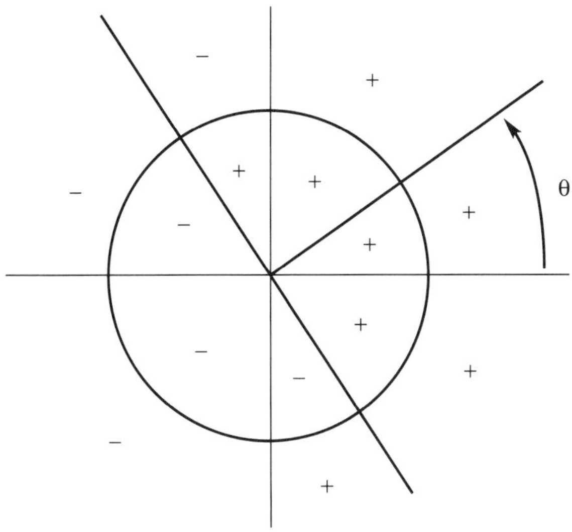

FIGURE 21

The inside integral on the right-hand side is none other than the area of the $(n-3)$-dimensional sphere of radius $\sqrt{1-x_1^2-x_2^2}$. Therefore,

$$\begin{aligned}\mathscr{I} &= \iint_{u^2+v^2\leq 1} f(u, v)\sigma_{n-3}(1-u^2-v^2)^{(n-4)/2}\,du\,dv \\ &= \sigma_{n-3}\int_0^{2\pi}\left\{\int_0^1 f(r\cos\varphi, r\sin\varphi)(1-r^2)^{(n-4)/2} r\,dr\right\} d\varphi.\end{aligned}$$

Obviously, $f(r\cos\varphi, r\sin\varphi) = f(\cos\varphi, \sin\varphi)$, and hence

$$\begin{aligned}\mathscr{I} &= \sigma_{n-3}\int_0^{2\pi} f(\cos\varphi, \sin\varphi)\,d\varphi\cdot\int_0^1 (1-r^2)^{(n-4)/2} r\,dr \\ &= \frac{\sigma_{n-3}}{n-2}\int_0^{2\pi} f(\cos\varphi, \sin\varphi)\,d\varphi \\ &= \frac{\sigma_{n-3}}{n-2}\int_0^{2\pi} (\operatorname{sgn}\cos\varphi)\cdot\operatorname{sgn}\cos(\varphi-\theta)\,d\varphi.\end{aligned}$$

We leave it to the reader to show that the last integral is equal to $2\pi - 4\theta$ (see Figure 21, where the sign of $\cos\varphi$ is indicated outside the disk and the sign of $\cos(\varphi-\theta)$ is indicated inside it). Therefore,

$$\mathscr{I} = \frac{2\pi}{n-2}\sigma_{n-3}\left(1-\frac{2}{\pi}\theta\right),$$

and since $\sigma_{n-1} = (2\pi/(n-2))\sigma_{n-3}$, we get the required result.

3.33. The assertion a) is a special case of problem 3.34. Using a) and the formula given before problem 3.24, we get that

$$I_n = \frac{1}{2}(2\pi)^{-n}(n\alpha_n)^2\int_0^\infty\int_0^\infty |r-\rho|(r\rho)^{n-1}e^{-(r^2+\rho^2)/2}\,dr\,d\rho.$$

Passing to polar coordinates in the integral on the right-hand side of the equality, we arrive at the equality

$$\begin{aligned}
I_n &= \frac{2}{2^n\Gamma^2(n/2)} \int_0^{\pi/2} |\cos\varphi - \sin\varphi|(\sin\varphi\cos\varphi)^{n-1}\,d\varphi \int_0^\infty u^{2n}\cdot e^{-u^2/2}\,du \\
&= \frac{\sqrt{2}\Gamma(n+1/2)}{2^n\Gamma^2(n/2)} \int_0^\pi |\cos\frac{\varphi}{2} - \sin\frac{\varphi}{2}| \sin^{n-1}\varphi\,d\varphi \\
&= \frac{2\sqrt{2}\Gamma(n+1/2)}{2^n\Gamma^2(n/2)} \int_0^{\pi/2} \frac{\sin^{n-1}\varphi\cos\varphi}{\sin(\varphi/2)+\cos(\varphi/2)}\,d\varphi\,.
\end{aligned}$$

It remains to observe that

$$\int_0^{\pi/2} \frac{\sin^{n-1}\varphi\cos\varphi}{\sin(\varphi/2)+\cos(\varphi/2)}\,d\varphi \underset{n\to\infty}{\sim} \frac{1}{n\sqrt{2}}$$

(see problem VI.2.8) and to use Stirling's formula.

3.34. Since $A \supset B(0,\varepsilon)$ for some $\varepsilon > 0$, it follows that $q_A(x) \le \|x\|/\varepsilon$, and hence

$$\begin{aligned}
\int_{\mathbb{R}^n}\int_{\mathbb{R}^n} |q_A(x) - q_A(y)|\,d\mu(x)\,d\mu(y) &\le \frac{1}{\varepsilon}\int_{\mathbb{R}^n}\int_{\mathbb{R}^n} (\|x\| + \|y\|)\,d\mu(x)\,d\mu(y) \\
&\le \frac{2}{\varepsilon}\int_{\mathbb{R}^n}\left(\sum_{1\le k\le n} |x_k|\right)d\mu(x) < \infty.
\end{aligned}$$

Let χ_{tA} be the characteristic function of the set tA $(t > 0)$. Then

$$\begin{aligned}
\int_0^\infty \mu(tA)[1-\mu(tA)]\,dt &= \int_0^\infty \left\{\int_{\mathbb{R}^n}\int_{\mathbb{R}^n} \chi_{tA}(x)[1-\chi_{tA}(y)]\,d\mu(x)\,d\mu(y)\right\}dt \\
&= \int_{\mathbb{R}^n}\int_{\mathbb{R}^n}\left\{\int_0^\infty \chi_{tA}(x)[1-\chi_{tA}(y)]\,dt\right\}d\mu(x)\,d\mu(y).
\end{aligned} \tag{1}$$

If $q_A(x) > q_A(y)$, then $\chi_{tA}(x)[1-\chi_{tA}(y)] = 0$; if $q_A(x) < q_A(y)$, then $\chi_{tA}(x)[1-\chi_{tA}(y)] = 1$ for $q_A(x) \le t \le q_A(y)$, and $\chi_{tA}(x)[1-\chi_{tA}(y)] = 0$ for $t \notin [q_A(x), q_A(y)]$. Therefore,

$$\int_0^\infty \chi_{tA}(x)[1-\chi_{tA}(y)]\,dt = \max(q_A(y) - q_A(x), 0).$$

With the help of the last equality we get from (1) that

$$\begin{aligned}
\int_0^\infty \mu(tA)[1-\mu(tA)]\,dt &= \int_{\mathbb{R}^n}\int_{\mathbb{R}^n} \max(q_A(y) - q_A(x), 0)\,d\mu(x)\,d\mu(y) \\
&= \frac{1}{2}\int_{\mathbb{R}^n}\int_{\mathbb{R}^n} |q_A(x) - q_A(y)|\,d\mu(x)\,d\mu(y).
\end{aligned}$$

3.35. Let $E \subset \mathbb{R}^m$ be a measurable set generating a tesselation (see problem 1.22). Prove that $\int_E f(x)\,dx = \int_Q f(x)\,dx$ (from this we get the assertions a) and b) for $E = a + Q$ and $E = A(Q)$). To verify this equality consider the sets $Q_l = l + Q$ and $E_l = -l + E \cap Q_l$ ($l \in \mathbb{Z}^m$) and show that

$$\bigcup_{l \in \mathbb{Z}^m} E_l = Q, \quad E_l \cap E_{l'} = \varnothing \qquad \text{for } l' \neq l.$$

Since f is periodic, it is clear that $\int_{E_l} f(x)\,dx = \int_{E \cap Q_l} f(x)\,dx$. It remains to use the countable additivity of the integral.

In the case when A is not an integer matrix consider the example

$$A = \begin{pmatrix} 1/2 & 0 \\ 0 & 2 \end{pmatrix}, \qquad f(x, y) = \sin(2\pi x) \quad (x, y \in \mathbb{R}).$$

3.36. Let $d = \operatorname{diam}(T)$. It can be assumed that the points $O = (0, \dots, 0)$ and $\mathscr{D} = (0, \dots, 0, d)$ belong to the set T. Let Q be the projection of T on the hyperplane $x_n = 0$. Consider the cone K with vertex at $\mathscr{D}$ and base Q. Since

$$\lambda_n(K) = \frac{d}{n}\lambda_{n-1}(Q) \geq \frac{d}{n}S,$$

it suffices to prove that $V \geq \lambda_n(K)$. To do this, show that the intersection of K with any straight line l through Q parallel to the x_n-axis is a segment of length not exceeding the length of the segment obtained by intersecting l with T. Constructing the plane passing through l and the x_n-axis, consider a point $M \in T$ of it that is farthest away from this axis and is located on the same side of it as l. Prove that segments of equal length are obtained by intersecting l with K and the triangle $O\mathscr{D}M$.

3.37. Verify that the convex hull of two ellipsoids obtained from each other by a translation contains an ellipsoid of larger volume. Deduce from this that the ellipsoid is centrally symmetric with respect to zero. It can be assumed without loss of generality that $\mathrm{Э} = \overline{B^n}$. In this case

$$\lambda_n(\mathrm{Э}) \leq \alpha_n \quad \text{if } \mathscr{E} \subset T \text{ and } \mathscr{E} \text{ is an ellipsoid.} \tag{1}$$

Assume that $x_0 \in T$, $\|x_0\| > \sqrt{n}$. It can obviously be assumed that $x_0 = (c, 0, \dots, 0)$, where $c = \|x_0\|$. Consider the convex hull T_1 of the ball $\overline{B^n}$ and the points $\pm x_0$. (The two-dimensional section S of the set T_1 by a plane through the x_1-axis is pictured in Figure 22.) Obviously, $T_1 \subset T$. Take a point a lying on the x_1-axis, $0 < a < c$, and consider the ellipse E determined by the inequality $(x_1^2/a^2) + (y^2/b^2) \leq 1$, where $b^2 = (c^2 - a^2)/(c^2 - 1)$, and the y-axis is taken in the plane of the section perpendicular to the x_1-axis. Verify that $E \subset S$. From this conclusion it follows that T_1 contains the ellipsoid $\mathscr{E}_a$ obtained by rotating E about the x_1-axis

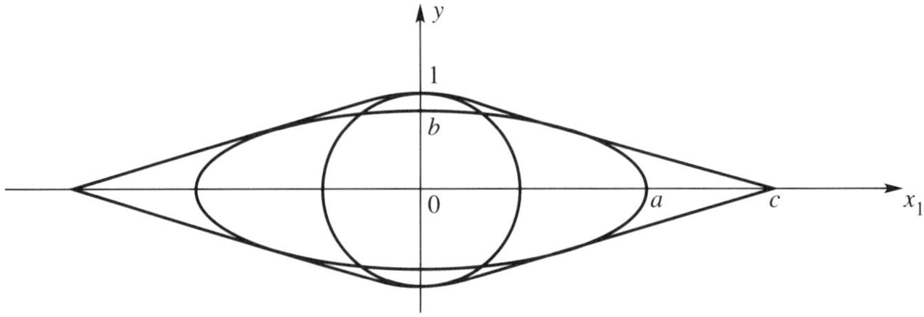

FIGURE 22

in the space $\mathbb{R}^n$. Let $V(a) = \lambda_n(\mathscr{E}_a)$, and

$$(\mathscr{E}_a)_x = \{(x_2, \dots, x_n) \in \mathbb{R}^n | (x, x_2, \dots, x_n) \in \mathscr{E}_a\}$$
$$= \left\{(x_2, \dots, x_n) \in \mathbb{R}^{n-1} \left| \frac{x_2^2 + \cdots + x_n^2}{b^2} \le 1 - \frac{x^2}{a^2}\right.\right\}.$$

Then

$$V(a) = \int_{-a}^{a} \lambda_{n-1}((\mathscr{E}_a)_x)\,dx$$
$$= \alpha_{n-1} \int_{-a}^{a} b^{n-1}\left(1 - \frac{x^2}{a^2}\right)^{(n-1)/2} dx = \alpha_n \left(\frac{c^2 - a^2}{c^2 - 1}\right)^{(n-1)/2} \cdot a.$$

It is clear that $V(1) = \alpha_n$ and $V'(1) = \alpha_n(c^2 - n)/(c^2 - 1) > 0$. Therefore, for a close to 1 and $a > 1$ the inequality $V(a) > \alpha_n$ holds although $\mathscr{E}_a \subset T$, which contradicts (1).

3.38. The continuity and boundedness of φ and its holomorphicity outside K can be established with the help of the following theorem (see, for example, [1], p. 169).

THEOREM. *Suppose that $p > 1$, $E \subset \mathbb{R}^m$, $\lambda_m(E) < \infty$, and $\{f_n\}$ is a sequence of measurable functions on E. If $f_n \to f_0$ in measure on E and $\sup_n \int_E |f_n(x)|^p\,dx < \infty$, then the functions f_0, f_n $(n = 1, 2, \dots)$ are integrable on E, and $\int_E |f_n(x) - f_0(x)|\,dx \to 0$; in particular, $\int_E f_n(x)\,dx \to \int_E f_0(x)\,dx$.*

If K is totally disconnected, then φ supplies an example (due to Pompeiu) showing that Liouville's theorem ceases to be valid if instead of entire functions we consider functions that are continuous and bounded on $\mathbb{C}$ and holomorphic on an open dense set. See problem 4.8 (an example of Urysohn) for a refinement of this result.

3.39. a) Let $p, q > -1$, and $\mathscr{I}_{p,q} = \int_{S^3} |\zeta_1|^{2p}|\zeta_2|^{2q}\, d\mu_3(\zeta_1, \zeta_2)$. Then

$$\begin{aligned}
\mathscr{I}_{p,q} &= \int_{x^2+y^2+u^2+v^2=1} (x^2+y^2)^p(u^2+v^2)^q\, d\mu_3(x, y, u, v) \\
&= 2\iiint_{x^2+y^2+u^2\le 1} (x^2+y^2)^p(1-x^2-y^2)^q \frac{dx\,dy\,du}{\sqrt{1-x^2-y^2-u^2}} \\
&= 4\int_{x^2+y^2\le 1} (x^2+y^2)^p(1-x^2-y^2)^q \\
&\quad \times \left\{\int_0^{\sqrt{1-x^2-y^2}} \frac{du}{\sqrt{1-x^2-y^2-u^2}}\right\} dx\,dy \\
&= (2\pi)^2 \int_0^1 r^{2p+1}(1-r^2)^q\, dr \\
&= 2\pi^2 \int_0^1 t^p(1-t)^q\, dt = 2\pi^2 \cdot \frac{\Gamma(p+1)\Gamma(q+1)}{\Gamma(p+q+2)}.
\end{aligned}$$

c) Note that the series $\sum_{n\ge 0}(n+1)(\zeta_1\overline{z}_1+\zeta_2\overline{z}_2)^n$ is uniformly convergent in the ball $|z_1|^2+|z_2|^2 \le 1$, because $|\zeta_1\overline{z}_1+\zeta_2\overline{z}_2| \le a\sqrt{|z_1|^2+|z_2|^2} \le a < 1$, where $a = \sqrt{|\zeta_1|^2+|\zeta_2|^2}$. Therefore,

$$\begin{aligned}
&\int_{S^3} f(z_1, z_2)\frac{d\mu_3(z_1, z_2)}{(1-\zeta_1\overline{z}_1-\zeta_2\overline{z}_2)^2} \\
&\quad = \sum_{n\ge 0}(n+1)\int_{S^3} f(z_1, z_2)(\zeta_1\overline{z}_1+\zeta_2\overline{z}_2)^n\, d\mu_3(z_1, z_2) \qquad (1) \\
&\quad = \sum_{n\ge 0}(n+1)\sum_{0\le k\le n}\binom{n}{k}\zeta_1^k\zeta_2^{n-k}\int_{S^3} f(z_1, z_2)\overline{z}_1^k\overline{z}_2^{n-k}\, d\mu_3(z_1, z_2).
\end{aligned}$$

We compute the integrals on the right-hand side of (1):

$$\begin{aligned}
&\int_{S^3} f(z_1, z_2)\overline{z}_1^k\overline{z}_2^{n-k}\, d\mu_3(z_1, z_2) \\
&\quad = \sum_{j,l\ge 0} a_{jl}\int_{S^3} z_1^j z_2^l \overline{z}_1^k \cdot \overline{z}_2^{n-k}\, d\mu_3(z_1, z_2) \\
&\quad = a_{k,n-k}\int_{S^3} |z_1|^{2k}\cdot|z_2|^{2(n-k)}\, d\mu_3(z_1, z_2) \\
&\quad = a_{k,n-k}\cdot\frac{2\pi^2}{n+1}\binom{n}{k}^{-1}.
\end{aligned}$$

Substituting this result in the right-hand side of (1), we get that

$$\int_{S^3} f(z_1, z_2)\frac{d\mu_3(z_1, z_2)}{(1-\zeta_1\bar{z}_1-\zeta_2\bar{z}_2)^2}$$

$$= \sum_{n\geq 0}(n+1)\sum_{0\leq k\leq n}\binom{n}{k}\zeta_1^k\zeta_2^{n-k}a_{k,n-k}\cdot\frac{2\pi^2}{n+1}\binom{n}{k}^{-1}$$

$$= 2\pi^2\sum_{n\geq 0}\sum_{0\leq k\leq n}a_{k,n-k}\zeta_1^k\zeta_2^{n-k} = 2\pi^2 f(\zeta_1, \zeta_2).$$

§4. The Stieltjes integral

4.1. Suppose first that $b < \infty$. We carry out the solution in this case in several steps.

Step 1. The function h is continuous on $[a, b]$. Break up the interval $[a, b]$ into parts by the points $t_0 = a < t_1 < \cdots < t_{n-1} < t_n = b$. Let $F(t) = \lambda_m([-t, t]^m) = (2t)^m$ $(t \geq 0)$. Then

$$\int_E h(\max|x_k|)\,dx = \sum_{0\leq k<n}\int_{t_k\leq\max|x_k|<t_{k+1}} h(\max|x_k|)\,dx$$

$$= \sum_{0\leq k<n} h(\bar{t}_k)[F(t_{k+1}) - F(t_k)]$$

$$= \sum_{0\leq k<n} h(\bar{t}_k)F'(\bar{\bar{t}}_k)(t_{k+1} - t_k),$$

where $\bar{t}_k, \bar{\bar{t}}_k \in [t_k, t_{k+1}]$ $(k = 0, 1, \ldots, n-1)$. Thus,

$$\int_E h(\max|x_k|)\,dx$$

$$= m2^m\sum_{0\leq k<n}(\bar{\bar{t}}_k)^{m-1}h(\bar{t}_k)(t_{k+1}-t_k) \xrightarrow[\max(t_{k+1}-t_k)\to 0]{} m2^m\int_a^b t^{m-1}h(t)\,dt.$$

Step 2. The function h is bounded on $[a, b]$. Consider a sequence $\{h_n\}$ of nonnegative continuous functions on $[a, b]$ that converges to h almost everywhere on $[a, b]$. It can be assumed without loss of generality that $h_n \leq \sup h$. Show that $h_n(\max|x_k|) \to h(\max|x_k|)$ a.e. on E. Using Lebesgue's theorem on taking the limit under the integral sign, we see that the equalities

$$\int_E h_n(\max|x_k|)\,dx = m2^m\int_a^b t^{m-1}h_n(t)\,dt \qquad (n = 1, 2, \ldots)$$

lead to the required result.

Step 3. The function h is arbitrary (measurable, nonnegative). The required result is obtained with the help of approximation by a sequence of bounded measurable functions.

The case $b = +\infty$ is exhausted by passing to the limit from finite intervals to an infinite interval.

b) The arguments are according to the same scheme as in a). Here the function $F(t) = (2t)^m$ is replaced by the function

$$F_1(t) = \lambda_m\{(x_1, \dots, x_m) \in \mathbb{R}^m | 1 \le \min x_k, \max x_k \le t\} \\ = (t-1)^m \quad (t-1).$$

4.2. The arguments are according to the same scheme as in the solution of problem 4.1a). The function $F(t) = (2t)^m$ is replaced by the function

$$F_1(t) = \lambda_m(O_m(t)) = \frac{(2t)^m}{m!} \qquad (t \ge 0).$$

4.3. The arguments are according to the same scheme as in the solution of problem 4.1a).

4.4. The arguments are according to the same scheme as in the solution of problem 4.1a).

4.5. It is clear that $\int_{\mathbb{R}^2} e^{-\rho(x)}\,dx = \lambda_2(A) + \int_{\mathbb{R}^2\setminus A} e^{-\rho(x)}\,dx$. Arguing according to the same scheme as in the solution of problem 4.1a), show that $\int_{\mathbb{R}^2\setminus A} e^{-\rho(x)}\,dx = \int_0^\infty e^{-t}\,dF(t)$, where $F(t) = \lambda_2(A_t)$, $A_t = \bigcup_{x\in A} B(x, t)$. Use the result of problem 1.16.

4.6. a) Show that the integral is the limit of the sums

$$2^{-n} \sum_{\varepsilon_1, \dots, \varepsilon_n \in \{0, 2\}} \sum_{1\le k\le n} \frac{\varepsilon_k}{3^k}.$$

b) Use the same method as in a);

c) Use the fact that the graph of the function σ is centrally symmetric with respect to the point $(1/2, 1/2)$.

d) Prove that $\int_0^t \sigma(x)\,d\sigma(x) = \sigma^2(t)/2$ $(t > 0)$ or use the same device as in the solution of a).

e) Use the equality

$$\sigma(x) + \sigma(1-x) = 1 \ (x \in \mathbb{R}).$$

4.7. a) Let $Q_j = [0, 3^{-j}] \times [0, 3^{-j}]$ and $E_j = Q_j \setminus Q_{j+1}$ $(j = 0, 1, 2, \dots)$. It is clear that

$$\int_0^1 \int_0^1 \frac{d\sigma(x)\,d\sigma(y)}{(x^2+y^2)^{p/2}} = \sum_{j\ge 0} \iint_{E_j} \frac{d\sigma(x)\,d\sigma(y)}{(x^2+y^2)^{p/2}}. \tag{$*$}$$

Since $2^{-p/2} \cdot 3^{jp} \le (x^2+y^2)^{-p/2} \le 3^{(j+1)p}$ on E_j, it follows that

$$2^{-p/2} \cdot 3^{jp} \cdot \frac{3}{4} \cdot 4^{-j} \le \iint_{E_j} \frac{d\sigma(x)\,d\sigma(y)}{(x^2+y^2)^{p/2}} \le 3^{(j+1)p} \cdot \frac{3}{4} \cdot 4^{-j}. \tag{$**$}$$

Therefore, the series $(*)$ converges simultaneously with the series $\sum (3^p/4)^j$.

b) Let $\widetilde{Q}_j = [u - 3^{-j}, u + 3^{-j}] \times [v - 3^{-j}, v + 3^j]$ and $\widetilde{E}_j = \widetilde{Q}_j \backslash \widetilde{Q}_{j+1}$ $(j = 0, 1, \dots)$. Arguing as in a), show that the integrals

$$\iint_{\widetilde{E}_j} \frac{d\sigma(x)\, d\sigma(y)}{((x-u)^2 + (y-v)^2)^{p/2}}$$

admit an estimate analogous to $(**)$.

4.8. The solution of this problem is based on the result of problem 4.7 and is carried out accordingly to the same scheme as the solution of problem 3.38. In essence, the function φ was first considered in [28], p. 93, where it was constructed as a limit of functions that are integral sums for the integral $\int_0^1 \int_0^1 (z - (\xi + in))^{-1} d\sigma(\xi) d\sigma(\eta)$.

§5. The ε-entropy and Hausdorff measures

5.1. Let E_0 be an ε-distinguishable subset of A with a maximal number of points. Then it forms an ε-net, and hence $N(A, \varepsilon) \le \operatorname{card}(E_0) = M(A, \varepsilon)$. On the other hand, distinct points in E_0 can be approximated to within $\varepsilon/2$ only by distinct points, and thus $M(A, \varepsilon) \le N(A, \varepsilon/2)$.

5.2. a). Use the results of problem 5.1 in computing the minimal number of points in an ε-net for the cube.

b) Let $\varepsilon > 0$, and let the set E be a minimal ε-net for A, that is, $\operatorname{card}(E) = N(A, \varepsilon)$. Then $A \subset \bigcup_{x \in E} B(x, \varepsilon)$ and $\lambda_n^*(A) \le \sum_{x \in E} \lambda(B(x, \varepsilon)) = N(A, \varepsilon)\alpha_n \varepsilon^n$, which yields the desired inequality with $C = \alpha_n^{-1} \lambda_n^*(A)$, where α_n is the n-dimensional volume of the unit ball.

In the solution of problems 5.3–5.12 we omit for brevity the expression "$\varepsilon \to +0$" when we use the symbols O, o, and $\sim$.

5.3. a) Suppose that $\varepsilon > 0$ and $2^{-(m+1)} < \varepsilon \le 2^{-m}$, that is $m = [\log_2(1/\varepsilon)]$. It is obvious that the points $1, 2^{-1}, \dots, 2^{-m}$ are ε-distinguishable and form a 2ε-net. Therefore (see problem 5.1), $N(A, 2\varepsilon) \le m + 1 \le N(A, \varepsilon/2)$, which implies that $H(A, \varepsilon) \sim \log_2 \log_2(1/\varepsilon)$.

d) Suppose that $\varepsilon > 0$ and $m^{-\alpha} - (m+1)^{-\alpha} < \varepsilon \le (m-1)^{-\alpha} - m^{-\alpha}$, that is, $m - 1 \le (\alpha/\varepsilon)^{1/(\alpha+1)} \le m + 1$. It is clear that the points $1, 2^{-\alpha}, \dots, m^{-\alpha}$ are ε-distinguishable. If we add the points of the form $k\varepsilon$ to them, where $k = 1, 2, \dots, p$ and $p\varepsilon \le m^{-\alpha} < (p+1)\varepsilon$, then we get an ε-net of the set $A = \{n^{-\alpha} | n \in \mathbb{N}\}$. Since $p = [1/(\varepsilon m^\alpha)]$, it follows that $p = O(m)$. Therefore, $m \le N(A, \varepsilon/2)$ and $N(A, \varepsilon) = O(m)$. Thus, $(\alpha/\varepsilon)^{1/(\alpha+1)} \le N(A, \varepsilon/2) + 1$ and $N(A, \varepsilon) = O(\varepsilon^{-1/(\alpha+1)})$, which implies that

$$H(A, \varepsilon) \sim \frac{1}{\alpha+1} \log_2 \frac{1}{\varepsilon}.$$

Problems 5.4, 5.5, and 5.6 can be solved similarly. We present a solution of problem 5.4b).

5.4. b) Together with the given number $\varepsilon > 0$ we consider a positive number δ whose choice will be made more specific below. Let us break up the graph A of the function $\sin \pi/x$ on $(0, 1]$ into parts B_δ and C_δ, where

B_δ is the part of A contained in the rectangle $P_\delta = (0, \delta) \times [-1, 1]$, and $C_\delta = A \backslash B_\delta$. The length of C_δ is

$$l_\delta = \int_\delta^1 \sqrt{1 + \frac{\pi^2}{t^4} \cos^2 \frac{\pi}{t}}\, dt = O(\delta^{-1}).$$

Therefore, the number $N(C_\delta, \varepsilon)$ can be estimated by the quantity $l_\delta/\varepsilon = O((\varepsilon\delta)^{-1})$. The number $N(B_\delta, \varepsilon)$ does not exceed $N(P_\delta, \varepsilon)$, and hence $N(B_\delta, \varepsilon) = O(\delta\varepsilon^{-2})$. Let δ be chosen such that the estimates for $N(B_\delta, \varepsilon)$ and $N(C_\delta, \varepsilon)$ have the same order; for this we take $\delta = \sqrt{\varepsilon}$. We thereby obtain the estimate $N(A, \varepsilon) = O(\varepsilon^{-3/2})$. To see that the number of ε-distinguishable points in A also has order $\varepsilon^{-3/2}$ we consider the intervals $\Delta_k = [1/(2k+1), 1/2k]$ for $k = 1, 2, \dots, [1/(2\sqrt{\varepsilon})]$ and the corresponding parts Γ_k of the graph. Since $f(\Delta_k) = [0, 1]$ for any k, by using ε-distinguishable points in $[0, 1]$ it is possible to determine at least $[1/\varepsilon]$ distinguishable points in Γ_k. Since the distance between neighboring intervals Δ_k is not less than ε, points lying on different arcs Γ_k will be ε-distinguishable. The total number of ε-distinguishable points constructed in A is $[1/\varepsilon] \cdot [1/2\sqrt{\varepsilon}]$. Thus, $H(A, \varepsilon) \sim \frac{3}{2} \log_2(1/\varepsilon)$.

5.7. Suppose that $\varepsilon > 0$ and $3^{-(m+1)} < \varepsilon < 3^{-m}$, i.e., $m = [\log_3(1/\varepsilon)]$. The endpoints of the intervals $\Delta_{\varepsilon_1 \cdots \varepsilon_m}$ of rank m (see the definition of the Cantor set before problem I.2.27) are obviously ε-distinguishable and form an ε-net. Therefore, $N(C, \varepsilon) \le 2 \cdot 2^m \le N(C, \varepsilon/2)$. This implies that $H(C, \varepsilon) \sim \log_3(1/\varepsilon) = (\log_3 2) \cdot \log_2(1/\varepsilon)$.

5.8. Suppose that $\varepsilon > 0$ and $1/(m+1)! < \varepsilon \le 1/m!$. The points of the form $\sum_{0 \le k \le m} \varepsilon_k/k!$ with $\varepsilon_k = 0$ or 1 are ε-distinguishable and form a 2ε-net. Therefore, $N(A, 2\varepsilon) \le 2^{m+1} \le N(A, \varepsilon/2)$. It follows from the definition of m that $\ln(1/\varepsilon) \sim m \ln m$, that is,

$$m \sim \frac{\ln(1/\varepsilon)}{\ln\ln(1/\varepsilon)}.$$

Consequently,

$$H(A, \varepsilon) \sim \frac{\ln(1/\varepsilon)}{\ln\ln(1/\varepsilon)}.$$

5.9. The assertion a) follows from the result of problem 5.10 if it is taken into account that in this case $\psi(x) = x + m!$ for $m! \le x < (m+1)!$ and that (as is easily seen) $\underline{\lim}_{y\to+\infty} \frac{1}{y}\psi^{-1}(y) = \frac{1}{2}$ and $\overline{\lim}_{y\to+\infty} \frac{1}{y}\psi^{-1}(y) = 1$.

See problem 1.11 about assertion b).

5.10. Suppose that $\varepsilon > 0$ and $2^{-\psi(m+1)} < \varepsilon \le 2^{-\psi(m)}$, i.e., $m = [\psi^{-1}(\log_2(1/\varepsilon))]$. Obviously, the points $\sum_{1 \le k \le m} \varepsilon_k 2^{-\psi(k)}$, where $\varepsilon_k = 0$ or 1, are ε-distinguishable and form a 2ε-net. Therefore, $N(A, 2\varepsilon) \le 2^m \le N(A, \varepsilon/2)$. Consequently,

$$H(A, 2\varepsilon) \le m \le H(A, \varepsilon/2). \tag{1}$$

It can be assumed without loss of generality that ψ is linear on each interval $[n-1, n]$ $(n \in \mathbb{N})$. Then, as is easy to see, $\psi(x+h) - \psi(x) \geq h$ for $x, h \geq 0$, and hence

$$\psi^{-1}(y+\eta) - \psi^{-1}(y) \leq \eta \qquad (y, \eta \geq 0). \tag{2}$$

It follows form (1) that $H(A, 2\varepsilon) - 1 \leq \psi^{-1}(\log_2 1/\varepsilon) \leq H(A, \varepsilon/2) + 1$. Together with (2) this leads to the required result.

5.11. Suppose that $\varepsilon > 0$, $3^{-(m+1)} < \varepsilon \leq 3^{-m}$, and $2^{-(p+1)} < \varepsilon \leq 2^{-p}$. We consider the points $(x, y) \in A$ with coordinates $x = \sum_{1 \leq k \leq p} \varepsilon_k 2^{-k}$ and $y = \sum_{1 \leq k \leq m} \eta_k 3^{-k} + \sum_{m < k \leq p} \varepsilon_k 3^{-k}$, where $\varepsilon_k = 0$ or 1 and $\eta_k = 0, 1,$ or 2, and ε_k and η_k have the same parity for $k \leq m$. It is clear that these points are ε-distinguishable and form a 4ε-net. The number of these points is equal to $3^m 2^{p-m}$. Therefore,

$$N(A, 4\varepsilon) \leq 3^m 2^{p-m} \leq N(A, \tfrac{\varepsilon}{2}),$$

that is,

$$H(A, 4\varepsilon) \leq [\log_3 \tfrac{1}{\varepsilon}] \log_2 \tfrac{3}{2} + [\log_2 \tfrac{1}{\varepsilon}] \leq H(A, \tfrac{\varepsilon}{2}).$$

Consequently, $H(A, \varepsilon) \sim (2 - \log_3 2) \log_2 1/\varepsilon$.

5.12. a) Suppose that $\varepsilon > 0$ and the set E is a minimal ε/m-net for A, and $\operatorname{card} E = N(A, \varepsilon/m)$. Then the set $E_m = E + \cdots + E$ (m terms) is an ε-net for A_m, and $\operatorname{card}(E_m) \leq (N(A, \varepsilon/m))^m$. Consequently, $H(A_m, \varepsilon) \leq mH(A, \varepsilon/m) = o(\ln 1/\varepsilon)$, which is possible only if $\lambda(A) = 0$ (see problem 5.2).

b) The solution is analogous to the preceding.

c) No. The indicated set serves as a counterexample.

5.13. a) If A is countable, $A = \{x_1, x_2, \ldots\}$, then to compute $\mu_p(A, \varepsilon)$ consider the covering of A by the balls $B(x_k, \varepsilon 2^{-k})$.

c) Let $\varepsilon > 0$, and let $\{B(x_k^l, r_k^l)\}_{k=1}^\infty$ be a covering of A such that $r_k < \varepsilon$ for all $l, k \in \mathbb{N}$ and $|\sum_{k \geq 1} (r_k^l)^p - \mu_p(A_l)| < \varepsilon/2^l$. Then

$$\begin{aligned} \mu_p(A, \varepsilon) &\leq \sum_{l \geq 1} \sum_{k \geq 1} (r_k^l)^p \\ &\leq \sum_{l \geq 1} \left(\mu_p(A_l) + \frac{\varepsilon}{2^l} \right) = \sum_{l \geq 1} \mu_p(A_l) + \varepsilon. \end{aligned}$$

Since $\varepsilon > 0$ is arbitrary, we get what is required.

5.14. a) Let $\varepsilon > 0$, and let $\{B(x_k, r_k)\}_{k=1}^\infty$ be a covering of A such that $r_k < \varepsilon$ for all $k \in \mathbb{N}$ and $\sum_{k \geq 1} r_k^p \leq \mu_p(A) + 1$. Then

$$\sum_{k\geq 1}\lambda_n(B(x_k\,,\,r_k)) = \sum_{k\geq n}\alpha_n r_k^n \leq \alpha_n\varepsilon^{n-p}\sum_{k\geq 1} r_k^p$$
$$\leq \alpha_n\varepsilon^{n-p}(\mu_p(A)+1).$$

Therefore, $\lambda_n^*(A) \leq \sum_{k\geq 1}\lambda_n(B(x_k\,,\,r_k)) \leq \alpha_n\varepsilon^{n-p}(\mu_p(A)+1)$, which yields what was required, since $\varepsilon > 0$ is arbitrary.

b) If $\mu_n(A) = 0$, then $\lambda_n(A) = 0$, because A can be covered by a sequence $\{B(x_k\,,\,r_k)\}_{k=1}^\infty$ of balls such that the sum $\sum_{k\geq 1} r_k^n$, hence also the sum $\sum_{k\geq 1}\lambda_n(B(x_k\,,\,r_k))$, is arbitrarily small.

If $\lambda_n(A) = 0$, then the set A can be covered by a sequence $\{Q_j\}_{j=1}^\infty$, $Q_j = \prod_{k=1}^n (a_k^j - h_j\,,\,a_k^j + h_j)$, of cubes such that the sum $\sum_{j\geq 1}\lambda_n(Q_j) = \sum_j (2h_j)^n$ is arbitrarily small. Then it is clear that the balls $B(a_j\,,\,\sqrt{n}h_j)$ "circumscribed about the cubes" (where $a_j = (a_1^j\,,\,\dots\,,\,a_n^j)$) also form a covering for A, and the sum $\sum_{j\geq 1}(\sqrt{n}h_j)^n$ differs from the sum $\sum_{j\geq 1}(2h_j)^n$ only by the factor $(\sqrt{n}/2)^n$.

5.15. Since (for $r \leq 1$)

$$\nu_{n-1}(S^{n-1}\cap B(a\,,\,r)) = \int_{B^{n-1}(\delta)}\frac{d\lambda_{n-1}(x)}{\sqrt{1-\|x\|^2}}\,,$$

where $a = (0\,,\,\dots\,,\,0\,,\,1)$ and $B^{n-1}(\delta)$ is the projection of the set $S^{n-1}\cap B(a\,,\,r)$ on the plane $x_n = 0$, it follows that there exist constants $m > 0$ and M (depending on the dimension) such that $mr^{n-1} \leq \nu_{n-1}(S^{n-1}\cap B(x\,,\,r)) \leq Mr^{n-1}$ for all $x\in S^{n-1}$. By using this circumstance it is possible to get the required results by a simple modification of the solution of problem 5.14.

5.16. a) This assertion is obvious, because if $r_k \leq 1$, then $\sum r_k^p \geq \sum r_k^q$.

b) The proof that $\mu_q(A) = 0$ if $q > p$ and $\mu_p(A) < +\infty$ is analogous to the solution of problem 5.14a). If $\mu_p(A) > 0$ and $1 < q < p$, then $\mu_q(A) = \infty$, because otherwise we would come to a contradiction of the part of the assertion already proved.

5.17. a) Suppose that $r_j > 0$, $r_j \to 0$, and $H(A\,,\,r_j)/\log_2(1/r_j) \to r(A)$. Further, let $q > r(A)$, and let E_j be an r_j-net for A such that $\operatorname{card}(E_j) = N(A\,,\,r_j)$. Then $A \subset \bigcup_{x\in E_j} B(x\,,\,r_j)$, and $H(A\,,\,r_j) < q\log_2 1/r_j$ for sufficiently large j. Therefore, $\mu_q(A\,,\,2r_j) \leq N(A\,,\,r_j)r_j^q \leq r_j^{-q}\cdot r_j^q = 1$. Consequently, $\mu_q(A) < \infty$ for $q > r(A)$, which implies (see problem 5.16b)) that $\dim_H(A) \leq r(A)$.

b) The inequality $\sup_k \dim_H(A_k) \leq \dim_H(\bigcup_{k=1}^\infty A_k)$ is obvious. On the other hand, if $q > \sup_k \dim_H(A_k)$, then for any $\varepsilon > 0$ there is a covering $\{B(x_k^j\,,\,r_k^j)\}_{k=1}^\infty$ of A_j such that $\sum_{k\geq 1}(r_k^j)^q < \varepsilon/2^j$, $r_k^j < \varepsilon$ $(j\,,\,k \in \mathbb{N})$. Consequently, $\bigcup_{j=1}^\infty A_j \subset \bigcup_{j,k=1}^\infty B(x_k^j\,,\,r_k^j)$, $r_k^j < \varepsilon$ $(j\,,\,k\in\mathbb{N})$, and $\sum_{j,k\geq 1}(r_k^j)^q < \sum_{j\geq 1}\varepsilon 2^{-j} = \varepsilon$. Thus, $\mu_q(\bigcup_{j=1}^\infty A_j) = 0$, which by the choice of q implies that $\dim_H(\bigcup_{j=1}^\infty A_j) \leq \sup_k \dim_H(A_k)$.

5.18. It is clear that

$$\begin{aligned}\mathscr{I}_\varepsilon &= \inf\left\{\sum_{k\geq 1}(\operatorname{diam}(C_k))^p \,|A\subset\bigcup_{k=1}^\infty C_k\,,\ \operatorname{diam}(C_k)<\varepsilon\right\}\\ &\leq \inf\left\{\sum_{k\geq 1}(2r_k)^p\,|A\subset\bigcup_{k=1}^\infty B(x_k\,,\,r_k)\,,\ 2r_k\leq\varepsilon\right\}\\ &= 2^p\mu_p(A\,,\,\varepsilon/2).\end{aligned}$$

Passing to the limit as $\varepsilon\to+0$, we get the inequality

$$\tilde{\mu}_p(A)\leq 2^p\mu_p(A).$$

On the other hand, let C_k^0 $(k\in\mathbb{N})$ be sets such that $A\subset\bigcup_{k=1}^\infty C_k^0$, $r_k=\operatorname{diam}(C_k^0)<\varepsilon$, and $\mathscr{I}_\varepsilon+\varepsilon>\sum_k r_k^p$. If $x_k\in C_k^0$, then $C_k^0\subset\overline{B}(x_k\,,\,r_k)\subset B(x_k\,,\,(1+\varepsilon)r_k)$, and hence $\mathscr{I}_\varepsilon+\varepsilon\geq(1/(1+\varepsilon)^p)\mu_p(A\,,\,\varepsilon(1+\varepsilon))$. From this we get that $\tilde{\mu}_p(A)\geq\mu_p(A)$ by passing to the limit as $\varepsilon\to+0$.

5.19. a) If $[a\,,\,b]\subset\bigcup_{k=1}^\infty(x_k-r_k\,,\,x_k+r_k)$, then $b-a\leq\sum_k 2r_k$ by the countable subadditivity one-dimensional Lebesgue measure λ. Consequently, $\mu_1([a\,,\,b])\geq(b-a)/2$. On the other hand, by dividing $[a\,,\,b]$ into equal parts of arbitrarily small length it is easy to see that $\mu_1([a\,,\,b])\leq(b-a)/2$.

b) Fix a number $\varepsilon>0$, small enough that the difference $l-h$ between the length l of the arc and the length h of the chord subtending it is less than εl if $l<\varepsilon$. Suppose that $A\subset\bigcup_{k=1}^\infty B(x_k\,,\,r_k)$, $r_k<\varepsilon$, $L_k=A\cap B(x_k\,,\,r_k)$, and let h_k be the length of the chord subtending the arc L_k and y_k the midpoint of this chord. It is clear that $h_k\leq 2r_k$ and $L_k=A\cap B(y_k\,,\,h_k/2)$. Therefore,

$$\sum_{k\geq 1}\frac{h_k}{2}\geq\frac{1}{2}\sum_{k\geq 1}l_k(1-\varepsilon),$$

where l_k is the length of the arc L_k. Thus,

$$\frac{1}{2}\sum_{k\geq 1}h_k\geq\frac{1-\varepsilon}{2}\sum_{k\geq 1}l_k\geq\pi(1-\varepsilon),$$

and hence $\mu_1(A\,,\,\varepsilon)\geq\pi(1-\varepsilon)$.

On the other hand, dividing the circle into n equal parts of length $2\pi/n<\varepsilon$ and covering these arcs by discs with centers at the midpoints of the chords subtending them and with radii equal to π/n, we see that $\mu_1(A\,,\,\varepsilon)\leq n\cdot\pi/n=\pi$, and hence $\mu_1(A)\leq\pi$.

c) This is a special case of d).

d) Let G be an open subset of $\mathbb{R}^n$, and let $\varepsilon>0$. Find balls $B(x_k\,,\,r_k)$ $(k\in\mathbb{N})$ such that

$$G\subset\bigcup_{k=1}^\infty B(x_k\,,\,r_k),\qquad r_k<\varepsilon,\qquad \sum_{k\geq 1}r_k^n<\mu_n(G\,,\,\varepsilon)+\varepsilon.$$

Then

$$\lambda_n(G) \le \sum_{k\ge 1} \lambda_n(B(x_k, r_k))$$
$$= \alpha_n \sum_{k\ge 1} r_n^k < \alpha_n(\mu_n(G, \varepsilon) + \varepsilon).$$

Since ε is arbitrary,

$$\lambda_n(G) \le \alpha_n \mu_n(G) \qquad (*)$$

Now take balls $B(y_k, \rho_k)$ $(k \in \mathbb{N})$ such that $\rho_k < \varepsilon$, $B(y_k, \rho_k) \cap B(y_j, \rho_j) = \varnothing$ for $k \ne j$, and $G = e \cup \bigcup_{k=1}^\infty B(y_k, \rho_k)$, where $\lambda_n(e) = 0$ (see problem 1.17). Further, fix balls $B(z_k, \delta_k)$ covering the set e and satisfying the conditions $\sum_{k\ge 1} \delta_k^n < \varepsilon$ and $\delta_k < \varepsilon$ $(k \in \mathbb{N})$ (see problem 5.14b)). Then $G \subset \bigcup_{k=1}^\infty (B(z_k, \delta_k) \cup B(y_k, \rho_k))$ and

$$\mu_n(G, \varepsilon) \le \sum_{k\ge 1} \rho_k^n + \sum_{k\ge 1} \delta_k^n < \alpha_n^{-1} \sum_{k\ge 1} \alpha_n \rho_k^n + \varepsilon$$
$$= \alpha_n^{-1} \sum_{k\ge 1} \lambda_n(B(y_k, \rho_k)) + \varepsilon$$
$$= \alpha_n^{-1} \lambda_n \left(\bigcup_{k=1}^\infty B(y_k, \rho_k) \right) + \varepsilon = \alpha_n^{-1} \lambda_n(G) + \varepsilon.$$

Since $\varepsilon > 0$ is arbitrary, this gives us the inequality $\mu_n(G) \le \alpha_n^{-1} \lambda_n(G)$, which together with $(*)$ leads to the final result: $\mu_n(G) = \alpha_n^{-1} \lambda_n(G)$. In particular, $\mu_n([0, 1]^n) = \mu_n((0, 1)^n) = \alpha_n^{-1}$.

5.20. Note that if $p = \log_3 2$, then $(1 + 2t)^p \le 1 + t^p$ for $0 \le t \le 1$. Therefore, if some interval of length l is divided into three parts with lengths l_1, l_2, and l_3 such that $l_1 \le l_2 \le l_3$, then

$$l^p \le l_2^p + l_3^p. \qquad (*)$$

We use this inequality to compute $\mu_p(K, 1/2)$. By definition,

$$\mu_p\left(K, \frac{1}{2}\right) = \inf\left\{ \sum_{k\ge 1} \left(\frac{b_k - a_k}{2}\right)^p \mid K \subset \bigcup_{k=1}^\infty (a_k, b_k),\ b_k - a_k < 1 \right\}.$$

It is clear that the condition $b_k - a_k < 1$ can be discarded, and the coverings can be assumed to be finite and made up of closed intervals. Accordingly,

$$\mu_p\left(K, \frac{1}{2}\right) = \inf\left\{ \sum_{1\le k\le N} \left(\frac{b_k - a_k}{2}\right)^p \mid K \subset \bigcup_{k=1}^N [a_k, b_k] \right\}.$$

Suppose that the intervals $[a_1, b_1], \ldots, [a_N, b_N]$ form a covering of K. Making them smaller if necessary, we can assume that no two of them have common interior points. Let

$$\min\{b_k - a_k \mid k = 1, 2, \ldots, N\} = b_m - a_m = \delta, \qquad \Delta = [a_m, b_m].$$

At least one of the intervals neighboring Δ, call it Δ', has distance from Δ not exceeding δ. Let Δ_0 be the smallest interval containing Δ and Δ'. In view of the inequality $(*)$ we get that $\delta_0^p \le \delta^p + (\delta')^p$, where δ_0 and δ' are the lengths of the respective intervals Δ_0 and Δ'. Thus, replacing Δ and Δ' by Δ_0, we again get a covering of K consisting of $N-1$ closed intervals $[a_1', b_1'], \ldots, [a_{N-1}', b_{N-1}']$, and $\sum_{1\le k<N}(b_k' - a_k')^p \le \sum_{1\le k\le N}(b_k - a_k)^p$. Continuing this process, we arrive at the covering of K consisting of the single interval $[0, 1]$, without increasing the sum of the pth powers of the lengths of the intervals making up the covering of K at each step. This gives us that $\mu_p(K, 1/2) \ge 2^{-p}$. Since the converse inequality is obvious, we get that $\mu_p(K, 1/2) = 2^{-p}$. It follows from similarity considerations that $\mu_p(K \cap [0, 3^{-n}]; 3^{-n}/2) = 2^{-p}3^{-np} = 2^{-(p+n)}$ for any $n \in \mathbb{N}$. Let $\Delta_{\varepsilon_1\cdots\varepsilon_n}$ be one of the intervals of rank n in the definition of the Cantor set. Obviously,

$$\mu_p(K \cap \Delta_{\varepsilon_1\cdots\varepsilon_n}, 3^{-n}/2) = \mu_p(K \cap [0, 3^{-n}], 3^{-n}/2) = 2^{-(p+n)}.$$

Since the distances between the intervals of rank n are $\ge 3^{-n}$, it follows that (see problem 5.18d))

$$\begin{aligned}\mu_p(K, 3^{-n}/2) &= \mu_p\left(\bigcup_{\varepsilon_1, \ldots, \varepsilon_n \in \{0,1\}} K \cap \Delta_{\varepsilon_1\cdots\varepsilon_n}, \frac{1}{2}3^{-n}\right)\\ &= \sum_{\varepsilon_1, \ldots, \varepsilon_n \in \{0,1\}} \mu_p(K \cap \Delta_{\varepsilon_1\cdots\varepsilon_n}, \frac{1}{2}3^{-n}) = 2^n \cdot 2^{-(p+n)} = 2^{-p}.\end{aligned}$$

Thus, $\mu_p(K) = \lim \mu_p(K, 3^{-n}/2) = 2^{-p}$.

To find the function φ we observe that it is continuous on $[0, 1]$ and is constant on the open intervals complementary to K, while its increment on $\Delta_{\varepsilon_1\cdots\varepsilon_n}$ is $2^p\mu_p(K \cap \Delta_{\varepsilon_1\cdots\varepsilon_n}) = 2^{-n}$. It is now easy to prove by induction that the values of φ at the endpoints of $\Delta_{\varepsilon_1\cdots\varepsilon_n}$ coincide with the value of the Cantor function.

5.21. a) See problem 5.20.

b) See problem 5.3e) and 5.13a).

5.22. Let $\varphi(x) = \mu_p(E \cap [0, x])$, $0 \le x \le 1$, and let $\Delta_{\varepsilon_1\cdots\varepsilon_n}$, where $\varepsilon_k = 0$ or 1, be the intervals of rank n corresponding to the set E. Since all the sets $E \cap \Delta_{\varepsilon_1\cdots\varepsilon_n}$ are congruent, it follows that $\mu_p(E \cap \Delta_{\varepsilon_1\cdots\varepsilon_n}) = 2^{-n}\mu_p(E)$. Consequently, the increment of φ on each interval $\Delta_{\varepsilon_1\cdots\varepsilon_n}$ is proportional to the increment of the function σ_E. Moreover, φ, like σ_E, is constant on the open intervals complementary to E. It is now easy to prove by induction that the values of the functions $\mu_p(E)\sigma_E$ and φ coincide at the endpoints of the intervals $\Delta_{\varepsilon_1\cdots\varepsilon_n}$. Since the set of such points is dense in E, we get what is required.

5.23. Suppose that σ_E is the Cantor function corresponding to the set E, $a = \underline{\lim}\, 2^n l_n^p$, and $\{n_k\}$ is a sequence of positive integers such that $2^{n_k} l_{n_k}^p \to a$. Let ε be an arbitrary positive number. Further, assume that the indices n_k are such that $l_{n_k} < \varepsilon$ and $2^{n_k} l_{n_k}^p < a + \varepsilon$. Now consider a covering of E by closed intervals of rank n_k (for fixed k). Then we get that $\mu_p(E, \varepsilon) \le 2^{n_k} l_{n_k}^p < a + \varepsilon$. In particular, we see that if $a = 0$, then $\mu_p(E) = 0$, while if $a < \infty$, then also $\mu_p(E) < \infty$.

Now let $0 < c < a < \infty$. Fix an index m such that $l_m < \varepsilon$ and $2^n l_n^p > c$ for $n \ge m$, and consider an arbitrary covering of E by open intervals $(x_j - r_j, x_j + r_j)$ satisfying the condition $r_j < l_m$ $(j = 1, 2, \dots,)$. Choose indices s_j such that $l_{s_j} \le r_j < l_{s_j - 1}$. Then $2r_j < l_{s_j-2}$, $s_j > m$, and

$$\begin{aligned}
\sum_{j\ge1} r_j^p \ge \sum_{j\ge1} l_{s_j}^p &\ge c \sum_{j\ge1} 2^{-s_j} \\
&= \frac{c}{4}\sum_{j\ge1} 2^{-s_j+2} \\
&= \frac{c}{4}\sum_{j\ge1} \sigma_E(l_{s_j-2}) \\
&\ge \frac{c}{4}\sum_{j\ge1} (\sigma_E(x_j - r_j + l_{s_j-2}) - \sigma_E(x_j - r_j)) \\
&\ge \frac{c}{4}\sum_{j\ge1} (\sigma_E(x_j + r_j) - \sigma_E(x_j - r_j)) \ge \frac{c}{4}\sigma_E(1) = \frac{c}{4}.
\end{aligned}$$

The last of the inequalities above is valid because the measure generated by the function σ_E is countably subadditive. Accordingly, we have proved that $\sum_{j\ge1} r_j^p \ge c/4$ for any covering of E by open intervals $(x_j - r_j, x_j + r_j)$ of sufficiently small length, which implies that $\mu_p(E) \ge c/4 > 0$. In particular, if $a = +\infty$, then since $c < a$ is arbitrary, $\mu_p(E) = +\infty$.

5.24. Verify first that $2^{-n_k} \le l_k \le 2 \cdot 2^{-n_k}$ $(k = 1, 2, \dots)$, and prove the equalities $\underline{\lim}(k/n_k) = 1/2$ and $\overline{\lim}(k/n_k) = 1$.

5.25. Use the result of problem 5.23.

5.26. Let $\varepsilon > 0$, and let N be a positive integer such that $n^{-(N+1)} \le \varepsilon < n^{-N}$. Then the points $\sum_{1\le k\le N} a_k n^{-k}$, where $a_k \in E$ $(k = 1, \dots, N)$, form an ε-net for A with cardinality $(\operatorname{card} E)^N = m^N$. Therefore, $H(A, \varepsilon) \le \log_2 m^N = N \log_2 m \le p \log_2 1/\varepsilon$, where $p = \log_2 m / \log_2 n$. Consequently (see problem 5.17), $\dim_H(A) \le r(A) \le p$. On the other hand, if $A \subset \bigcup_{j=1}^{\infty} (x_j - r_j, x_j + r_j)$, then the sum $\sum_{j\ge1} r_j^p$ can be estimated from below as follows. Let σ be the Cantor function corresponding to A, and let k_j be indices such that $n^{-(k_j+1)} \le 2r_j < n^{-k_j}$ $(j \in \mathbb{N})$. Then

$$\begin{aligned}\sum_{j\ge1} r_j^p &\ge \sum_{j\ge1}\left(\frac{1}{2n^{k_j+1}}\right)^p = \frac{1}{(2n)^p}\sum_{j\ge1} m^{-k_j}\\ &= \frac{1}{2^p m}\sum_{j\ge1}\sigma(n^{-k_j})\\ &\ge \frac{1}{2^p m}\sum_{j\ge1}\left(\sigma\left(x_j - r_j + \frac{1}{n^{k_j}}\right) - \sigma(x_j - r_j)\right)\\ &\ge \frac{1}{2^p m}\sum_{j\ge1}(\sigma(x_j + r_j) - \sigma(x_j - r_j))\\ &\ge \frac{1}{2^p m}(\sigma(1) - \sigma(0)) = \frac{1}{2^p m}.\end{aligned}$$

The last inequality is valid because the measure generated by the function σ is countably subadditive. Thus, $\mu_p(A) \ge 1/(2^p m) > 0$ and $\dim_H(A) \ge p$.

5.27. To compute $\dim_H(A)$ and $\dim_H(A+A)$ use the result of the preceding problem.

5.28. Consider the function

$$\varphi(z) = \int_0^1\int_0^1 \frac{d\sigma(u)\,d\sigma(v)}{z-(u+iv)} \qquad (z \in \mathbb{C}),$$

where σ is the Cantor function corresponding to the set K (see problem III.3.20). Use the same idea as in problem 4.8.

It is known (see, for example, [10]) that a function with the indicated properties does not exist if $\dim_H(K) < 1/2$. For the case $\dim_H(K) = 1/2$ see [8], pp. 346–348.

§6. Asymptotics of integrals of higher multiplicity

6.1. c) Using Fubini's theorem, show that

$$\int_{B^n(r)} |x_1|^p\,dx = 2\alpha_{n-1}\int_0^r t^p(r^2-t^2)^{(n-1)/2}\,dt.$$

Express the last integral with the help of the function Γ and use its properties.

6.4. Using Fubini's theorem, we see that

$$\lambda_n\{(x_1, \dots, x_n) \in B^n \mid |x_1| < \varepsilon\} = \int_{-\varepsilon}^{\varepsilon} \alpha_{n-1}(1-x^2)^{(n-1)/2}\,dx,$$

$$\alpha_n = \int_{-1}^{1} \alpha_{n-1}(1-x^2)^{(n-1)/2}\,dx.$$

The right-hand sides of these equalities are equivalent as $n \to \infty$ (see problem VI.2.4). Therefore, $\alpha_n^{-1}\lambda_n\{(x_1, \dots, x_n) \in B^n \mid |x_1| < \varepsilon\} \underset{n\to\infty}{\longrightarrow} 1$.

6.5. By definition, $L_n = \sigma_{n-1}^{-2}\int_{S^{n-1}}\int_{S^{n-1}} \|x' - x\|\,d\mu_{n-1}(x)\,d\mu_{n-1}(x')$.

Since the measure μ_{n-1} is spherically invariant, we get that $L_n = \sigma_{n-1}^{-1} \times \int_{S^{n-1}} \|e_1 - x\| \, d\mu_{n-1}(x)$, where $e_1 = (1, 0, \ldots, 0)$. Reduce the last integral to an integral over $[-1, 1]$ and use the asymptotic theorem of Laplace. The result obtained ($\lim_{n\to\infty} L_n = \sqrt{2}$) attests that radii drawn to two randomly chosen points on the sphere are almost orthogonal, "as a rule."

6.6. Assume that $0 \le a < b \le \sqrt{n}$. Then

$$\begin{aligned} p_n &= P_n\{(x_1, \ldots, x_n) \in S^{n-1}(\sqrt{n}) | a \le x_n \le b\} \\ &= \frac{n^{-(n-1)/2}}{\sigma_{n-1}} \int \cdots \int_{n-b^2 \le x_1^2 + \cdots + x_{n-1}^2 \le n-a^2} \frac{\sqrt{n}\, dx_1 \cdots dx_{n-1}}{\sqrt{n - x_1^2 - \cdots - x_{n-1}^2}}. \end{aligned}$$

Reduce this integral to the form

$$\frac{(n-1)\alpha_{n-1}}{n\alpha_n} \int_{\sqrt{1-b^2/n}}^{\sqrt{1-a^2/n}} u^{n-2} \frac{du}{\sqrt{1-u^2}}.$$

Making the change of variable $u = \sqrt{1 - t^2/n}$ in the last integral, we get that

$$p_n = \frac{(n-1)\alpha_{n-1}}{n^{3/2}\alpha_n} \int_a^b (1 - t^2/n)^{(n-3)/2} \, dt.$$

It remains to pass to the limit under the integral sign and to find the limit of the factor in front of it.

6.7. We represent the integral over the ball $\alpha_n^{-1} \int_{B^n} f(x)\, dx$ in the form $\alpha_n^{-1} \int_{B^n} f(x/\|x\|)\, dx + \mathcal{I}_1 + \mathcal{I}_2 + \mathcal{I}_3$, where

$$\begin{aligned} \mathcal{I}_1 &= \alpha_n^{-1} \int_{B^n(1-\varepsilon)} f(x)\, dx, \\ \mathcal{I}_2 &= \alpha_n^{-1} \int_{1-\varepsilon \le \|x\| \le 1} \left(f(x) - f\left(\frac{x}{\|x\|}\right) \right) dx, \\ \mathcal{I}_3 &= -\alpha_n^{-1} \int_{B^n(1-\varepsilon)} f\left(\frac{x}{\|x\|}\right) dx. \end{aligned}$$

It is clear that $|\mathcal{I}_1| \le M(1-\varepsilon)^n$, $|\mathcal{I}_3| \le M(1-\varepsilon)^n$, and $\mathcal{I}_2 \le h(\varepsilon)$. Moreover,

$$\begin{aligned} \alpha_n^{-1} \int_{B^n} f\left(\frac{x}{\|x\|}\right) dx &= \alpha_n^{-1} \int_0^1 r^{n-1} \left\{ \int_{S^{n-1}} f(w)\, d\mu_{n-1}(w) \right\} dr \\ &= \frac{1}{n\alpha_n} \int_{S^{n-1}} f(w)\, d\mu_{n-1}(w). \end{aligned}$$

Since $n\alpha_n = \sigma_{n-1}$, it follows that

$$\begin{aligned} \left| \alpha_n^{-1} \int_{B^n} f(x)\, dx - \sigma_{n-1}^{-1} \int_{S^{n-1}} f(x)\, d\mu_{n-1}(x) \right| &\le |\mathcal{I}_1| + |\mathcal{I}_2| + |\mathcal{I}_3| \\ &\le 2M(1-\varepsilon)^n + h(\varepsilon). \end{aligned}$$

6.8. Represent the given integral in the form

$$\left(\frac{1}{\sqrt{2\pi}}\int_{-\infty}^{+\infty}|x|^{q/n}e^{-x^2/2}\,dx\right)^n$$

and use the fact that

$$\frac{1}{\sqrt{2\pi}}\int_{-\infty}^{+\infty}|x|^{q/n}e^{-x^2/2}dx = 1+\frac{1}{\sqrt{2\pi}}\int_{-\infty}^{+\infty}\frac{q}{n}\ln|x|\cdot e^{-x^2/2}\,dx+O\left(\frac{1}{n^2}\right).$$

6.9. Use the formula $(*)$ from the beginning of §6.

6.10. a) Use the result of problem V.2.3.

b) Use the result of problem VI.2.16.

c) Let $\mathcal{I}=\int_{\|x\|\geq 1+\varepsilon}d\tilde{\gamma}_n(x)$. Then

$$\begin{aligned}\mathcal{I}&=\left(\frac{n}{2\pi}\right)^{n/2}\cdot\frac{2\pi^{n/2}}{\Gamma(n/2)}\cdot\int_{1+\varepsilon}^{\infty}t^{n-1}\cdot e^{-nt^2/2}\,dt\\&=\frac{n^{n/2}}{\Gamma(n/2)}2^{1-n/2}\cdot e^{-(n-1)/2}\int_{1+\varepsilon}^{\infty}e^{-(n-1)(t^2/2-\ln t-1/2)}e^{-t^2/2}\,dt.\end{aligned}$$

Since

$$\frac{t^2}{2}-\ln t-\frac{1}{2}\geq\frac{(1+\varepsilon)^2}{2}-\ln(1+\varepsilon)-\frac{1}{2}\geq\frac{\varepsilon^2}{2}$$

and $\Gamma(n/2)\geq(n/2)^{(n-1)/2}e^{-n/2}\sqrt{2\pi}$, it follows that

$$\begin{aligned}\mathcal{I}&\leq\sqrt{\frac{ne}{\pi}}\cdot e^{-(n-1)\varepsilon^2/2}\int_{1+\varepsilon}^{\infty}e^{-t^2/2}\,dt\\&\leq\sqrt{\frac{n}{\pi}}\cdot e\cdot e^{-n\varepsilon^2/2}\int_{1+\varepsilon}^{\infty}e^{-t^2/2}\,dt.\end{aligned}$$

We get similarly that

$$\int_{\|x\|\leq 1-\varepsilon}d\tilde{\gamma}_n(x)\leq\sqrt{\frac{n}{\pi}}\cdot e\cdot e^{-n\varepsilon^2/2}\int_0^{1-\varepsilon}e^{-t^2/2}\,dt.$$

Consequently,

$$\begin{aligned}\int_{|\,\|x\|-1|\geq\varepsilon}d\tilde{\gamma}_n(x)&\leq\sqrt{\frac{n}{\pi}}\cdot e\cdot e^{-n\varepsilon^2/2}\int_0^{\infty}e^{-t^2/2}\,dt\\&=\frac{e}{\sqrt{2}}\sqrt{n}\cdot e^{-n\varepsilon^2/2}<2\sqrt{n}e^{-n\varepsilon^2/2}.\end{aligned}$$

6.11. Using the formula $(*)$ from the beginning of §6, we get that

$$\begin{aligned}\int_{\mathbb{R}^n}\|x\|_p^q\,d\tilde{\gamma}_n(x)&=\left(\frac{n}{2\pi}\right)^{n/2}\int_{\mathbb{R}^n}\|x\|_p^q\cdot e^{-n\|x\|^2/2}\,dx\\&=\left(\frac{n}{2\pi}\right)^{n/2}\int_0^{\infty}r^{n-1+q}e^{-nr^2/2}\left(\int_{S^{n-1}}\|w\|_p^q\,d\mu_{n-1}(w)\right)dr.\end{aligned}$$

Moreover,

$$\left(\frac{n}{2\pi}\right)^{n/2}\int_0^\infty r^{n-1+q}e^{-nr^2/2}\,dr$$
$$=\frac{2^{q/2-1}}{\pi^{n/2}}\cdot\frac{1}{n^{q/2}}\cdot\Gamma\left(\frac{n}{2}+\frac{q}{2}\right)\sim\frac{2^{q/2-1}}{\pi^{n/2}}\cdot\frac{\Gamma\left(\frac{n}{2}\right)\cdot\left(\frac{n}{2}\right)^{q/2}}{n^{q/2}}=\sigma_{n-1}^{-1}.$$

6.12. Use the formula $(*)$ and the relation $\Gamma(a+c)\underset{a\to+\infty}{\sim}a^c\cdot\Gamma(a)$ (see problem VI.2.15).

6.13. Let

$$F(t)=\int_{\|x\|_\infty\le t}d\gamma_n(x)=\left(1-\sqrt{\frac{2}{\pi}}\int_t^\infty e^{-u^2/2}\,du\right)^n\qquad(t\ge0).$$

Then

$$\int_{\mathbb{R}^n}\|x\|_\infty^q\,d\gamma_n(x)=\int_0^\infty t^q\,dF(t)$$
$$=n\sqrt{2/\pi}\int_0^\infty t^q\left(1-\sqrt{2/\pi}\int_t^\infty e^{-u^2/2}\,du\right)^{n-1}\cdot e^{-t^2/2}\,dt. \tag{1}$$

Let $z(t)=\sqrt{2/\pi}\int_t^\infty e^{-u^2/2}\,du$. It is clear that $z'(t)=-\sqrt{2/\pi}\cdot e^{-t^2/2}$, $z(0)=1$, $z(t)\underset{t\to\infty}{\to}0$. Therefore, making the change of variable $z=z(t)$ in the integral on the right-hand side of (1), we get that

$$\int_{\mathbb{R}^n}\|x\|_\infty^q\,d\gamma_n(x)=n\int_0^1 t^q(z)(1-z)^{n-1}\,dz,$$

where $t(z)$ is the function inverse to $z(t)$. Since $z(t)\underset{t\to+\infty}{\sim}\sqrt{2/\pi}\cdot(1/t)\cdot e^{-t^2/2}$, it follows that (see problem VI.4.1) $t(z)=\sqrt{2\ln(1/z)}\cdot(1+\alpha(z))$, where $\alpha(z)\underset{z\to+0}{\to}0$. Consequently,

$$\int_{\mathbb{R}^n}\|x\|_\infty^q\,d\gamma_n(x)=n\int_0^1(2\ln(1/z))^{q/2}(1+\alpha(z))^q\cdot(1-z)^{n-1}\,dz. \tag{2}$$

Using the results of problems VI.2.8 and VI.2.12, we get that

$$\int_0^1(\ln(1/z))^{q/2}(1+\alpha(z))^q(1-z)^{n-1}\,dz\sim\int_0^1(\ln(1/z))^{q/2}(1-z)^{n-1}\,dz$$
$$\sim\frac{1}{n}(\ln n)^{q/2}.$$

Together with (2) this leads to the required result.

6.14. Arguing as in the solution of problem 6.11, replace the given integral by the integral with respect to Gaussian measure. Then use the result of problem 6.13.

6.15. Using the inequality $\|x\|_p\ge n^{1/p}|x_1\cdot\dots\cdot x_n|^{1/n}$ and the result of problem 6.8, we get a lower estimate for $\mathcal{I}_n(p,q)$ in the case $q>0$, and an upper estimate in the case $q<0$.

a) We estimate $\mathcal{I}_n(p, q)$ from above. Since $\|x\|_p \le n^{1/p-1/2} \cdot \|x\|_2$, it follows that $\mathcal{I}_n(p, q) \le n^{q/p-q/2} \cdot \int_{\mathbb{R}^n} \|x\|_2^q \, d\gamma_n(x)$. Using the result of problem 6.12, we get what is required.

b) In this case

$$\mathcal{I}_n(p, q) \le \left(\int_{\mathbb{R}^n} \|x\|_p^p \, d\gamma_n(x)\right)^{q/p} = n^{q/p} \left(\int_{-\infty}^{+\infty} |x|^p \, d\gamma_1(x)\right)^{q/p}.$$

c) Since $\|x\|_p \le n^{1/p-1/q} \cdot \|x\|_q$, it follows that

$$\mathcal{I}_n(p, q) \le n^{q/p-1} \int_{\mathbb{R}^n} \|x\|_q^q \, d\gamma_n(x) = n^{q/p} \int_{-\infty}^{+\infty} |x|^q \, d\gamma_1(x).$$

We estimate $\mathcal{I}_n(p, q)$ from below for $q < 0$. Let $a = -q$. Then

$$1 = \int_{\mathbb{R}^n} \|x\|_p^{a/2} \cdot \|x\|_p^{q/2} \, d\gamma_n(x) \le \mathcal{I}_n^{1/2}(p, a) \cdot \mathcal{I}_n^{1/2}(p, q).$$

Therefore, $\mathcal{I}_n(p, q) \ge \mathcal{I}_n^{-1}(p, a)$, and it remains to use the upper estimate for $\mathcal{I}_n(p, q)$.

6.16. Use the results of problems 6.11 and 6.15.

6.17. Using the equality

$$1 = \frac{\pi^{n/2} \cdot r_n^n}{\Gamma(n/2+1)}$$

and Stirling's formula, we get that

$$r_n = \sqrt{\frac{n}{2\pi e}} \cdot e^{((\ln n)/2 + (\ln \pi)/2 + o(1))/n} = \sqrt{\frac{n}{2\pi e}} \left(1 + \frac{\ln n}{2n} + \frac{\ln \pi}{2n} + o\left(\frac{1}{n}\right)\right).$$

6.18. a) Show that the radius of the ball is equal to $\sqrt{n-1}$.

b) Show that the radius of the ball is equal to $\sqrt{n}/2 - 1$; use Stirling's formula.

6.19. Show that $\gamma_n(B^n(\rho_n)) = \frac{1}{\Gamma(n/2)} \cdot \int_0^{\rho_n^2/2} t^{n/2-1} \cdot e^{-t} \, dt$, and use the result of problem VI.2.17a) to show that $\rho_n^2/2 = n/2 + o(n^{1/2})$.

6.20. For $t > 0$ let $O_n(t) = \{(x_1, \dots, x_n) \in \mathbb{R}^n \mid \sum_{1 \le k \le n} |x_k| \le t\}$ and $F(t) = \lambda_n(O_n(t)) = (2t)^n/n!$. Find the mean value $a_n(t)$ of the function $\|x\|_1^{-1}$ on $O_n(t)$. It is clear that

$$\begin{aligned} a_n(t) &= \frac{n!}{(2t)^n} \int_0^t \frac{dF(u)}{u} = \frac{n}{t^n} \int_0^t u^{n-2} \, du \\ &= \frac{n}{n-1} t^{-1} \le \frac{2}{t} \qquad (n \ge 2). \end{aligned}$$

Let us estimate b_n. Since $\|x\|_1 \le \sqrt{n} \cdot \|x\|$, it follows that

$$\begin{aligned} b_n &\ge \alpha_n^{-1} n^{-1/2} \int_{B^n} \frac{dx}{\|x\|} \\ &= \alpha_n^{-1} n^{-1/2} \alpha_n \cdot n \int_0^1 t^{n-2} \, dt = \frac{\sqrt{n}}{n-1}. \end{aligned}$$

To get an upper estimate of b_n we fix a number $\varepsilon > 0$ to be made more precise later. Then

$$\begin{aligned}\alpha_n b_n &\le \int_{O_n(\varepsilon\sqrt{n})} \frac{dx}{\|x\|_1} + \int_{B^n \setminus O_n(\varepsilon\sqrt{n})} \frac{dx}{\|x\|_1} \\ &\le \frac{(2\varepsilon\sqrt{n})^n}{n!} \cdot \frac{2}{\varepsilon\sqrt{n}} + \frac{\alpha_n}{\varepsilon\sqrt{n}}.\end{aligned}$$

Thus,

$$b_n \le \frac{1}{\varepsilon\sqrt{n}} + \frac{2}{\varepsilon} \cdot \frac{(2\varepsilon)^n}{\pi^{n/2}} \cdot n^{(n-1)/2} \cdot \frac{\Gamma(n/2+1)}{n!}.$$

By using Stirling's formula it is easy to see that the second term is

$$O\left(\frac{1}{\sqrt{n}}\left(\varepsilon\sqrt{\frac{2e}{\pi}}\right)^n\right).$$

Taking $\varepsilon = 1/\sqrt{2}$, we get that $b_n \le \sqrt{2/n} + o(1/n)$.

We proceed to the estimation of c_n. Let $G(t) = \lambda_n([-t, t]^n)$. Then $G(t) = (2t)^n$, and since $|x_1| + \cdots + |x_n| \le n \max_{1\le k \le n} |x_k|$, it follows that

$$c_n \ge \frac{1}{2^n n} \int_0^1 \frac{dG(t)}{t} = \int_0^1 t^{n-2}\, dt = \frac{1}{n-1}.$$

On the other hand, using the same device as in the determination of an upper estimate for the numbers b_n, we have that

$$\begin{aligned}c_n &\le 2^{-n} \int_{O_n(\varepsilon n)} \frac{dx}{\|x\|_1} + 2^{-n} \int_{[-1,1]^n \setminus O_n(\varepsilon n)} \frac{dx}{\varepsilon n} \\ &\le \frac{(\varepsilon n)^n}{n!} a_n(\varepsilon n) + \frac{1}{\varepsilon n} \\ &= \frac{n}{n-1} \frac{(\varepsilon n)^{n-1}}{n!} + \frac{1}{\varepsilon n}.\end{aligned}$$

Stirling's formula gives us easily that the first term on the right-hand side of the last inequality is $O((\varepsilon e)^n)$. Taking $\varepsilon = 1/3$, we get that $c_n \le 3/n + o(1/n)$.

6.21. a) Let

$$\frac{n!}{2^n} \int_{O_n} \frac{\|x\|_2^2}{\|x\|_1}\, dx = \mathscr{I}_n.$$

Then

$$\begin{aligned}\mathscr{I}_n &= \frac{n!n}{2^n} \int_{O_n} \frac{x_1^2}{\|x\|_1}\, dx \\ &= \frac{n!n}{2^n} \int_{-1}^{+1} u^2 \left(\int \cdots \int_{O_{n-1}(1-|u|)} \frac{dx_2 \cdots dx_k}{|u| + |x_2| + \cdots + |x_n|} \right) du \\ &= \frac{n!n}{2^n} \int_{-1}^{1} u^2 \left(\int_0^{1-|u|} \frac{dF(t)}{|u|+t} \right) du,\end{aligned}$$

where $F(t) = \lambda_{n-1}(O_{n-1}(t)) = (2t)^{n-1}/(n-1)!$. Therefore,

$$\mathcal{J}_n = n^2(n-1)\int_0^1 u^2\left(\int_0^{1-u}\frac{t^{n-2}\,dt}{u+t}\right)du$$
$$= n^2(n-1)\int_0^1 t^{n-2}\left(\int_0^{1-t}\frac{u^2\,du}{t+u}\right)dt.$$

Since $u^2 \le u^2/(u+t) \le u^2/t$, it follows that

$$n^2(n-1)\int_0^1 t^{n-2}\frac{(1-t)^3}{3}\,dt \le \mathcal{J}_n \le n^2(n-1)\int_0^1 t^{n-3}\frac{(1-t)^3}{3}\,dt \sim \frac{2}{n}.$$

b) Let $\alpha_n^{-1}\int_{B^n}\|x\|^2/\|x\|_1\,dx = \mathcal{J}_n$. Since $\|x\|_1 \le \sqrt{n}\|x\|_2$, it follows that

$$\mathcal{J}_n \ge \frac{\alpha_n^{-1}}{\sqrt{n}}\int_{B^n}\|x\|\,dx = \frac{\alpha_n^{-1}}{\sqrt{n}}n\alpha_n\int_0^1 t^n\,dt = \frac{\sqrt{n}}{n+1}.$$

The upper estimate for $\mathcal{J}_n$ follows from the inequality $\mathcal{J}_n \le \alpha_n^{-1} \times \int_{B^n} 1/\|x\|_1\,dx$ and the estimate for b_n in problem 6.20.

c) It is clear that

$$K_n = n\cdot 2^{-n}\int\cdots\int_{[-1,1]^n}\frac{x_1^2\cdot dx_1\cdots dx_n}{|x_1|+|x_2|+\cdots+|x_n|}.$$

Using the inequality $|x_1|+\cdots+|x_n| \le n\cdot\max_{1\le k\le n}|x_k|$, we get that

$$K_n \ge n2^{-n}\int\cdots\int_{[-1,1]^n}\frac{|x_1|^2}{n}\,dx_1\cdots dx_n = \frac{1}{3}.$$

The upper estimate of K_n follows at once from the inequality

$$K_n \le 2^{-n}\int_{[-1,1]^n} n/\|x\|_1\,dx$$

and the estimate for c_n in problem 6.20.

6.22. b) We remark first that

$$\frac{\lambda_n(O_n)}{\alpha_n} = \frac{2^{n-1}\cdot n\cdot\Gamma(n/2)}{\pi^{n/2}\cdot n!} \underset{n\to\infty}{\sim} \frac{1}{\sqrt{2}}\left(\frac{2e}{\pi n}\right)^{n/2}. \qquad (*)$$

Show that

$$O_n(1+\varepsilon\sqrt{n})\cap[-1,1]^n \subset O_n+\varepsilon B^n \subset O_n(1+\varepsilon\sqrt{n}).$$

It follows from these inclusions that

$$\frac{2^n(1+\varepsilon\sqrt{n})^n}{n!} - n\frac{(2\varepsilon\cdot\sqrt{n})^n}{n!} \le V_n \le \frac{2^n}{n!}(1+\varepsilon\sqrt{n})^n,$$

where $V_n = \lambda_n(O_n+\varepsilon B^n)$, that is,

$$\frac{(2\varepsilon\sqrt{n})^n}{n!}\left(\left(1+\frac{1}{\varepsilon\sqrt{n}}\right)^n - 1\right) \le V_n \le \frac{(2\varepsilon\sqrt{n})^n}{n!}\left(1+\frac{1}{\varepsilon\sqrt{n}}\right).$$

Using $(*)$, we get that

$$\frac{V_n}{\alpha_n} \underset{n\to+\infty}{\sim} \frac{\lambda_n(O_n)}{\alpha_n}(\varepsilon\sqrt{n})^n e^{\sqrt{n}/\varepsilon}$$
$$\underset{n\to\infty}{\sim} \frac{\varepsilon^n}{\sqrt{2}}\left(\frac{2e}{\pi}\right)^{n/2} e^{\sqrt{n}/\varepsilon - 1/(2\varepsilon^2)}.$$

From this it follows, in particular, that $(V_n/\alpha_n)^{1/n} \underset{n\to+\infty}{\to} \varepsilon\sqrt{2e/\pi}$.

6.23. For $0 \le t \le 1$ let $A_n(t) = \{(x_1, \dots, x_n) \in A_n | 0 \le x_n \le t\}$ $(n \ge 2)$, $A_1(t) = [0, t]$, $V_n(t) = \lambda_n(A_n(t))$ for $n \ge 1$, and $V_0(t) \equiv 1$. Then $V_n(t) = \int_0^t V_{n-1}(1-u)\,du$ for $n \ge 1$, and hence

$$V_n'(t) = V_{n-1}(1-t), \quad V_n''(t) = -V_{n-2}(t) \qquad (n \ge 2).$$

Deduce from this that

$$\sum_{0\le j\le k} \frac{(-1)^j}{(2j)!} V_{2(k-j)}(1) = 0, \qquad \sum_{0\le j\le k} \frac{(-1)^j}{(2j)!} V_{2(k-j)+1}(1) = \frac{(-1)^k}{(2k+1)!}. \qquad (*)$$

Let $F(x) = \sum_{k\ge0} V_{2k}(1)\cdot x^{2k}$ and $G(x) = \sum_{k\ge0} V_{2k+1}(1)x^{2k+1}$. Using $(*)$, verify that $F(x)\cos x \equiv 1$ and $G(x)\cos x \equiv \sin x$, that is, $F(x) = 1/\cos x$ and $G(x) = \tan x$. Use the expansions

$$\frac{1}{\sin x} = \sum_{n\in\mathbb{Z}} \frac{(-1)^n}{x+n\pi}, \qquad \operatorname{cotan} x = \sum_{n\in\mathbb{Z}} \frac{1}{x+n\pi}.$$

Replacing x by $\pi/2 - x$ in them, we get that

$$\frac{1}{\cos x} = \sum_{k\ge1} \frac{(-1)^{k-1}(2k-1)\pi}{(2k-1)^2\pi^2/4 - x^2}, \qquad \tan x = \sum_{k\ge1} \frac{2x}{(2k-1)^2\pi^2/4 - x^2}.$$

Consequently (for $|x| < \pi/2$),

$$\frac{1}{\cos x} = \sum_{k\ge1} \frac{4(-1)^{k-1}}{\pi(2k-1)} \sum_{m\ge0}\left(\frac{2x}{\pi(2k-1)}\right)^{2m}$$
$$= \frac{4}{\pi}\sum_{m\ge0}\left(\frac{2x}{\pi}\right)^{2m} \sum_{k\ge1} \frac{(-1)^{k-1}}{(2k-1)^{2m+1}},$$

$$\tan x = \sum_{k\ge1} \frac{8x}{\pi^2(2k-1)^2} \sum_{m\ge0}\left(\frac{2x}{\pi(2k-1)}\right)^{2m}$$
$$= \frac{4}{\pi}\sum_{m\ge0}\left(\frac{2x}{\pi}\right)^{2m+1} \sum_{k\ge1} \frac{1}{(2k-1)^{2m+2}}.$$

Thus,

$$V_{2m}(1) = \frac{4}{\pi}\left(\frac{2}{\pi}\right)^{2m} \sum_{k\geq 1} \frac{(-1)^{k-1}}{(2k-1)^{2m+1}},$$

$$V_{2m+1}(1) = \frac{4}{\pi}\left(\frac{2}{\pi}\right)^{2m+1} \sum_{k\geq 1} \frac{1}{(2k-1)^{2m+2}}.$$

From this, $V_n(1) = 2(2/\pi)^{n+1}(1 + O(1/3^n))$.

6.24. Since $\lambda_n(2V_n) > 2^n(3/2)^n$, the set $2V_n$ contains at least $(3/2)^n$ points with integer coordinates (see problem 1.3b)). We remark that the nonzero coordinates of these points can be equal only to $+1$ or -1, because $2V_n \subset (-2, 2)^n$. Let K_n be the number of points with coordinates 0 and ± 1 such that the number of nonzero coordinates does not exceed $m = [n/10]$. It is clear that $K_n = 1+2\binom{n}{1}+2^2\cdot\binom{n}{2}+\dots+2^m\cdot\binom{n}{m}$. Verify that $K_n < (3/2)^n$ for large n; this implies that the set $2V_n$ contains the desired point.

6.25. a) Fix an arbitrary positive number ε and consider the set

$$E_n(\varepsilon) = \left\{(x_1, \dots, x_n) \in [0, 1]^n \,\middle|\, \left|\frac{x_1^2 + \dots + x_n^2}{n} - \frac{1}{3}\right| > \varepsilon\right\}.$$

By Chebyshev's inequality and the pairwise orthogonality of the functions $x_k^2 - 1/3$ $(k = 1, 2, \dots, n)$ on $[0, 1]^n$,

$$\begin{aligned}\lambda_n(E_n(\varepsilon)) &\leq \frac{1}{n^2\varepsilon^2}\int_{[0,1]^n}\left(\sum_{1\leq k\leq n}\left(x_k^2 - \frac{1}{3}\right)\right)^2 dx \\ &= \frac{1}{n^2\varepsilon^2}\int_{[0,1]^n}\sum_{1\leq k\leq n}\left(x_k^2 - \frac{1}{3}\right)^2 dx = O\left(\frac{1}{n}\right).\end{aligned}$$

(Thus, $\|x\|/\sqrt{n}$ "almost coincides" with $1/\sqrt{3}$ on a set of measure as close as desired to 1.) Consequently,

$$\begin{aligned}\left|\frac{1}{\sqrt{n}}\int_{[0,1]^n}\|x\|\,dx - \frac{1}{\sqrt{3}}\right| &\leq \int_{[0,1]^n}\left|\frac{\|x\|}{\sqrt{n}} - \frac{1}{\sqrt{3}}\right| dx \\ &\leq \int_{[0,1]^n}\left|\frac{\|x\|^2/n - 1/3}{1/\sqrt{3}}\right| dx \\ &\leq \sqrt{3}\left(\int_{E_n(\varepsilon)} dx + \int_{[0,1]^n\setminus E_n(\varepsilon)} \varepsilon\,dx\right) \\ &\leq \sqrt{3}(\lambda_n(E_n(\varepsilon)) + \varepsilon).\end{aligned}$$

Since $\lambda_n(E_n(\varepsilon)) \to 0$, for sufficiently large n

$$\left|\frac{1}{\sqrt{n}}\int_{[0,1]^n}\|x\|\,dx - \frac{1}{\sqrt{3}}\right| \leq 2\varepsilon\sqrt{3}.$$

b) Consider the set

$$E_n(\varepsilon) = \left\{ (x_1, \dots, x_n) \in [0, 1]^n \,\middle|\, \left| \frac{x_1 + \dots + x_n}{n} - \frac{1}{2} \right| > \varepsilon \right\}$$

and prove that $\lambda_n(E_n(\varepsilon)) \to_{n\to\infty} 0$. Use the continuity of the function f at the point $1/2$.

c) Since $\sqrt[n]{x_1 \cdots x_n} = e^{(1/n)\sum_{1\le k\le n} \ln x_k}$ and $\int_0^1 \ln x\, dx = -1$, consider the set

$$E_n(\varepsilon) = \left\{ (x_1, \dots, x_n) \in [0, 1]^n \,\middle|\, \left| \frac{1}{n} \sum_{1\le k\le n} \ln x_k + 1 \right| > \varepsilon \right\}$$

and use the same considerations as in a) and b).

d) Remarking that

$$\frac{1}{\sqrt{2\pi}} \int_{-\infty}^{+\infty} |x| e^{-x^2/2}\, dx = \sqrt{\frac{2}{\pi}},$$

consider the set

$$E_n(\varepsilon) = \left\{ (x_1, \dots, x_n) \in \mathbb{R}^n \,\middle|\, \left| \frac{1}{n} \sum_{1\le k\le n} |x_k| - \sqrt{\frac{2}{\pi}} \right| > \varepsilon \right\}.$$

6.26. c) It obviously suffices to consider only the case when $nw_n \to \infty$. Let $A_n = \{(x_1, \dots, x_n) \in \mathbb{R}^n \mid \sum_{1\le k\le n} |x_k|^p \le nw_n\}$ and

$$\mathcal{I}_n = \int \cdots \int_{A_n} f(x_1) \cdots f(x_n)\, dx_1 \cdots dx_n.$$

There is a number $\delta > 0$ such that $\int_{-\delta}^{\delta} f(x)\, dx < e^{-1}$. For $x = (x_1, \dots, x_n) \in A_n$ let $E(x) = \{j \in \mathbb{N} \mid |x_j| \ge \delta\}$. It is clear that $\operatorname{card} E(x) \le [nw_n/\delta^p]$. Let $N = [nw_n/\delta^p]$, $\mathcal{B} = \{B \subset \{1, 2, \dots, n\} \mid \operatorname{card} B = N\}$, and $C_B = \{(x_1, \dots, x_n) \in \mathbb{R}^n \mid |x_k| < \delta \text{ for } k \notin B\}$. Then $A_n \subset \bigcup_{B\in\mathcal{B}} C_B$. Since $\int \cdots \int_{C_B} f(x_1) \cdots f(x_n)\, dx_1 \cdots dx_n \le (\int_{-\delta}^{\delta} f(x)\, dx)^{n-N} < e^{-(n-N)}$, it thus follows that

$$\mathcal{I}_n \le \sum_{B\in\mathcal{B}} \int \cdots \int_{C_B} f(x_1) \cdots f(x_n)\, dx_1 \cdots dx_n \le \binom{n}{N} e^{-(n-N)}. \qquad (1)$$

Estimating $\binom{n}{N}$, we get that

$$\binom{n}{N} \le \frac{n^N}{N!} \le \left(\frac{n}{N}\right)^N e^N \le \left(\frac{2\delta^p}{w_n}\right)^N e^N.$$

Together with (1) this gives us that

$$\mathcal{I}_n \le e^{-n} \cdot e^{N(2+\ln(2\delta^p/w_n))} \le e^{-n(1-(2w_n/\delta^p)-(w_n/\delta^p)\ln(2\delta^p/w_n))}.$$

Since $w_n \ln(2\delta^p/w_n) \to 0$ as $n \to \infty$, we get for sufficiently large n that $\mathcal{I}_n \le e^{-n/2}$.

6.27. It follows from the Hölder inequality that the quantity $S_p(a) = (2^{-n} \int_{Q_n} |(x, a)|^p\, dx)^{1/p}$ is increasing as p increases. For $p = 2$ it coincides with $\|a\|/3$. Thus, the right-hand inequality is nontrivial only for $p > 2$, and the left-hand inequality is nontrivial for $p < 2$. Let us prove the right-hand inequality. Obviously, it suffices to do this in the case when $\|a\| = 1$. It will be assumed first that $p = 2m$ is an even number. Using the inequality $t^{2m}/(2m)! \le \cosh t$, we get that

$$\int_{Q_n} (x, a)^{2m}\, dx \le (2m)! \int_{Q_n} \cosh(x, a)\, dx = 2^n (2m)! \prod_{1\le k\le n} \frac{\sinh a_k}{a_k},$$

where a_k $(k = 1, \dots, n)$ are the coordinates of the vector a. Using the inequality $(\sinh t)/t \le e^{t^2/2}$, we get that

$$2^{-n} \int_{Q_n} (x, a)^{2m}\, dx \le (2m)! e^{(1/2)\sum_{1\le k\le n} a_k^2} = (2m)! e^{1/2}.$$

This implies that $S_{2m}(a) \le 2m\|a\|$. In the case of an arbitrary $p > 2$ we find a positive integer m such that $2m \le p < 2m + 2$. Then

$$S_p(a) \le S_{2m+2}(a) \le (2m+2)\|a\| \le (p+2)\|a\|.$$

Let us proceed to the proof of the left-hand inequality for $p \in (0, 2)$. Consider a number $\theta \in (0, 1)$ such that $2 = \theta p + 4(1-\theta)$. Using the Hölder inequality with the exponent $1/\theta$, we get that

$$\begin{aligned}\frac{1}{3}\|a\|^2 &= 2^{-n} \int_{Q_n} |(x, a)|^2\, dx = 2^{-n} \int_{Q_n} |(x, a)|^{\theta p} \cdot |(x, a)|^{4(1-\theta)}\, dx \\ &\le \left(2^{-n} \int_{Q_n} |(x, a)|^p\right)^{\theta} \cdot \left(2^{-n} \int_{Q_n} |(x, a)|^4\, dx\right)^{1-\theta}.\end{aligned}$$

Consequently,

$$\begin{aligned}\left(2^{-n} \int_{Q_n} |(x, a)|^p\, dx\right)^{\theta} &\ge \frac{\|a\|^2}{3(S_4(a))^{4(1-\theta)}} \\ &\ge \frac{\|a\|^2}{3B_4^{4(1-\theta)} \cdot \|a\|^{4(1-\theta)}} = \frac{\|a\|^{p\theta}}{3B_4^{4(1-\theta)}},\end{aligned}$$

that is, $S_p(a) \ge (3B_4^{4(1-\theta)})^{-1/p\theta} \cdot \|a\|$.

CHAPTER IX

Sequences of Measurable Functions

§1. Convergence in measure and almost everywhere

1.2. b) Show that the series $\sum_{k\geq 1}\int_0^1 f_{[c^k]}(x)\,dx$ converges.

1.3. Consider the functions $g_n = (-1)^n \cdot n \cdot f_n$, where the f_n are the functions in problem 1.2.

1.5. Let $A_n \in \mathbb{N}$. Then the functions $|\sin(A_n x + \varphi_n)|$ and $|\sin x|$ are equidistributed on $(0, 2\pi)$. Therefore,

$$\begin{aligned}&\lambda\{x \in (0, 2\pi) \mid |\sin(A_n x + \varphi_n)|^{p_n} > \varepsilon\}\\ &\quad= \lambda\{x \in (0, 2\pi) \mid |\sin x| > \varepsilon^{1/p_n}\} \underset{n\to\infty}{\longrightarrow} 0.\end{aligned}$$

If $A_n \notin \mathbb{N}$, then consider the smallest interval containing $(0, 2\pi)$ and an integer number of periods of the function $|\sin(A_n x+\varphi_n)|$, and use analogous considerations.

1.6. To prove the first of the required equalities consider the sets

$$E_n(\varepsilon) = \{x \in (0, 2\pi) \mid \sin(A_n x + \varphi_n) < 1 - \varepsilon\} \qquad (\varepsilon > 0)$$

and prove that for any $\varepsilon > 0$ there is a sequence $\{n_k\}_{k\geq 1}$ of indices such that the intersection $\bigcap_{k\geq p} E_{n_k}(\varepsilon)$ has zero measure for each $p \in \mathbb{N}$.

1.7. Let $I_n = \int_0^\infty e^{-x}|\sin(nx + \varphi_n)|^n\,dx$. It is clear that

$$\begin{aligned}I_n &= \frac{1}{n}\int_0^\infty e^{-t/n}|\sin(t + \varphi_n)|^n\,dt = \frac{1}{n}\sum_{k\geq 0}\int_{k\pi}^{(k+1)\pi} e^{-t/n}|\sin(t + \varphi_n)|^n\,dt\\ &= \frac{1}{n}\sum_{k\geq 0}\int_0^\pi e^{-\frac{s+k\pi}{n}}|\sin(s + \varphi_n)|^n\,ds\\ &= \frac{1}{n}\sum_{k\geq 0}e^{-\frac{k\pi}{n}}\int_0^\pi \left(1 + O\left(\frac{s}{n}\right)\right)|\sin(s + \varphi_n)|^n\,ds\\ &= \frac{1}{n}\left(1 + O\left(\frac{1}{n}\right)\right)\sum_{k\geq 0}e^{-\frac{k\pi}{n}}\int_0^\pi \sin^n s\,ds = \frac{1 + O(\frac{1}{n})}{n(1 - e^{-\pi/n})}\int_0^\pi \sin^n s\,ds.\end{aligned}$$

Since $\int_0^\pi \sin^n s\,ds \sim \sqrt{\frac{2\pi}{n}}$ (see VI.2.9a)), it follows that

$$I_n \sim \frac{1}{\pi}\sqrt{\frac{2\pi}{n}} = \sqrt{\frac{2}{\pi n}}.$$

1.9. a) Use the result of problem 1.8.

b) Consider the set $e = \bigcup_{n\geq p}\{x \in \mathbb{R} \mid |f_n(x)| > \varepsilon_n\}$ for sufficiently large p.

1.10. It can be assumed without loss of generality that the sequence $\{f_n\}_{n\geq 1}$ is decreasing. Consider a subsequence $\{f_{n_k}\}_{k\geq 1}$ such that $\lambda\{x \in (0, 1) \mid f_{n_k} \geq 1/k^2\} \leq 1/k^2$ $(k = 1, 2, \dots)$. Using the result of problem 1.9a), show that the sequence $\{A_n\}_{n\geq 1}$, where $A_n = k$ for $n_k \leq n < n_{k+1}$, is the desired sequence.

1.11. Let $\{A_n\}_{n\geq 1}$ be a sequence such that $A_n \to \infty$ and $A_n(f_n(x) - f(x)) \to 0$ a.e. on $(0, 1)$. Prove that $g = \sup_n |A_n(f_n - f)|$ is the desired function.

1.12. Consider the set $e = \{x \in (0, 1) \mid g(x) > C\}$, where g is the function in problem 1.11 and C is a sufficiently large positive number.

1.15. Consider a sequence $\{k_n\}_{n\geq 1}$ such that

$$\lambda\{x \in (0, 1) \mid |f_{n,k_n}(x) - f_n(x)| > 1/n\} < 1/n,$$

and use the inclusion

$$\begin{aligned}\{x \in (0, 1) \mid |f_{n,k_n}(x) - f_0(x)| > \varepsilon\} \\ \subset \{x \in (0, 1) \mid |f_{n,k_n}(x) - f_n(x)| > \varepsilon/2\} \\ \cup \{x \in (0, 1) \mid |f_n(x) - f_0(x)| > \varepsilon/2\}.\end{aligned}$$

b) Use Egorov's theorem (problem 1.12).

c) Use the result of problem III.1.6.

§2. Convergence in the mean. The law of large numbers

2.1. Use the Hölder inequality.

2.2. a) Use the inequality

$$\lambda\{x \in E \mid |f_n(x) - f_0(x)| > \varepsilon\} \leq \frac{1}{\varepsilon^r}\int_E |f_n(x) - f_0(x)|^r\,dx.$$

2.3. Use the Cauchy-Schwartz-Bunyakovskiĭ inequality and the absolute continuity of the integral.

2.4. Prove that $\int_0^1 (f_n(x) - f_0(x))f_0(x)\,dx \underset{n\to\infty}{\to} 0$.

2.6. a) To estimate the integral $\int_E |f(x)|\,dx$ represent $|f|$ as the product $|f|^t|f|^{1-t}$ and choose the numbers $t > 0$ and $s > 1$ such that $st = r$ and $(1-t)s' = 2$, where $s' = s/(s-1)$; use the Hölder inequality.

b) For $\alpha > 0$ let $E(\alpha) = \{x \in E \mid |f(x)| > \mathcal{J}_\alpha(f)\}$. It is clear that $\lambda_m(E(\alpha)) \leq \alpha$ and

$$\begin{aligned}\|f\|_1 &\leq \int_{E(\alpha)} |f(x)|\,dx + \lambda_m(E)\mathcal{J}_\alpha(f) \\ &\leq \sqrt{\alpha}\|f\|_2 + \lambda_m(E)\mathcal{J}_\alpha(f) \\ &\leq C\sqrt{\alpha}\|f\|_1 + \lambda_m(E)\mathcal{J}_\alpha(f).\end{aligned}$$

Consequently, if $C\sqrt{\alpha} \le 1/2$, then $\|f\|_1/2 \le (1-C\sqrt{\alpha})\|f\|_1 \le \lambda_m(E)\mathscr{J}_\alpha(f)$.

2.7. We fix a number $\varepsilon > 0$ whose choice is made more precise below, and we let $f = \sum_{n\ge 1} |a_n f_n|$. It is clear that there is a number $t > 0$ such that $\lambda_m(E_t) < \varepsilon$, where $E_t = \{x \in E \mid f(x) > t\}$. Then $\int_{E_t} |f_n(x)|\,dx \le \sqrt{\lambda_m(E_t)}C \le C\sqrt{\varepsilon}$. The number $\varepsilon > 0$ is chosen so that $C\sqrt{\varepsilon} < \delta/2$. In this case $\int_{E\setminus E_t} |f_n(x)|\,dx > \delta/2$. Consequently,

$$t\lambda_m(E) \ge \int_{E\setminus E_t} f(x)\,dx = \sum_{n\ge 1} |a_n| \int_{E\setminus E_t} |f_n(x)|\,dx$$
$$\ge \sum_{n\ge 1} |a_n|\delta/2,$$

from which

$$\sum_{n\ge 1} |a_n| \le \frac{2t}{\delta}\lambda_m(E).$$

2.8. Prove that $\varliminf \int_E |\sin(nx+\varphi_n)|\,dx > 0$ for any (measurable) set $E \subset (0, 2\pi)$, and use the result of problem 2.7.

2.9. Use the result of problem 2.2a).

2.10. Use the result of problem 2.9.

2.11. Prove that the measure of the set

$$\left\{x \in (-\pi, \pi) \mid \frac{1}{n} \sum_{1\le k\le n} k \sin^2 kx < \frac{n}{4}\right\}$$

tends to zero as $n \to \infty$.

2.12. b) Prove that $|\sigma_n| \le |\sigma_{k^2}| + 2M/k$, where $k = [\sqrt{n}]$.

2.13. Let $A_k = \sum_{k^2<n\le(k+1)^2} \|f_n\|_2^2$ $(k = 0, 1, 2, \ldots)$. Then

$$\sum_{n\ge 1} n^{-3/2}\|f_n\|_2^2 \ge \sum_{k\ge 1} k^{-3} \sum_{k^2<n\le(k+1)^2} \|f_n\|_2^2 = \sum_{k\ge 1} k^{-3} A_k.$$

Therefore,

$$\sum_{k\ge 1} \|\sigma_{k^2}\|_2^2 \le \sum_{k\ge 1} k^{-4} \sum_{0\le j<k} A_j$$
$$\le \sum_{j\ge 0} A_j \sum_{k\ge j+1} k^{-4}$$
$$\le 2A_0 + \sum_{j\ge 1} j^{-3} A_j < \infty.$$

Thus, $\sigma_{k^2}(x) \to 0$ a.e. on E. Suppose now that $n = k^2+p$, where $k = [\sqrt{n}]$ and $0 \le p \le 2k$, and let $g = \sum_{n\ge 1} n^{-3/2} f_n^2$. It is clear that $g \in \mathscr{L}^1(E)$, and hence $\omega_k(x) = \sum_{n\ge k} n^{-3/2} f_n^2(x) \to 0$ a.e. on E. Moreover, $|\sigma_n| \le$

$|\sigma_n - (k^2/n)\sigma_{k^2}| + |\sigma_{k^2}|$, and

$$\left|\sigma_n(x) - \frac{k^2}{n}\sigma_{k^2}(x)\right| \le \frac{1}{k^2}\left|\sum_{k^2<j\le k^2+p} f_j(x)\right|$$
$$\le \frac{1}{k^2}\sum_{k^2<j\le(k+1)^2} j^{3/4}\frac{|f_j(x)|}{j^{3/4}}$$
$$\le \frac{1}{k^2}\left(\sum_{k^2<j\le(k+1)^2} j^{3/2}\right)^{1/2}\omega_k^{1/2}(x)$$
$$\le \frac{1}{k^2}(k+1)^{3/2}(2k+1)^{1/2}\omega_k^{1/2}(x) \underset{k\to\infty}{\longrightarrow} 0$$

almost everywhere on E.

2.14. b) As established in problem VIII.3.16, $\int_0^1(\sigma_n(x) - 1/2)^4\,dx = O(1/n^2)$. Therefore, the series $\sum_{n\ge1}(\sigma_n(x) - 1/2)^4$ converges almost everywhere, and hence $\sigma_n(x) - 1/2 \to 0$ a.e. on $(0, 1)$.

2.15. The solution is analogous to that of the preceding problem.

2.17. Use Chebyshev's inequality and the pairwise orthogonality of the functions $f(x_1), f(x_2), \dots, f(x_n)$ in the cube $(0, 1)^n$.

2.18. The solution is analogous to that of the preceding problem.

2.19. Consider the function $f(x) = x^2 - 1/2\pi$ $(x \in \mathbb{R})$ and the measure γ with density $e^{-\pi x^2}$. Prove that the $\gamma\times\cdots\times\gamma$-measure ($n$ factors) of the set

$$E_n(\varepsilon) = \left\{(x_1, \dots, x_n) \in \mathbb{R}^n \,\middle|\, \frac{1}{n}\left|\sum_{1\le k\le n} f(x_k)\right| > \varepsilon\right\}$$

tends to zero, and hence the integral $\int_{E_n(\varepsilon)}\varphi(\|x\|^2/n)e^{-\pi\|x\|^2}\,dx$ is small for large n. The integral $\int_{\mathbb{R}^n\setminus E_n(\varepsilon)}\varphi(\|x\|^2/n)e^{-\pi\|x\|^2}\,dx$ is close to $\varphi(1/2\pi)$ for small $\varepsilon > 0$.

2.20. a) Since $\|S - S_k\|_2^2 = \sum_{j>k} a_j^2$, it follows that

$$\sum_{k\ge1}\|S - S_{k^2}\|_2^2 = \sum_{k\ge1}\sum_{j>k^2} a_j^2 \le \sum_{j\ge2}\sqrt{j}\cdot a_j^2 < \infty.$$

b) It is clear that $S_{k^2}(x) \underset{k\to\infty}{\to} S(x)$ a.e. on E. Let $n = k^2 + p$, where $k = [\sqrt{n}]$ and $0 \le p \le 2k$. Then

$$|S_n - S_{k^2}| = \left|\sum_{k^2<j\le n} a_j f_j\right|^2 \le p\sum_{k^2<j\le n} a_j^2 f_j^2$$
$$\le 2\sum_{j>k^2}\sqrt{j}a_j^2 f_j^2 = 2R_k.$$

Since the series $\sum_{j\geq 1}\sqrt{j}\cdot a_j^2 f_j^2(x)$ converges almost everywhere, it follows that $R_k(x) \underset{k\to\infty}{\to} 0$ almost everywhere. Thus,

$$\begin{aligned}|S_n(x)-S(x)| &\leq |S_n(x)-S_{k^2}(x)|+|S(x)-S_{k^2}(x)| \\ &\leq \sqrt{2R_k(x)}+|S(x)-S_{k^2}(x)| \to 0 \quad \text{a.e.}\end{aligned}$$

c) It follows from the preceding inequality that $|S_n - S| \leq g$, where the function $g = \sqrt{2\sum_{j\geq 1}\sqrt{j}a_j^2 f_j^2} + \sqrt{\sum_{k\geq 1}|S-S_{k^2}|^2}$ is square-integrable. It is clear that $|S_n| \leq |S| + |S_n - S| \leq |S| + g$, $|S| + g \in \mathscr{L}^2(E)$.

§3. The Rademacher functions. Khintchine's inequality

3.4. a)–c) Use the result of problem 3.2b).

d) Use the fact that the infinite products $\prod_{n\geq 1}\cos tc_n$ and $\prod_{n\geq 1}(1 - t^2c_n^2/2)$ converge only simultaneously.

3.5. Show with the help of 3.3b) that for real c_k

$$\int_0^1 \left(\sum_{1\leq k\leq n} c_k r_k(x)\right)^4 dx = \sum_{1\leq k\leq n} c_k^4 + 6\sum_{1\leq j<k\leq n} c_j^2 c_k^2 \leq 3\left(\sum_{1\leq k\leq n} c_k^2\right)^2.$$

b) Use the idea of the solution of problem 2.6a).

3.6. Represent the double integral as an iterated integral and use the result of problem 3.5.

3.7. The solution is analogous to that of problem 3.5.

3.8. The solution is analogous to that of problem 3.5.

3.10. Use the results of problems 3.9a) and 3.5a).

3.11. Use the equality $(C = (c_{jk})_{1\leq j,k\leq n})$

$$\sum_{x,y\in A} |(Cx, y)|^p = 2^{2n}\int_0^1\int_0^1 \left|\sum_{1\leq j,k\leq n} c_{jk} r_j(s) r_k(t)\right|^p ds\,dt$$

and the result of problem 3.6.

3.12. Compute $\widehat{F}(s)$ by using the equality

$$\widehat{F}(s) = \int_{-\infty}^{+\infty} e^{-ist}\,dF(t) = \int_0^1 e^{-isf(x)}\,dx,$$

where F is the desired distribution function, along with the results of a) and b) in problems 3.3 and 3.4.

3.13. a) Let $S_n = \sum_{1\leq k\leq n} a_k r_k$. Use the equality $S_m = \frac{1}{2}(S_m + a_{m+1}r_{m+1}) + \frac{1}{2}(S_m - a_{m+1}r_{m+1})$, the triangle inequality, and induction on n $(n \geq m)$.

b) Use the convexity of the functions $s \mapsto e^{xs}$ and induction on n $(n \geq m)$.

c), d) Represent E as a union of intervals of the form $\Delta_{m,j}$ $(j = 0, 1, \ldots, 2^m - 1)$ and use a) and b).

3.14. a) It will be assumed without loss of generality that $A = 1$ (otherwise, t can be replaced by At'). We introduce the sets

$$E_+ = \left\{ x \in (0, 1) \mid \sum_{1 \le k \le n} a_k r_k(x) > t \right\},$$

$$E_- = \left\{ x \in (0, 1) \mid \sum_{1 \le k \le n} a_k r_k(x) < -t \right\}.$$

Let s be a positive number to be made more precise below. Then

$$\begin{aligned} \lambda(E_+) &\le \int_{E_+} e^{s(\sum_{1 \le k \le n} a_k r_k(x) - t)}\, dx \\ &\le e^{-st} \int_0^1 e^{s \sum_{1 \le k \le n} a_k r_k(x)}\, dx \\ &= e^{-ts} \prod_{1 \le k \le n} \cosh(s a_k) \le e^{-st} \cdot e^{s^2/2}. \end{aligned}$$

Choosing s so that the right-hand side of the inequality is minimal $(s = t)$, we get that $\lambda(E_+) \le e^{-t^2/2}$. The measure of the set E_- can be estimated analogously.

b) Use a) and the inclusion

$$\begin{aligned} &\left\{ x \in (0, 1) \,\middle|\, \left| \sum_{1 \le k \le n} c_k r_k(x) \right| > t \right\} \\ &\quad \subset \left\{ x \in (0, 1) \,\middle|\, \left| \sum_{1 \le k \le n} \alpha_k r_k(x) \right| > \frac{t}{\sqrt{2}} \right\} \\ &\quad \cup \left\{ x \in (0, 1) \,\middle|\, \left| \sum_{1 \le k \le n} \beta_k r^k(x) \right| > \frac{t}{\sqrt{2}} \right\}, \end{aligned}$$

where $\alpha_k = \mathrm{Re}(c_k)$ and $\beta_k = \mathrm{Im}(c_k)$.

c) Represent E as a union of disjoint intervals of the form $\Delta_{m,j}$ $(1 \le m \le n,\ 0 \le j < 2^m)$, chosen so that

$$\max_{1 \le l \le n} \left| \sum_{1 \le k \le l} a_k r_k(x) \right| = \left| \sum_{1 \le k \le m} a_k r_k(x) \right| \quad \text{for } x \in \Delta_{m,j}.$$

Then it follows from the inequalities a) and b) of the preceding problem that

$\lambda(E) \leq t^{-1} \int_E |\sum_{1\leq k\leq n} a_k r_k(x)|\, dx \leq At^{-1}$ and

$$\lambda(E) \leq 2 \cdot e^{-st} \int_E \cosh\left(s \sum_{1\leq k\leq n} a_k r_k(x)\right) dx$$

$$\leq 2 \cdot e^{-st} \int_0^1 \cosh\left(s \sum_{1\leq k\leq n} a_k r_k(x)\right) dx.$$

The last integral was estimated in the proof of a).

d) Let F be the distribution function of g. Using the inequality $1 - F(t) \leq 2 \cdot e^{-t^2/2A^2}$ established in c) and assuming that $s < 1/2A^2$, we get that

$$\int_0^1 e^{sg^2(x)}\, dx = \int_0^\infty e^{st^2}\, dF(t) \leq 4s \int_0^\infty te^{-(1/2A^2-s)t^2}\, dt < \infty.$$

But if $s \geq 1/2A^2$, then we consider the function

$$g_1(x) = \sup_{n\geq N} \left| \sum_{N\leq k\leq n} a_k r_k(x) \right|,$$

where N is chosen so that $A_1^2 = \sum_{k>N} a_k^2 < 1/(4s)$. As already shown above, $\int_0^1 e^{2sg_1^2(x)}\, dx < \infty$. Moreover, $g(x) \leq K + g_1(x)$, where $K = \sum_{1\leq j\leq N} |a_j|$, and hence $g^2(x) \leq 2K^2 + 2g_1^2(x)$. Consequently,

$$\int_0^1 e^{sg^2(x)}\, dx \leq e^{2sK^2} \int_0^1 e^{2sg_1^2(x)}\, dx < \infty.$$

3.15. Obviously, all the numbers a_k can be assumed to be real. Let $S_n(x) = a_1 r_1(x) + \cdots + a_n r_n(x)$. Using the result of problem 3.14c), prove that for any number $\varepsilon > 0$ the measure of the set

$$\{x \in (0, 1) \mid \sup_{p\in\mathbb{N}} |S_{n+p}(x) - S_n(x)| > \varepsilon\}$$

does not exceed $\varepsilon^{-1}(\sum_{k>n} a_k^2)^{1/2}$. Derive from this that

$$\lambda\left\{x \in (0, 1) \middle| \varlimsup_{n\to\infty} S_n(x) - \varliminf_{n\to\infty} S_n(x) > \varepsilon\right\} = 0.$$

3.16. By shrinking the set E if necessary the problem can be reduced to the case when the sum f of the series is bounded on E. Suppose that $|f| \leq C$ on E. Consider a subsequence $\{S_{n_k}\}_{k\geq 1}$ of partial sums of the series that converge uniformly to f on E. Obviously, we can assume that $|S_{n_k}(x)| \leq 2C$ for any $x \in E$ and $k \in \mathbb{N}$. Fix a positive integer m to be made more precise below, and let

$$T^2 = \sum_{1\leq j\leq m} a_j^2, \quad u_k = \sum_{m<j\leq n_k} a_j^2 \qquad \text{for } n_k > m.$$

We prove that the sequence $\{u_k\}_{k\geq 1}$ is bounded, which suffices for the solution of the problem. It is clear that

$$2C\sqrt{\lambda(E)} \geq \left(\int_E (S_{n_k}(x))^2\,dx\right)^{1/2} \geq \left(\int_E \left(\sum_{m<j\leq n_k} a_j r_j(x)\right)^2 dx\right)^{-1/2} - T.$$

Consequently,

$$\begin{aligned}(T+2C\sqrt{\lambda(E)})^2 &\geq \int_E \sum_{m<j,\,l\leq n_k} a_j a_l r_j(x) r_l(x)\,dx \\ &= \lambda(E)u_k + 2\sum_{m<j<l\leq n_k} a_j a_l \int_E r_j(x) r_l(x)\,dx \qquad (1)\\ &\geq \lambda(E)u_k - 2\sqrt{\sum_{m<j<l\leq n_k} a_j^2 a_l^2}\cdot\sqrt{\sum_{m<j<l\leq n_k}\left(\int_E r_j(x) r_l(x)\,dx\right)^2}.\end{aligned}$$

Since the family $\{r_j r_l\}_{1\leq j<l}$ is an orthonormal system (see problem 3.7a)), Bessel's inequality gives us that

$$\theta_m = \sqrt{\sum_{m<j<l}\left(\int_E r_j(x) r_l(x)\,dx\right)^2} \underset{m\to\infty}{\longrightarrow} 0.$$

We take m large enough that $4\theta_m < \lambda(E)$. Then it follows from (1) that

$$(T+2C\sqrt{\lambda(E)})^2 \geq \lambda(E)u_k - 2u_k\theta_m \geq \frac{\lambda(E)}{2}u_k.$$

Thus, $u_k \leq 2(\lambda(E))^{-1}(T+2C\sqrt{\lambda(E)})^2$ for all $n_k > m$ for this choice of m, as required.

3.17. It is clear that a) $\Rightarrow$ b) $\Rightarrow$ c) (see problems 2.1 and 2.2). The implications c) $\Rightarrow$ a) $\Rightarrow$ d) were established in problems 3.16 and 3.15. Finally, d) $\Rightarrow$ c) by Lebesgue's theorem.

3.18. The solution is analogous to that of problem 3.16. Holding to the notation introduced there and assuming that $E \subset (0, 2\pi)$, we get instead of (1) that

$$(\pi T + 2C\sqrt{\lambda(E)})^2 \leq \sum_{m<j\leq n_k} a_j^2 \int_E \sin^2 2^j x\,dx - 2\sqrt{\sum_{m<j<l\leq n_k} a_j^2 a_k^2} \cdot\sqrt{\sum_{m<j<l\leq n_k}\left(\int_E \sin 2^j x \sin 2^l x\,dx\right)^2}.$$

Since

$$\int_E \sin^2 2^j x\,dx = \frac{\lambda(E)}{2} - \frac{1}{2}\int_E \cos 2^{j+1} x\,dx \underset{j\to\infty}{\longrightarrow} \frac{\lambda(E)}{2},$$

m can be taken large enough that $\int_E \sin^2 2^j x\,dx > \frac{1}{4}\lambda(E)$ for $j > m$. To estimate the sum $\sum_{m<j<l\le n_k}(\int_E \sin 2^j x \sin 2^l x\,dx)^2$ use the equality

$$\int_E \sin 2^j x \sin 2^l x\,dx = \frac{1}{2}\left(\int_E \cos(2^l - 2^j)x\,dx - \int_E \cos(2^l + 2^j)x\,dx\right)$$

and Bessel's inequality for the trigonometric system.

3.19. a) Show that the boundedness of the sum of the series $\sum_{n\ge1} a_n r_n$ on Δ implies its boundedness on $(0, 1)$. Suppose that $|\sum_{n\ge1} a_n r_n| \le C$ on $(0, 1)$. Fix an arbitrary positive integer n and consider an interval

$$\Delta_{n,k} = \left(\frac{k}{2^n}, \frac{k+1}{2^n}\right) \subset (0, 1),$$

such that $r_j(x) = \operatorname{sgn} a_j$ for $x \in \Delta_{n,k}$, $j = 1, 2, \dots, n$. Since $\int_{\Delta_{n,k}} r_j(x)\,dx = 0$ for $j > n$, it follows that

$$C2^{-n} \ge \int_{\Delta_{n,k}} \sum_{j\ge1} a_j r_j(x)\,dx = \sum_{1\le j\le n} a_j \int_{\Delta_{n,k}} r_j(x)\,dx = \sum_{1\le j\le n} |a_j| 2^{-n}.$$

Thus, $\sum_{1\le j\le n} |a_j| \le C$ for any $n \in \mathbb{N}$.

b) Use the idea of the solution of problem IV.6.28.

3.20. b) Use the idea of the solution of problem 3.13c).

c) Use the idea of the solution of problem 3.14c).

d) Use the idea of the solution of problem 3.15.

3.21. a) Let S_n be the nth partial sum of the Fourier series of f, and let ε be an arbitrary positive number. Prove that

$$\lambda\left\{x \in (0, 1) \,\middle|\, \max_{1\le p\le N} |S_{n+p}(x) - S_n(x)| > \varepsilon\right\} \le \varepsilon^{-1} \int_0^1 |S_{n+p}(x) - S_n(x)|\,dx.$$

Derive from this that

$$\lambda\left\{x \in (0, 1) \,\middle|\, \sup_{p\in\mathbb{N}} |S_{n+p}(x) - S_n(x)| > \varepsilon\right\} \le \varepsilon^{-1} \int_0^1 |f(x) - S_n(x)|\,dx.$$

To complete the proof it remains to repeat the arguments used in the solutions of problems 3.15 and 3.20d).

b) Let $E_n = \{x \in (0, 1) \mid \sup_{1\le m\le n} |S_m(x)| > t\}$. Using as a model the arguments in the proof of c) in the preceding problem, prove that $\lambda(E_n) \le \|S_n\|_1 t^{-1}$, after which it remains only to pass to the limit as $n \to \infty$.

3.22. Prove that $\|\sum_{1\le k\le n} a_k r_k\|_p \le B_p(\sum_{1\le k\le n} a_k^2)^{1/2}$. The second inequality can be obtained with the help of the method in the solution of problem 2.6a). Let F be the distribution function of the function $|\sum_{1\le k\le n} a_k r_k|$.

Using the estimate $1 - F(t) \le 2e^{-t^2/2A^2}$ obtained in problem 3.14a), where $A^2 = \sum_{1\le k\le n} a_k^2$, show that for any $p > 0$

$$\int_0^1 \left| \sum_{1\le k\le n} a_k r_k(x) \right|^p dx = \int_0^\infty t^p\, dF(t) \le 2p \int_0^\infty t^{p-1} e^{-t^2/2A^2}\, dt$$
$$= 2pA^p \int_0^\infty y^{p-1} e^{-y^2/2}\, dy\,.$$

Expressing the last integral in terms of the function Γ and using Stirling's formula, we get the required estimate for B_p.

A proof of Khintchine's inequality without making explicit use of the properties of the distribution function can be obtained by analogy with the solution of problem VIII.6.27.

3.23. The implication e) $\Rightarrow$ b) is trivial, and the implication a) $\Rightarrow$ e) follows from Khintchine's inequality.

3.24. It will be assumed that $\sum_{|k|\le n} |c_k|^2 = 1$. Consider the function

$$P(x, t) = \sum_{|k|\le n} c_k r_{n+k+1}(t) e^{-ikx} \qquad (x \in \mathbb{R},\ t \in (0, 1)).$$

a) Use the inequality $\int_0^1 |P(x, t)|\, dt \ge \alpha$ (where $\alpha > 0$ is an absolute constant), which follows from problem 3.22.

b) If we show that the measure of the set

$$E = \{t \in (0, 1) \mid \max_{x\in\mathbb{R}} |P(x, t)| > 10\sqrt{\ln n}\}$$

is less than 1, then by taking a point t_0 in $(0, 1)\backslash E$ (not dyadic rational) and letting $\varepsilon_k = r_{n+k+1}(t_0)$ we get the desired trigonometric polynomial. To estimate the measure of E we observe that $\max_{x\in\mathbb{R}} |P(x, t)|$ is "almost realized" at one of the points $x_j = 2\pi j/n^2$, $j = 0, 1, \dots, n^2$. Indeed, if $|x - x_j| \le \pi/n^2$, then

$$|P(x, t) - P(x_j, t)| \le \sum_{|k|\le n} |c_k| \cdot 2 \cdot \left| \sin \frac{k\pi}{2n^2} \right| \le \frac{\pi}{\sqrt{n}}\,.$$

Therefore, $\max_{x\in\mathbb{R}} |P(x, t)| \le \pi/\sqrt{n} + \max_{1\le j\le n^2} |P(x_j, t)|$. Consequently,

$$E \subset \left\{ t \in (0, 1) \Big| \max_{1\le j\le n^2} |P(x_j, t)| > \left(10 - \frac{\pi}{\sqrt{n \ln n}}\right) \sqrt{\ln n} \right\}$$
$$\subset \widetilde{E} = \left\{ t \in (0, 1) \Big| \max_{1\le j\le n^2} |P(x_j, t)| > 6\sqrt{\ln n} \right\}.$$

Finally, it is clear that

$$\widetilde{E} \subset \bigcup_{1\le j\le n^2} \{t \in (0, 1) \mid |P(x_j, t)| > 6\sqrt{\ln n}\}\,.$$

As established in problem 3.14b), the measure of each of the sets on the right-hand side of the last inclusion does not exceed $4e^{-9\ln n} = 4n^{-9}$. Consequently,

$$\lambda(E) \leq \lambda(\widetilde{E}) \leq n^2 \cdot 4 \cdot n^{-9} = 4n^{-7} < 1 \quad \text{for } n \geq 2.$$

As is clear from the last inequality, $\sup_{x\in\mathbb{R}} |P(x, t)| \leq 10\sqrt{\ln n}$ for all t belonging to the set $(0, 1)\backslash E$, the measure of which is close to 1. Thus, arranging the signs ε_k randomly by means of coin tosses, we get polynomials satisfying the required estimate in the overwhelming majority of the cases.

We remark that numerous results relating to the questions touched on this problem can be found in [11].

3.25. Let $p_n = 2\ln n$. With the help of Khintchine's inequality show that the series

$$\sum_{n\geq 2} \int_0^1 \left(\frac{|r_1(x) + \cdots + r_n(x)|}{\sqrt{an\ln n}}\right)^{2\ln n} dx$$

converges for any $a > 2$. In the proof of this use the estimate $B_p \leq \sqrt{p(1+\varepsilon)/e}$, valid for any $\varepsilon > 0$ and sufficiently large p (see problem 3.22). It follows from the convergence of this series that the series

$$\sum_{n\geq 2} \left(\frac{|r_1(x) + \cdots + r_n(x)|}{\sqrt{an\ln n}}\right)^{2\ln n}$$

converges almost everywhere, which is possible only if

$$\varlimsup_{n\to\infty} \frac{|r_1(x) + \cdots + r_n(x)|}{\sqrt{an\ln n}} \leq 1$$

almost everywhere.

3.26. Use the same idea as in the solution of the preceding problem.

§4. Fourier series and the Fourier transform

4.2. It is clear that for $h > 0$ and any $x \in \mathbb{R}$

$$|\varphi(x+h) - \varphi(x)| \leq L \int_{-\pi}^{\pi} |g_\alpha(y+h) - g_\alpha(y)|\, dy,$$

where $L = \operatorname{ess\,sup}_{\mathbb{R}} |f(x)|$. Show that the integral on the right-hand side of the inequality does not exceed $(4/(1-\alpha))h^{1-\alpha}$.

4.3. a) Use the equality

$$\hat{h}(n) = \lim_{k\to\infty} \frac{1}{2\pi} \int_{-\pi}^{\pi} S_k(x)\widetilde{S}_k(x) e^{-inx}\, dx,$$

where S_k and $\widetilde{S}_k$ are the kth partial sums of the Fourier series of the respective functions f and g. Prove first that $\|h - S_k\widetilde{S}_k\|_1 \underset{k\to\infty}{\to} 0$ (see problem 2.2b)).

b) Represent Φ as the sum of the series $\sum_{k\geq 0} f^k/k!$ and use a).

4.4. Use the equality $\hat{f}(n) = 2\pi \hat{g}(n)\hat{h}(n) \quad (n \in \mathbb{Z})$.

4.5. Compute the Fourier coefficients $\hat{\varphi}(n)$ of the function $\varphi = f * g_{1/2}$ (see problem 4.2) and show that for $n \neq 0$

$$\hat{\varphi}(n) = a\frac{\hat{f}(n)}{\sqrt{|n|}} + O\left(\frac{1}{n^2}\right), \qquad a \neq 0.$$

4.6. Use the equality $\hat{f}(n) = -\frac{1}{2\pi}\int_{-\pi}^{\pi} f(x+\pi/n)e^{-inx}\,dx$.

4.7. Use integration by parts.

4.8. Express $\hat{f}(n)$ in terms of the integrals

$$\int_0^\pi xe^{\pm i(nx\pm 1/x)}\,dx = O\left(\frac{1}{n}\right) \pm \int_{2/\sqrt{n}}^{\pi} \frac{x}{n \mp 1/x^2}\,d\left(\frac{1}{i}e^{\pm i(nx\pm 1/x)}\right).$$

Estimate the last integral by integrating by parts.

4.9. Suppose that $|a_n| \leq C/n^{1+\alpha} \quad (n = 1, 2, \ldots)$. Show that

$$|f(x) - f(x')| \leq C \sum_{1\leq n\leq |x-x'|^{-1}} \frac{|x-x'|}{n^\alpha} + 2C \sum_{n>|x-x'|^{-1}} \frac{1}{n^{1+\alpha}}.$$

Prove that each of the sums on the right-hand side of the inequality is $O(|x-x'|^\alpha)$.

For $\alpha = 1$ consider the function $f(x) = \sum_{n\geq 1} e^{inx}/n^2$ and verify that its derivative is unbounded on $(0, \pi)$.

4.10. Let

$$S_j(x) = \sum_{|n|\leq j} \hat{f}(n)e^{inx} \quad \text{and} \quad f_k(x) = \sum_{1\leq n\leq k} \frac{\sin(2^{n!}|x|)}{n^2}.$$

Show that

$$S_{m_k}(0) = \frac{1}{\pi}\int_0^\pi f_k(x)\frac{\sin(m_k+1/2)x}{\sin(x/2)}\,dx, \quad \text{where } m_k = 2^{k!}.$$

Prove that

$$\int_0^\pi f_k(x)\frac{\sin(m_k+1/2)}{\sin(x/2)}\,dx$$
$$\geq \frac{2}{k^2}\int_0^\pi \frac{\sin^2 m_k x}{x}\,dx - \sum_{1\leq n<k}\frac{1}{n^2}\int_0^\pi \frac{|\sin m_n x|}{x}\,dx + O(k).$$

Use the result of problem VI.1.10.

4.11. Use the equality

$$\sigma_n(x) - f(x) = \frac{1}{2\pi n}\int_{-\pi}^{\pi} (f(x-t) - f(x))\frac{\sin^2(nt/2)}{\sin^2(t/2)}\,dt. \qquad (*)$$

Prove that

$$\theta_n = \frac{1}{\pi n}\int_\delta^\pi \frac{\sin^2(nt/2)}{\sin^2(t/2)}\,dt \underset{n\to\infty}{\longrightarrow} 0$$

for any $\delta > 0$, and deduce from this that for any $x \in \mathbb{R}$

$$|\sigma_n(x) - f(x)| \le \omega_f(\delta) + 2\theta_n \|f\|_\infty,$$

where ω_f is the modulus of continuity of f.

4.12. a) Using the formula $(*)$ in the solution of the preceding problem, prove that

$$|\sigma_n(x) - f(x)| \le \frac{c}{n} \int_0^\pi \frac{\sin^2(nt/2)}{t^{2-\alpha}}\, dt \qquad (x \in \mathbb{R}),$$

where c is a Lipschitz constant L for f.

b) Use the result of a) and the extremal property of the partial sums of the Fourier series.

4.13. Using the Cauchy-Schwarz-Bunyakovskiĭ inequality and the assertion b) of the preceding problem, prove that $\sum_{2^k<|n|\le 2^{k+1}} |\hat{f}(n)| = O(2^{(1/2-\alpha)k})$.

4.14. With the help of the inequality $|\hat{\varphi}(k)| \le \|\varphi\|_1/2\pi$, applied to the function $\varphi = f - \sigma_n$, show that for $k \ne 0$ the Fourier coefficients $\hat{f}(k)$ of f are equal to zero.

4.15. Use the equality $\hat{f}(-n) = \overline{\hat{f}}(n)$ $(n \in \mathbb{Z})$.

4.16. It is clear that $|F_\varepsilon(x)| \le \sum_{k\ge 1} |f_k(x)| e^{-\varepsilon \sum_{j\ge k} |f_j(x)|}$. Therefore, a) follows from the result of problem IV.2.12c). To prove b) show that

$$\widehat{F}_\varepsilon(n_q) - a_q = \frac{1}{2\pi} \int_{-\pi}^{\pi} \left(F_\varepsilon(x) - \sum_{k\ge p} f_k(x) \right) e^{-in_q x}\, dx.$$

With the help of the result of problem 4.3b) derive from this that

$$\widehat{F}_\varepsilon(n_q) - a_q = \frac{1}{2\pi} \int_{-\pi}^{\pi} \sum_{k\ge p} f_k(x)(e^{-\varepsilon \sum_{j\ge k} h_j(x)} - 1) e^{-in_q x}\, dx,$$

and hence

$$\begin{aligned} |\widehat{F}_\varepsilon(n_q) - a_q| &\le \frac{\varepsilon}{\pi} \int_{-\pi}^{\pi} \sum_{k\ge p} |f_k(x)| \sum_{j\ge k} |h_j(x)|\, dx \\ &= \frac{\varepsilon}{\pi} \sum_{k\ge p} \sum_{j\ge k} \int_{-\pi}^{\pi} |f_k(x)||h_j(x)|\, dx. \end{aligned}$$

It remains to use the Cauchy-Schwarz-Bunyakovskiĭ inequality.

4.17. Use the inequality

$$\int_{-\pi}^{\pi} \left| \sum_{1\le j\le m} c_j e^{in_j x} \right| dx \ge \varepsilon \left| \int_{-\pi}^{\pi} \left(\sum_{1\le j\le m} c_j e^{in_j x} \right) F_\varepsilon(x)\, dx \right|,$$

where ε is a sufficiently small positive number, and F_ε is the function constructed in the preceding problem with $a_j = j^{-1} c_j/|c_j|$ $(j = 1, 2, \ldots, m)$.

The inequality in 4.17 was conjectured (in the particular case when $c_1 = c_2 = \cdots = 1$) by J. Littlewood. It was proved half a century later by S. Konyagin (On the Littlewood problem, Izv. Akad. Nauk SSSR Ser. Math. **45** (1981), no. 2) and simultaneously by O. McGehee, L. Pigno, and B. Smith (see [41]).

4.18. a) Show that the function $\hat{f}_c(s)$ is a solution of the differential equation $y'(s) + csy(s) = 0$.

4.19. We get by n-fold integration by parts that

$$\hat{h}_n(s) = (-1)^n e^{s^2/2} \int_{-\infty}^{+\infty} e^{-x^2} \frac{d^n}{dx^n}(e^{(x-is)^2/2})\, dx.$$

Using the equality

$$\frac{d}{dx}\varphi(x - is) = \frac{1}{-i}\frac{d}{ds}\varphi(x - is),$$

prove that

$$\hat{h}_n(s) = \frac{1}{i^n} e^{s^2/2} \frac{d^n}{ds^n}\left(e^{-s^2/2} \int_{-\infty}^{+\infty} e^{-x^2/2} \cdot e^{isx}\, dx\right).$$

To compute the last integral use the result of problem 4.18.

4.21. Since $\sigma(2/3 + t) = 1/2 + \sigma(t)$ for $0 \le t \le 1/3$, it follows that

$$\begin{aligned}\hat{\sigma}(s) &= \int_0^{1/3} e^{-ist}\, d\sigma(t) + \int_{2/3}^{1} e^{-ist}\, d\sigma(t) \\ &= \int_0^{1/3} e^{-ist}\, d\sigma(t) + \int_0^{1/3} e^{-is(t+2/3)}\, d\sigma(t) \\ &= (1 + e^{-2is/3}) \int_0^{1/3} e^{-ist}\, d\sigma(t).\end{aligned}$$

Using induction, show that

$$\hat{\sigma}(s) = \left(\prod_{1\le k\le n} (1 + e^{-2is/3^k})\right) \int_0^{3^{-n}} e^{-ist}\, d\sigma(t).$$

Derive from this that

$$\hat{\sigma}(s) = e^{-is/2} \prod_{k\ge 1} \cos(s3^{-k}).$$

It is clear from the last equality that

$$|\hat{\sigma}(2\pi 3^n)| = \frac{1}{2} \prod_{k\ge 2} \cos(2\pi 3^{-k}) \ne 0$$

for any $n \in \mathbb{N}$.

4.22. a) Let $0 \le x < y \le 2$. Since

$$\begin{aligned}\sigma_1(y) &= \int_0^y \sigma(y - t)\, d\sigma(t) \ge \int_0^x \sigma(y - t)\, d\sigma(t) \\ &\ge \int_0^x \sigma(x - t)\, d\sigma(t) = \sigma_1(x),\end{aligned}$$

the function σ_1 is nondecreasing. To prove that it is strictly monotone, show that there exist closed intervals $[k3^{-n}, (k+1)3^{-n}]$ and $[l3^{-n}, (l+1)3^{-n}]$ of rank n (see the definition of the Cantor set before problem I.1.27) such that

$$0 \le k, l < 3^n, \qquad x < \frac{k+l}{3^n} < \frac{k+l+1}{3^n} < y$$

(cf. problem I.1.28b)). Then

$$\begin{aligned}\sigma_1(y) - \sigma_1(x) &\ge \sigma_1\left(\frac{k+l+1}{3^n}\right) - \sigma_1\left(\frac{k+l}{3^n}\right)\\ &\ge \int_{k3^{-n}}^{(k+1)3^{-n}} \left(\sigma\left(\frac{k+l+1}{3^n} - t\right) - \sigma\left(\frac{k+l}{3^n} - t\right)\right) d\sigma(t)\\ &= \int_0^{3^{-n}} \left(\sigma\left(\frac{l+1}{3^n} - \tau\right) - \sigma\left(\frac{l}{3^n} - \tau\right)\right) d\sigma(\tau)\\ &\ge \int_0^{3^{-n}/2} \left(\sigma\left(\frac{l+1}{3^n} - \frac{1}{2}3^{-n}\right) - \sigma\left(\frac{l}{3^n}\right)\right) d\sigma(t)\\ &= \frac{1}{2}\cdot 2^{-n}\cdot\frac{1}{2}\cdot 2^{-n} > 0.\end{aligned}$$

b) Show that $\hat{\sigma}_1(s) = (\hat{\sigma}(s))^2$, and use the result of the preceding problem.

4.23. Use the fact that the Fourier transform of a convolution of integrable functions is equal to the product of their Fourier transforms.

4.24. a) Show that the range of g is dense in $[0, 1]$ and use monotonicity.

b) Study the intervals of constancy of g and find the measure of their union.

c) Use the same device as in the solution of problem 4.21.

d) Study the Fourier transform of the measure corresponding to the function

$$h(x) = \int_{-\infty}^{+\infty} g(x-t)\, dg(2t) \qquad (x \in \mathbb{R}).$$

4.25. The solution is analogous to the solution of the preceding problem.

4.26. Prove that

$$\hat{\mu}(s) = \int_{-\infty}^{+\infty} e^{-iu}\, dF_s(u),$$

where $F_s(u) = \mu\{t \in \mathbb{R}^n \mid (s, t) \le u\}$ $(u \in \mathbb{R})$.

4.27. Let μ be a measure satisfying the conditions of the problem. Then $\mu = \mu_1 \times \mu_2$, where μ_1 and μ_2 are Borel probability measures on subspaces L and M, respectively. Since μ is rotation-invariant, it follows that $\dim L + \dim M = n$. Again since μ is rotation-invariant, we see that $\hat{\mu}(s) = \hat{\mu}(\|s\|a)$, where a is an arbitrary normalized vector. The invariance of μ with respect to orthogonal transformations implies that μ_1 and μ_2 are also invariant with respect to such transformations (in L and M, respectively). Therefore, $\hat{\mu}_1(u) = \hat{\mu}_1(\|u\|u_0)$ and $\hat{\mu}_2(v) = \hat{\mu}_2(\|v\|\cdot v_0)$, where $u \in L$, $v \in M$, and

u_0, v_0 are arbitrary normalized vectors in L and M, respectively. With the help of Fubini's theorem we get that

$$\hat{\mu}(u+v) = \hat{\mu}_1(u)\cdot\hat{\mu}_2(v), \quad \text{where } u \in L,\ v \in M.$$

Since $\|u+v\| = \sqrt{\|u\|^2 + \|v\|^2}$, this implies the equality

$$\hat{\mu}(\sqrt{\|u\|^2 + \|v\|^2}a) = \hat{\mu}_1(\|u\|u_0)\cdot\hat{\mu}_2(\|v\|v_0). \tag{1}$$

Let $\varphi(t) = \hat{\mu}(ta)$ for $t > 0$. Setting $v = 0$ and $u = 0$ in turn in (1), we see that

$$\varphi(\|u\|) = \hat{\mu}(\|u\|a) = \hat{\mu}_1(\|u\|u_0),$$
$$\varphi(\|v\|) = \hat{\mu}(\|v\|a) = \hat{\mu}_2(\|v\|v_0).$$

Thus, it follows that (1) that

$$\varphi(\sqrt{t_1^2 + t_2^2}) = \varphi(t_1)\varphi(t_2) \quad \text{for any } t_1,\ t_2 \geq 0.$$

Using the result of problem III.5.9, we get that $\varphi(t) = e^{-ct^2}$ for $t \geq 0$. Thus, $\hat{\mu}(s) = e^{-c\|s\|^2}$, and $c \geq 0$ because $|\hat{\mu}(s)| \leq 1$. If $c = 0$, then μ is the unit measure concentrated at the origin of coordinates. The case $c > 0$ corresponds to the Gaussian measure with density $(4\pi c)^{-n/2}e^{-\|x\|^2/4c}$.

CHAPTER X

Iterates of Transformations of an Interval

§1. Topological dynamics

1.1. Prove by induction that $n \le f(n) < \min\{f(k) | k > n\}$ for each $n \in \mathbb{N}$.

1.2. a) Prove that $f(0) = 0$ and investigate the sign of $f(x)$ for $x > 0$.

b) Prove that f is strictly increasing and that $\inf f = a \in (-\infty, 0)$, $f(a) = 0$. Fix an arbitrary strictly increasing function $f_0 \in C([a, 0])$ satisfying the conditions $f_0(a) = 0$ and $f_0(0) = e^a$, and extend it to $\mathbb{R}$ by using the equation.

c) Prove that f must be strictly increasing on $[0, +\infty)$ and strictly decreasing on $(-\infty, 0]$ and that $f(0) \ge 0$, which is impossible, because F, and hence f, takes negative values.

1.4. Let $f : X \to X$ be a bijection. For an arbitrary point $x \in X$ let $x_0 = x$, $x_1 = f(x)$, $x_{-1} = f^{-1}(x)$, $x_2 = f(x_1)$, $x_{-2} = f^{-1}(x_{-1})$, and so on. Denote by $\mathrm{Orb}(x)$ the set $\{x_i\}_{i \in \mathbb{Z}}$ and call it the orbit of the element x. Two cases are possible. a) There is a number $n \in \mathbb{N}$ such that $x_i \ne x_0$ for $i = 0, \dots, n-1$, and $x_n = x_0$. In this case $x_i = x_{n+i}$ for any $i \in \mathbb{Z}$, and the set $\mathrm{Orb}(x)$ consists of precisely n points. b) $x_n \ne x_0$ for all $n > 0$. Then $x_i \ne x_j$ for all $i \ne j$, i, $j \in \mathbb{Z}$, and the set $\mathrm{Orb}(x)$ is infinite.

It is not hard to see that the orbits of two points x and y either are disjoint (if $y \notin \mathrm{Orb}(x)$) or coincide. Thus, the whole set X splits into disjoint subsets X_α, each of which is the orbit of any of its points and is thus invariant with respect to f. It suffices to define the desired involutions (denote them by h and k) for each set X_α separately, taking an arbitrary point x_0 in it. This can be done as follows: for $x_i \in X_\alpha = \mathrm{Orb}(x_0)$ let $h(x_i) = x_{n+1-i}$ and $k(x_i) = x_{n-i}$, in case a), and $h(x_i) = x_{1-i}$ and $k(x_i) = x_{-i}$ in case b).

1.5. Show that the difference $x - f(x)$ cannot have constant sign, and use the Bolzano-Cauchy theorem.

1.6. The sequence $\{x_n\}$ is monotone. The assertion is false for a decreasing function (for example, $f(x) = 1 - x$, $x \in [0, 1]$).

1.7. The absence of a fixed point implies that the sequence $\{x_n\}$ is monotone. The rest goes as in the preceding problem. In the complex case a fixed point can fail to exist: consider the mapping f given by the formula $f(z) = (|z| + 1)e^{i(|z|+1)}$ $(z \in \mathbb{C})$.

1.8. Consider the arithmetic mean of the images of an arbitrary point with respect to all the transformations of the group. 1.13. a) To construct a conjugating mapping h use the method in the solution of problem III.5.12 (see also Problem 1.16).

b) Suppose that $h(\sin x) = \cos h(x)$, where $h : \mathbb{R} \to \mathbb{R}$ is a continuous bijection. Prove that $h([-1, 1]) = [-1, 1]$ and use the strict monotonicity of h.

c), e) Choose a linear function h.

d) See the hint for b).

1.14. a) Considering the results of problems 1.11 and 1.12, show that since the point $x = 0$ is a minimum point and a fixed point of f, the point $x = -a/2$ must play the same role for g in the case when $f \sim g$, and this implies the necessity of the condition $b = (a^2 - 2a)/4$. If this equality holds, then $g(x)$ takes the form $(x + a/2)^2 - a/2$, and it is not hard to choose a linear function conjugating f and g.

b) Comparing the number of fixed points for f and g, we arrive at the conclusion that if $f \sim g$, then $a \geq -1/4$. For all such nonzero a and for $b = 1 \pm \sqrt{1+4a}$ (let $b = 0$ for $a = 0$) there exists a linear function conjugating f and g $(h(x) = (4x-2)/(b-2))$. Note that here b can be anything except $b = 2$. For $b = 2$ the maximum point of the mapping g is simultaneously a fixed point, and this does not happen for f, whatever the value of a.

c) Assuming that $g \sim \tilde{g}$ and $g \circ h = h \circ \tilde{g}$, we arrive at the conclusion (see problem 1.10) that h carries the fixed points of $\tilde{g}$ into the fixed points of g (that is, $h(0) = 0$ and $h(1/3) = 2/3$), and the intervals of monotonicity of $\tilde{g}$ into the intervals of monotonicity of g. However, g is in fact monotone on $[0, 1/3]$, while $\tilde{g}$ is not monotone on $[0, 2/3] = h([0, 1/3])$. Contradiction.

d) First, it follows from the condition $g([0, 1]) \subset [0, 1]$ that $b \in [0, 4]$, it follows from the condition $f(\Delta) \subset \Delta$ that $a \leq 2$, it follows from the condition $g(0) = g(1)$ that $f(p) = f(q)$, that is, $p = -q$, and it follows from the condition $g(0) = 0$ that p or q is a fixed point of f. Moreover, the constraints $a \geq -1/4$ and $b \neq 2$ hold for the same reasons as in b).

Let x_1, x_2 $(x_1 \leq x_2)$ be the roots of the equation $1 - ax^2 = x$. We consider the following cases:

α) $a \in [-1/4, 0]$, $\Delta = [-x_1, x_1]$;
β) $a \in [-1/4, 0)$, $\Delta = [-x_2, x_2]$;
γ) $a \in (0, 2]$, $\Delta = [x_1, -x_1]$;
δ) $a \in (0, 2]$, $\Delta = [-x_2, x_2]$;
A) $b \in [0, 1]$;
B) $b \in [1, 2)$;
Γ) $b \in (2, 4]$.

Comparing the number of fixed points of f and g, we arrive at the conclusion that if $f \sim g$, then case α is incompatible with B and Γ, and β and γ are incompatible with A. Arguing by analogy with c), we discover also that α and B are incompatible, and β and A are incompatible. In the case δ the condition $f(\Delta) \subset \Delta$ is not satisfied. Mappings f and g relating to cases α and A (respectively, β and B, γ and Γ) are conjugate if $a = b(b-2)/4$. The conjugating mapping is the same as in b).

1.15. a) This is the special case of b) for $n = 2$. A conjugating mapping is $h(x) = \sin \frac{\pi}{2} x$.

c) A conjugating mapping is odd, and is given for $x > 1$ by the formula $h(x) = \cosh a(x-1)$.

1.16. Obviously, the differences $f(x) - x$ and $g(x) - x$ have constant signs inside the intervals. Choose arbitrary points $c_0 \in (a_1, b_1)$ and $d_0 \in (a_2, b_2)$, and let $\{c_n\}$ and $\{d_n\}$ be the trajectories of these points under the action of f and g, that is, $c_{n+1} = f(c_n)$, $d_{n+1} = g(d_n)$, $n \in \mathbb{Z}$. These sequences are monotone, and they converge (since f and g are continuous) to fixed points, that is, to the endpoints of the intervals, as $n \to \pm\infty$. For definiteness assume that f and g are increasing.

We define a homeomorphism $h : [a_2, b_2] \to [a_1, b_1]$ as follows. Let $h(a_2) = a_1$, $h(b_2) = b_1$, and $h(d_n) = c_n$ $(n \in \mathbb{Z})$. On $[d_0, d_1]$ define h so that it is a homeomorphism of this interval onto $[c_0, c_1]$, and on the remaining intervals $[d_n, d_{n+1}]$ let $h(x) = f^n(h(g^{-n}(x)))$, where $f^0(x) = x$, $f^1(x) = f(x)$, $f^n(x) = f(f^{n-1}(x))$ for $n > 1$, and $f^n = (f^{-n})^{-1}$ for $n < 0$. Verify that the definition is unambiguous and that h is a homeomorphism conjugating f and g.

1.17. Suppose that f is a homeomorphism of $[a, b]$, and $a = x_0 < x_1 < \cdots < x_n = b$ are all its fixed points, and let $\varepsilon_i(f) = \operatorname{sgn}(f(x) - x)$, where x is an arbitrary point in (x_{i-1}, x_i), $i = 1, \dots, n$. With the help of problems 1.11 and 1.12 show that if $f \sim g$, then they have the same number of fixed points, and the collections of numbers $\varepsilon_i(f)$ and $\varepsilon_i(g)$ either coincide or (in the case of a decreasing conjugating function) satisfy the relation $\varepsilon_{n+1-i}(f) = -\varepsilon_i(g)$, $i = 1, \dots, n$. To verify that this condition suffices for f and g to be topologically conjugate observe that f implements a homeomorphism of each interval $[x_{i-1}, x_i]$, $i = 1, \dots, n$, onto itself. If $[y_{i-1}, y_i]$ is the ith closed interval bounded by fixed points of g, then $f|_{[x_{i-1}, x_i]} \sim g|_{[y_{i-1}, y_i]}$ (see the preceding problem). Verify that if $\varepsilon_i(f) = \varepsilon_i(g)$, then the conjugating mappings h_i satisfy the conditions $h(y_{i-1}) = x_{i-1}$ and $h(y_i) = x_i$, which enables us to "glue together" from them a conjugating mapping defined on $[a, b]$. If $\varepsilon_{n+1-i}(f) = -\varepsilon_i(g)$, then this means that the restriction of g to the ith interval (starting from the point a) is conjugate to the restriction of f to the ith interval (starting from b), and the conjugating homeomorphism changes the orientation, which again enables us to construct a single conjugating homeomorphism h on $[a, b]$ (moreover, $h(a) = b$, $h(b) = a$).

1.18. By the monotone sequence theorem, the limit $\lim x_n = c$ exists. By the continuity of f, the relation $x_{n+1} = f(x_n)$ leads to the equality $f(c) = c$. The behavior of the iterates is illustrated in Figure 23.

1.19. To determine the character of the behavior of the sequence $\{x_n\}$ it is useful to picture the process of its construction graphically. This is done in Figure 24 for the functions under consideration. In the cases a)–d) use problem 1.18. In e) the point x_1 falls in $[1, \infty)$ whatever the position of x_0, and problem 1.18 can be used on this interval. In f) (see page 322) the interval $(0, 1]$ plays the same role.

In g) (see page 322) verify that the only fixed point is $x^* = \sqrt{5} - 1$. The inequality $x^* < f^2(x) = f(f(x)) < x$ holds for $x > x^*$, and the inequality $x < f^2(x) < x^*$ holds for $x < x^*$. Therefore, $\lim x_{2n} = l$ and $\lim x_{2n+1} = l'$ exist and satisfy the relations $f^2(l) = l$ and $f^2(l') = l'$, which implies that $l = l' = x^*$.

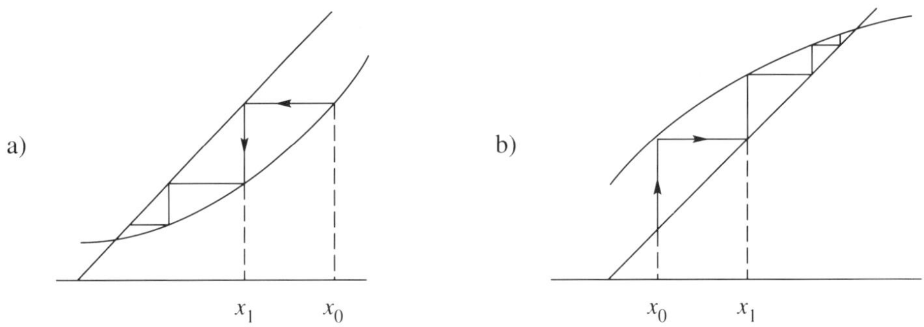

FIGURE 23

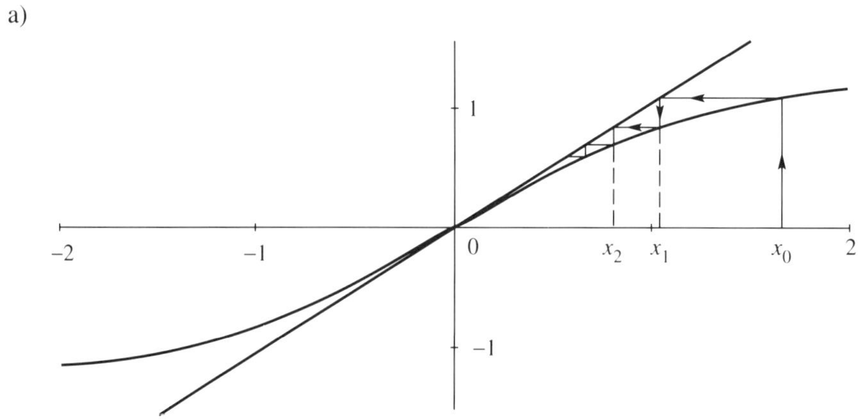

FIGURE 24a)

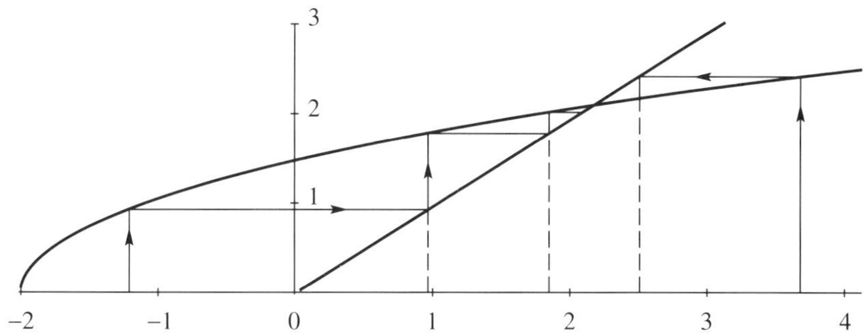

FIGURE 24b)

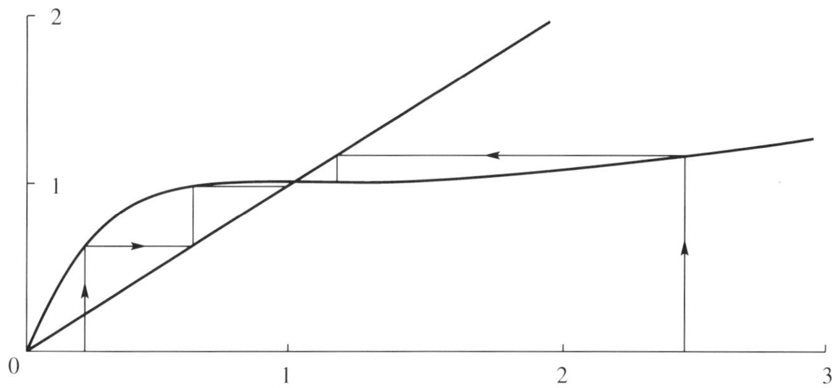

FIGURE 24d)

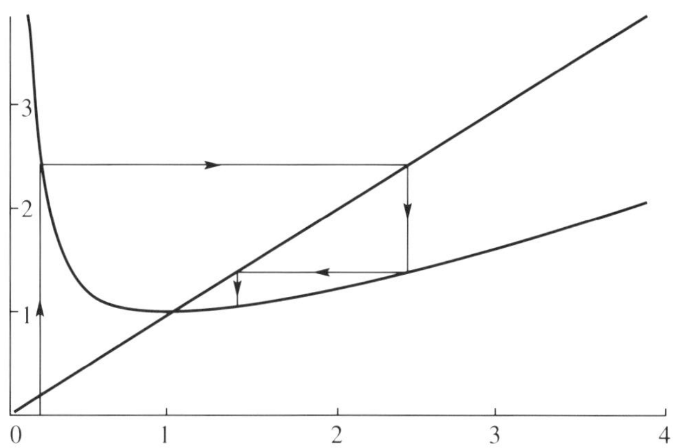

FIGURE 24e)

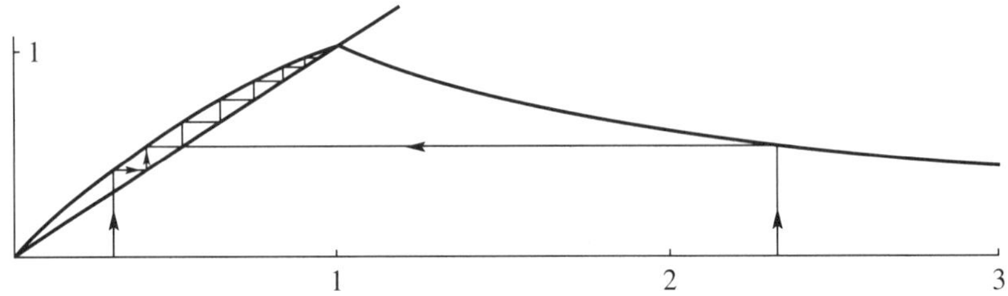

FIGURE 24f)

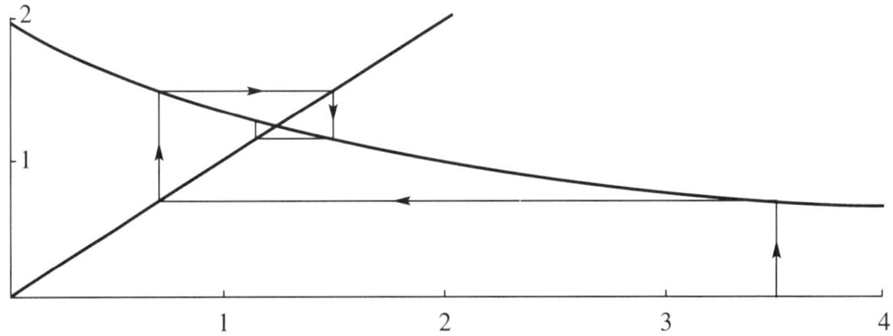

FIGURE 24g)

1.20. The solutions of these problems are analogous to the solution of the problems in 1.19. In b)–d) the interval of values of x_0 such that convergence holds has the form $[-p, p]$, where $p = \max(|t_1|, |t_2|)$, and t_1 and t_2 are the roots of the equation $f(t) = t$. In d), e), and h) it is useful to consider the sequences $\{x_{2n}\}$ and $\{x_{2n+1}\}$ (see the solution of problem 1.19g)). In the case of the mapping in h) the sequence $\{x_{2n}\}$ is monotonically increasing for $x_0 \in (1/\sqrt{5}, 1/\sqrt{3})$, while $\{x_{2n+1}\}$ is decreasing for n less than $n_0 = \min\{n||x_n| > 1/\sqrt{3}\}$, and for $n > n_0$, x_{2n} is in one of the intervals $(-\infty, -1)$ or $(1, +\infty)$, depending on the parity of n_0. Consequently, $\lim x_n$ is 1 or -1. For $x_0 \in (-1/\sqrt{3}, -1/\sqrt{5})$ the situation is symmetric. If $|x_0| = 1/\sqrt{5}$, then $x_{2n} = x_0$ and $x_{2n+1} = -x_0$, and the limit does not exist. If $|x_0| < 1/\sqrt{5}$, then $x_n \to 0$.

1.21. a) It follows from the relation

$$(f(x) - x^*)/(x - x^*) \underset{x \to x^*}{\longrightarrow} f'(x^*)$$

that $|x_{n+1} - x^*| \leq c^n |x_0 - x^*|$, $0 < c < 1$, for x_0 sufficiently close to x^*.

b) The proof is analogous.

c) Consider the mappings $f(x) = 1 - x$ $(x \in \mathbb{R})$ and $f(x) = 1 + x^2/4$ $(x \in \mathbb{R})$.

1.22. Note first that if f is continuous, and the iterates x_n converge to a limit x^*, then x^* is a fixed point for f.

a), b) Show with the help of problem 1.21 that for $|a| > 1$ the fixed points are repelling, and hence x_n does not have a limit unless x_n stabilizes beginning from some n_0. Considering the graphs, show that if $|a| < 1$, then for any x_0 the unique fixed point is the limit of x_n. The cases $a = \pm 1$ are trivial.

We study the mapping in d) (the same considerations can be used in the case c)). Let $a > 1$. Investigating the sign of the minimum (which is equal to $(1 + \ln\ln a)/\ln a)$ of the convex function $\varphi(x) = a^x - x$, we see that the fixed points $\bar{x}$, $\bar{\bar{x}}$ $(\bar{x} \le \bar{\bar{x}})$ of f exist only for $a \in (1, e^{1/e}]$. Further (see problem 1.18), $x_n \to \bar{x}$ for $x_0 < \bar{\bar{x}}$, $x_n \to \infty$ for $x_0 > \bar{\bar{x}}$, and $x_n \equiv \bar{\bar{x}}$ for $x_0 = \bar{\bar{x}}$.

If $a < 1$, then f has a unique fixed point x^*. If $a < e^{-e}$, then $f'(x^*) < -1$, and hence (see problem 1.21) x^* cannot serve as the limit of x_n if $x_0 \ne x^*$. For $a \in [e^{-e}, 1)$ prove that the function $f^2 = f \circ f$ is increasing and satisfies the inequality $f^2(x) > x$ for $x < x^*$, and $f^2(x) < x$ for $x > x^*$ (for this consider $(f^2)'$ and $(f^2)''$), and, using problem 1.18, check that $x_{2n} = f^{2n}(x_0) \to x^*$ and $x_{2n+1} = f^{2n+1}(x_0) \to x^*$ for all $x_0 \in \mathbb{R}$.

1.23. a) Verify that $N'_\varphi(c) = 0$ and use problem 1.21a).

b) By Taylor's formula,

$$0 = \varphi(c) = \varphi(x) + \varphi'(x)(c - x) + \frac{1}{2}\varphi''(\xi)(c - x)^2,$$

where ξ lies between x and c, which implies that

$$-\frac{\varphi(x)}{\varphi'(x)} = c - x + \frac{1}{2}\frac{\varphi''(\xi)}{\varphi'(x)}(c - x)^2.$$

Setting $x = x_k$ in this formula and writing x_{k+1} in the form $x_k - \varphi(x_k)/\varphi'(x_k)$, we obtain the relation

$$x_{k+1} - c = \frac{1}{2}\frac{\varphi''(\xi)}{\varphi'(x_k)}(x_k - c)^2,$$

which implies the estimate $|x_{k+1} - c| \le L|x_k - c|^2 \le \cdots \le L^{2^k}|x_0 - c|^{2^{k+1}}$.

1.24. Verify that $N'_\varphi(c) = 1 - 1/n$.

1.26. Verify that on the interval between any two neighboring extremum points of the polynomial P it has a root and an inflection point. Then use problem 1.25.

1.27. Use the fact that $(f^n)'(x_0) = f'(f^{n-1}(x_0))f'(f^{n-2}(x_0))\cdots f'(x_0) = \prod_{0\le j<n} f'(x_j) = (f^n)'(x_k)$, $k = 0, \ldots, n-1$, along with problem 1.21.

1.28. a) Use the following simple assertion:

(A) *If I, J are closed intervals and $J \subset f(I)$, then there is a closed interval $K \subset I$ such that $f(K) = J$.*

Indeed, let $J = [m, M]$, and let $x, y \in I$ be points with $x < y$ such that $f(x) \le m$ and $f(y) \ge M$. Setting $w = \min\{t \in [x, y] | f(t) \ge M\}$, $z = \max\{t \in [x, w] | f(t) \le m\}$, and $K = [z, w]$, we get the desired closed interval.

We set $K_0 = I_0$, and define the remaining closed intervals by induction: if K_{n-1} has already been constructed and has the properties that $K_{n-1} \subset K_{n-2}$ and $f^{n-1}(K_{n-1}) = I_{n-1}$, then, applying (A) to the closed intervals I_{n-1}, $I_n \subset f(I_{n-1}) = f^n(K_{n-1})$ and to the mapping f^n, we find a closed interval $K_n \subset K_{n-1}$ such that $f^n(K_n) = I_n$.

b) For any point x in the nonempty intersection $\bigcap_{n\ge 0} K_n$ we have that $f^n(x) \in I_n$.

1.29. Suppose that the points $w < x < y < z$ form a 4-cycle. Verify that, whatever the order in which these points are mapped one to the other, two of the intervals $[w, x]$, $[x, y]$, and $[y, z]$ (call them I and J) satisfy the relations $f(I) \supset J$ and $f(J) \supset I$. Prove that there is a closed interval $K \subset I$ not containing the endpoints of I such that $f^2(K) \supset K$, and use problem 1.29.

1.30. It suffices to prove that c) follows from a) (the implications c) $\Rightarrow$ b) $\Rightarrow$ a) are obvious). Assume for definiteness that the points a, b, c, and d are in the order $d \le a < b < c$. Denote $[a, b]$ by L and $[b, c]$ by R, and for arbitrary $m \ge 2$ let $I_0 = \cdots = I_{m-2} = R$, $I_{m-1} = L$, and $I_m = R'$, where $R' \subset R$ is a closed interval not containing the endpoints of R. It follows from a) that $f(R) \supset L$, $f(R) \supset R$, and $f(L) \supset R'$, and hence $f(I_n)$ always contains I_{n+1}. By the assertion in 1.28, there is a closed interval $K_m \subset R'$ such that $f^m(K_m) = I_m \supset K_m$. If we apply to f^m the assertion formulated in Problem 1.5b), there is a point $x_0 \in K_m$ such that $x_m = f^m(x_0) = x_0$. It remains to verify that $f^n(x) = x_n \ne x_0$ for $0 < n < m$.

If $x_n = x_0$ for $0 < n < m$, then all the points x_k, $k = 0, 1, 2, \dots$, have the number n as their period. Then since $m - 1$ is representable in the form $pn + q$ $(p \in \mathbb{N},\ 0 \le q < n)$, $x_{m-1} = x_q \in R'$, contradicting the inclusion $x_{m-1} \in L$, which follows from the definition of x_0. The case $m = 1$ (that is, the existence of a fixed point) is trivial (see problem 1.5a)).

The assertion proved is due to T. Li and J. Yorke [39].

1.31. Consider the closed intervals $\Delta_1 = [\cos(2\pi/n),\ \cos(\pi/n)]$ and $\Delta_2 = [\cos(\pi/n), 1]$, which are mapped by $f = T_n$ in a one-to-one fashion onto $[-1, 1]$. Take c to be the point $\cos 2\pi/n$, b to be the unique point in Δ_1 such that $f(b) = c$, a to be the unique point in Δ_2 such that $f(a) = b$, and d to be 1. Then $f(c) = d$ and $c < b < a < d$, which implies (see problem 1.30) that f has cycles of any order.

1.32. b) Only the case when $a > 0$ is interesting. Removing the absolute value signs on the eight intervals into which the line is divided by the points

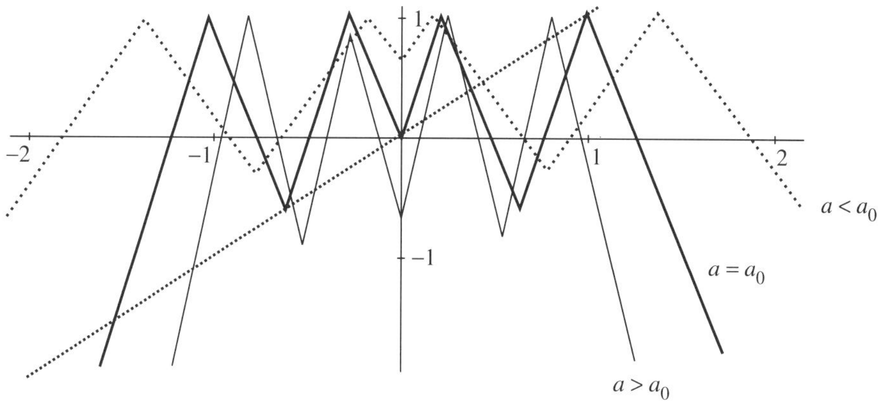

FIGURE 25

$$0 \quad \pm\frac{1}{a}, \quad \pm\left(\frac{a-1}{a^2}\right), \quad \pm\left(\frac{a+1}{a^2}\right),$$

solve the equation

$$1 - a|1 - a|1 - a|x||| = x$$

and verify that its solutions are contained among the numbers of the form $x = (1 + \varepsilon_1 a + \varepsilon_2 a^2)/(1 + \varepsilon_3 a^3)$, where $\varepsilon_1, \varepsilon_2, \varepsilon_3 = \pm 1$; moreover, for $1 < a < a_0 = (1+\sqrt{5})/2$ only two of them, $p = 1/(1-a)$ and $q = 1/(1+a)$, turn out to be roots of the equation (these are obviously fixed points of f). For $a \geq a_0$ all eight numbers of the indicated form are roots, and if we renumber them in increasing order, then $x_1 = p$, $x_6 = q$, and the triples $\{x_2, x_4, x_8\}$ and $\{x_3, x_5, x_7\}$ form cycles that coincide for $a = a_0$. The fact that they are repelling follows from the inequality $|f'(x)| > 1$ for $x = x_i$ $(i = 1, \dots, 8)$ (cf. the solution of problem 1.21). In the case $-1 < a \leq 1$ only the root q remains, and for $a \leq -1$ there are no roots. The character of the graphs for the three values of the parameter is shown in Figure 25.

1.33. a), b) Since each fixed point of f is a fixed point of any of its iterates f^n, the fourth-order polynomial $f^2(x) - x$ is divisible by the quadratic trinomial $f(x) - x$. As a result of the division we obtain the polynomial $a^2x^2 - ax + 1 - a$, which has real roots for a $\geq 3/4$. If $a \leq 3/4$, then the fixed points of f^2 reduce to the two fixed points of f. For $a > 3/4$ there are two points x_1 and x_2 satisfying the relations $f^2(x) = x$ and $f(x) \neq x$. It is easy to see that $f(x_1) = x_2$ and $f(x_2) = x_1$, that is, $\{x_1, x_2\}$ is a 2-cycle. The character of the graph of f^2 is pictured in Figure 26 for $a < 3/4$, $a = 3/4$, and $a > 3/4$.

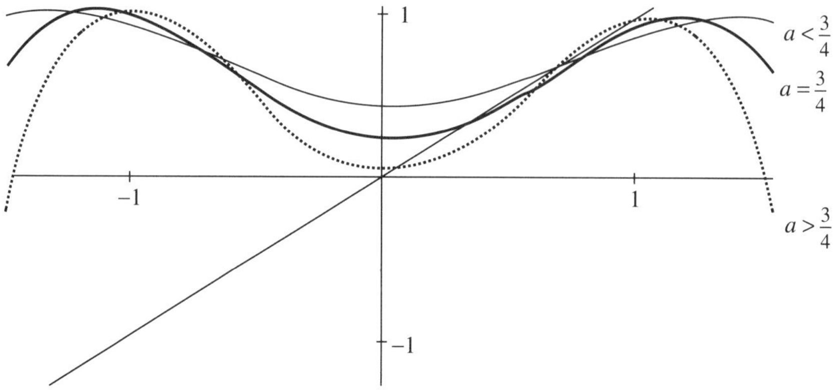

FIGURE 26

Since $(f^2)'(x_1) = f'(x_1)f'(x_2) = (-2ax_1)(-2ax_2) = 4a^2x_1x_2 = 4(1-a)$, the cycle turns out to be attracting in the interval $3/4 < a < 5/4$, while it is repelling for $a > 5/4$ (see problem 1.27).

c) Prove that f does not have 4-cycles for $a \leq 5/4$. The case $a \leq 1$ is almost obvious, and we assume below that $a > 1$. Let $P(x) = f^2(x) = 1 - a(1 - ax^2)^2$ $(x \in \mathbb{R})$. To a 4-cycle of f there correspond two 2-cycles of P. It is these 2-cycles we are looking for. To find a 2-cycle of P means to find a solution of the system

$$\begin{cases} y = P(x), \\ x = P(y), \end{cases} \tag{1}$$

satisfying the condition $x \neq y$.

Let L and L_s be the graphs of the functions $y = P(x)$ and $x = P(y)$, respectively (Figure 27). It is clear that the curves L and L_s are symmetric to each other with respect to the line $y = x$, and it is not hard to get the answers of interest to us from accurately drawn sketches of these graphs.

Now let us give a rigorous argument. Show that to verify that the system (1) does not have nontrivial solutions it suffices to prove the absence of solutions with abscissae in the intervals $[1-a, x_0]$ and $[x_1, 1]$, where x_0 and x_1 $(x_0 < x_1)$ are points forming a 2-cycle of f. Consider first the interval $[x_1, 1]$. Let Q be the function on $[x_1, 1]$ whose graph coincides with part of the curve L_s. In other words, Q is the function inverse to the restriction of P to $[1/\sqrt{a}, x_0]$. Our goal is to show that $P(x) \neq Q(x)$ for $x \in (x_0, 1]$.

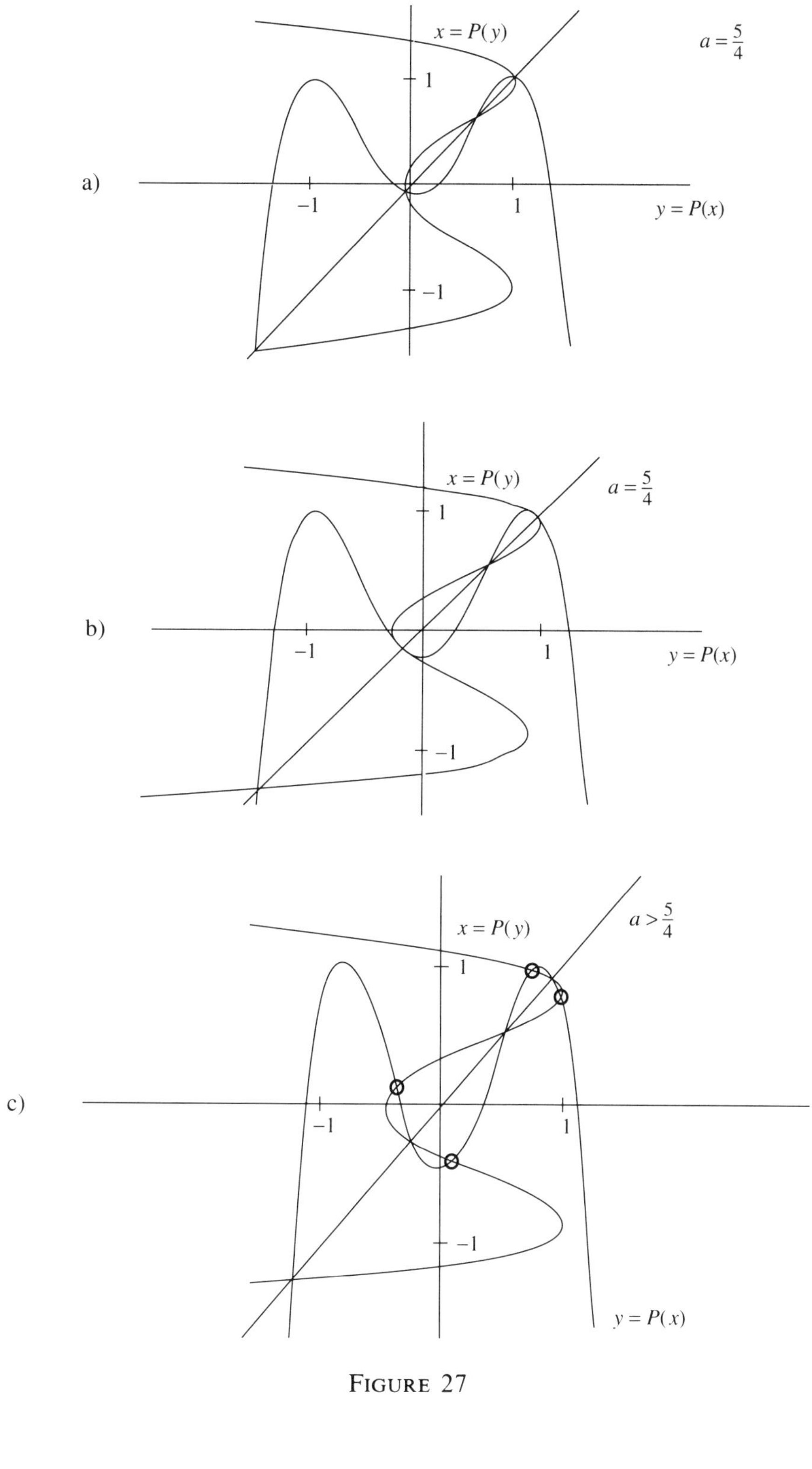
a)
$x = P(y)$
$a = \frac{5}{4}$
1
−1
1
−1
$y = P(x)$
b)
$x = P(y)$
$a = \frac{5}{4}$
1
−1
1
−1
$y = P(x)$
c)
$x = P(y)$
$a > \frac{5}{4}$
1
−1
1
−1
$y = P(x)$

FIGURE 27

First note that

$$P'(x) < 0, \qquad P''(x) < 0, \qquad P'''(x) < 0$$

for $x > 1/\sqrt{a}$, and, moreover, $|P'(x_0)| \leq 1$ (it is here that the condition $a \leq 5/4$ plays a role). Using the rule for differentiating an inverse function and denoting $Q(x)$ by t, we get that

$$Q'(x) = \frac{1}{P'(t)}, \qquad Q''(x) = -\frac{P''(t)}{(P'(t))^3},$$
$$Q'''(x) = -\frac{P'''(t)}{(P'(t))^4} + 3\frac{(P''(t))^2}{(P'(t))^5}.$$

Obviously,

$$P(x_1) = Q(x_1) = x_1, \qquad P'(x_1) - Q'(x_1) \geq 0,$$
$$P''(x_1) - Q''(x_1) = P''(x_1) + \frac{P''(x_1)}{(P'(x_1))^3} \geq 0.$$

Finally, let us see that $P'''(x) - Q'''(x) > 0$ for $x \in (x_1, 1)$. Indeed, since $|P'(t)| \leq 1$ and $at^2 \geq 1$ for $x \in (x_1, 1)$, we have that

$$\begin{aligned} P'''(x) - Q'''(x) &= -24a^3x - \frac{24a^3t}{|P'(t)|^4} + 3\frac{(4a^2(3at^2-1))^2}{|P'(t)|^5} \\ &\geq -\frac{48a^3}{|P'(t)|^4} + 3\frac{(8a^2)^2}{|P'(t)|^5} > 0. \end{aligned}$$

It follows from what has been proved that the difference $P(x) - Q(x)$ is strictly increasing on $[x_1, 1]$. Thus, the system (1) does not have solutions with abscissae in $(x_1, 1]$. Then it does not have solutions with positive abscissae at all. But if $(\bar{x}, \bar{y})$ is assumed to be a solution of (1) such that $1 - a < \bar{x} < x_0$, then the pair $(f(\bar{x}), f(\bar{y}))$ is also a solution of (1), which, however, is impossible, because $f(\bar{x}) > f(1-a) > 0$ (the condition $a \leq 5/4$ is again used in verifying the last inequality).

If $a > 5/4$, then $P'(x_0) < -1 < Q'(x_0) < 0$, and the existence of nontrivial solutions of (1) follows from considerations connected with the Bolzano-Cauchy theorem.

We remark that for $a = 2$ the system (1) has 12 nontrivial solutions (see Figure 28), which form three 4-cycles (the points generating a particular cycle are marked by the same symbols).

d)([3]) By the assertion in a) and problem 1.29, f does not have cycles for $a \leq 3/4$. Suppose that $a > 0$ and let p and q be the fixed points of f.

([3])R. Sibilev helped in the solution of the this part of the problem.

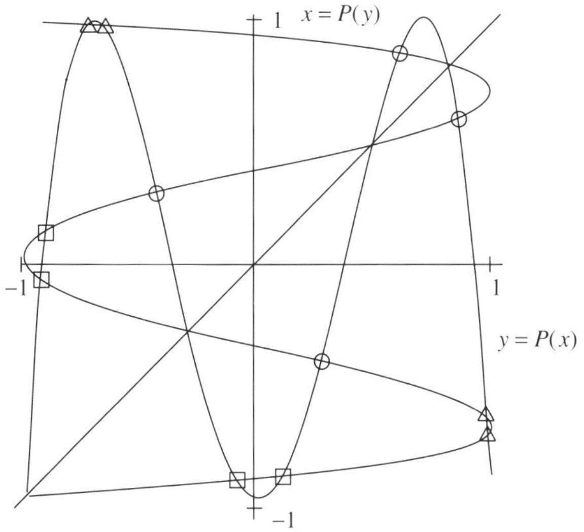

FIGURE 28

With the help of a sketch of the graph of $y = f^3(x)$ it is not hard to see that the number of fixed points of f^3 can only increase as the parameter a increases (follow the movement of the points corresponding to the extrema of f^3 on Figure 29, and then try to make a rigorous argument). Verify that the fixed points of f^2 different from p and q generate a 3-cycle. For $a \leq 3/4$ only p and q are fixed points of f^3, but f^3 has eight fixed points for larger values of a. Thus, there exists a minimal value a_0 of the parameter for which the number of fixed points of f^3 is greater than two. Suppose that $a = a_0$, $f^3(x_1) = x_1$, $x_2 = f(x_1)$, $x_3 = f(x_2)$, and $x_1 \neq p$, q. The points x_1, x_2, and x_3 form a 3-cycle, and each of them is a root of the equation $f^3(x) = x$; moreover, x_1, x_2, and x_3 must be multiple roots, because the graphs of $y = f^2(x)$ and $y = x$ are tangent at these points. Considering the number of roots and the degree of the polynomial $P(x) = f^3(x) - x$, we conclude that $P(x) = A(x-p)(x-q)(x-x_1)^2(x-x_2)^2(x-x_3)^2$. Dividing $P(x)$ by $x - f(x) = a^2x^2 + x - 1 = a^2(x-p)(x-q)$, we arrive at the equality

$$\begin{aligned}-a^6x^6 + a^5x^5 + (3a^5 - a^4)x^4 + (a^3 - 2a^4)x^3 + (3a^3 - 3a^4 - a^2)x^2 \\ + a^3 - 2a^2 + a - 1 = -a^6(x-x_1)^2(x-x_2)^2(x-x_3)^2 \\ = -a^6(x^3 + bx^2 + cx + d)^2.\end{aligned}$$

Equating coefficients of like powers of x, we get a system of six equations in a, b, c, and d. Elimination of b, c, and d from this system leads to the

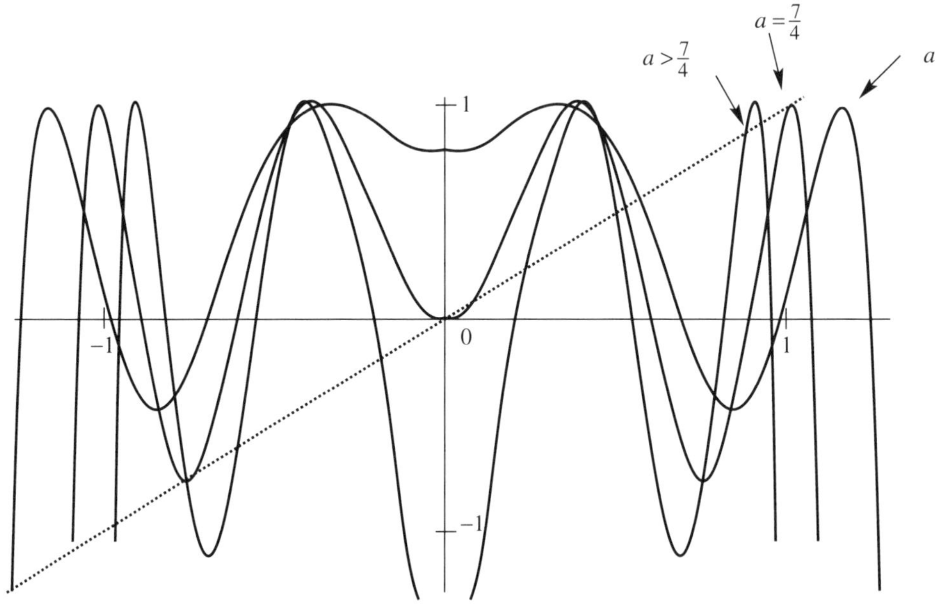

FIGURE 29

equation $(4a-7)^2=0$, which gives us that a must be equal to $7/4$, and a check confirms that the system is consistent in this case. Thus, there are no 3-cycles if $a<7/4$, f^3 has 5 fixed points if $a=7/4$ (p, q, and three points generating a cycle), and f^3 has eight fixed points if $a>7/4$ (namely, p and q and six more points, each of which, as already mentioned, must be in a 3-cycle). Consequently, there are exactly two 3-cycles for $a>7/4$.

1.34. Use problem 1.14d).

1.35. Let L be the curve symmetric to the graph of f over the interval $[x_0, 1]$ with respect to the line $y=x$, and let

$$\gamma(x)=(f(x), x), \qquad x_0 \le x \le 1,$$

be its parametrization. Show that by virtue of the conditions on the one-sided derivatives there is a number $\delta>0$ such that the points $\gamma(x)$ lie under the graph of f if $x_0<x<x_0+\delta$. On the other hand, the point $\gamma(1)$ lies above the graph of f (or on it), and hence the graph of f intersects the curve L. The abscissa of the intersection point generates a 2-cycle.

1.36. Consider the mapping $f=g^{2^{n-1}}$ and prove that f has a fixed point satisfying the condition of the preceding problem. Verify that the points of a 2-cycle of f generate a 2^n-cycle of g. To construct an example, consider the continuous function on $[0, 1]$ whose graph is the polygonal curve with vertices at the points

$$\left(1-\frac{1}{n}, 1-\frac{7+6(-1)^n}{16n}\right).$$

1.37. Let $g = f^n$. This mapping also belongs to the class $S(\Delta)$ (see problem VII.2.24c)). The assumption that $g(x) = x$ for infinitely many points implies by the mean value theorem that $g'(x) = 1$ also on an infinite set. As a consequence of this, $|g'|$ has infinitely many local minimum points. By the "minimum modulus principle" (problem VII.2.24f)), it follows that g, hence also f, has infinitely many critical points.

1.38. There exist points u and v such that $x < u < y < v < z$ and $g'(u) = g'(v) = 1$. The assertion now follows from the "minimum modulus principle" (problem VII.2.24f)) for the interval $[u, v]$.

1.39. Suppose that $x \in (a, b)$ is an attracting fixed point for $g = f^n$ ($n \in \mathbb{N}$) (if $x = a$ or $x = b$, then there is nothing to prove). Then $|g'(x)| \leq 1$.

a) Suppose first that $|g'(x)| < 1$. Consider the maximal interval $\langle r, s\rangle$ containing x such that the trajectory $\{g^m(y)\}$ of any point y in it is attracted to x. If $r > a$, then for all m we have that $g^m(r) \notin (r, s)$ (a consequence of the maximality of the interval), and if $s < b$, then $g^m(s) \notin (r, s)$ for $m \in \mathbb{N}$. Thus, this interval has the form $[a, s)$, $(r, b]$, $[a, b]$, or (r, s). The first three possibilities lead at once to the assertion to be proved. In the fourth case it follows from the obvious relations $g((r, s)) \subset (r, s)$ and $g(r)$, $g(s) \notin (r, s)$ that $g(r)$, $g(s) \in \{r, s\}$. In view of assertion 1.38, the assumption that $g(r) = r$ and $g(s) = s$ leads to the inequality $g'(x) > 1$, which is impossible. If $g(r) = s$ and $g(s) = r$, then $g^2(r) = r$ and $g^2(s) = s$, which is impossible for an analogous reason. Thus, we arrive at the conclusion that $g(r) = g(s)$; but then, by Rolle's theorem, the interval (r, s) contains a critical point p of g that is attracted to x. Since $g = f^n$, p must have the form $f^k(c)$, $k \in \mathbb{N}$, which at once implies the assertion to be proved.

b) Suppose now that $|g'(x)| = 1$. Replacing g by g^2 if necessary, we can assume that $g'(x) = 1$. Let (r, s) be the maximal interval containing x and not containing other fixed points of g. It is nonempty, because the set of fixed points of g is finite (problem 1.37). Then either $g(y) > y$ for all $y \in (r, x)$ or $g(y) < y$ for $y \in (x, s)$, since otherwise each of the intervals (r, x) and (x, s) contains infinitely many points t at which $g'(t) > 1$, which leads to a contradiction with the minimum modulus principle (VII.2.24f)) and the finiteness of the set of critical points of g.

Assume for definiteness that $g(y) > y$ for $y \in (r, x)$. If $r = a$ and g is increasing on $[a, x]$, then the trajectory of a is attracted to x (see problem 1.18). If $r = a$ and g is not monotone on $[a, x]$, then some critical point p of g falls in (a, x), and g is increasing on $[p, x]$. Then the trajectory $\{g^m(p)\}$ is attracted to x. If $r > a$, then $g(r) = r$, and by the mean value theorem the interval (r, x) contains a point z such that $g'(z) = 1$. Using the facts that $g(z) > z$, $g(x) = x$, and $g'(x) = 1$ and again using the minimum modulus principle, we arrive at the conclusion that some critical point p lies between z and x and is attracted to x. The argument concludes in the same way as in a). The assertion proved is due to D. Singer [43](4).

1.40. Let $p < 0$ be a fixed point of f, and let $q = -p$. Then $f([p, q]) \subset [p, q]$ for $a \leq 2$, and p is a repelling fixed point, $f^n(q) = p$ for $n \geq 1$, and an attracting cycle must attract the trajectory of the point 0 in view of assertions VII.2.25a) and 1.39. For $a = 2$ the mapping f is the Chebyshev polynomial T_2, and we can use problems 1.31 and 1.39.

1.41. (5) a) Choose the direction of traverse of the circle so that upon such a circuit the fixed point t and the points of the cycle are encountered in the order $t \to a \to b \to c \to t$, where $\varphi(a) = b$, $\varphi(b) = c$, and $\varphi(c) = a$. If x and y are two points of the circle, denote by $[x, y]$ the arc which in the chosen direction begins at x and ends at y. We say that the condition A_0 holds if $\varphi([t, a]) \supset [a, b]$, and that the condition A_1 holds if $\varphi([t, a]) \supset [b, a]$ (obviously, one of these conditions always holds). This will be written symbolically as follows: $A_0 = (\varphi([t, a]) \supset [a, b])$ and $A_1 = (\varphi([t, a]) \supset [b, a])$. In exactly the same way, one of the two conditions in each of the following pairs is always valid:

$$B_0 = (\varphi([a, b]) \supset [b, c]), \qquad B_1 = (\varphi([a, b]) \supset [c, b]);$$
$$C_0 = (\varphi([b, c]) \supset [c, a]), \qquad C_1 = (\varphi([b, c]) \supset [a, c]);$$
$$D_0 = (\varphi([c, t]) \supset [t, a]), \qquad D_1 = (\varphi([c, t]) \supset [a, t]).$$

We always have at least one of the sixteen propositions of the form $A_{\varepsilon_1} \& B_{\varepsilon_2} \& C_{\varepsilon_3} \& D_{\varepsilon_4}$, $\varepsilon_i = 0, 1$.

Observe that the intervals with endpoints t, a, b, or c always contain either two intervals I and J with the properties that $\varphi(I) \supset J$ and $\varphi(J) \supset I \cup J$, or three intervals I, J, and K with the properties that $\varphi(I) \supset J$, $\varphi(J) \supset K$, and $\varphi(K) \supset I \cup K$. Modifying the arguments used in the solution of problem 1.30, verify that there is always a point x whose iterates visit successively the intervals $\underbrace{J, J, \dots, J}_{n-1}, I, J$ or $\underbrace{K, K, \dots, K}_{n-2}, I, J, K$ and which is such that $\varphi^n(x) = x$ but $\varphi^k(x) \neq x$ for $k < n$.

b) Parametrizing the circle in the usual way, we identify it with the interval $[0, 2\pi)$. Consider the mapping $f : [0, 2\pi) \to [0, 2\pi)$ given by the formula

$$f(t) = \begin{cases} t + 2|\frac{\pi}{3} - t|, & 0 \leq t \leq \frac{2\pi}{3}, \\ (t + \frac{2\pi}{3}) \bmod 2\pi, & \frac{2\pi}{3} \leq t < 2\pi \end{cases}$$

and carry it over to the circle. Corresponding to the point $\pi/3$ is a fixed point of the circle, and corresponding to the points 0, $2\pi/3$, and $4\pi/3$ is a 3-cycle, but it is easy to see that there are no 2-cycles.

(4)The book [33] contains extensive material about iterates of transformations of a closed interval.

(5)The solution of this problem was communicated to us by Yu. Khokhlov.

1.42. b) Verify that the inequality $m \le x - x_0 < m+1$ $(m \in \mathbb{Z})$ implies that $m \le f^n(x) - f^n(x_0) < m+1$ for any $n \in \mathbb{N}$.

c) Let $\varphi^m(z_0) = z_0 = e^{2\pi i x_0}$. Then $f^m(x_0) = x_0 + k$ for some integer k. Using the fact that $f^n(x_0)$ can be represented in the form $f^r(x_0) + lk$ for $n = lm + r$ $(l \in \mathbb{N},\ 0 \le r < m)$, prove that $\alpha = k/m$.

d) In this case $f^m(x) - x$ cannot be an integer, whatever the values of $m \ge 1$ and $x \in \mathbb{R}$. Then for a fixed m and some $k \in \mathbb{Z}$

$$k < f^m(x) - x < k + 1.$$

Setting x equal to 0, $f^m(0), \dots, f^{(l-1)m}(0)$ in this equality and adding the inequalities obtained, we get that

$$lk < f^{lm}(0) < l(k+1) \qquad (l \in \mathbb{N}),$$

in particular,

$$k < f^m(0) < k + 1.$$

These inequalities give us that

$$\left| \frac{f^{lm}(0)}{lm} - \frac{f^m(0)}{m} \right| < \frac{2}{m}.$$

Interchanging the roles of l and m and adding the inequalities obtained, we get that

$$\left| \frac{f^m(0)}{m} - \frac{f^l(0)}{l} \right| < \frac{2}{m} + \frac{2}{l},$$

which at once implies the existence of a limit α for $x_0 = 0$, and hence for all $x \in \mathbb{R}$.

It remains to verify that for rational α some power of φ has a fixed point.

Suppose first that $\alpha = 0$. If it is assumed that there are no fixed points, then $\{f^n(0)\}$ is a monotone sequence (say increasing). The assumption that $f^m(0) \ge 1$ for some $m \ge 1$ implies that $f^{km}(0) \ge k$, $k \in \mathbb{N}$, which contradicts the stipulation that $\alpha = 0$. But then the sequence $\{f^m(0)\}$ is bounded, and its limit is a fixed point, contrary to assumption.

The case $\alpha = k/m$ can be reduced to the preceding case by passing to the function $g(x) = f^m(x) - k$, which gives the mapping φ^m.

e) Show that f_1 and f_2 differ by an integer.

§2. Transformations with an invariant measure

2.1. It suffices to compute the measures of the inverse images of open intervals (or arcs).

2.2. It is convenient to represent the baker's transformation as the composition of the mappings T_1 and T_2 represented in Figure 30 (which justifies the name of the transformation). It is obvious that each of T_1 and T_2 is area-preserving.

FIGURE 30

2.3. a) If $T(x_1, x_2) = T(x_1', x_2')$, then $z_i = a_{i1}x_1 + a_{i2}x_2 - (a_{i1}x_1' + a_{i2}x_2') \in \mathbb{Z}$ $(i = 1, 2)$. Solving this system, we get (thanks to the condition $|\det A| = 1$) that $x_1 - x_1' \in \mathbb{Z}$ and $x_2 - x_2' \in \mathbb{Z}$, but this is possible only for $x_1 = x_1'$ and $x_2 = x_2'$.

b) Consider the linear transformation $\widetilde{T}_A : \mathbb{R}^2 \to \mathbb{R}^2$ given by the matrix A. It is invertible and preserves Lebesgue measure. In particular, $\lambda_2(\widetilde{T}^{-1}(B)) = \lambda_2(B)$ for $B \subset X$. The intersections $C_v = \widetilde{T}_A^{-1}(B) \cap (X + v)$ $(v \in \mathbb{Z}^2)$ are nonempty only for finitely many vectors v, and translations of them lie in X, are pairwise disjoint (see a)), and make up the set $T^{-1}(B)$. See also the solution of the following problem.

2.4. Consider the measure $\nu = T_A\lambda_2$ on $X = \mathbb{R}^2/\mathbb{Z}^2$. Like Lebesgue measure, it is translation-invariant (all sums are assumed to be modulo 1):

$$\begin{aligned} \nu(B + y) &= \nu(B + T_A(x)) = \lambda_2(T_A^{-1}(B + T_A(x))) \\ &= \lambda_2(T_A^{-1}(B) + x) = \lambda_2(T_A^{-1}(B)) = \nu(B) \qquad (B \subset X,\ y \in X) \end{aligned} \tag{1}$$

Therefore (see the assertion before the statement of the problem), $\nu = c\lambda$. Setting $B = X$ in (1), we get that $c = 1$. Thus, $\lambda_2(T_A^{-1}(B)) = \lambda_2(B)$.

2.5. The measure μ is a probability measure when $c = 1/\ln 2$. It suffices to verify that $\mu(T^{-1}(A)) = \mu(A)$ for all sets A of the form $(0, a]$, $0 < a \leq 1$. Since

$$T^{-1}(A) = \bigcup_{k \in \mathbb{N}} \left(\frac{1}{k+a}, \frac{1}{k} \right],$$

the measure of the inverse image is equal to

$$\begin{aligned} \sum \frac{1}{\ln 2} \int_{1/(k+a)}^{1/k} \frac{dx}{1+x} &= \frac{1}{\ln 2} \sum (\ln(1+k) - \ln k + \ln(1+k+a) - \ln(k+a)) \\ &= \frac{1}{\ln 2} \lim \left(\ln(1+a) - \ln\left(1 + \frac{a}{n+1}\right)\right) \\ &= \frac{\ln(1+a)}{\ln 2} = \frac{1}{\ln 2} \int_0^a \frac{dx}{1+x} = \mu(A). \end{aligned}$$

2.6. a) Use the topological conjugacy of T_n and the sawtooth mapping f_n (problem 1.15b)), which preserves Lebesgue measure (this is obvious). We find the image of Lebesgue measure with respect to a conjugating mapping: $\mu = h\lambda$, where $h(x) = \sin \frac{\pi}{2}x$ $(x \in [-1, 1])$ (it will clearly be the desired invariant measure). For intervals,

$$\mu([a, x]) = \lambda([h^{-1}(a), h^{-1}(x)]) = \frac{2}{\pi}(\arcsin x - \arcsin a),$$

and this implies that $d\mu = 2\,dx/(\pi\sqrt{1-x^2})$ $(|x| < 1)$. The measures λ and μ can be multiplied by an arbitrary constant, of course.

b) Use problem 1.15c).

2.7. To prove sufficiency consider the measure $\nu(A) = \frac{1}{n}(\mu(A) + \cdots + \mu(T^{-(n-1)}(A)))$, where μ is a measure invariant with respect to T^n.

2.8. Suppose first that T is increasing. Denote by F the set of fixed points of T, and by G the complement of F. If $\Delta = (\alpha, \beta)$ is some maximal interval in G, then $\alpha, \beta \in F$, and hence $T(\Delta) = \Delta$. Therefore, it suffices to prove the assertion for each such interval.

Let c be an arbitrary point in Δ. If we let $x_n = T^n(c)$, $n \in \mathbb{Z}$ (T is invertible), then the closed intervals Δ_n having x_n and x_{n+1} as endpoints form a covering of Δ. For any positive function $h_0 \in \mathscr{L}^1(\Delta_0)$ we define the measure $\nu_0 = h_0\,dx$ on Δ_0 and extend it by zero to $\Delta\backslash\Delta_0$. The measures $\nu_n = T^n\nu_0$ are nonzero on Δ_n and are also absolutely continuous. The equality $\nu = \sum_{n\in z}\nu_n$ defines the desired (infinite) measure.

If T is decreasing, then consider T^2 and use the assertion in 2.5.

2.9. Suppose that T has an attracting n-cycle. Then the mapping $S = T^n$ has an attracting fixed point x^*. Let $f > 0$ be the density of some measure μ. We show that there is a set A such that $S^{-1}(A) \supset A$ and $\mu(S^{-1}(A)\backslash A) > 0$. These relations contradict either the invariance with respect to S or the finiteness of the measure μ.

If $S'(x^*) > 0$, then S is monotonically increasing in some neighborhood of x^*, and then there exist points c and d such that $x^* \in (c, d)$, and $A = (c, d)$ satisfies the condition above. If $S'(x^*) < 0$, then we replace S by S^2. If $S'(x^*) = 0$, then $|S(x) - S(x^*)| \le |x - x^*|/2$, for $|x - x^*| < \varepsilon$ (ε sufficiently small), which implies the existence of the desired set A.

It follows from problem 2.7 that there is no measure with the desired properties with respect to T.

2.10. a) Suppose first that $a > 0$. The mapping H_a, considered separately on each half-line $(-\infty, 0)$ and $(0, +\infty)$, is topologically conjugate to the translation $x \mapsto x + \ln a$ on $\mathbb{R}$ ($h(t) = -e^t$ is a conjugating homeomorphism in the first case, and $h(t) = e^t$ in the second). Verify that the measure on $\mathbb{R}\backslash\{0\}$ determined by the density $f(x) = 1/|x|$ is the image of Lebesgue measure with respect to h, and that this measure is invariant with respect to all the H_a.

b) Prove that in this case the desired measure μ does not exist. Since μ must be, in particular, translation-invariant, its values on intervals must be proportional to the length, that is, $\mu = k\lambda$ $(k > 0)$. But such a measure is obviously not invariant with respect to homotheties (the cases $k = 0$ or $k = \infty$ lead to the trivial measures).

2.11. Using the fact that these transformations are topologically conjugate to suitable transformations in problem 2.10, verify that in the case a) the invariant measure has the form $(x|\ln x|)^{-1}\,dx$, while in the case b) there is no nontrivial invariant measure.

2.12. a) Let $A \subset \mathbb{R}^n$ be such that $\mu(A) = 0$. Then $\lambda_n(A)$, $\lambda_n(A+y)$, and $\mu(A+y)$ $(y \in \mathbb{R}^n)$ are also equal to zero.

b) The quasi-invariance of μ means the existence of a function f on $\mathbb{R}^n \times \mathbb{R}^n$ such that for all $y \in \mathbb{R}^n$ and $x \subset \mathbb{R}^n$

$$\mu(y + A) = \int_A f(x, y)\, d\mu(x) \tag{1}$$

and $f(x, y) > 0$ for almost all x, y. It suffices to verify just for bounded sets that the equalities $\mu(A) = 0$ and $\lambda_n(A) = 0$ are equivalent. We integrate (1) over a set Y such that $\lambda_n(Y) > 0$. Let χ_E denote the characteristic function of a set E. On the one hand,

$$\begin{aligned} I &= \int_Y \mu(y + A)\, dy = \int_Y dy \int_{\mathbb{R}^n} \chi_A(x - y)\, d\mu(x) \\ &= \int_{\mathbb{R}^n} dy \int_{\mathbb{R}^n} \chi_Y(y)\chi_A(x - y)\, d\mu(x) = \int_{\mathbb{R}^n} dt \int_{\mathbb{R}^n} \chi_Y(x - t)\chi_A(t)\, d\mu(x) \\ &= \int_A dt \int_{\mathbb{R}^n} \chi_Y(x - t)\, d\mu(x). \end{aligned}$$

On the other hand,

$$\int_Y \mu(y + A)\, dy = \int_Y dy \int_A f(x, y)\, d\mu(x) = \int_A d\mu(x) \int_Y f(x, y)\, dy.$$

Equating the right-hand sides of these equalities, we conclude that if $\lambda_n(A) = 0$, then $I = 0$, and then we get that $\mu(A) = 0$, because $f(x, y) > 0$ and $\lambda_n(Y) > 0$.

By the Radon-Nikodým theorem, the measure μ is determined by some density g. We show that the set $e = \{x | g(x) = 0\}$ has zero Lebesgue measure. Assuming the opposite, consider the family of translates $e + x_k$ of e by vectors x_k forming a countable dense subset of $\mathbb{R}^n$. The set $C = \mathbb{R}^n \setminus \bigcup_k (e + x_k)$ has zero Lebesgue measure (see problem VIII.1.23), and then also $\mu(C) = 0$. But $\mu(e + x_k) = \mu(e) = 0$, that is, $\mu(\mathbb{R}^n) = 0$. Contradiction.

2.13. a) If $f \geq 0$, then $A \mapsto \nu(A) = \int_{T^{-1}(A)} f(x)\, dx$ $(A \subset X)$ is an absolutely continuous measure on X, and the existence of f^* follows from the Radon-Nikodým theorem. If f is arbitrary, then it can be decomposed into a difference of nonnegative functions.

b) Let $f \geq 0$, and let U and V be neighborhoods of x and $y = T(x)$ with $U = T^{-1}(V)$. Then

$$\int_U f(x)\, dx = \int_V f(y)\, dy = \int_U f(T(x))|\det T'(x)|\, dx.$$

Using "differentiation with respect to the domain," that is, dividing both sides by $\lambda_n(U)$ and passing to the limit as U shrinks to the point x, we get what is required.

c) This is proved similarly.

d) Verify (1) for the sets $A = [0, a]$, $0 \le a \le 1$.

2.14. a)

$$d_i = \frac{1}{\lambda(\Delta_i)} \int_{\Delta_i} f_d(x)\,dx = \frac{1}{\lambda(\Delta_i)} \int_{T^{-1}(\Delta_i)} f_c(x)\,dx$$
$$= \sum_{1\le j\le n} \frac{c_j \lambda(\Delta_j \cap T^{-1}(\Delta_i))}{\lambda(\Delta_i)},$$

which implies that

$$p_{ij} = \frac{\lambda(\Delta_j \cap T^{-1}(\Delta_i))}{\lambda(\Delta_i)} = n\lambda(\Delta_j \cap T^{-1}(\Delta_i)).$$

c) Note that if $d = \pi c$, then $\sum_i d_i = \sum_i \sum_j p_{ij} c_j = \sum_j c_j \sum_i p_{ij} = \sum_j c_j$. Therefore, the transformations

$$u \mapsto \pi u \quad \text{and} \quad u \mapsto v^{(m)} = \frac{1}{m} \sum_{0\le k\le m} \pi^k u$$

map the compact set $K = \{u \in \mathbb{R}^n_+ \mid \sum_{1\le i\le n} u_i = C\}$ $(C > 0)$ into itself. Consequently (for fixed $u \in \mathbb{R}^n_+$, $u \ne 0$), the sequence $\{v^{(m)}\}$ has a limit point $v = \lim v^{(m_i)}$. Since π is continuous,

$$\pi v = \lim_i \pi v^{(m_i)} = \lim \frac{1}{m_i} \sum_{1\le k\le m_i} \pi^k u$$
$$= \lim \left(v^{(m_i)} + \frac{1}{m_i}(\pi^{m_i} u - u) \right) = \lim v^{(m_i)} = v = (v_1, v_2, \dots, v_n).$$

It follows from Perron's theorem that if $p_{ij} > 0$ for all i and j, then such a limit point is unique and independent of $u \in K$, and also $v_i > 0$ for all i. Frobenius extended this result to certain nonnegative matrices (see, for example, Gantmakher's book, "The theory of matrices").

2.15. a) It is easy to find (for even n) by algebraic methods a fixed point of the discrete Perron-Frobenius operator (see problem 2.14), that is, an eigenvector v of the matrix π that corresponds to the eigenvalue 1. Furthermore, it turns out that $v_k = 0$ for $k \le n/2$, and $v_k = v_{3n/2-k}$ for $k > n/2$. This suggests the following answer: if $Pf = f$ (P is the Perron-Frobenius operator; see problem 2.13), then $f(x) = 0$ for $x < 1/2$, and $f(x) = f(3/2 - x)$ for $x > 1/2$. To verify it we write the equality

$$Pf(y) = \begin{cases} \frac{1}{2} f(\frac{1}{2}y) & \text{for } y < \frac{1}{2}, \\ \frac{1}{2} f(\frac{1}{2}y) + f(\frac{3}{2} - y) & \text{for } y > \frac{1}{2} \end{cases}$$

(see problem 2.13c)), which implies that

$$\int_0^{2^{-n}} f(y)\,dy = \int_0^{2^{-n}} Pf(y)\,dy = \int_0^{2^{-(n+1)}} f(z)\,dz, \qquad n = 1, 2, \ldots,$$

which is possible only for

$$\int_{2^{-(n+1)}}^{2^{-n}} f(y)\,dy = 0 (n \in \mathbb{N}).$$

Consequently, $f(y) = 0$ almost everywhere on $[0, 1/2]$. It follows from this immediately that $f(y) = Pf(y) = f(3/2 - y)$ for $y > 1/2$.

b) For $n = 3$ the discrete Perron-Frobenius operator is written by the matrix

$$\pi = \begin{pmatrix} 0 & 0 & 1/2 \\ 1 & 0 & 1/2 \\ 0 & 1 & 0 \end{pmatrix}$$

(see problem 2.14a)). The vector $v = (1/5, 2/5, 2/5)$ satisfies the equality $\pi v = v$. Let $f(x) = 1/5$ for $x < 1/3$ and $f(x) = 2/5$ for $x > 1/3$, and show that the measure $d\mu = f\,dx$ is invariant with respect to T.

c) The transformation T does not have an absolutely continuous invariant measure, because it has an attracting fixed point (see problem 2.9). The fixed points $v \in \mathbb{R}^n$ for the operator π have the following form: $v_1 = c$, $v_k = 0$ for $k > 1$. The measures on the interval that correspond to them are determined by the densities $f_v(x) = n$ for $x \le 1/n$ and $f_v(x) = 0$ for $x > 1/n$. If μ_n is the measure determined by such a density, then it is clear that $\mu_n(A) \to 0$ if $A = (\varepsilon, 1)$, $\varepsilon > 0$, and $\mu_n([0, 1]) = 1$. This enables us to compute the limit measure, which is concentrated at the point 0 (it is invariant with respect to T, but does not have a density).

2.16. Let f be a function on Ω satisfying the condition $Pf = f$ (problem 2.13b), e)). Denoting $f(0, 1)$ by C and writing the equation $Pf = f$ for the mappings $z \mapsto yz$ $(y > 0)$, we get that $f(0, y) = C/y^2$. The application of the same device to the mapping $z \mapsto z + x$ $(x \in \mathbb{R})$ leads to the equality $f(x, y) = f(0, y) = C/y^2$. It remains to see that an arbitrary automorphism T is a composition of automorphisms of the form considered, and hence $Pf = f$ for all T.

2.17. By the results in 2.13b) and e), the density f_L of the measure μ_L must satisfy the relation

$$f_L(ax, ay + b) = \frac{1}{a^2} f_L(x, y). \tag{1}$$

Setting $x = 1$, $y = 0$, and $A = f_L(1, 0)$, we get from this that $f_L(a, b) = A/a^2$ for any $(a, b) \in P$. Obviously, such a function satisfies (1). The density f_R of the measure μ_R is found similarly from the equation

$$f_R(ax, bx + y) = \frac{1}{a} f_R(x, y).$$

2.18. Let f be the density of an invariant measure for T. Consider the function

$$h(x) = a + (b-a)\left(\int_a^b f(t)\,dt\right)^{-1} \int_a^x f(t)\,dt$$

and verify that the mapping $S = h \circ T \circ h^{-1}$ has the required property.

2.19. It suffices for us to confine ourselves to the case when $\alpha \in [0, 1)$ (see 1.42e)). Let μ be a finite measure on S^1 that preserves T (it exists by the Bogolyubov-Krylov theorem). It can be assumed that $\mu(S^1) = 1$. Identifying S^1 with the interval $[0, 1)$, consider the function $h(x) = \mu([0, x])$. The assertion 1.42c) ensures that T does not have periodic points. Consequently, the trajectory of each point is infinite, and then the measure of each point is equal to zero, the function h is continuous, $h(0) = 0$, and $h(x) \to 1$ as $x \to 1$. Let $\beta = \mu([0, T(0)])$. Then

$$\begin{aligned} h(T(x)) &= \mu([0, T(x)]) \\ &= (\mu([0, T(0)]) + \mu([T(0), T(x)])) \bmod 1 \\ &= (\beta + \mu([0, x])) \bmod 1 = (h(x) + \beta) \bmod 1. \end{aligned}$$

From this it is clear that $h \circ T = T_\beta \circ h$, where T_β is the rotation of the circle through the angle $\beta \in [0, 1)$.

To prove that $\beta = \alpha$ we consider some function $f : \mathbb{R} \to \mathbb{R}$ such that $f(x) = T(x)$ for $x \in [0, 1)$ and $f(x+1) = f(x) + 1$ for all $x \in \mathbb{R}$, and verify that $\lim f^n(0)/n = \beta$ (see 1.42b), e)).

Let ν be a measure on $\mathbb{R}$ whose restriction to each interval of the form $[k, k+1)$ $(k \in \mathbb{Z})$ is obtained from the measure μ by translation by k. Then it follows from the invariance of μ that $\nu([f^n(0), f^{n+1}(0)]) = \nu([0, f(0)]) = \beta$ for $n \in \mathbb{Z}$. Therefore, $\beta = \frac{1}{n}\nu([0, f^n(0)])$. But since $|\nu([0, y]) - y| < 1$ for all y,

$$\beta = \lim \frac{1}{n}\nu([0, f^n(0)]) = \lim \frac{1}{n} f^n(0) = \alpha.$$

2.20. If T is not ergodic, then there is an invariant set A such that $\mu(A) > 0$ and $\mu(\overline{A}) > 0$, where $\overline{A} = X \backslash A$. Then a function f taking the value a on A and b $(\neq a)$ on $\overline{A}$ is invariant and not constant. Conversely, if f is an essentially nonconstant function, then there is a number c such that the sets $A = \{x \in X \mid f(x) \geq c\}$ and $\overline{A}$ have positive measure. If f is also invariant, then A and $\overline{A}$ are invariant, and T is not ergodic.

2.21. a) Use the preceding problem and note that to prove ergodicity it suffices to check the functions f in $\mathscr{L}^2([0, 1])$. As is known, every such function can be expanded in a Fourier series:

$$f(x) \sim \sum_{n \in \mathbb{Z}} c_n e^{2\pi i n x} \qquad (c_n = \hat{f}(n)),$$

and the sequence $\{c_n\}_{n \in \mathbb{Z}}$ of coefficients determines f uniquely (up to values on a set of measure zero). Corresponding to the function $g(x) = f(T(x))$ is the Fourier series $\sum_{n\in\mathbb{Z}} c_n e^{2\pi i\alpha n} e^{2\pi i x n}$. If $f(x) = g(x)$ almost everywhere, then $c_n = c_n e^{2\pi i\alpha n}$ for all $n \in \mathbb{Z}$. For irrational α this implies that $c_n = 0$ for $n \neq 0$, that is, the Fourier series for f reduces to a constant. It follows that $f(x) = c_0$ almost everywhere, which means ergodicity. If $\alpha = k/l \in \mathbb{Q}$, then the coefficients c_{ml} $(m \in \mathbb{Z})$ can be nonzero, that is, there exist nonconstant invariant functions.

b) Arguing as in a), we take the function $f(x) \sim \sum_{n\in\mathbb{Z}} c_n e^{2\pi i n x}$ and get from the equality $f(x) = f(T(x)) \sim \sum_{n\in\mathbb{Z}} c_n e^{2\pi i 2nx}$ the relations $c_n = c_{2n} = c_{4n} = \cdots$ for all $n \in \mathbb{Z}$. The condition $f \in \mathscr{L}^2(X)$, or $\sum_{n\in\mathbb{Z}} |c_n|^2 < \infty$, leads to the fact that $c_n = 0$ for $n \neq 0$, which implies ergodicity of T as in a).

c) This is analogous to b), since the transformation T can be interpreted as the mapping $x \mapsto nx \bmod 1$ of the interval $[0, 1)$.

d), e) The functions in $\mathscr{L}^2(X)$ can be "decomposed" in a double Fourier series: $f(x, y) \sim \sum_{n_1, n_2\in\mathbb{Z}} c_{n_1 n_2} e^{2\pi i n_1 x} e^{2\pi i n_2 y} = \sum_{n_1, n_2\in\mathbb{Z}} c_{n_1 n_2} e^{2\pi i(n_1 x + n_2 y)}$. Then

$$f(T_A(x, y)) \sim \sum_{n_1, n_2\in\mathbb{Z}} c_{n_1 n_2} e^{2\pi i((a_{11}n_1 + a_{21}n_2)x + (a_{12}n_1 + a_{22}n_2)y)} = \sum_{n_1, n_2\in\mathbb{Z}} c_{n_1 n_2} e^{2\pi i(m_1 x + m_2 y)}.$$

Denote the vectors (n_1, n_2) and (m_1, m_2) by n and m, respectively. Then $m = A^* n$, where A^* is the matrix obtained by transposing A. If f is invariant, then $c_n = c_{A^* n} = c_{(A^*)^2 n} = \cdots$, and the convergence of the series $\sum_{n\in\mathbb{Z}^2} |c_n|^2$ leads to the fact that if the trajectory $\{(A^*)^k n\}$ of the vector n is infinite, then $c_n = c_{A^* n} = \cdots = 0$.

If the eigenvalues of A are not roots of unity, then $(A^*)^k n \neq n$ for all $k \in \mathbb{N}$ and $n \in \mathbb{Z}^2 \backslash \{0\}$. In this case $f(x) = c_{00}$ almost everywhere as above, that is, T_A is ergodic. In the opposite case there exist a number $k \in \mathbb{N}$ and a vector $v \in \mathbb{R}^2$ such that $(A^*)^k v = v$. Since A^* is an integer matrix, a vector v satisfying this relation can be assumed to be an integer vector. As a nonzero invariant function we can take the sum of the exponentials corresponding to the vectors v, $A^* v, \ldots, (A^*)^{k-1} v$. Consequently, T_A is not ergodic.

2.22. Suppose that the trajectory $\{T^m(x)\}$ of the point x is not dense in X. This means that there exists an open ball B (with rational coordinates of the center and with rational radius) such that $x \notin \bigcup_{m\in\mathbb{N}} T^{-m}(B) = G_B$. Since G_B is open, it follows that $\lambda_n(G_B) > 0$. It follows from ergodicity that $\lambda_n(X \backslash G_B) = 0$ (consider the invariant set $\bigcap_{n\in\mathbb{N}} T^{-n}(G_B) \subset G_B$). Since the balls with rational parameters form a countable family $\{B_k\}$, the union $e = \bigcup_k (X \backslash G_{B_k})$ has zero measure. And every point in $X \backslash e$ has a dense trajectory.

2.23. If S is ergodic, then in view of the preceding problem the trajectory of almost every point is dense, and this, by the result in VII.4.5a), implies that $\alpha_1, \dots, \alpha_n$ are linearly independent over $\mathbb{Q}$ modulo 1. The sufficiency of this condition can be proved with the help of Fourier series (modify the solution of problem 2.21a)).

2.24. Show that the transformations T and S admit the same "symbolic representation." Let D be the set of dyadic rational numbers in $I = [0, 1]$, and let $X = I \backslash D$. Since $\lambda(D) = 0$, it suffices to present a measurable invertible transformation $h : X \to X$ that preserves Lebesgue measure and is such that $S \circ h = h \circ T$.

Let $A_0^1 = [0, 1/2] \cap X$ and $A_1^1 = [1/2, 1] \cap X$. Define successively the sets $A_0^k = S^{-1}(A_0^{k-1})$ and $A_1^k = S^{-1}(A_1^{k-1})$, $k = 2, 3, \dots$. Prove that these sets have the following properties:

1) $\lambda(A_{i_1}^{k_1} \cap A_{i_2}^{k_2} \cap \dots \cap A_{i_n}^{k_n}) = 2^{-n}$ for any $k_1 < k_2 < \dots < k_n$ and $i_1, \dots, i_n \in \{0, 1\}$ (can be verified by induction);

2) the set $\bigcap_{k=1}^{\infty} A_{i_k}^k$ consists of a single point for any sequence $\{i_k\}$, $i_k \in \{0, 1\}$, and each point in X is uniquely representable as such an intersection.

The last property allows us to identify the points in X with binary sequences, and then $1, i_2, \dots) = (i_2, i_3, \dots)$.

Let $x = 0.x_1x_2\dots$ be the binary expansion of a point in X. Setting $i_k = x_k$, we get a binary sequence determining (with the help of 2)) a point $y \in X$. Let $h(x) = y$. Obviously, $T(x) = 0.x_2x_3\dots$, which yields the equality $S \circ h = h \circ T$.

It remains to see that h is measurable and preserves Lebesgue measure (the invertibility of h is obvious). This follows from the fact that for any interval of the form $\Delta_{\varepsilon_1\varepsilon_2\cdots\varepsilon_n} = \{x = \sum x_k/2^k \,|\, x_1 = \varepsilon_1, \dots, x_n = \varepsilon_n\}$ where $\varepsilon_k \in \{0, 1\}$, the equality $h(\Delta_{\varepsilon_1\cdots\varepsilon_n} \cap X) = A_{\varepsilon_1}^1 \cap A_{\varepsilon_2}^1 \cap \dots \cap A_{\varepsilon_n}^1$ and property 1) are satisfied. The measurability of h follows from the fact that the inverse image $h^{-1}(\Delta \cap X)$ of any dyadic rational interval $\Delta = [k/2^n, (k+1)/2^n]$ of rank n coincides with $\Delta' \cap X$, where $\Delta' = [l/2^n, (l+1)/2^n]$.

2.25. a) Use the fact that by a linear substitution T can be transformed into the transformation S in the preceding problem, and hence is metrically conjugate to the mapping $x \mapsto (2x) \bmod 1$ $(x \in [0, 1))$, which is ergodic (see problem 2.21b)).

b) The problem can be reduced to the preceding with the help of the solution of problem 2.6a), in which it is established that the mappings $x \mapsto 1 - 2x^2$ and $x \mapsto 1 - 2|x|$ are metrically conjugate.

c) With the help of problems 1.15b) and 2.6a) replace the mapping T_n by the saw-tooth mapping f_n. Then, modifying the assertion in problem 2.24, replace f_n by the mapping $x \mapsto (nx) \bmod 1$ $(x \in [0, 1))$, which was actually proved to be ergodic in 2.21c).

2.26. For every sufficiently large $k \in \mathbb{N}$ the set $A_k = \{x \in X \,|\, f(x) \geq 1/k\}$

has positive measure. Consider the set B of points x in A_k such that $T^n(x) \notin A_k$, $n \in \mathbb{N}$ (nonrecurrent points). Verify that $T^{-m}(B) \cap B = T^{-(m+n)}(B) \cap T^{-n}(B) = \varnothing$ for all m, $n \in \mathbb{N}$. The finiteness of the measure of X leads to the conclusion that $\mu(B) = 0$, and this implies that almost every point $x \in A_k$ returns to A_k infinitely many times, that is, $f(T^n(x)) \geq 1/k$ for infinitely many indices, and $\sum f(T^n(x)) = +\infty$.

To finish the proof, exhaust X by the sets A_k.

2.27. Apply the ergodic theorem to the characteristic function of A.

2.28. Let $B \subset X$ be a set such that $\mu_1(B) \neq \mu_2(B)$, and let χ_B be its characteristic function. By the ergodic theorem, for $i = 1$ or 2

$$\varphi(x) = \lim \frac{1}{n} \sum_{0 \leq k < n} \chi_B(T^k(x)) = \mu_i(B)$$

for all x in a set A_i with $\mu_i(A_i) = 1$. Since $A_1 \cap A_2 = \varnothing$, this implies the assertion to be proved.

2.29. a) First suppose that $f(0) = f(1)$. Then f can be thought of as defined on the circle, where it is uniformly continuous. For any $\varepsilon > 0$ there exists a $\delta > 0$ such that if $|(x - x')| \bmod 1| < \delta$, then

$$|(S^k(x) - S^k(x')) \mod 1| < \delta \quad \text{and} \quad |f(S^k(x)) - f(S^k(x'))| < \varepsilon (k \in \mathbb{N}).$$

Let $\sigma_n^f(x) = \frac{1}{n} \sum_{0 \leq k < n} f(S^k(x))$, and suppose that x' is a point at which, in addition, (1) holds. Then $|\sigma_n(x) - \sigma_n(x')| < \varepsilon$ for all n, which implies that

$$\int_0^1 f(x)\, dx - \varepsilon \leq \underline{\lim}\, \sigma_n^f(x) \leq \overline{\lim}\, \sigma_n^f(x) \leq \int_0^1 f(x)\, dx + \varepsilon.$$

Since ε is arbitrary, this implies (1) for x.

If $f(0) \neq f(1)$, then we can select functions g, $h \in C([0, 1])$ such that $g(0) = g(1)$, $h(0) = h(1)$, $g \leq f \leq h$, and $\int_0^1 (h(x) - g(x))\, dx < \varepsilon$. It then remains to use the inequalities

$$\sigma_n^g(x) \leq \sigma_n^f(x) \leq \sigma_n^h(x), \quad \sigma_n^h(x) - \sigma_n^g(x) < \varepsilon \qquad (x \in [0, 1]).$$

To prove b) replace f in this argument by the function χ_Δ.

See problem I.3.9 for a proof of a) without use of the ergodic theorem.

2.30. Let $T(x) = (2x) \bmod 1$. For those $x \in [0, 1)$ that are not dyadic rational numbers (as always, we can neglect a set of measure zero when considering questions involving measure) the equality $x_k = 1$ means that $T^k(x) \in [1/2, 1)$. Since T is ergodic (problem 2.21b)), the ergodic theorem gives us that the desired limit is 1/2 for almost all x (problem 2.27). It can be proved in exactly the same way that the corresponding frequency is equal to $1/p$ for the p-ary number system. The following fact can be deduced at once from this: the set of all x such that for any p all the digits in the p-ary expansion of x occur identically often (such numbers are said to be *normal*) has measure 1. See also problems IX.2.14 and IX.3.10.

2.31. If A is invariant, then $T^{-n}(A) \cap A = A$ for all $n \in \mathbb{N}$, and the relation $(*)$ leads to the equality $\mu(A) = (\mu(A))^2$, that is, $\mu(A)$ is equal to 0 or 1.

2.32. Let $x = \sum x_k/2^k$ and $y = \sum y_k/2^k$ $(x_k, y_k \in \{0, 1\})$ be the binary expansions of numbers x and y in $[0, 1)$. If $T(x) = y$, then $y_k = x_{k+1}$ $(k \in \mathbb{N})$. We prove first the relation $(*)$ from the previous problem for sets of the form $\Delta_{\varepsilon_1 \cdots \varepsilon_m} = \{x \in [0, 1) | x_k = \varepsilon_k, \ k = 1, \dots, m\}$ $(\varepsilon_k \in \{0, 1\})$ (we call them *cylinders*). If $A = \Delta_{\varepsilon_1 \cdots \varepsilon_m}$ and $B = \Delta_{\eta_1 \cdots \eta_p}$, then (for $n > m$)

$$A \cap T^{-n}(B) = \{x | x_1 = \varepsilon_1, \dots, x_m = \varepsilon_m, \ x_{n+1} = \eta_1, \dots, x_{n+p} = \eta_p\}.$$

Computing the measures of these sets, we get

$$\lambda(A) = \frac{1}{2^m}, \quad \lambda(B) = \frac{1}{2^p}, \quad \lambda(A \cap T^{-n}(B)) = \frac{1}{2^{m+p}},$$

and hence $(*)$ holds. In almost the same way $(*)$ can be verified for sets A and B representable as finite unions of cylinders (in particular, for all intervals with dyadic rational endpoints). Every measurable set A can be approximated to any accuracy ε by a set A' representable in such a form (this means that $\lambda(A \setminus A') + \lambda(A' \setminus A) < \varepsilon$). In this case we write $A \overset{\varepsilon}{=} A'$. If $B \overset{\varepsilon}{=} B'$, then also $T^{-n}(B) \overset{\varepsilon}{=} T^{-n}(B')$ and $A \cap T^{-n}(B) \overset{\varepsilon}{=} A' \cap T^{-n}(B')$. From this, $\overline{\lim} |\lambda(A \cap T^{-n}(B)) - \lambda(A)\lambda(B)| \le 2\varepsilon$. Since ε is arbitrary, $(*)$ is proved.

2.33. a) The proof is the same as in the case of Lebesgue measure (see the preceding problem).

b) Verify that the measures coincide on intervals of the form $[k/2^n, (k+1)/2^n]$.

c), d) Use the ergodicity of μ_p (see problem 2.31) and assertion 2.28.

2.34. Suppose that the first digit of the number 2^k is p. This means that for some $m \in \mathbb{N}$

$$m + \log_{10} p \le k \log_{10} 2 < m + \log_{10}(p + 1). \tag{1}$$

Consider the transformation $S(x) = (x + \alpha) \bmod 1$ on $[0, 1)$, with $\alpha = \log_{10} 2$. The relation (1) means that $S^k(0) \in [\log_{10} p, \log_{10}(p+1))$. Since α is irrational, the desired limit is equal to $\log_{10}(1 + 1/p)$ (see 2.21a), 2.29b)).

2.35. a) Consider the sets $\Delta_{a_1 \cdots a_n} = \{x \in (0, 1] | a_1(x) = a_1, \cdots, a_n(x) = a_n\}$, which are intervals of monotonicity of T^n. Let $\psi = \psi_{a_1 \cdots a_n}$ be the function on $[0, 1)$ that is the inverse of T^n on $\Delta = \Delta_{a_1 \cdots a_n}$. Using the explicit expression

$$\psi(t) = \cfrac{1}{a_1 + \cfrac{1}{a_2 + \cdots \cfrac{1}{a_n + t}}} = \frac{p_n + t p_{n-1}}{q_n + t q_{n-1}}, \qquad t \in [0, 1)$$

for ψ (we omit the elementary computations), we can compute the length of Δ:

$$\lambda(\Delta) = 1/(q_n(q_n + q_{n+1})).$$

If $A = [\alpha, \beta) \subset [0, 1]$, then

$$\lambda(T^{-n}(A) \cap \Delta) = \psi(\beta) - \psi(\alpha) = (\beta - \alpha)\frac{1}{(q_n + \alpha q_{n-1})(q_n + \beta q_{n-1})}.$$

Since the quotient on the right-hand side is between $\lambda(\Delta)/2$ and $2\lambda(\Delta)$, we arrive at the inequality

$$\frac{1}{2}\lambda(A)\lambda(\Delta) \le \lambda(T^{-n}(A) \cap \Delta) \le 2\lambda(A)\lambda(\Delta)$$

for an arbitrary interval $A \subset [0, 1)$, hence also for an arbitrary measurable set A. If we replace λ by a Gauss measure μ with density between $1/(2\ln 2)$ and $1/\ln 2$, then, replacing the constant 2 by $C = 4/\ln 2$, we can write the inequality

$$\frac{1}{C}\mu(A)\mu(\Delta) \le \mu(T^{-n}(A) \cap \Delta) \le C\mu(A)\mu(\Delta). \tag{1}$$

Since the value of the measure of any interval can be recovered from its values on all possible intervals of the form $\Delta_{a_1 \cdots a_n}$ (this follows from the fact that $\lambda(\Delta_{a_1 \cdots a_n}) \le 2^{-n+1}$), we can replace Δ in (1) by an arbitrary set. Suppose now that A is an invariant set, and take the $\overline{A} = (0, 1]\backslash A$ in place of Δ. Then $\frac{1}{C}\mu(A)\mu(\overline{A}) \le \mu(A \cap \overline{A}) = 0$, which implies that $\mu(A) = 0$ or $\mu(\overline{A}) = 0$.

b1) Use assertion 2.27.

b2) Since

$$\frac{1}{n}\sum_{0 \le k < n} a_k = \frac{1}{n}\sum_{0 \le k < n} a_1(T^k(x)),$$

the ergodic theorem should be applied to the function $f(x) = a_1(x)$. However, this function is not integrable, because $f(x) = k$ for $x \in (1/(k+1), 1/k)$. Therefore, we replace f by the smaller function $f_N = \min(f, N)$, for which

$$\lim \frac{1}{n}\sum_{0 \le k < n} f_N(T^k(x)) = \frac{1}{\ln 2}\int_0^1 \frac{f_N(x)}{1+x}\,dx.$$

The expression obtained increases unboundedly as $N \to \infty$, and hence the limit of interest to us is infinite.

b3) Apply the ergodic theorem to the function $f(x) = \ln a_1(x)$.

b4) Verify first the equality $p_n(x) = q_{n-1}(T(x))$, which implies that

$$\frac{p_n(x)}{q_n(x)} \cdot \frac{p_{n-1}(T(x))}{q_{n-1}(T(x))} \cdots \frac{p_1(T^{n-1}(x))}{q_1(T^{n-1}(x))} = \frac{1}{q_n(x)},$$

and hence

$$-\frac{1}{n}\ln q_n(x) = \frac{1}{n}\sum_{0\le k<n} \ln\left(\frac{p_{n-k}(T^k(x))}{q_{n-k}(T^k(x))}\right).$$

If we set $f(x) = \ln x$, then the last expression differs from the sum $\frac{1}{n}\sum_{0\le k<n} f(T^k(x))$ by the quantity

$$\sigma_n = \frac{1}{n}\sum_{0\le k<n}\left(\ln(T^k(x)) - \ln\frac{p_{n-k}(T^k(x))}{q_{n-k}(T^k(x))}\right),$$

which tends to zero, because for all $x \in (0, 1)$

$$\left|\ln\frac{x}{p_k/q_k}\right| \le \left|\frac{x}{p_k/q_k} - 1\right| = \frac{q_k}{p_k}\left|x - \frac{p_k}{q_k}\right| \le \frac{q_k}{p_k}\cdot\frac{1}{q_k q_{k+1}} = \frac{1}{p_k q_{k+1}} \le \frac{1}{2^{k-1}}$$

(the last inequality can be proved by induction, for which it is useful to write out the recursion relations connecting p_n and q_n with p_{n-1}, q_{n-1} and p_{n-2}, q_{n-2}; Khintchine's book "Continued fractions" contains a wealth of information about continued fractions).

Finally,

$$\frac{1}{n}\sum_{0\le k<n} f(T^k(x)) \to \frac{1}{\ln 2}\int_0^1 \frac{\ln x}{1+x}\,dx = -\frac{\pi^2}{12\ln 2}.$$

Answers

Chapter I

1.6. Yes. 1.7. Yes. 1.8. Yes. 1.14. $N \le 11a^2$. 1.20. a) $\{0, 1, 1/2, 1/3, \ldots\}$; b) c) $\mathbb{R}$; d) $[-1, 1]$. 1.25. Yes. 1.28 b) $[0, 2]$, $[-1, 1]$. 1.29. No. 1.30. For example, the set of vertices of regular n-polygons inscribed in circles of radii $1 - 1/n$ with center at a particular point. On the line—the set of midpoints of the intervals complementary to the Cantor set.

2.10. If and only if $\{a_k\}_{k=1}^n$ is an arithmetic progression. 2.15. c) $\Rightarrow$ b). 2.18. f) No. 2.19. c) Reverse the inequality for $p < 0$; d) for $p \in (-1, 0)$ the inequality is satisfied for all $x \in (0, 1)$.

3.6. a) $[-1, 1]$; b) $\mathbb{R}$; c)–e) $[-1, 1]$. 3.7. a) $\{0, 1/q, \ldots, (q-1)/q\}$ if $x = p/q$ is an irreducible fraction; b)–d) $[0, 1]$. 3.8. Yes.

Chapter II

1.1. $1/2$. 1.2. $\sup x_n = \overline{\lim}\, x_n = \sqrt{\pi/2}$, $\inf x_n = x_2 = \sqrt{1/2}$, $\underline{\lim}\, x_n = \sqrt{2/\pi}$. 1.3. a), b) 2. 1.4. $e^{-1/18}$. 1.5. $e^{a+1}/(e-1)$. 1.6. a) $e/(e-1)$; b) $S_1 < S_2 < \cdots < S_6 > S_7 > S_8 > \cdots$ c) $S_6 = 2.002\ldots$; $S_n < 2$ for $n \ne 6$. 1.7. $4e^{-1}$. 1.8. $-p \ln p - (1-p)\ln(1-p)$ if $p \in (0, 1)$; 0 if $p = 0$ or $p = 1$. 1.9. a) $(1/2)\ln 2$; b) $\pi/8$; c) $(1/2)\ln 5$; d) $\pi \ln 2$; e) π. 1.10. a) 0; b) 2π; c) $+\infty$; d) 0. 1.12. b) $C_2 = \pi^2/4$.

3.2. a) $e^{-1}(x_0 + (e-1)x_1)$; b) $e^{-1/2}((\sqrt{e} - 1)x_0 + x_1)$. 3.3. a) $x_n \sim n2^{n-1}(x_1 - 1 + 2\ln 2)$ for $x_1 \ne 1 - 2\ln 2$; b) $x_n \to 1$ for $x_1 = 1 - 2\ln 2$.

Chapter III

1.1. a), e) The empty set; b) the set of all functions; c), d) the set of constant functions; f) the set of functions bounded on each finite interval; g) the set of functions uniformly continuous on $\mathbb{R}$; h) the set of functions bounded on $\mathbb{R}$; i) the set of nondecreasing functions uniformly continuous on $\mathbb{R}$. 1.2. No. 1.5. Yes. 1.6. b) No. 1.7. Yes. 1.12. The assertion is true for a separable space and is not true for a nonseparable space. 1.14. a) No; b) yes. 1.23. The functions of the form $ax + b$. 1.25. Yes.

2.3. No. 2.5. For a finite family of functions it is true, for a countable family it is false.

3.9. a) $\Leftrightarrow$ b), c) does not imply a). 3.12. Yes, $g'(0) = 2/\pi$. 3.18. 2. 3.19. $\alpha \le \log_3 2$.

4.1. No. 4.2. $\varphi^{-1}(t) = (1-t^2)/(1+t^2) + i(2t)/(1+t^2)$. 4.3. Yes. 4.4. c) For example, $\xi = 2/9$, $\eta = 1/3$.

5.1. a) The constants; b) $f(x) = g(\ln x)$, where g is an arbitrary continuous function on $\mathbb{R}$ having period $\ln 2$.

5.3. $f(x) = A\ln x$. 5.4. $f(x) = Ax\ln|x|$ for $x \in \mathbb{R}$, $x \neq 0$, $f(0) = 0$. 5.5. $f(x) \equiv 0$ or $f(x) = x$. 5.6. $f(x) \equiv 0$, $f(x) = x$, and, moreover, $f(x) = -x$ if n is odd. 5.10. $f(x) = A\ln x + h(\ln x)$, where h is an arbitrary continuous function on $(-\infty, +\infty)$ with period $T = \ln(a/b)$ and $A = T^{-1}(f(a) - f(b))$. 5.11. a) $\varphi(1) = 0$ and $\varphi(x) = (\ln x)H(\ln|\ln x|)$ for $0 < x < 1$, where H is an arbitrary continuous function on $\mathbb{R}$ having period $\ln 2$; b) $f(x) \equiv 0$, $f(x) \equiv 1$, or $f(x) = x^{e^{\omega(\ln|\ln x|)}}$ for $0 < x < 1$, where ω is a continuous function on $\mathbb{R}$ having period $\ln 2$ and satisfying the condition $\omega(t+h) - \omega(t) > -h$ for $t, h \in \mathbb{R}$, $h > 0$. 5.12. For example, $f(x) = (\operatorname{sgn} x)|x|^{|\ln|x||}$ for $x \neq 0$, $f(0) = 0$. 5.13. $f(z) = z$ or $f(z) = \overline{z}$ for $|z| = 1$, $f(0) = 0$, and $f(z) = |z|^{e^{\omega(\ln|\ln|z||)}} \cdot e^{i(\ln|z|)H(\ln|\ln|z||)} f(z/|z|)$ for $0 < |z| < 1$, where ω and H are continuous real functions with period $\ln 2$, and $\omega(t+h) - \omega(t) > -h$ for $t, h \in \mathbb{R}$ and $h > 0$. 5.15. $f(x, y) = a(x^2 + 4xy + y^2) + b(x+y) + c$, where a, b, and c are arbitrary real numbers. 5.16. If $f \neq \text{const}$, then g exists if and only if f is strictly monotone on $\mathbb{R}$ or f is even and strictly monotone on $[0, +\infty)$.

In the answers to problems 5.17 and 5.18 the letter φ stands for a character.

5.17. a) $\varphi(n) = \zeta^n$ $(n \in \mathbb{Z})$, where $\zeta \in S^1$; b) $\varphi((k, l)) = \zeta_1^k \xi_2^l$ $((k, l) \in \mathbb{Z}^2)$, where $\zeta_1, \zeta_2 \in S^1$; c) $\varphi(n) = \zeta^n$ $(n \in \mathbb{Z}_m)$, where ζ is an mth root of unity. 5.18. a) $\varphi(x) = e^{itx}$ $(x \in \mathbb{R})$, where $t \in \mathbb{R}$; b) $\varphi(x) = e^{i(x, t)}$, where (x, t) is the inner product of the vectors x and t in $\mathbb{R}^n$; c) $\varphi(x, n) = e^{itx}\zeta^n$ $(x \in \mathbb{R}, n \in \mathbb{Z})$, where $t \in \mathbb{R}$ and $\zeta \in S^1$; d) $\varphi(z) = e^{i(t \operatorname{Re} z + s \operatorname{Im} z)}$ $(z \in \mathbb{C})$, where $t, s \in \mathbb{R}$; e) $\varphi(\zeta) = \zeta^n$ $(\zeta \in S^1)$, where $n \in \mathbb{Z}$; f) $\varphi(x) = e^{it\ln x}$ $(x > 0)$, where $t \in \mathbb{R}$; g) $\varphi(z) = e^{it\ln|z|}(z/|z|)^n$ $(z \in \dot{\mathbb{C}})$, where $t \in \mathbb{R}$ and $n \in \mathbb{Z}$. 5.19. An arbitrary continuous homomorphism φ is defined by the respective equalities: a) $\varphi(x) = ax$; b) $\varphi(x) = e^{ax}$; c) $\varphi(x) = a\ln x$; d) $\varphi(x) = x^a$; e) $\varphi(x) = e^{iax}$; f) $\varphi(z) = |z|^a e^{ib\ln|z|}(z/|z|)^n$, where a and b are arbitrary real numbers, and $n \in \mathbb{Z}$.

Chapter IV

1.1. Converges. 1.2. a)–c) Diverges. 1.4. a) $p > 0$; b) $p > 1/2$; c) $p > 2$. 1.5. a) Converges; b–d) diverges. 1.7. a), b), e), g) Diverges; c), d), f), h) converges.

2.5. For nonmonotone sequences the assertion is false. 2.6. a) Yes; b) no. 2.9. Not always. 2.10. Not always. 2.12. d) Not always. 2.14. For $a = +\infty$ either convergence or divergence can occur. 2.18. Yes. 2.19. Monotonicity is essential. 2.20. b) The series $\sum \lambda_n a_n$ can converge, but if $\lambda_n \uparrow +\infty$, then it diverges. 2.21. b) True.

3.1. Yes. It is impossible to choose a positive sequence. 3.2. Yes. 3.4. The converse assertion is true if $a_n \downarrow 0$. Monotonicity is essential.

4.1. $\frac{3}{4}(x - \sin x)$, 4.2. $\frac{3}{4}\cos x - \cos^3 x$, 4.3. $1/r$. 4.4. $\int_0^1 4/(4 - z + zt^2)\,dt$. 4.5. $\ln m$. 4.6. $e/2$. 4.7. $1 - \gamma$. 4.8. $(\gamma - (\ln 2)/2)\ln 2$. 4.9. $\gamma + \ln 2$. 4.10. $\pi(1 - (\ln(2\pi))/2)$. 4.11. $\frac{\pi}{2}(1 - \ln \pi)$.

5.1. $2y < x^2$, $q > 2p$. 5.2, 5.3. Yes. 5.4. The series converges nonuniformly. 5.9. The condition $\sum a_n^2 < \infty$ is essential.

6.3. For example, $E = \{\sum \varepsilon_k/k! | \varepsilon_k = 0 \text{ or } 1\}$, $B_n = 2\pi n!$. 6.6. The limits are equal to $2/\pi$ and $1/2$; $L_n \sim (8/\pi)\ln n$, $\widetilde{L}_n \sim 2(\ln n)$. 6.8. a) $(\pi - x)/2$ for $0 < x < 2\pi$; b) $\frac{1}{2}\cos\frac{x}{2}\ln\cot\frac{x}{4} - \frac{\pi}{4}\sin\frac{x}{2}$ for $0 < x < 2\pi$; c) $-x/2$ for $|x| < \pi$; d) $\frac{1}{2}\sin\frac{x}{2}\ln\cot\frac{x}{4} + \frac{\pi}{4}\cos\frac{x}{2}$ for $0 < x < 2\pi$.

Chapter V

1.1. The integral converges. 1.2. The integral converges for $p > 2$. 1.3. The integral converges in the following three cases: $q < -1$, r and p arbitrary; $q \geq -1$, $r \leq 1$, $1 + q < p$; $q \geq -1$, $r > 1$, $1 + q < p/r$. 1.4. The integral converges in the following three cases: $q < -1$, r and p arbitrary; $q \geq -1$, $r \leq 0$, $p > 0$; $q \geq -1$, $r > 0$, $1 + q < p/r$. 1.5. The integral converges if $q > 1$, $p \leq 0$ or $q > 1 + p$, $p < 0$. 1.6. The integral converges for $q < -1$ and $p > 0$. 1.7. The integral converges for $p < 1$ and $q > 1$. 1.8. The integral converges if $p < -1$, $q \leq 0$ or $q > 0$, $2(p+1) < q$. 1.9.–1.11. The integrals converge. 1.12. a), b) No. 1.13. For any $\varepsilon > 0$. 1.14. a) $f(0)\ln(b/a)$; b) $(f(0) - \frac{1}{T}\int_0^T f(x)\,dx)\ln(b/a)$; c) $(f(0) - l)\ln(b/a)$. 1.16. 0. 1.17. $-(\pi/2)\ln 2$. 1.18. $-\pi^2/6$. 1.19. $-\pi^2/12$. 1.20. $\pi^2/6$. 1.21. $\pi/2 - \arctan a$. 1.22. $\pi/2$. 1.23. $\pi/4$. 1.24. 0. 1.25. $\pi(|a| + |b| - |a + b|)$. 1.26. $-\gamma$. 1.27. $\gamma^2 + \pi^2/6$. 1.28. γ. 1.29. $(1/q - 1/p)\gamma$. 1.30. $(1/q - 1/p)\gamma$. 1.31. $\ln 4$. 1.32. $-(1/2)\ln^2 2$. 1.33. $\sqrt{\pi}$. 1.34. $3/4$; $\frac{4-\sqrt{2}}{8}\sqrt{\pi}$; $\frac{1}{2}$; $(\sqrt{2} - 1)\sqrt{\frac{\pi}{2}}$; $\ln 2$; $2(\sqrt{2} - 1)\pi$. 1.35. $\sqrt{\pi}$. 1.36. $(\sqrt{\pi}/2)e^{-2a}$. 1.37. a), b) $\sqrt{\pi/2}$. 1.38. $\ln(\pi/\sqrt{6})$.

2.1. $2\ln 2$. 2.2. a) n/(n+1) ; b) 1/(n+1) ; c) $n!\varepsilon^{-n}$. 2.3. $(2\pi)^{n/2}$. 2.4. a) $2\sqrt{2\pi(a^2 + b^2)}$, b) $2^{(p+1)/2}(2\pi)^{(n-1)/2}\Gamma((p + 1)/2)$. 2.5. $2\pi - 2\pi\|x\|^2/3$ for $\|x\| \leq 1$; $4\pi/(3\|x\|)$ for $\|x\| \geq 1$. 2.6. $\pi^2(\cosh 1 - 1)$. 2.7. $\pi^2(\det A)^{-1/2}$. 2.8. $t \in \operatorname{Int}(A)$; $K(t) = 8\pi(t_1^2 - t_2^2 - t_3^2 - t_4^2)^{-2}$ for $t = (t_1, t_2, t_3, t_4) \in \operatorname{Int}(A)$. 2.9. $M = (b - a)/3$, $P = 4/9$. 2.10. $1/2$.

2.11. a) $P_1(a) > 1/2$ for any a, $P_1(a) \underset{a\to\infty}{\to} 1$; b) $1/2$. 2.12. $\pi/4$. 2.13. a) $4R/\pi$; b) $5\pi/6$. 2.14. $\Psi(x, y) = \pi(x^2+y^2-1)$ for $x^2+y^2 \le 1$ and $\Psi(x, y) = \pi\ln(x^2+y^2)$ for $x^2+y^2 \ge 1$. 2.17. $(n!\,a_1a_2\cdots a_n)^{-1}$. 2.19.

$$\frac{1}{24}\int_0^1\int_0^1 (f''_{x^2}(x, y) + f''_{y^2}(x, y))\,dx\,dy.$$

Chapter VI

1.2. a) $a^{-1}|\ln(1-t)|$; b) $2^{-1}|\ln(1-t)|$; c) $\frac{\pi}{2}(1-t)^{-1}$. 1.3. a) $A/\ln A$; b) A; c) $2A/\pi$; d) $e^{-A^2}/2A$; e) A^ae^{-A}. 1.4. a) $(\ln A)/p$; b) $e^A/2A$; c) $\sqrt{\pi A}e^{A/4}$; d) $A\ln^p A$. 1.5. a) $(2\pi n^2)^{-1}$; b) $(-1)^np/(\pi n)^{p+1}$; c) $(-1)^n(\pi n)^{-p}$. 1.10.

$$\text{a)}\ \frac{1}{2}\frac{A^{1-p}}{1-p}\quad\text{for } p<1,\qquad \frac{\ln A}{2}\quad\text{for } p=1;$$

$$\text{b)}\ \frac{2}{\pi}\frac{\ln A}{1+p};$$

$$\text{c)}\ \frac{2}{\pi}\frac{A^{1-p}}{1-p}\quad\text{for } p<1,\qquad \frac{2}{\pi}\ln A\quad\text{for } p-1;$$

$$\text{d)}\ \frac{2}{\pi}\frac{A^{1-p}}{p-1}.$$

1.16. a), b) $(\varepsilon|\ln\varepsilon|)^{-1}$.

2.1. $f(0)$. 2.3. $\sqrt{2/A}\int_0^\theta e^{-t^2}\,dt$. 2.5. a), b) $\Gamma(1+1/p)A^{-1/p}$; c) $\Gamma((1+p)/2)(2/A)^{(p+1)/2}$. 2.9. a) $\sqrt{2\pi/A}$; b) $\sqrt{\pi/A}/2$; c) $\sqrt{\pi/2}A^{A+1/2}$; d) $\Gamma(2-p)A^{p-2}$; e) $\sqrt{\pi/A}/2$; f) A^{-2}; g) $C_p(e^{(1-1/p)x_pA})/\sqrt{A}$, where $x_p = p^{1/(1-p)}$, $C_p = \sqrt{2\pi x_p/(p-1)}$; h) A^{-1}. 2.10. a) $(a+1)^{n+1/2}\sqrt{\pi/2}$; b) $\sqrt{3\pi/(2n)}$; c) $\frac{1}{2}\sqrt{\pi/n}(1+(-1)^n\sqrt{2})$. 2.11. a) $e^{-\sqrt{A}}\sqrt{2\pi e/A}$; b) $A^p/(p\ln A)$; c) C_pA^A, where $C_p = \int_0^\infty e^{x^p-x}\,dx$.

3.11. a) $\pi/(2t)$; b) $\sqrt{\pi t}/2$. 3.27. a), c) $\Gamma(1+1/p)(1-t)^{-1/p}$; b)

$$\sqrt{\frac{2\pi}{p-1}S(t)\exp\left(\left(1-\frac{1}{p}\right)S(t)\right)},$$

$$\text{where } S(t) = (p|\ln t|)^{1/(1-p)} \underset{t\to1-0}{\sim} (p(1-t))^{1/(1-p)};$$

d) $|\ln(1-t)|/\ln a$; e), f) $|\ln(1-t)|/\ln|\ln(1-t)|$. 3.28.

$$\text{a)}\ \frac{\Gamma(1+p)}{(1-t)^{1+p}};\qquad \text{b)}\ \frac{|\ln(1-t)|^p}{1-t}.$$

3.29.

$$
\text{a)}\ \frac{\ln 2}{1-t};\quad \text{b)}\ \frac{\pi^2}{12(1-t)^2};\quad \text{c)}\ \frac{1}{2(1-t)};
$$

$$
\text{d)}\ \frac{|\ln(1-t)|}{\ln a};\quad \text{e)}\ \frac{|\ln(1-t)|}{1-t};\quad \text{f) } g)\ \frac{\pi^2}{6(1-t)^2};
$$

$$
\text{h)}\ \frac{|\ln(1-t)|}{(1-t)^2};\quad \text{i)}\ -\frac{\ln 2}{1-t};\quad \text{j)}\ -\frac{1}{2(1-t)};
$$

$$
\text{k)}\ -\frac{\pi^2}{12(1-t)^2};\quad \text{l)}\ -\frac{\ln 2}{(1-t)^2}.
$$

3.30.

$$
\text{a)}\ \frac{\pi}{4(1-t)};\quad \text{b)}\ \frac{\pi+2}{8}(1-t)^{-1};\quad \text{c)}\ \frac{\pi^2}{8}(1-t)^{-2};
$$

$$
\text{d)}\ \frac{1}{2}\frac{|\ln(1-t)|}{1-t};\quad \text{e)}\ \frac{\pi^2}{24}(1-t)^{-2}.
$$

3.31. a) $|\ln(1-t)|/(1-t)$; b) $(\pi^2/6)(1-t)^{-2}$.
4.3.

$$
\text{a) } y = x - x^2 + \frac{3}{2}x^3 + \cdots;
$$

$$
\text{b) } y = 1 + \frac{x-1}{2} - \frac{1}{4}(x-1)^2 + \frac{5}{16}(x-1)^3 + \cdots;
$$

$$
\text{c) } y = \frac{x-1}{2} - \frac{9}{16}(x-1)^2 + \frac{109}{192}(x-1)^3 + \cdots;
$$

$$
\text{d) } y = 1 + x - \frac{x^2}{2} + \frac{2x^3}{3} + \cdots;
$$

$$
\text{e) } y = -x^2 + x^3 - x^4 + x^5 - \frac{3}{2}x^6 + \cdots;
$$

$$
\text{f) } y = -\frac{x^3}{3} + \frac{8}{15}x^5 + a_7x^7 + \cdots.
$$

4.8. a) $1/n$; b) $\sqrt{3/n}$; c) $(3n)^{-1/3}$; d) $\sqrt{3/(2n)}$; e), f) $2/n$; g) $4/n^2$; h) $(2n)^{-1/4}$. 4.10. For example, $f(x) = x - x^2e^{-1/x}$ for $0 < x \le 1$, $f(x) = f(1)$ for $x > 1$. 4.11. If the sequence $\{x_n\}$ does not become stationary, then $x_n \sim (-1)^n/\sqrt{2n}$ or $(-1)^{n-1}/\sqrt{2n}$, depending on the choice of x_0. In particular, the first case is realized if $0 < x_0 < 1$. 4.13. $a_n \sim K_p n^{(1-p)/(1+p)}$, where

$$
K_p = \left(\frac{1+p}{1-p}\left(\frac{2}{1+p}\right)^p\right)^{1/(p+1)}.
$$

4.14. $a_n \sim ((p+1)\ln n)^{1/(p+1)}$ for $p > -1$, $a_n \sim \sqrt{\pi n}$ for $p = -1$, and $a_n \sim C_p n$ for $p < -1$, where C_p is a constant depending on p.

Chapter VII

1.33. No. 1.36. a) $x^2/2$; b) $|x|^q/q$, where $q = p/(p-1)$; c) $+\infty$ outside the interval $[\min f', \max f']$; the piecewise linear function having as break points the values of f' on the interval $[\min f', \max f']$; the values of $(f^*)'$ are the break points of the function f; d) $+\infty$ for $x \geq 0$, $-|x|^q/q$ for $x < 0$, where $q = p/(p-1)$; e) $a\sqrt{1+x^2}$; f) $f(-x)$; g) $x(\ln x) - x$ for $x > 0$, 0 for $x = 0$, and $+\infty$ for $x < 0$.

2.12. $3a^2$.

3.3. a) $B_n(f_0, x) \equiv 1$, $B_n(f_1, x) \equiv x$, $B_n(f_2, x) \equiv x^2 + x(1-x)/n$; b) $(1 + x(a^{1/n} - 1))^n$.

4.3. For arbitrary λ_i. 4.4. a), b) $\lambda_1, \dots, \lambda_k$ are linearly independent over $\mathbb{Q}$. 4.5. a), b) $\lambda_1, \dots, \lambda_k$ are linearly independent over $\mathbb{Q}$ modulo 1. 4.19. Only if the sequence $\{\varepsilon_k\}$ is periodic. 4.20. No.

Chapter VIII

1.4. a) $1/10$; b) $4/10$; c) $1/100$; d) $1/2$. 1.5. 1. 1.6. a) Yes; b) no. 1.11. a) if $\overline{\lim}(n_{k+1} - n_k) \geq 2$. 1.12. Yes. 1.13. a) $l_n = 3^{-n}$; b) $l_n = \sum_{k>n} 1/k!$; c) if $n_{k+1} > n_k + 1$ infinitely many times; in this case $l_m = \sum_{k>m} 2^{-n_k}$. 1.16. $a = \lambda_2(A)$, $b = L$, $c = \pi$, where L is the length of the curve bounding the set A.

2.2. For example, $\sum_{n\geq 1} 2^{-n}\chi_{(0,+\infty)}(t - r_n)$, where $\{r_n\}_{n\geq 1}$ is the sequence of all rational numbers. 2.3. Yes. 2.5. If and only if $\lambda_m\{x \in E \mid f(x) = t_0\} = 0$. 2.6. a), b), d) $F(x) = \pi + 2\arcsin x$ for $|x| \leq 1$, $F(x) = 0$ for $x < -1$, $F(x) = 2\pi$ for $x \geq 1$; c) $F(x) = 0$ for $x < -1$, $F(x) = \frac{2}{3}(\pi + 2\arcsin x)$ for $-1 \leq x \leq 0$, $F(x) = \frac{2}{3}(\pi + 4\arcsin x)$ for $0 \leq x \leq 1$, and $F(x) = 2\pi$ for $x > 1$. 2.7. a) Yes; b) no. 2.8. a) $F(t-c)$, $F(t/c)$, $F(\sqrt[3]{t})$; b) $F(t)$. 2.9. c) No in general. 2.10. $f^*(\tau) = F^{-1}(\lambda_m(E) - \tau)$, $0 \leq \tau \leq \lambda_m(E)$. 2.11. a) $\sin((\pi - \tau)/2)$ $(0 \leq \tau \leq 2\pi)$; b) $\cos(\tau/4)$ $(0 \leq \tau \leq 2\pi)$; c) $\cot\tau$ $(0 < \tau < \pi)$. 2.12. $f^*(\tau) = \sqrt{1 - \tau/\pi}$ $(0 \leq \tau \leq \pi)$.

3.1. $6\binom{N}{3} - (6\binom{N-1}{2} - 1)S_1 + 6(N-1)S_2$. 3.6. Yes. 3.7. a) $p > 1$; b) $1 < p < 2$; c) $p < 1$; d) $p > 2$; e) $-1 < p < 2$. 3.8. a), b) Yes; 0, 0; c) no; 0, 0; d) no; $-\pi$, π. 3.9. a) $p < 2$; b) $p < 1$; c) $p < 3$. 3.15. No, as is clear from the example $E = (0, 1)$, $g \equiv 1$, $f(x) = e^{1/\sqrt{x}}$. 3.16. a) $1/2$, $1/4$; b) $1/4n$. 3.17. a) $1/10$; b) $1/100$. 3.18. $1/11$. 3.19. a) 3; b) $F(t) = 0$ for $t \leq 1$, $F(t) = 1 - (\frac{2}{3})^n$ for $n < t < n+1$; $f^*(\tau) = n$ for $(\frac{2}{3})^n \leq \tau < (\frac{2}{3})^{n-1}$. 3.25.

$$\frac{1}{\Gamma(n/2)} \int_0^{r^2/2} t^{n/2-1} e^{-t}\, dt.$$

3.26. $\frac{2}{\pi}(\arctan a - a/(1+a^2))$. 3.27. $\pi^2 R^2(4a^2+R^2)$. 3.30. $2^n\Gamma^n(1+1/p)/\Gamma(1+n/p)$.

3.31. a)
$$\frac{\Gamma((p+1)/2)\Gamma^2(n/2)}{2\sqrt{\pi}\Gamma((n-1)/2)\Gamma((n+p+1)/2)}\|a\|.$$

3.35. b) No.

3.39. a)
$$\frac{2\pi^2}{m+n+1}\binom{m+n}{n}^{-1}.$$

4.2. b)
$$\frac{1}{ea_1a_2\cdots a_m}\left(1+1+\frac{1}{2!}+\cdots+\frac{1}{(m-1)!}\right).$$

4.5. $\lambda_2(A)+L+2\pi$, where L is the length of the curve bounding A. 4.6. a) $1/2$; b) $3/8$; c) $1/2$; d) $1/2$; e) $1/2$. 4.7. a), b) $p<2\log_3 2$.

5.3. a), b) $\log_2\log_2\frac{1}{\varepsilon}$; c) $\frac{1}{2}\log_2\log_2\frac{1}{\varepsilon}$; d) $\frac{1}{\alpha+1}\log_2\frac{1}{\varepsilon}$; e) $\log_2\frac{1}{\varepsilon}$. 5.4. a) 1; b) $3/2$. 5.5. a), b) $5/3$; c) $4/3$. 5.6. a), b), c) 2. 5.7. $\log_3 2$. 5.8. $(\ln\frac{1}{\varepsilon})/\ln\ln\frac{1}{\varepsilon}$. 5.11. $2-\log_3 2$. 5.12 c) No. 5.19. a) $p=1$, $\mu_1(A)=(b-a)/2$; b) $p=1$, $\mu_1(A)=\pi$; c) $p=2$, $\mu_2(A)=\pi^{-1}$; d) $p=n$, $\mu_n(A)=\alpha_n^{-1}$. 5.20. $p=\log_3 2$, $\mu_p(K)=2^{-p}$. 5.21. a) $\dim_H(A)=M\text{-}\dim(A)=\log_3 2$; b) $\dim_H(A)=0$, $M\text{-}\dim(A)=1$; c) $\dim_H(A)=\log_3 2$, $M\text{-}\dim(A)=1$. 5.24. a) $l_m=\sum_{k>m}2^{-n_k}$; b) $1/2$; c) ∞. 5.26. $(\ln m)/\ln n$. 5.27. $\dim_H(A)=(\ln 5)/\ln 10$, $\dim_H(A+A)=(\ln g)/\ln 10$.

6.2. a) About 14%, b) more than 99%. 6.3. Less than 0.25%. 6.4. The probability that $|x_1|<\varepsilon$ tends to 1 as $n\to\infty$ for any $\varepsilon>0$. 6.5. $\sqrt{2}$. 6.10. b) $1/2$. 6.18. The ratio of the volumes of the cube and the ball is equivalent to: a) $\sqrt{\pi en}(2/(\pi e))^{n/2}$; b) $\sqrt{\pi en}(8/(\pi e))^{n/2}e^{\sqrt{n}}$. In case b) the ball is contained in the cube only for $n\le 9$. 6.19. $\rho_n=\sqrt{n}+o(1)$.

6.22. b)
$$\frac{\varepsilon^n}{\sqrt{2}}\left(\frac{2e}{\pi}\right)^{n/2}e^{\sqrt{n}/\varepsilon-1/(2\varepsilon^2)}.$$

Chapter IX

1.1. Convergence in measure holds in all cases except e). There is an integrable majorant only in cases b) and d). 1.10, 1.11. The assertion is true also for $\mathbb{R}$. 1.12. The theorem ceases to be true when $(0, 1)$ is replaced by $\mathbb{R}$. 1.14. If the condition is violated, then the assertion is false. 1.15. a), b) Yes; c) no.

2.11. π^2. 2.19. $\varphi(1/2\pi)$.

3.1. d) $1-2x$. 3.4. a) $(\sin t)/t$; b) $e^{-it/2}\int_0^1 e^{ixt}\,d\sigma(x)$; c) $\prod_{k=1}^n\cosh(zc_k)$. 3.12. a) F is the Cantor function; b) $F(t)=(t+1)/2$ for $|t|\le 1$, $F(t)=(1+\operatorname{sgn} t)/2$ for $|t|\ge 1$.

4.1. a), b) $f_{\sqrt{c_1^2+c_2^2}}$; c) $f_{c_1+c_2}$. 4.9. No. 4.18. a), b) $\hat{f}_c = (2\pi/c)^{n/2} f_{1/c}$; c) $e^{-|cs|}$. 4.27. The measure concentrated at zero or the measure with density $(2\pi c)^{-n/2} e^{-\|x\|^2/(2c)}$ $(c > 0)$.

Chapter X

1.2. a) No; b) yes; c) no. 1.3. $h(x) = 1 - x$, $k(x) = -x$. 1.13. a) Yes; b) no; c) yes; d) no; e) yes. 1.14. a) $b = (a^2 - 2a)/4$, $a \in \mathbb{R}$; b) for $a \geq -1/4$, $a \neq 0$: $b = 1 \pm \sqrt{1+4a}$, and for $a = 0 : b \in \mathbb{R}\backslash\{2\}$; d) for $a = b(b-2)/4$ with $b \in [0, 2) \cup (2, 4)$; moreover, $\Delta = [-x_1, x_1]$ for $a \in [-1/4, 0]$ and $b \in [0, 1]$, $\Delta = [-x_2, x_2]$ for $a \in [-1/4, 0]$ and $b \in [1, 2)$, and $\Delta = [x_1, -x_1]$ for $a \in (0, 2]$ and $b \in (2, 4)$, where x_1, x_2 $(x_1 \leq x_2)$ are the roots of the equation $f(x) = x$. 1.19. a) 0; b) 2; c) -2; d) 1; e) 1; f) 1; g) $\sqrt{5}-1$. 1.20. b) $\lim x_n = 0$ for $x_0 \in (-1, 2)$, $\lim x_n = 2$ for $x_0 \in \{-1, 2\}$, $\lim x_n = +\infty$ for the remaining values of x_0; c) the limit is equal to $(1-\sqrt{1-4a})/2a$ for $|x_0| < (1+\sqrt{1-4a})/2a$, $(1+\sqrt{1-4a})/2a$ for $|x_0| = (1+\sqrt{1-4a})/2a$, and $+\infty$ for the remaining x_0; d) the limit is equal to $(-1+\sqrt{1+4a})/2a$ for $|x_0| < (1+\sqrt{1+4a})/2a$, $(-1-\sqrt{1+4a})/2a$ for $|x_0| = (1+\sqrt{1+4a})/2a$, and $+\infty$ for the remaining x_0; e) the limit is equal to $(1-\sqrt{1-4c})/2$ for $|x_0| < (1+\sqrt{1-4c})/2$, $(1+\sqrt{1-4c})/2$ for $|x_0| = (1+\sqrt{1-4c})/2$, and $+\infty$ for the remaining x_0; f) the limit is equal to c for $x_0 \in [c-1, c]$ and $+\infty$ for the remaining x_0; g) $\lim x_n$ is equal to $1/3$ when $x_0 \neq 1$ and 1 when $x_0 = 1$; h) the limit is equal to 0 for $|x_0| < 1/\sqrt{5}$, ± 1 for $|x_0| > 1/\sqrt{5}$, and does not exist for $|x_0| = 1/\sqrt{5}$. 1.22. a), b) $x_n \to 1/(1+a)$ for $|a| < 1$ and any x_0, and for the remaining a the sequence either diverges or stabilizes; c) see the answer to problem 1.20c) for $a \leq 3/4$, and the sequence either diverges or stabilizes for $a > 3/4$; d) for $a \in [e^{-e}, 1]$ and for all $x_0 \in \mathbb{R}$ the sequence $\{x_n\}$ converges to the unique fixed point of f, for $a \in (1, e^{1/e})$ we have that $\lim x_n = \overline{x}$ for $x_0 < \overline{\overline{x}}$, $\lim x_n = \overline{\overline{x}}$ for $x_0 = \overline{\overline{x}}$, and $\lim x_n = +\infty$ for $x_0 > \overline{\overline{x}}$, where $\overline{x}, \overline{\overline{x}}$ $(\overline{x} < \overline{\overline{x}})$ are the fixed points of f, and for the remaining a the sequence $\{x_n\}$ either diverges or stabilizes. 1.32. a), b) For $a = 1$ there is a continuum of 2-cycles $\{t, 1-t\}$ $(t \in [0, 1/2))$, for $a > 1$ there is one repelling 2-cycle

$$\left\{\frac{1-a}{1+a^2}, \frac{1+a}{1+a^2}\right\}$$

and from one to three repelling 4-cycles; c) for $a > (1+\sqrt{5})/2 = 1.618\ldots$ there are two repelling 3-cycles

$$\left\{\frac{1-a-a^2}{1+a^3}, \frac{1+a-a^2}{1+a^3}, \frac{1+a+a^2}{1+a^3}\right\}$$

and

$$\left\{\frac{1-a+a^2}{1-a^3}, \frac{1+a-a^2}{1-a^3}, \frac{1-a-a^2}{1-a^3}\right\},$$

and for $a = (1+\sqrt{5})/2$ they merge into one. 1.33. a) $a = 3/4$; b), c) $a = 5/4$; d) $a = 7/4$. 1.34. a) $b = 3$; b), c) $b = 1+\sqrt{6} = 3.449\ldots$; d) $b = 1+2\sqrt{2} = 3.828\ldots$. 1.41. b) No.

2.6. a) $d\mu = A\,dx/\sqrt{1-x^2}$; b) $d\mu = A\,dx/\sqrt{1-x^2}$ for $|x| < 1$, $d\mu = B\,dx/\sqrt{x^2-1}$ for $|x| > 1$ $(A, B > 0)$. 2.10. a) Yes, $d\mu = dx/|x|$ $(x \neq 0)$; b) no. 2.11. a) Yes, $d\mu = dx/(x|\ln x|)$ $(x \neq 0, 1)$; b) no. 2.14. a) $p_{ij} = \lambda(\Delta_j \cap T^{-1}(\Delta_i))/\lambda(\Delta_i)$. 2.15. $d\mu = f\,dx$, where a) $f(x) = 0$ for $x < 1/2$, $f(x) = f(3/2 - x)$ for $x > 1/2$; b) $f(x) = 1/5$ for $x < 1/3$, $f(x) = 2/5$ for $x > 1/3$; c) the limit measure is concentrated at the point 0. 2.16. $d\mu = dx\,dy/y^2$. 2.17. $d\mu_L = dx\,dy/x^2$, $d\mu_R = dx\,dy/x$. 2.21. a) Yes, if and only if α is irrational; b) yes; c) yes; d), e) yes, if and only if there are no roots of unity among the eigenvalues of the matrix A. 2.23. $\alpha_1, \ldots, \alpha_n$ must be linearly independent over $\mathbb{Q}$ modulo 1. 2.28. $\mu(A)$. 2.30. $1/2$, $1/p$. 2.34. $\log_{10}(1+1/p)$.

Appendix I

We establish here an interesting fact showing that the arithmetic properties of a real number that were discussed in Chapter I can unexpectedly play a decisive role in investigations far removed at first glance from this topic.

A trigonometric polynomial in two variables $x, y \in \mathbb{R}$ is defined to be a sum

$$T(x, y) = \sum_{n, m \in \mathbb{Z}} c_{n,m} e^{i(nx+my)}$$

in which only finitely many coefficients $c_{n,m} \in \mathbb{C}$ are nonzero. We consider the projection P_Ω acting in the set of all such polynomials as follows:

$$P_\Omega T(x, y) = \sum_{(n, m) \in \Omega} c_{n,m} e^{i(nx+my)}$$

(Ω is a subset of $\mathbb{Z}^2$). We say that the projection P_Ω is bounded if there exists a number $C_\Omega > 0$ such that

$$\max_{x,y} |P_\Omega T(x, y)| \le C_\Omega \max_{x,y} |T(x, y)| \qquad (*)$$

for every polynomial T.

We consider the simple case when Ω is a strip, that is,

$$\Omega = \{(n, m) \in \mathbb{Z}^2 \,|\, |m - \lambda n| \le \Delta\} \qquad (\lambda \in \mathbb{R},\ \Delta > 0).$$

It turns out that this projection is bounded if and only if the coefficient λ is rational (see the paper of É. S. Belinskiĭ in the collection "Theory of mappings and approximation of functions," "Naukova Dumka," Kiev, 1983, pp. 18–20).

To prove this assertion, consider first of all the analogous problem in the one-dimensional case: for any trigonometric polynomial $Q(t) = \sum_{k \in \mathbb{Z}} c_k e^{ikt}$ (only finitely many of the coefficients $c_k \in \mathbb{C}$ are nonzero), let $PQ(t) = \sum_{k \in \mathbb{Z}, |k| \le N} c_k e^{ikt}$. It will be assumed that $N \ge 2$, $N \in \mathbb{N}$. Since PQ is a partial sum of the Fourier series of Q, it follows that

$$PQ(t) = \frac{1}{\pi} \int_{-\pi}^{\pi} Q(t+s) \frac{\sin(N + 1/2)s}{2\sin(s/2)}\, ds.$$

Therefore, the projection P is bounded: $|PQ(t)| \le C\max_t|Q(t)|$, and

$$C \le \frac{1}{\pi}\int_{-\pi}^{\pi}\left|\frac{\sin(N+1/2)s}{2\sin(s/2)}\right|\,ds.$$

It follows from the result of problem IV.6.6 that $C \le \text{const}\cdot\ln N$. The example of the polynomial

$$Q_0(t) = e^{iNt}\sum_{1\le k\le N}\frac{1}{k}\sin kt = \frac{1}{2i}\sum_{\substack{0\le k\le 2N\\ k\ne N}}\frac{1}{k-N}e^{ikt}$$

shows that the estimate for C cannot be essentially improved: on the one hand (see IV.6.15) $|Q_0(t)| \le A$ (the number A is independent of t and N), and, on the other hand,

$$PQ_0(t) = \frac{1}{2i}\sum_{1\le k<N}\frac{1}{k-N}e^{ikt},$$

hence

$$\max_t|PQ_0(t)| = |PQ_0(0)| = \frac{1}{2}\sum_{1\le k<N}\frac{1}{N-k} > \frac{1}{2}\ln N.$$

Thus, $C \asymp \ln N$ in the one-dimensional case.

For a better understanding of the subsequent arguments we point out the principle difference between a strip with rational slope and a strip with irrational slope. It is this difference that gives rise to the effect under discussion. In the case of a rational slope the following alternative holds for any line $L \subset \mathbb{R}^2$: either $L \subset \Omega$ or the set $L \cap \Omega$ contains a bounded number N_L of integer points (the upper bound for N_L depends on the strip Ω). But if $\lambda \notin \mathbb{Q}$, then the situation is different: it is possible to draw a line L such that the set $L\cap\Omega$ contains an arbitrarily long series of integer points with coordinates forming an arithmetic progression.

Now consider the projection on a strip with rational slope. Let $\lambda = p/q$, where $p \in \mathbb{Z}$, $q \in \mathbb{N}$, and p and q are relatively prime, and choose integers p_0 and q_0 such that $pp_0 + qq_0 = 1$. The integer matrix $\left(\begin{smallmatrix} p & -q \\ q_0 & p_0\end{smallmatrix}\right)$ determines a bijective linear mapping of $\mathbb{Z}^2$ onto itself (since the determinant is equal to 1, the inverse matrix is an integer matrix). Let us make the substitution

$$\begin{cases} n' = np - mq, \\ m' = nq_0 + mp_0, \end{cases}\qquad \begin{cases} x' = xp_0 - yq_0, \\ y' = xq + yp. \end{cases}$$

Then

$$\begin{aligned} T(x,y) &= \sum_{n,m\in\mathbb{Z}} c_{n,m}e^{i(nx+my)} \\ &= \sum_{n',m'\in\mathbb{Z}} c'_{n',m'}e^{i(n'x'+m'y')} = T'(x',y') \end{aligned}$$

(here $c'_{n',m'} = c_{n'p_1+m'q,\,m'p-n'q_0}$). Since the inequality $|m - \lambda n| \leq \Delta$ is equivalent to the inequality $|n'| \leq q\Delta$, the projection P_Ω is written as follows in the new coordinates:

$$P_\Omega T(x, y) = \sum_{n',m'\in\mathbb{Z},\,|n'|\leq q\Delta} c'_{n',m'} e^{i(n'x'+m'y')}.$$

Consequently, the action of the projection P_Ω reduces to the action of the one-dimensional projection with respect to the variable x', and thus

$$C_\Omega \leq \mathrm{const}\cdot\ln(2 + q\Delta).$$

Let us now consider the case when $\lambda \notin \mathbb{Q}$. Choose an irreducible fraction p/q such that $|\lambda - p/q| \leq q^{-2}$. Let L be the line through the points $(0, 0)$ and (q, p). Since p and q are relatively prime, all the integer points on L have the form (kq, kp), $k \in \mathbb{Z}$. It is clear that $\Omega \cap L = \{(kq, kp)|k \in \mathbb{Z}, |k| \leq N\}$ for some $N > 0$. We show that $N \geq q\Delta$. To do this we represent the number λ in the form $\lambda = p/q + \theta/q^2$, $|\theta| \leq 1$. For $|k| \leq q\Delta$ we have that $|kp - \lambda kq| = |k|\,|p - (p/q + \theta/q^2)q| = |k\theta|/q \leq |k|/q \leq \Delta$, that is, $(kq, kp) \in \Omega \cap L$ for such k.

Consider the action of the projection P_Ω on polynomials T with coefficients $c_{n,m}$ equal to zero for $(n, m) \notin L$, that is, on the polynomials $T(x, y) = \sum_{k\in\mathbb{Z}} a_k e^{ik(qx+py)}$. It is clear that

$$P_\Omega T(x, y) = \sum_{k\in\mathbb{Z},\ |k|\leq N} a_k e^{ik(qx+py)}.$$

It follows from the treatment of the one-dimensional case that P_Ω is bounded, that is, the inequality $(*)$ holds, only if

$$C_\Omega \geq \mathrm{const}\cdot\ln N \geq \mathrm{const}\cdot\ln(q\Delta).$$

Since the numerator q can take arbitrarily large values, it follows that $C_\Omega = +\infty$ for a strip with irrational slope.

Appendix II

Suppose that the partial sums $S_n(x)$ of a trigonometric series $a_0/2 + \sum(a_k \cos kx + b_k \sin kx)$ with real coefficients are nonnegative on a set $E \subset [0, 2\pi)$ with $\lambda(E) > 0$. Then $S_n(x) = o(n^2)$ almost everywhere on E (in particular, a_n, $b_n = o(n^2)$). Thus, if the partial sums of a trigonometric series are bounded below then they are not too large above. This result, obtained by W. Darsow (see J. London Math. Soc. **35** (1960), 237–238) was essentially strengthened by S. V. Konyagin (see Math. Notes **44** (1988), 910–920), who proved that if the partial sums are nonnegative on a set of positive measure, then they are bounded almost everywhere on this set. Consequently, a trigonometric series cannot diverge to $+\infty$ on a set of positive measure. Moreover, for an arbitrary real trigonometric series with partial sums not necessarily positive on a set E with $\lambda(E) > 0$ the following alternative holds: either the sequence $\{S_n(x)\}$ is bounded for almost all $x \in E$ or $\underline{\lim}_{n\to+\infty} S_n(x) = -\infty$ and $\overline{\lim}_{n\to+\infty} S_n(x) = +\infty$ for almost all $x \in E$. But if $\lambda(E) = 0$ and E is closed, then, as shown by Olevskiĭ (see Colloq. Math. **60/61** (1990), pp. 709–712), there exists a Fourier series with sum $+\infty$ on E and finite off E.

We prove only the simpler assertion formulated at the beginning of this appendix. For the proof we use the well-know fact that almost all points of a measurable set E are points of density: $\lim_{\varepsilon\to+0} \frac{1}{2\varepsilon}\lambda(E \cap (x-\varepsilon, x+\varepsilon)) = 1$ for almost all $x \in E$.

Let x be some point of density of E. Then 0 is a point of density of the set $F = \{t - x | t \in E\} \cap \{x - t | t \in E\}$. Fix numbers u and v with $\pi/2 < u < v < \pi$; their choice is made more precise below. Since 0 is a point of density of F, for sufficiently large n the set $F \cap [u/n, v/n]$ is not empty, that is, there is an index N such that for all $n \geq N$ there exists a $\varphi_n \in F$ with $u \leq n\varphi_n \leq v$. Using the fact that $x \pm \varphi_n \in E$ for $n \geq N$, we get that $S_n(x \pm \varphi_n) \geq 0$. Therefore,

$$0 \leq (S_n(x+\varphi_n)+S_n(x-\varphi_n))/2 = a_0/2+\sum_{1\leq k\leq n}(a_k \cos kx+b_k \sin kx)\cos k\varphi_n$$

$$= S_n(x)\cos n\varphi_n + \sum_{0\leq k<n} S_k(x)(\cos k\varphi_n - \cos(k+1)\varphi_n).$$

Thus, for $n \geq N$

$$\begin{aligned} S_n(x) &\leq \frac{-1}{\cos n\varphi_n} \sum_{0\leq k<n} S_k(x) \cdot 2\sin\frac{\varphi_n}{2} \cos\left(k+\frac{1}{2}\right)\varphi_n \\ &\leq \frac{-v}{n\cos u} \sum_{0\leq k<n} S_k(x). \end{aligned}$$

Let $w = -v/\cos u$ $(w > 1)$, and $\sigma_n = S_0(x)+\cdots+S_n(x)$. Then $\sigma_n - \sigma_{n-1} \leq \frac{w}{n}\sigma_{n-1}$, and hence $\sigma_n \leq (1+w/n)\sigma_{n-1} \leq (1+1/n)^w \sigma_{n-1}$. Consequently,

$$\begin{aligned} \sigma_n &\leq \left(1+\frac{1}{n}\right)^w \left(1+\frac{1}{n-1}\right)^w \sigma_{n-2} \\ &\leq \cdots \leq \left(\frac{n+1}{n}\frac{n}{n-1}\cdots\frac{N+1}{N}\right)^w \sigma_{N-1} \leq (n+1)^w \sigma_{N-1}. \end{aligned}$$

Thus,

$$S_n(x) = \sigma_n - \sigma_{n-1} \leq \frac{w}{n}\sigma_{n-1} \leq wn^{w-1}\sigma_{N-1} = O(n^{w-1}).$$

It remains to see that $\inf_{(\pi/2,\pi)}(-u/\cos u) < 3$, and hence the parameters u and v can be chosen so that $w = -u/\cos v < 3$.

Appendix III

Consider the chord obtained by intersecting a convex closed curve Γ and a line L. What is the rate of decrease of the chord length as the line approaches some support line without changing direction? This is clearly determined by the order of contact of the support line and the curve Γ. The length of the chord can fail to tend to zero in general if the curve Γ contains a segment parallel to L. It is intuitively clear that this is an exceptional case. The same can be surmised about contact of order greater than two. In other words, the order of contact of "most" support lines with Γ is at most two, and hence the lengths of the corresponding chords have the estimate $O(\sqrt{\varepsilon})$ (ε is the distance between L and the support line parallel to it). A possible interpretation of this vague assertion, one which gives a quantitative estimate, is the inequality

$$\int_{-\pi}^{\pi} \mu^2(\varphi, \varepsilon)\, d\varphi \leq \text{const} \cdot \varepsilon .$$

Here $\mu(\varphi, \varepsilon)$ is the length of a chord at a distance ε from the support line; the angle φ gives the direction of the support line (the vector $e_\varphi = (\cos\varphi, \sin\varphi)$ is perpendicular to it and directed toward the curve Γ); the constant depends only on Γ. Thus, in the mean over all rotations of the support line the rate of decrease of the chord length for the curve Γ is not less than for a circle.

We prove that the constant on the right-hand side of the inequality is determined only by the two quantities r and R if the curve Γ is contained in a disk of radius R and contains a disk of radius r. In proving such an inequality it can be assumed that Γ is a smooth strictly convex curve. It suffices to prove that for small ε (for example, $\varepsilon < r$)

$$\int_{-\pi}^{\pi} (\mu_2'(\varphi, \varepsilon))^2\, d\varphi \leq C_{r,R}/\varepsilon .$$

This yields the required inequality, because

$$\sqrt{\int_{-\pi}^{\pi} \mu^2(\varphi, \varepsilon)\, d\varphi} = \sqrt{\int_{-\pi}^{\pi} \left(\int_0^{\varepsilon} \mu_2'(\varphi, \delta)\, d\delta \right)^2 d\varphi}$$
$$\leq \int_0^{\varepsilon} \sqrt{\int_{-\pi}^{\pi} (\mu_2'(\varphi, \delta))^2\, d\varphi}\, d\delta$$
$$\leq \int_0^{\varepsilon} \sqrt{C_{r,R}/\delta}\, d\delta = 2\sqrt{C_{r,R}\varepsilon}.$$

Denote by y_φ the point on Γ at which the vector e_φ is the inner normal to Γ. For each φ denote by $\varphi + \Delta(\varphi)$ (respectively, $\varphi - \delta(\varphi)$) the slope of the normal to Γ at the endpoint (respectively, at the initial point) of the chord of length $\mu(\varphi, \varepsilon)$ orthogonal to the vector e_φ (the angles $\Delta(\varphi)$ and $\delta(\varphi)$ depend on ε). It is not hard to see that $0 < \delta(\varphi)$, $\Delta(\varphi) \leq \theta_{r,R} < \pi$ for $\varepsilon \in (0, r)$. Since $\mu_2'(\varphi, \varepsilon) = \cot\Delta(\varphi) + \cot\delta(\varphi)$, it follows that

$$\int_{-\pi}^{\pi} (\mu_2'(\varphi, \varepsilon))^2\, d\varphi \leq 2\int_{-\pi}^{\pi} (\cot^2\Delta(\varphi) + \cot^2\delta(\varphi))\, d\varphi.$$

To estimate the resulting integral we inscribe in Γ a polygonal curve with each component subtending an arc of "height" ε. This polygonal curve is not closed in general, and hence its first and last components intersect. We consider three successive components. Suppose that they subtend arcs of lengths m_-, m_0, and m_+. Assume that the vector e_{φ_0} is perpendicular to the middle chord (that is, y_{φ_0} is the "vertex" of the middle arc). The point $y_{\varphi_0-\delta(\varphi_0)}$ divides the arcs of length m_- and m_0, while the point $y_{\varphi_0+\Delta(\varphi_0)}$ divides the arcs of length m_0 and m_+. We prove that

$$\int_{\varphi_0-\delta(\varphi_0)}^{\varphi_0-\Delta(\varphi_0)} (\cot^2\Delta(\varphi) + \cot^2\delta(\varphi))\, d\varphi \leq A_{r,R}(m_- + m_0 + m_+)/\varepsilon. \qquad (*)$$

It suffices to estimate the integral over the interval $[\varphi_0, \varphi_0+\Delta(\varphi_0)]$. Further, it can be assumed without loss of generality that $\varphi_0 = 0$. Let $\Delta = \Delta(0)$. We note first the estimates

$$|\cot\Delta(\varphi)| \leq B_{r,R}(m_0 + m_+)/\varepsilon, \qquad |\cot\delta(\varphi)| \leq B_{r,R}(m_- + m_0)/\varepsilon,$$

which are uniform with respect to $\varphi \in [0, \Delta]$. We verify only the first one. For this we introduce a rectangular system of coordinates u, v with center at the point y_φ and such that the positive direction of the OU-axis coincides with the vector e_φ. In the new coordinates the curve Γ can be represented as the union of the graph of a concave function M and the graph of a convex function N, both defined on some interval $[0, \overline{u}]$, $\overline{u} \geq 2r$. It is not hard to see that $|N(u)|$, $|M(u)| \leq \mathscr{D}_{r,R}\,\mu(u, \varphi)$ for $u \in [0, r]$. If $\Delta(\varphi) \leq \pi/2$, then, using the concavity of the function M, we get that

$$0 < \cot\Delta(\varphi) = M'(\varepsilon) < M(\varepsilon)/\varepsilon \leq \mathscr{D}_{r,R}\,\mu(\varphi, \varepsilon)/\varepsilon$$
$$\leq \mathscr{D}_{r,R}(m_0 + m_+)/\varepsilon.$$

But if $\Delta(\varphi) > \pi/2$, then, as is not hard to see from a sketch,

$$\begin{aligned}|\cot\Delta(\varphi)| = |M'(\varepsilon)| &\le (|N(\varepsilon)| + \mu(\varphi, \varepsilon))/\varepsilon \\ &\le (1+\mathscr{D}_{r,R})(m_0 + m_+)/\varepsilon.\end{aligned}$$

To estimate the integral of $\cot^2\Delta(\varphi)$, note that the function $\alpha(\varphi) = \varphi + \Delta(\varphi)$ is increasing, and hence $\alpha(\varphi) \ge \alpha(0)$, that is, $\Delta(\varphi) \ge \Delta - \varphi$. This implies that $\cot\Delta(\varphi) = O(1/(\Delta-\varphi))$. Thus

$$|\cot\Delta(\varphi)| \le L_{r,R}\min(1/(\Delta-\varphi),\ (m_0+m_+)/\varepsilon).$$

Consequently,

$$\int_0^\Delta \cot^2\Delta(\varphi)\,d\varphi \le 2L_{r,R}(m_0+m_+)/\varepsilon.$$

To estimate the integral of $\cot^2\delta(\varphi)$ we observe that $\cot\delta(\varphi) = O(1/\varphi)$. Indeed, since $\delta(\varphi) \le \theta_{r,R} < \pi$, it suffices to consider the case $\delta(\varphi) < \pi/2$, and hence $0 < \cot\delta(\varphi) = |N'(\varepsilon)| \le |N(\varepsilon)|/\varepsilon$. Since $|N(\varepsilon)|\sin\varphi \le 2\varepsilon$, we have that $0 < \cot\delta(\varphi) \le 2/\sin\varphi = O(1/\varphi)$. Thus,

$$|\cot\delta(\varphi)| \le L_{r,R}\min(1/\varphi,\ (m_0+m_-)/\varepsilon).$$

Consequently,

$$\int_0^\Delta \cot^2\delta(\varphi)\,d\varphi \le 2L_{r,R}(m_0+m_-)/\varepsilon.$$

The inequality $(*)$ is proved. Summing all the inequalities corresponding to the different components of the polygonal curve, we get (m_Γ is the length of the curve Γ):

$$\begin{aligned}\int_{-\pi}^{\pi}(\mu_2'(\varphi, \varepsilon))^2\,d\varphi &\le 2\int_{-\pi}^{\pi}(\cot^2\Delta(\varphi) + \cot^2\delta(\varphi))\,d\varphi \\ &\le \frac{8}{\varepsilon}A_{r,R}m_\Gamma \le \frac{C_{r,R}}{\varepsilon}.\end{aligned}$$

Bibliography

1. G. P. Akilov, B. M. Makarov, and V. P. Khavin, *Elementary introduction to the theory of the integral*, Izdat. Leningrad. Gos. Univ., Leningrad, 1969.
2. S. N. Bernstein, *On the convergence of certain sequences of polynomials*, Collected works of S. N. Bernstein, Vol. 2, 1954, pp. 187–196. (Russian)
3. P. Billingsley, *Ergodic theory and information*, Wiley, 1965.
4. B. Z. Vulikh, *A short course in the theory of functions of a real variable*, "Nauka," Moscow, 1973. (Russian)
5. A. O. Gel′fond and Yu. V. Linnik, *Elementary methods in analytic number theory*, Fizmatgiz, Moscow, 1962; English transl., Rand McNally, Chicago, Ill., 1965, and Pergamon Press, N. Y., 1966.
6. A. Ya. Dorogovtsev, *Mathematical analysis. A problem book*, "Vishcha Shkola," Golovnoe Izdat., Kiev, 1987. (Russian)
7. V. A. Zorich, *Mathematical analysis*, Parts 1 and 2, "Nauka," Moscow, 1981, 1984. (Russian)
8. L. D. Ivanov, *Variations of sets and functions*, "Nauka," Moscow, 1975. (Russian)
9. L. V. Kantorovich, *On convergence of a sequence of Bernstein polynomials outside the basic interval*, Otdel. Mat. Estestv. Nauk, Izv. Akad. Nauk, 1931, pp. 1103–1115. (Russian)
10. L. Carleson, *Selected problems on exceptional sets*, Van Nostrand, Princeton, N. J., 1967.
11. J.-P. Kahane, *Some random sequences of functions*, Heath, Lexington, Mass., 1968.
12. M. Kac, *Statistical independence in probability theory, analysis, and number theory*, Wiley, 1959.
13. J. L. Kelley, *General topology*, Van Nostrand, Princeton, N. J., 1955.
14. A. N. Kolmogorov and S. V. Fomin, *Elements of the theory of functions and of functional analysis*, 3rd ed., "Nauka," Moscow, 1972; English transl. of 1st ed., Vols. I, II, Graylock Press, Albany, N. Y., 1961.
15. I. P. Kornfel′d, Ya. G. Sinaĭ, and S. V. Fomin, *Ergodic theory*, "Nauka," Moscow, 1980; English transl., Springer-Verlag, 1982.
16. L. K. Kudryavtsev, *Mathematical analysis*, Vols. 1 and 2, "Vysshaya Shkola," Moscow, 1970. (Russian)
17. B. M. Levitan, *Almost periodic functions*, GITTL, Moscow, 1953. (Russian)
18. G. Ya. Lozanovskiĭ, *A characterization of a power function in terms of functional inequalities*, Qualitative and approximation methods in the investigation of operator equations, vyp. 2, Yaroslavl′, 1977, pp. 162–165. (Russian)
19. A. Ya. Dorogovtsev, editor, *Matematika sevodnya (Scientific compendium)*, "Visha Shkola," Kiev, 1982. (Russian)
20. K. R. Parthasarathy, *Introduction to probability theory and measure theory*, Springer-Verlag, 1978.
21. G. Pólya and G. Szegö, *Aufgaben und Lehrsätze aus der Analysis*, Vols. I, II, 3rd ed., Springer-Verlag, 1964.
22. Yu. G. Reshetnyak, *Problem book for a course in mathematical analysis*, Novosibir. Gos. Univ., Novosibirsk, 1979. (Russian)

23. W. Rudin, *Principles of mathematical analysis*, 2nd ed., McGraw-Hill, 1964.
24. V. A. Sadovnichiĭ, A. A. Grigor′yan, and S. V. Konyagin, *Problems at the student mathematical olympiads*, Izdat. Moskov. Univ., Moscow, 1987. (Russian)
25. V. A. Sadovnichiĭ and A. S. Podkolzin, *Problems at the student mathematical olympiads*, "Nauka," Moscow, 1980. (Russian)
26. V. N. Sergeev and G. A. Tonoyan, *Student mathematical olympiads*, Izdat. Erevan. Gos. Univ., Erevan, 1985. (Russian)
27. E. C. Titchmarsh, *The theory of functions*, 2nd ed., Oxford Univ. Press, Oxford, 1939.
28. P. S. Urysohn, *Works on topology and other areas of mathematics*, Part 1, GITTL, Moscow-Leningrad, 1951. (Russian)
29. G. M. Fikhtengol′ts, *A course in differential and integral calculus*, Vols. I, II, 7th ed., Vol. III, 5th ed., "Nauka," Moscow, 1969; German transl., VEB Deutscher Verlag, Berlin, 7th ed. of I, 1972, 6th ed. of II, 1966, 5th ed. of III, 1972.
30. A. N. Sharkovskiĭ, *The co-existence of cycles of a continuous mapping of the line onto itself*, Ukrain. Mat. Zh. **16** (1964), 61–71. (Russian)
31. L. Schwartz, *Méthodes mathématiques pour les sciences physiques*, Hermann, Paris, 1961; English transl., Hermann, Paris, and Addison-Wesley, Reading, Mass., 1966.
32. A. B. Shidlovskiĭ, *Transcendental numbers*, "Nauka," Moscow, 1987; English transl., de Gruyter, Berlin, 1989.
33. P. Collet and J.-P. Eckmann, *Iterated maps on the interval as dynamical systems*, Birkhäuser, Basel, 1980.
34. A. Connes, U. Haagerup, and E. Størmer, *Diameters of state spaces of type* III *factors*, Lecture Notes in Math., Vol. 1132, Springer-Verlag, Berlin and New York, 1985, pp. 91–116.
35. M. Dekking, M. Mendès-France, and A. J. van der Poorten, *Folds!*, Math. Intelligencer **4** (1982), no. 3, 134–138.
36. A. Grothendieck, *Résumé de la théorie métrique des produits tensoriels topologiques*, Bol. Soc. Mat. São Paulo **8** (1956), nos. 1-2, 1–79.
37. U. Haagerup, *The best constants in the Khintchine inequality*, Proc. Internat. Conf. on "Operator Algebras, Ideals and their Applications in Theoretical Physics" (Leipzig, 1977), Teubner Texte Math., Leipzig, 1978, pp. 69–79.
38. G. Klambauer, *Problems and propositions in analysis*, New York, 1979.
39. T.-Y. Li and J. A. Yorke, *Period three implies chaos*, Amer. Math. Monthly **82** (1975), no. 10, 985–992.
40. G. G. Lorentz, *Approximation of functions*, New York, 1966.
41. O. C. McGehee, L. Pigno, and B. Smith, *Hardy interval inequality and the* L^1 *norm of exponential sums*, Ann. of Math. **113** (1981), no. 3, 613–618.
42. M. Mendès-France and G. Tenenbaum, *Dimension des courbes planes, papiers pliés et suites de Rudin–Shapiro*, Bull. Soc. Math. France **109** (1981), no. 2, 207–215.
43. D. Singer, *Stable orbits and bifurcations of maps of the interval*, SIAM J. Appl. Math. **35** (1978), no. 2, 260–267.

Subject Index

Recent Titles in This Series

(*Continued from the front of this publication*)

72 **A. V. Pogorelev,** Bendings of surfaces and stability of shells, 1988
71 **A. S. Markus,** Introduction to the spectral theory of polynomial operator pencils, 1988
70 **N. I. Akhiezer,** Lectures on integral transforms, 1988
69 **V. N. Saliĭ,** Lattices with unique complements, 1988
68 **A. G. Postnikov,** Introduction to analytic number theory, 1988
67 **A. G. Dragalin,** Mathematical intuitionism: Introduction to proof theory, 1988
66 **Ye Yan-Qian,** Theory of limit cycles, 1986
65 **V. M. Zolotarev,** One-dimensional stable distributions, 1986
64 **M. M. Lavrent′ev, V. G. Romanov, and S. P. Shishat·skiĭ,** Ill-posed problems of mathematical physics and analysis, 1986
63 **Yu. M. Berezanskiĭ,** Selfadjoint operators in spaces of functions of infinitely many variables, 1986
62 **S. L. Krushkal′, B. N. Apanasov, and N. A. Gusevskiĭ,** Kleinian groups and uniformization in examples and problems, 1986
61 **B. V. Shabat,** Distribution of values of holomorphic mappings, 1985
60 **B. A. Kushner,** Lectures on constructive mathematical analysis, 1984
59 **G. P. Egorychev,** Integral representation and the computation of combinatorial sums, 1984
58 **L. A. Aĭzenberg and A. P. Yuzhakov,** Integral representations and residues in multidimensional complex analysis, 1983
57 **V. N. Monakhov,** Boundary-value problems with free boundaries for elliptic systems of equations, 1983
56 **L. A. Aĭzenberg and Sh. A. Dautov,** Differential forms orthogonal to holomorphic functions or forms, and their properties, 1983
55 **B. L. Roždestvenskiĭ and N. N. Janenko,** Systems of quasilinear equations and their applications to gas dynamics, 1983
54 **S. G. Kreĭn, Ju. I. Petunin, and E. M. Semenov,** Interpolation of linear operators, 1982
53 **N. N. Čencov,** Statistical decision rules and optimal inference, 1981
52 **G. I. Èskin,** Boundary value problems for elliptic pseudodifferential equations, 1981
51 **M. M. Smirnov,** Equations of mixed type, 1978
50 **M. G. Kreĭn and A. A. Nudel′man,** The Markov moment problem and extremal problems, 1977
49 **I. M. Milin,** Univalent functions and orthonormal systems, 1977
48 **Ju. V. Linnik and I. V. Ostrovskiĭ,** Decomposition of random variables and vectors, 1977
47 **M. B. Nevel′son and R. Z. Has′minskiĭ,** Stochastic approximation and recursive estimation, 1976
46 **N. S. Kurpel′,** Projection-iterative methods for solution of operator equations, 1976
45 **D. A. Suprunenko,** Matrix groups, 1976
44 **L. I. Ronkin,** Introduction to the theory of entire functions of several variables, 1974
43 **Ju. L. Daleckiĭ and M. G. Kreĭn,** Stability of solutions of differential equations in Banach space, 1974
42 **L. D. Kudrjavcev,** Direct and inverse imbedding theorems, 1974
41 **I. C. Gohberg and I. A. Fel′dman,** Convolution equations and projection methods for their solution, 1974
40 **D. P. Želobenko,** Compact Lie groups and their representations, 1973

(See the AMS catalog for earlier titles)